Battle in the Mind Fields

Battle in the Mind Fields

JOHN A. GOLDSMITH
AND BERNARD LAKS

THE UNIVERSITY OF CHICAGO PRESS CHICAGO AND LONDON

The University of Chicago Press, Chicago 60637
The University of Chicago Press, Ltd., London

Published 2019
Printed in the United States of America

28 27 26 25 24 23 22 21 20 19 1 2 3 4 5

ISBN-13: 978-0-226-55080-0 (cloth)
ISBN-13: 978-0-226-55094-7 (e-book)
DOI: https://doi.org/10.7208/chicago/9780226550947.001.0001

Library of Congress Cataloging-in-Publication Data
Names: Goldsmith, John A., 1951– author. | Laks, Bernard, author.
Title: Battle in the mind fields / John A. Goldsmith and Bernard Laks.
Description: Chicago ; London : The University of Chicago Press, 2019. | Includes bibliographical references and index.
Identifiers: LCCN 2018037456 | ISBN 9780226550800 (cloth : alk. paper) | ISBN 9780226550947 (ebook)
Subjects: LCSH: Linguistics—Philosophy. | Philosophy, Modern.
Classification: LCC P33 .G65 2018 | DDC 410.9—dc23
LC record available at https://lccn.loc.gov/2018037456

♾ This paper meets the requirements of ANSI/NISO Z39.48-1992 (Permanence of Paper).

Contents

Preface

First of all, a word about what this book is, and what it is not. It is a historical account of some central ideas in modern linguistics—an account of the ideas and some of the events surrounding their development, debate, and disposition. The book is *not*, appearances to the contrary, the history of modern linguistics or of any other period. It is far too selective in its choice of topics to be thought of as *the* history of anything. If it is historical, it is because we feel that this is the only way to narrate the story and the best way to hear it as well.

It is a study of rupture and continuity in linguistics. The primary lesson that we draw from the work we have studied here is that in the realm of ideas, continuity is overwhelmingly the way things work, while in the realm of personal interactions, acknowledgments, and jealousies, the degree of rupture that our scholars have described is great. We might even say that it is *astonishing*, but there is nothing to be surprised at, really, if we listen to what historians of ideas and historians of science have been telling us. Our goal in this book is to make clear *how* this pattern of continuity and of rupture has come to be and to shed a bit of light on *why* it is. In the end, we think that this situation has some regrettable sides to it, and we have not shied away from drawing some normative conclusions as well. But by and large we have subscribed to the eternally optimistic philosophy that the truth will set us free and so have tried to keep the moralizing to a minimum. Not to avoid it completely, but to keep it to a minimum.

We will have occasions in this book to remind ourselves, as well as the readers, what intellectual continuity means and what it does not mean. When we find intellectual continuity in the development of a new idea, we do not mean that the new idea was easy to come by, or that it was not novel, or that it was not a work of first-class originality. It is easy to misread a history such as ours in which the connections between new perspectives and older developments are emphasized. Continuity means that the new ideas were based on the present; it does not mean that this basis was trivial, or obvious, or less astonishing than anyone may have thought.

What *does* it mean, then? In our view, it is based on the notion (hardly controversial, in our day and age) that at any given moment, there are a range of ideas, opinions, and beliefs that comprise the current state of affairs. These ideas, these common beliefs, will vary with their degree of adhesion: some will be held by many, some by few. Some will have arisen recently, others will have been around for a long time. These ideas will not all be consistent with one another. (If they were, there would be no notion of controversy in a discipline.) These ideas form, in some respects, a large organic garden, or perhaps a zoo, in which change and variety is the principal constant. It is always the case that new creatures are descendants of other living organisms: new creatures do not come on the scene with no living, direct ancestors, or arise as the descendant of a long-extinct breed or race.

To put it slightly differently, when we look at the origin of *new ideas*, they are always the creative modification of several ideas that have been developed recently that no one has yet connected. There are three crucial elements in that: there is a connection that is made of several ideas; those ideas are current ideas of some recency; and this novel connection, once made, is developed and elaborated in a genuinely creative new way. That is the pattern that we find, over and over. And that is the pattern we will show our readers over the course of the growth and development of the mind sciences. Our view of intellectual history is thus both historical and variationist. It is historical in that we believe that there is no way to understand the ideas of a discipline at a particular moment in time without understanding the historical path which led the field from there to where it is today. It is variationist in that it explicitly denies the Kuhnian notion that a scientific discipline will subscribe to a core set of ideas which define a paradigm, a climate of opinion; a living discipline is a quiltwork of disagreements.

The discovery and the acknowledgment of continuity in the study of the mind in these fields is not an exercise in showing that for each idea traditionally attributed to one scholar, there was an earlier scholar who had pretty much said the same thing. That game is rarely of interest if it goes no further than that. The real lesson to be learned from studying the continuity of thought in this area is that all of these thinkers are engaging in a greater conversation, and that no single scholar is large enough to hold any single important idea: all of the ideas have developed over the course of generations of controversies in which people with different perspectives and prejudices have served and returned ideas in a great game.

We noted just above that at the level of personal interaction among scholars, the continuity of ideas seems to vanish, and instead we find all sorts of conflicts, of alliances, and of branding. The people whose work we study are, when all is said and done, just people, with all the baggage that they bring with them.[1]

It is both helpful and healthy to redouble our efforts to focus on the real intellectual substance in this story, but we have found that we are interested in both sides—both the idea side and the personal and institutional side of the story. Perhaps the most interesting part of the second side of the story is a phenomenon that we find ourselves up against throughout the story: a moment when a leading thinker decides that essentially all the work that has preceded him is no longer worth reading or taking seriously. This stratagem (for what else can we call it?) comes up on quite a few occasions, and there are quite a few more who adopt what the Voegelins once called an *eclipsing stance*. We are fascinated by the double fact that so many feel called to adopt that stance, and that it seems to work so often, for so long. In some instances, this stance is adopted explicitly, with a statement that what has preceded can be safely jettisoned, while in other cases, the message is passed on implicitly, by failing to state the obvious.

The reader is likely to have noticed already that in the pages that follow, there are many dates, places, and events. But do not be fooled by this: that is not what the book is really about. The dates and the events are there to allow us to reflect on questions with real intellectual depth, on hypotheses and the arguments developed for them, on the ways in which questions and positions may remain or return despite differences in their formulation. We care deeply about the ways in which we find conceptual continuity across the work of thinkers who were themselves not aware of the continuity. We care equally about the flip side of this coin: the ways in which change and rupture can emerge from underneath the cover of loyalty and common community.

What this means, in practical terms, is that we undertake a synchronic dialogue with the great writers of the past, and so we discuss their hypotheses and their arguments *not* as if they were archeological ruins but as if their hypotheses were *alive*, and as if they were colleagues whose offices were next door. It might take a bit of effort to see how their perspectives bear on our own questions, but that is a challenge that we always face in the real world. The point is that to unearth the continuities and the ruptures and to construct an internal history, what we must do is to engage in a dialogue which allows us to actually feel the agreements and disagreements as if they were ours today.

Our interest in rupture and continuity has led us to take more seriously certain aspects of external history as well. There are three kinds of external forces that play a major role in this story. The first is political, and in this book, the most striking case is the rise of Nazism in Central Europe during the 1930s and 1940s, a world historical fact that led to a major exodus of intellectuals out of Europe at critical moments of our story. From a larger perspective, that movement of scholars from Europe to the United States is part of a bigger picture which began when the United States was younger and not so rich, a time when the natural place for would-be American scholars to go for higher education was Western Europe. The present book is the first of two volumes telling a single story, and we will focus in this book on the events that brought the mind sciences up to World War II. It will be followed by a second volume that treats the three decades that followed the outbreak of the war.

The second kind of external force is quite simply death: a scholar's work stops abruptly at the time of his death, and if death does not stop his or her influence, it changes the character of that influence mightily. While ideas can survive the death of the people who championed them, people have no such longevity; their direct and personal influence vanishes with their death.

The third kind of force is the way in which economic resources are allotted in the creation of jobs, which in turn lead academics to leave some institutions and go to some others. We will see occasions when money that came from the Rockefeller Foundation (to take only one example) made it possible for European academics to leave their homes and avoid almost certain death, and also made it possible for academics to be invited to leave one university and come with all their students to another one. There are—not always, but often—stories that are of interest to us about why an academic institution decides it wants to hire significantly in an area, such as linguistics, psychology, or philosophy, and when that has a significant impact on the story here, we have every reason to look further into what those reasons were.

As we explore these questions, we are aware that we remain linguists, and we are deeply interested in the ideas themselves; we are not dependent on secondary sources to help us understand what is at stake. It is our strong belief, made more certain throughout the process of writing this book, that a deep account of a discipline cannot be neutral, cannot be so external that it rests on nothing but objective facts. If it is to deal both with ideas and with people, if it is to examine both the ideas that formed the people and the people who brought the ideas to life, then the histories of our disciplines must be internal histories which are capable of understanding the nature of the debates, the arguments, and the stakes. An internal history is not always a history as it was lived by the actors, each with his or her own particular point of view; in fact, it rarely is, and it may be the history that is constructed by partisans who attempt to put down their particular positions in order to reconstruct the underlying dynamics that are at play in the world of a given scientific domain at a particular time. It is less a history of events and more a history of ideas, a history whose primary aim is to bring to light the forces that act upon the growth and development of a discipline. These can include the strengths and the weaknesses of the actors themselves, the arguments and ideas both within the discipline and outside of it, as well as prestige, legitimacy, the strength of the orthodox, and the enthusiasm of the Young Turks—in short, everything that is at play in a disciplinary field and that makes it what it is.

We have naturally chosen particular incidents, schools, scholars, and coalitions in our discussions, and the fact that we have left a movement or a scholar out of our discussion does not mean that we think they are less worthy, important, or influential than those we have discussed. We have little discussion of Sigmund Freud in psychology, or of J. R. Firth in linguistics, and nothing to say about Kierkegaard or Bergson in philosophy. We talk more about Bloomfield than we do about Sapir, a fact that in no way reflects a view on their relative importance. We do not discuss Reichenbach's ideas of time and tense, which have had a great impact on current semantics. We barely mention sociology, anthropology, and economics. In all these cases, we were sorely tempted to include discussions. But we have done our best to maintain a tight coherence of the discussion that is to follow, and to do that, we have had to embrace the fact that an omission from our account should never be interpreted as a tacit message that whatever is left out is of less importance.

The particular story that we focus on in this book involves one part of the field of linguistics as we saw it when we embarked on our careers in linguistics some 40 years ago. Our own experiences begin roughly where the story leaves off, although we know (or knew) personally many of the principals whose careers extended into the 1970s and beyond. We have great admiration for all of the linguists we describe in this book (for some a bit more than for others, but that is only natural). Some of them are our teachers, and some our friends or professional colleagues, although of course many died before we were born, and those we only know through their writings. A large number of the people we discuss have set to paper their views about where their work comes from, or where the work of others comes from, and in quite a few cases, we aim to show that they are mistaken—sadly mistaken, if you will.

Our intention in this book is to help the reader better understand where our current beliefs in linguistics come from, and how they have been justified. We do not mean by this to criticize or dismiss any particular theory or framework, except insofar as a theory may have been offered to the public with an inaccurate pedigree. But each theory offers an answer to a set of questions which are more often implicit than explicit, and a historical perspective is sometimes the best, if not the only, way to come to understand what those questions are.

Both of us began our studies in linguistics in graduate school around the same time. We were drawn into the field because of the appeal of the questions and methods being explored and developed in generative grammar. If Chomsky had not come onto the scene when he did, it is highly unlikely that we would be here writing about linguistics. We, like so many of our generation, were inspired by the nature of the questions that generative grammar allowed us to explore. So just in case it is not clear, let us say it up front: we consider all of the thinkers and scholars that we write about in this book to be heroes. They are humans, but heroes nonetheless, and there are none of whom it cannot be said that they left the field better for having been there.

One reader of this book, a friend and participant from time to time in this book's story, was not happy by the occasional observation on our part that seemed to be suggesting that we were taking sides in a particular confrontation: at one point, we used the word "strident" to characterize a particular linguist's prose. We've left the word in; we have done our very best to remain sympathetic to all sides in these disagreements, which does not mean that we cannot call a sentence "strident" in tone when it is. As for our position, we are reminded of a statement almost certainly apocryphally attributed to John Lennon: we gave up being fans when we became professionals.

Needless to say, we have our own views on a number of subjects that we will discuss in this book, and we would not be unhappy if, as the result of reading it, some of our readers become convinced of our views. Still, that is not our primary aim, which is rather to show that among the great questions and ideas that have been central to the mind sciences over the last several centuries, there is more than one way to look at things. No matter how convinced you are of whatever you are convinced of, there is a good case to be made for other points of view. Progress generally comes from finding a new synthesis that brings together older ideas that seemed—but only *seemed*—to be in conflict.

This book is itself also the product of a debate, or a dialectic in the etymological sense of the term. It grew out of the pleasure that we found in discussion, in agreement and in disagreement, in the enjoyment of confronting ideas and arguments. Writing this book has been a project that began a decade ago, and the decision to write this book came only after years of extended discussions between us. It is the result of the agreements and disagreements shared by two linguists from two different continents, who grew up in two intellectual traditions and different material cultures, but who both share a great pleasure in debate, in arguing, and in encouraging controversy as a form of dialogue. We know full well that this is something that we learned from our teachers. Morris Halle, who advised one of us and greatly influenced the other, expressed what we feel: "Convince me," he would say. "Argue with me!"[2]

We have been sensitive to the extreme gender bias that leaps out at us as we tell this story. There are women who play important roles in the developments that we discuss, but there are not enough. In the early work on the mathematics of computation, there is Ada Lovelace, and in the story of the exodus of the psychologists from Central Europe to the United States, there is Charlotte Bühler, and there are a few more, such as Margaret Mead. But the academic world has not had a long history of encouraging and supporting women who sought a career at a research university. In our professional lifetime, we have seen the gender balance in linguistics come to parity or near it, but the same cannot be said for some of the other academic disciplines that we explore.

Our friends have warned us that this will not be an easy book to read. There are parts that are a bit dramatic, and there might even be some humor, but there are more parts that are difficult. Despite the tone, we do not offer a simplification of the issues. The reader who does not already have at least a smattering of knowledge of linguistics, philosophy, and psychology is going to be introduced to quite a number of unfamiliar characters and ideas. The reader who *does* have some knowledge of these fields is likely to have his assumptions challenged. We think, on the whole, that these issues have not been treated very well in the literature, and it has taken us decades to get to the point where we have been able to see some of these things.

It is often said that there are two ways to read the older literature in one's discipline: one either tries to force the earlier vocabulary into today's categories, translating as best one can into today's terminology, or else one tries to put oneself in the earlier mind-set, and read yesterday's articles from the point of view of a contemporary who was reading it for the first time. Over the course of writing this book, we have come to realize that for our purposes, *both* of these perspectives are necessary, and we do our best to help the reader come to grips with an older literature in both of these ways.

For that reason, we have made a special effort to include more snippets from writers than are typically found in studies of this sort, for the simple reason that the readers deserve to get a bit of a feel for themselves of how an earlier thinker chose to frame his thoughts and make his case.

Notes and Comments

Unless otherwise indicated, all the translations from French and German are our own. Russian names that occur have required a transliteration in English, and in some cases we have simply adopted the common transliterations that have been used, and when there is no common usage to fall back on, we have used a transliteration that makes the most sense, given familiar English orthography. We write Shpet, therefore, rather than Chpet or Špet, and Karchevsky rather than Karcevskij.

We have many people to thank for their help in the course of writing this book. There have been moments when we realized that just about anyone we have ever had a conversation with about linguistics has likely influenced this book in one way or another. Among those whose observations came at particularly important moments, we think of Farrell Ackerman, Daniel Andler, Robert Barsky, Hans Basbøll, Gabriel Bergounioux, Jackson Bierfeldt, Diane Brentari, Noam Chomsky, Katya Chvany, Jacques Durand, Pierre Encrevé, Lila Gleitman, Morris Halle, Chas Hockett, Fred Householder, Geoff Huck, Simon Jacobs, Bill Labov,

Chantal Lyche, Geoff Pullum, Robert Richards, Jason Riggle, Haj Ross, Jerry Sadock, Gillian Sankoff, Patrick Sériot, David Stampe, Guri Bordal Steien, and Atanas Tchobanov.

John Goldsmith wishes to express his gratitude to the University of Chicago, which has always been an ideal place for the kind of discussions that have gone into the writing of this book and whose deans have been generous over the last few years with helping him to find the time needed to work on this book. Bernard Laks expresses his gratitude to the Institut universitaire de France and the Université de Paris Nanterre for their support. The University of Vienna and the University of Chicago kindly provided funds to support a seminar organized by Elissa Pustka at the University of Vienna on April 6, 2017, which provided valuable feedback for us.

We both want to thank our wives, Jessie Pinkham and Claudie Laks, for their indulgence and support in this project, and we're especially delighted that Claudie's work could serve as the basis for the cover of this book.

Diagrams/Figures

The multicolored schemas we have included should be used with care. Each presents a number of actors in our story, in boxes that are color coded to roughly indicate what discipline the actors were involved in. Their placement in the schema is determined in part objectively: their height in the schema is a direct reflection of the year of their birth (we have shifted a few people up or down in interests of visual clarity). We have greatly simplified things by indicating relationships between various pairs of these people with colored lines, indicating roughly four relationships. One relationship is between colleagues, people who knew each other and influenced each other's work. The second relationship is one of important intellectual influence without personal influence or contact. The third is the most important, in a sense, represented in blue; it is the relationship between a mentor or dissertation advisor and the young scholar being advised. In the cases we look at here, there are a good number of secondary relations of just this sort, where a senior scholar plays a mentoring relationship of someone who was not officially his student (such as Sapir and Whorf), and we have indicated this with a dashed blue line. Finally, in a few cases, we wish to emphasize the hostile relationship between two scholars, and we have chosen to indicate these relationships in red. Bear in mind that restricting relationships to just these four kinds has led to some strange designations: for example, the relationship between Edward Sapir and Margaret Mead is represented with the color that indicates "colleagues," which is not a very good description, but it is better than any of the other choices. In some cases, we describe in the text a group of people who all influenced each other a good deal, but we have not made our figures more cluttered to include all of those pairwise connections. We have included a few mixed categories, notably "philosopher-psychologist," but that did not really help, because it is hardly a meaningful question to ask whether Brentano should be classified as a philosopher or as a philosopher-psychologist. Therefore, the reader should use the colors provided as a roadmap, but they cannot be relied upon in cases where the boundaries are blurred.

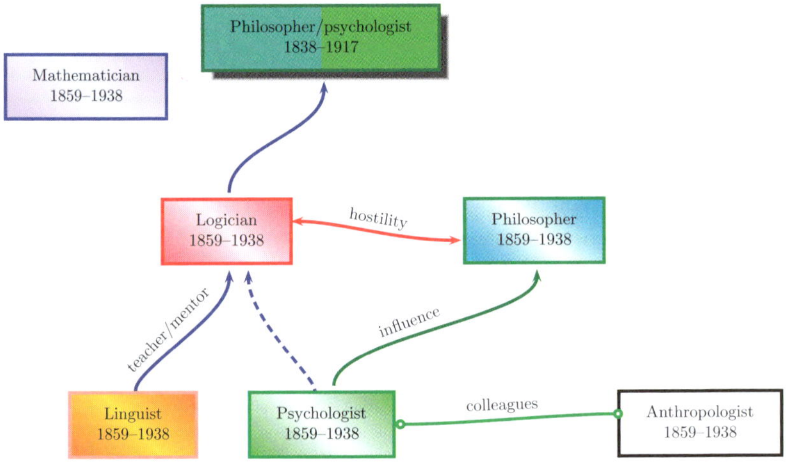

FIGURE 0.1. Sample schema. There are some guidelines needed to understand our figures. The information contained here is intended to serve as a visual reminder of who is who, and what they did. In all cases, a simplification is needed to do this, and the reader must bear in mind that the categorization here is in every instance a simplification of what we describe in the text. The decisions we have made here are simply what seems to us the most helpful and the least inaccurate. The vertical position is determined by date of birth—strictly, in most cases, with a very small amount of adjustment made for clarity. The colors of the individual boxes reflects the disciplines of the actors, but in most cases, some real simplification was needed. Quite a number of people are assigned to two categories, with two colors. The colors of the arrows connecting the boxes correspond to four kinds of relations: mentor (or teacher), colleague, influence, hostility. In many cases, it is hard to determine the relative importance of various teachers, and (as elsewhere) our choices represent an interpretation on our parts.

In order to help the reader organize the characters visually, we have included a number of ovals or rectangles of various sizes, usually with a label, such as "Prague Linguistic Circle." We caution the reader not to take these indications as claims about membership in the organizations or as some sort of Venn diagram that includes or excludes members. They are there purely to help the reader remember who is who, and should be thought of as pointers to the text, where more information is noted. In particular, the reader should not interpret our depictions as signifying something about the relationship between a school, a circle, or anything else. To repeat: the information presented in the diagrams is in most regards highly subjective, and on different days, we ourselves would make different choices in a few cases as to which color to use or whom to place inside a colored box.

CHAPTER ONE

Battle in the Mind Fields

In the Beginning

Battle in the mind fields: the characters in this story are, for the most part, a feisty and pugnacious cast. They come prepared for battle, they rarely take prisoners, and they enter the fray defending the faith. These are philosophers, psychologists, linguists, cognitive researchers of all stripes, the inheritors of the great classical questions that may live forever: What is thought? How is it that we are conscious of ourselves? How is it that humans are endowed with the gift of language? Is the multiplicity of languages in the world an indication that there are many ways of viewing the world, or are all the languages of mankind cut from a common cloth?

This book describes the evolution of some of these ideas and provides a rough snapshot of some of these people, with the goal of understanding the present, and with the certainty that the only way to understand the present is to understand where it has come from. A glance at what is to come may give the impression that we have wandered a bit through the pages of the past, but we promise that what we have included has reverberated in some fashion right down to the present day.

One of the best reasons to study the history of our disciplines is that everything we think we have learned was once an answer to a living, breathing question, and it was an answer provided at a time when alternative answers were also being taken every bit as seriously. But once an answer is certified as true and placed among our certainties, we forget the question to which it was the answer, and the consequence is that we forget what were the alternatives that once enjoyed some traction. In short, we become trapped by our beliefs—not always a bad thing, as long as it leads to no problems. But this phenomenon leads in a natural way to a sclerosis of the mind, a hardening of the mental arteries, and in the end a less adequate understanding of what the disciplines have learned the hard way.

Although much of our perspective in this book derives from personal experience, we have also gained a great deal from the sociologists and philosophers who have studied the evolution of thought in various disciplines. Pierre Bourdieu, for example, made the case for what he called "anamnesis," with a slight nod towards Plato, though using the term in his own way. He argued that a necessary condition for scientific progress was understanding explicitly the conditions (not to mention the context and the constraints) under which dominant scientific ideas had emerged. He was referring not just to science, but also to the vast range of social endeavors that constitute human society. Whether we call it change, or development, or evolution, the fact is that the moment that we live in is always one of confrontation and contestation, for all the reasons discussed in this book. Once that moment has passed, powerful forces enter into play to pretty up the past, to make it docile and submissive. Understanding and wisdom demand just the opposite, though; they demand that we know where we came from and how we got here.

Why? Because the sine qua non of scientific progress is what we might call the *disenchantment* of the scientific world. The student discovers a scientific world, ready-made and already endowed with simplified stories of the past. But the scholar who wants to understand must free herself of that thrall and be on a first-name basis with that world; the scientist must eventually become the master of those stories, and in most cases, that means knowing how we got to where we are. Know where you came from, and you will know where you are going.[1] And so we will have to begin in the past: not as far back as we might—in ancient Greece, say—but with a rapid introduction to the most relevant themes of the nineteenth century, when it seems that we can find the odd character here and there who is already contemporary and many others who are almost there.

People respond and react to what they read, what they hear, and what they are told. That's only human nature. No one locks himself in a closet and refuses to be influenced by other people. Yet it is not at all rare to encounter brilliant thinkers who try to wipe the historical slate clean—tabula rasa!—and start over, afresh. Of course they *themselves* never do start over afresh, *themselves* unaffected by all the ideas and scholarship of the past—that would be impossible—but they send forth the message that the work of the past is unimportant. This seems very odd, and so it is. There is some willful forgetting going on, and we would like to know why and to figure out what ought to be done to overcome it.

All thinking is a continuation of conversations that we have overheard or participated in. If we want to understand a book, we might need to have read perhaps not *everything* its author has ever read, but a quite a bit, and often what we find obscure in a difficult writer is obscure simply because we have to roll back some thought process that the writer had engaged in when presented with other questions, other possibilities, and other ideas.[2] Sometimes we engage in fast reading, just as we sometimes eat fast food, but just as there is for slow food, there is also a great need for slow reading, and we will engage the reader in such an activity in this book. We are tempted to say that a bibliography which goes back no more than five years is either unscientific or dishonest. That is too simple, and of course we could *imagine* papers where a slender bibliography was all that was needed. But as a generalization, it has lot going for it. When it comes to the central questions of the mind, the giants of human thought have preceded us, and we must remember that if we often disagree with them, we never leave them behind. It is critical that we remind ourselves that part of the essence of scientific work consists of confronting a vast library of ideas. When we know a field thoroughly, we find that nine times out of ten, we can summarize and on occasion even evaluate a book by doing nothing more than reading the bibliography carefully.[3]

The second half of the twentieth century saw the development of an overarching new view of mind which, despite its importance, has no simple name and which will be a major concern of both volumes of this book. This new view is tightly bound to the machine that has changed our lives: the computer. But the connection is not a simple one. Computers, the real thing, first appeared during World War II, largely as part of the war effort, in the United States, in England, and in Germany. Computers were needed at first to solve differential equations rapidly so that artillery could be more accurately aimed, then to break enemy codes and encryption systems, and eventually to help in the development of the atomic bomb. But computers were not the simple source of the new ideas about the mind. If anything, it was the other way around. People were able to invent and create computers because these new ideas about logic and computation were already being developed. Technology, philosophy, logic, mathematics: all these fields were tied together in a complex unity that is no less real today than it was in the beginning of the twentieth century.

Soft Mentalism, Hard Mentalism

A principal focus of our account is this transitional period and the change in the way the mind was understood. To give a name to this transition (though one that will need a good deal of spelling out over the course of the book), we will look at this shift from a *soft mentalism* to a *hard mentalism*. Soft mentalism focused on consciousness and self-awareness, while hard mentalism focused on representation, intension, and belief. Hard mentalism began as a fantasy: machines that could talk, play chess, and do sums. Pascal and Leibniz had some success with machines that could calculate. These fantasies began to take on form, if not life, and Charles Babbage came as near as anyone in the nineteenth century with his analytical engine. Hard mentalism sees Leibniz as its patron saint, while soft mentalism looks to Descartes.[4] And as logic is the science of what makes thought possible, there are two concepts of logic that correspond to these mentalisms: *hard logic* and *soft logic*.

The physical sciences over the past four centuries have been extraordinarily successful, as no thinking person could fail to see. Like a sharp investor looking for a place to put his money, many thoughtful people have looked to the physical sciences to try to figure out what they are doing so right and to see whether there are lessons to be learned that could be applied elsewhere. The crass might call this "physics envy"; others will see it as prudence and good common sense. We will see how the fascination with science and with measurement came to center stage in the nineteenth century, and well into the twentieth century, as more *kinds* of objects came to be placed under the scientific microscope: the sound changes in language studied in depth and detail by philologists and linguists, especially by German philologists and linguists, in the nineteenth century, for example. Taxonomic structures of cultural and social systems, of biological species, and of chemical elements all developed quickly during this period. Some of these systems were shaken up again at midcentury by the Darwinian revolution, the revolution that gave a new account, without recourse to divine intervention, of how change over long periods could be scientifically explained.

One of the messages that we expect our readers to take away is the idea that it is simply impossible to understand any of the mind fields—linguistics, philosophy, psychology, logic—over the past 100 years in isolation. Each field influenced, and was in turn influenced by, the others.

This interaction, on the rare occasions it is discussed, is usually presented as a quaint corner of dusty history. We will try to show how wrong this view is, and how much these disciplines have suffered from being unaware of the origins of many of the most important ideas and values that have shaped them. An important part of this intimate relation between the fields derives directly from the fact that these disciplines share deep historical roots, and in many ways these fields were once one. There is much to be learned, for example, from watching how psychology fought for its independence from philosophy after the middle of the nineteenth century and how linguistics continues to view its independence from psychology and to reflect on that independence.

We will frequently see an idea appear in one discipline as if it were new, when it actually migrated from another discipline, like a mole that dug under a fence and popped up on the other side. Disciplines may at times emphasize their limits; under most conditions this is a bad thing, but these limits help clarify for a wide range of workers what the questions are that they should be addressing. Still, there are always individuals who are passionately interested in issues that transcend a single discipline and whose work therefore becomes multidisciplinary. It may be possible to write a history of a single discipline, but it is not possible to research a history of a discipline and restrict oneself to that discipline: the reality, the boots on the ground, has always seen thinkers read and write across the disciplinary boundaries.[5]

We have found it useful to adopt some of Bourdieu's perspectives, as we noted just above.[6] Bourdieu generalizes the notion of capital from the economic domain to a wide range of social arenas, all the while recognizing that this capital can grow, diminish, accumulate, or even in some cases be wiped out in a crash. It is a banality to say that money is both a reality and a social construction. No one needs any explanation that money has its reality: it can be transformed into a sweater, a dinner, a car. And it is a social construction; without the force of a government behind it, a 10-dollar bill is just a slip of custom-made paper, not good for much at all. And while there is an arbitrariness to the units with which we measure monetary value, all capital has the possibility of accumulating, of being added to by its owner.

In various social arenas, which Bourdieu calls *fields*, individuals enter into different relations with one another; most of the relations discussed in in this book involve academic and scientific roles. In different fields, actors may work to accumulate capital, even though the capital is generally specific to each field. In the academic realm, the notion of capital corresponds to authority and influence, and under certain conditions it can transfer across fields; although the economic metaphor breaks down in such cases, since a transfer from one field to another need not involve a decrease in accumulated capital in the first. But transfer across fields, as Bourdieu underscores, is far from obvious and far from automatic: it is indeed a complex alchemy, which can involve far more than an explicit or pre-established set of rules; it may depend on a larger context, including ideas circulating on more extensive fields, or a sensitivity to the widest field of all, the zeitgeist.[7]

In the rest of this chapter, we will survey the principal themes that return frequently in the story that will capture our attention. We have cast a wide net, from a chronological point of view, so that we can see recurrences—and see them we will.

Liberation Moments

Here is the first noteworthy observation: new ideas that catch on are always perceived by the catchers-on to be liberating them not just from a set of ideas but from a *dogma* of an earlier generation. Each successful new way of looking at mind, language, and reasoning is viewed as a *notional liberation moment*. This way of putting it captures both the heady revolutionary fervor that comes along with a new scientific perspective and the sensation that a new perspective brings out explicitly what was wrong with the old conventional wisdom. Now, with the problem out in the open, we can get rid of it, put it behind us, and move forward with new vigor. We see the dogma of our elders and wonder how they could have failed to see it for what it was, as we see it now.

One of the ideas we will try to spell out is that we never completely drop old ideas: they remain with us, often getting harder and harder to see consciously, which is generally not a good thing. But one of the constants we will hear in the stories that are recounted by participants is this: each person, individually and in concert, felt that a great weight had been lifted from his or her shoulders, and that weight was the weight of a heavy past tradition. The behaviorists felt that way, as did the logical positivists, the early generative grammarians, and then the later generative semanticists. Yehoshua Bar-Hillel told of his similar conversion experience upon first encountering Carnap's and Reichenbach's work.

It follows from this that if you do not understand how a once dominant idea could have captured the imagination of smart, young people, then you simply do not understand it. All new ideas that grab the imagination of new people in a field do so because they are perceived as liberations from some kind of orthodoxy of the past.

Noam Chomsky expressed the heady emotion that we are talking about very well:

> The whole history of grammar, for thousands of years, had been a history of rules and constructions, and transformational grammar in the early days, generative grammar, just took that over. So the early generative grammar had a very traditional flair. There is a section on the Passive in German, and another section on the VP in Japanese, and so on: it essentially took over the traditional framework, tried to make it precise, asked new questions and so on. What happened in the Pisa discussions was that the whole framework was turned upside down.
>
> So, from that point of view, there is nothing left of the whole traditional approach to the structure of language, other than taxonomic artifacts, and that's a radical change, and it was a very liberating one. The principles that were suggested were of course wrong, parametric choices were unclear, and so on, but the way of looking at things was totally different from anything that had come before, and it opened the way to an enormous explosion of research in all sorts of areas, typologically very varied. It initiated a period of great excitement in the field. In fact I think it is fair to say that more has been learned about language in the last 20 years than in the preceding 2000 years.[8]

The last sentence is certainly a showstopper: either you believe it or you are stunned by its scientific immodesty. But immodesty (if that is what it is) aside, it illustrates the giddy feeling of liberation that so often comes along with being part of a movement that takes itself to be revolutionary. Martin Joos, an ornery member of the post-Bloomfieldian generation, must have had this in mind when he wrote that "linguistics has been preeminently a young man's pursuit ever since the 1920's."[9]

Sociology also reminds us that it is not always best to focus too much on the individual: as Bourdieu put it, it is not so much the heir that inherits the inheritance, in the world of ideas, as it is the inheritance that inherits the heir![10] We should not be too shocked to discover that systems of positions and dispositions are reborn in each individual in each new generation of scholars.

Here's another way to think of it. There is a *force* that we can feel when we read the work of giants who have preceded us, an energy that comes with it, an ability to *make us think today*. At the same time, the most profound contributions have always been the result of a thorough knowledge of orthodoxy and its dogma mixed with a passion for heterodoxy. There is no deep mystery why this should be so. It is the simple result of the fact that no one thinks alone or starts over from scratch.

Here is something else to keep in mind, something that we will state more than once, because it bears repetition: if the constant reminders of the sources of our ideas make the dead weight of the past seem inescapable, don't worry. Escaping the dead weight of the past is usually very simple: all that is necessary is to become aware, to become knowledgeable. The liberation is virtually instantaneous. There are grounds for hope and optimism.

Our Kind of Science

Any observer of the linguistic scene would notice that every generation has wanted its field to be *scientific*, and what's more, each generation thinks that it will be the very *first* generation to have succeeded in the quest to become a science. Within the mind sciences (linguistics, psychology, philosophy, logic), each generation rebukes the previous one for having wrongly thought that it had its hands on a legitimate scientific method and framework, and then it immediately goes on to offer what it takes to be a *truly* scientific vision.

It is much more interesting for the reader to see this directly. Here is a modest sample of moments when linguists observe that *finally* linguistics has become a science. We will begin here with a typographical convention that we employ in the rest of the book: within a quotation, added emphasis appears in boldface, and original emphasis appears in italics. Feel free to skim.[11]

> Since the commencement of the present century, and especially within the last fifteen years, the philosophy of language has been pursued with great ardor, and the learned on the continent of Europe, **by following the grand Baconian principle of induction, have placed this science on a solid basis**, and are in the way of most important discoveries. These discoveries are modifying the grammars and lexicons of every language. . . . The new method of

grammar has a thorough and proper unity, because it commences with the proposition, as the central point. The value of every word and of every form is made to depend on its relation to the proposition. This develops the organic relations of language, and gives to the new method a scientific form. . . . The new method . . . of course is the same for all languages. Different languages may all be analyzed in the same way. (Josiah Willard Gibbs 1838)

Another science, cultivated with great zeal and success in modern times, compares the languages of different countries and nations, and by an examination of their materials and structure, endeavours to determine their descent from one another: **this science has been termed *Comparative Philology*, or *Ethnography*; and by the French, *Linguistique***, a word which we might imitate in order to have a single name for the science, but the Greek derivative *Glossology* appears to be more convenient in its form. (William Whewell 1858)

In old classical usage, [philology] meant the love of literature; afterwards the scholastic mastery and exposition of language; more recently a sort of general amateur study of language, as a matter of mere pleasant curiousity; and last of all, **the scientific exploration and comprehension of its interior mechanism, in relation both to its original elements, and also to their varied transformations**, through a wide range of comparative analysis. (Benjamin W. Dwight 1859)

The science of language is a science of very modern date. We cannot trace its lineage much beyond the beginning of our century, and it is scarcely received as yet on a footing of equality by the elder branches of learning. We hear it spoken of as comparative philology, scientific etymology, phonology, and glossology. In France it has received the convenient, but somewhat barbarous, name of *Linguistique*. . . . We do not want to know languages, we want to know language; what language is, how it can form a vehicle or an organ of thought; we want to know its origin, its nature, its laws; and it is only in order to arrive at that knowledge that we collect, arrange, and classify all the facts of language that are within our reach. (Max Müller 1862)

In a course of lectures which I had the honour to deliver in this Institution two years ago, I endeavored to show that the language which we speak, and the languages that are and that have been spoken in every part of our globe since the first dawn of human life and human thought, **supply materials capable of scientific treatment**. . . . We can treat them, in fact, in exactly the same spirit in which the geologist treats his stones and petrications, nay in which the botanist

> treats the flowers of the field, and the astronomer the stars of heaven. **There *is* a Science of Language**, as there is a science of the earth, of its flowers and its stars; and though, as a young science, it is very far as yet from that perfection which . . . has been reached in astronomy, botany, and even in geology, it is, perhaps, for that very reason all the more fascinating. (Max Müller 1864)

> **Those who are engaged in the investigation of language have but recently begun to claim for their study the rank and title of a science.** Its development as such has been wholly the work of the present century, although its germs go back to a much more ancient date. It has had a history, in fact, not unlike that of the other sciences of observation and induction—for example, geology, chemistry, astronomy, physics—which the intellectual activity of modern times has built up upon the scanty observations and crude inductions of other days. . . . But to draw out in detail the history of growth of linguistic science down to the present time, with particular notice of its successive stages, and with due mention of the scholars who have helped it on, does not lie within the plan of these lectures. . . . Its execution would require more time than we can spare. (William Dwight Whitney 1867b)

In 1871, August Schleicher described linguistics in a way that seems so modern that we cannot present less than the first two paragraphs:

> Grammar forms one part of **the science of language**: **this science** is itself a part of the natural history of Man. Its **method is in substance that of natural science generally**; it consists in accurate investigation of our object and in conclusions founded upon that investigation. One of the chief problems of **the science of language** is the inquiry into, and description of the classes of languages or speech-stems, that is, of the languages which are derived from one and the same original tongue, and the arrangement of these classes according to a natural system. In proportion to the remainder but few speech-stems have hitherto been accurately investigated, so that the solution of this chief problem of the science must be looked for only in the future.
>
> By grammar we mean **the scientific comprehension and explanation of the sound, the form, the function of words and their parts, and the construction of sentences.** Grammar therefore treats of the knowledge of sounds, or phonology; of forms, or morphology; of functions, or the science of meaning and relation, and syntax. The subject of grammar may be language in general, or one particular language or group of languages; grammar may be universal or special: it will in most cases be concerned in explaining the language as a

product of growth, and will thus have to investigate and lay down the development of the language according to its laws. This is its exclusive province, and therefore its subject is the laying-down of the "life of language," generally called historical grammar, or history of language, but more correctly "science of the life of a language" (of sound, form, function, and sentence), and this again may be likewise as well general as more or less special. (August Schleicher 1871)

Great progress has been made in phonological science during the past score or two of years, and it is hardly too much to say that the mode of production of the ordinary articulate sounds composing human language is now understood in all its main features. (William Dwight Whitney 1865)

Here is the objection, which we take to be more or less well grounded: you transform the study of languages into the study of Language, of Language as considered as a human faculty, as one of the distinctive signs of its species, as an anthropological, or even zoological, character. . . . The most elementary phenomena of language will not be suspected, or clearly noticed, classified, and understood, if we do not insist on the study of languages from beginning to end. Language and languages [*langue* and *langage*] are one thing: one is the generalization of the other. If you want to study Language without undertaking the effort to study the quite evident diversity of what is found in languages, your effort will be in vain; on the other hand, if you want to study languages but lose track of the fact that in their very nature these languages are governed by certain principles of Language, your work will be *even more bereft of serious significance, and of all real scientific basis.* (Ferdinand de Saussure 1891)

A new science, called Phonetics or Phonology, has sprung up, and is now universally admitted to have created the modern science of language. In addition to this physiological and physical basis, **the superstructure of the science of language has likewise been stated to be no longer a historical or a philosophical, but to have become a physical, science.** It is true that, as with other natural sciences, so also in this case, the morphological, genetic, and biological aspects can be specially studied; also analogies can be drawn between geology and glossology as to their mode of inductive reasoning.

[Merz adds, in a footnote:] In the modern science of language we have one among the many cases where a historical or philosophical science is becoming an exact science by attaching itself to physics and physiology. . . . "It is phonology," says Prof. Sayce (*Introduction to the Science of Language*, 2 vols. 1880, chap. iv) "which has created the modern science of language, and phonology may therefore be forgiven if it has claimed more than rightfully belongs to it or forgotten that it is but one side and one branch of the master science itself. . . . It is when we pass from the outward vesture of speech to the meaning which it clothes, that the science of language becomes a historical one. The inner meaning of speech is the reflection of the human mind, and the development of the human mind must be studied historically." (John Theodore Merz 1903)

The essential point . . . is . . . that de Saussure has here first mapped out the world in which historical Indo-European grammar (the great achievement of the past century) is merely a single province; **he has given us the theoretical basis for a science of human speech**. (Leonard Bloomfield 1924)

In order to ascertain whether and to what extent linguistics is entitled to the name of a science, we must remember that in Modern English the term "science" may be understood in two different ways, viz: (1) in a broad sense . . . i.e., scholarly knowledge; (2) in a more modern and more technical sense, so as to be applied exclusively to branches of learning concerned with permanent and invariable relations, such as mathematics, chemistry, physics. These and similar sciences, it is claimed, are able to make predictions for the future. If interpreted in this way, the term would not be applicable even to the evolutionary branches of natural science, such as geology and biology. . . . **The science of linguistics is . . . concerned with uniformities and permanent or steadily recurring conditions in human speech generally.** We may count here, e.g., topics like the relation between language and dialects, the causes of phonetic change, the nature of phonetic laws, the mutual relation between appellatives and proper names, the various systems of counting, etc. . . . As branches of linguistics concerned with permanent conditions, we may claim, above all, general phonetics and general grammar. Phonetics nowadays has assumed such proportions as almost to constitute a science by itself. . . . In general or "philosophical" grammar, on the contrary, stress is laid principally on the relation between grammatical forms and mental categories. (Hermann Collitz 1924)

The layman—natural scientist, philologian, or man in the street—does not know that there is a science of language. **Such a science, however, exists**; its aims are so well defined, its methods so well developed, and its past results so copious, that students of language feel as much need for a professional society as do adherents of any other science. (Leonard Bloomfield 1925)

At the present time phonology is characterized by its structuralism and its systematic universalism. . . . This direction of thought can be seen in physics, in chemistry, in biology, in psychology, in economic sciences, etc. Phonology is thus not isolated. It is part of a broader scientific movement. (Nikolai Trubetzkoy 1933)

It is only within the last century or so that language has been studied in a scientific way, by careful and comprehensive observation. (Leonard Bloomfield 1933)

Dynamic philology [which is what Zipf called his approach] has the ultimate goal of **bringing the study of language more into line with the exact sciences**. To this end it views speech-production as a natural psychological and biological phenomenon to be investigated in the objective spirit of the exact sciences from which its methods have been taken. (George Zipf 1936)

The starting-point in **such a science** is to define (1) the universe of discourse and (2) the criteria which are used in making the classifications. (Charles Hockett 1942)

The native languages of our country had been studied by some very gifted men, but none had succeeded [before Boas] in **putting this study upon a scientific basis**. (Leonard Bloomfield 1943)

The Kimhian [David Kimhi, b. 1235?] theory placed the study of the Hebrew phonetics on a scientific basis. (William Chomsky 1945)

There can be no doubt that Bloomfield's greatest contribution to the study of language was to **make a science of it**. Others before him had worked scientifically in linguistics; but no one had so uncompromisingly rejected all prescientific methods, or had been so consistently careful, in writing about language, to use terms that would imply no tacit reliance on factors beyond the range of observation. (Bernard Bloch 1949)

Before the appearance of Bloomfield's *Language*, linguistics was usually treated as an essentially humanistic discipline, often fruitful but not completely amenable to scientific method to procedure by postulates, hypotheses, and verification. . . . Descriptive linguistics had, with few exceptions, remained on the level of our traditional West European normative grammar on the Graeco-Latin model; there had been scattered recognition of the need for improved methods of linguistic description (de Saussure, Boas, Sapir), but pre-Bloomfieldian efforts along this line had been relatively desultory and unsystematized. **Bloomfield was the first to demonstrate the possibility and to exemplify the means of a unified scientific approach to all aspects of linguistic analysis**: phonemic, morphological, syntactical; synchronic and diachronic. (Robert A. Hall Jr. 1949)

Modern scientific study has forced us to abandon many of the older commonly held views of language and has provided us with new principles and new assumptions which underlie new methods of analysis and verification. But the cultural lag in assimilating the results of **this modern scientific study of language has been so great that the views and practices of a prescientific era still dominate the schools**. (C. C. Fries 1952)

Crossland: Linguistics is still rather a young discipline. It's only in the last 25 years or so that really serious attempts have been made to put the study of languages on something like a scientific basis. And the people who've been making them have been mainly occupied so far—and I'd say quite well occupied—in working at suitable procedures for recording and analyzing individual languages. . . .
Halliday: I agree. . . . I think that in a sense what enables linguistics now to combine usefully with other disciplines is that it has been freed from entanglement with other subjects, such as psychology, and **its establishment as an independent scientific study**. (BBC Cambridge Language Research Unit 1952)

If one wanted to characterize in a word the direction in which linguistics seems to prolong [the views developed by Meillet], one could say that they mark the beginning of a linguistics conceived of as a *science*, by its coherence, its autonomy, and its goals. To say that linguistics moves in a direction of making itself scientific is not only to insist on a need for rigor which is common to all disciplines—it concerns first of all a change of attitude towards its object, which is defined by an effort to *formalize it*. (Emile Benveniste 1954)

All those who knew Bloomfield best seem to agree that his chief professional concern was to develop *linguistics as a science*. . . . There has been considerable difference of opinion from time to time as to the demands of "scientific" linguistics, but concerning the label of the ultimate goal itself there has been unanimity. (C. C. Fries 1962)

The essence of Chomsky's revolution in linguistics was **his gift to the field of a truly scientific perspective**. (Frederick Newmeyer 1986)

[Minimalist grammar] is well on its way to becoming a **full-blown natural science**, offering a serious promise of an advanced field of scientific inquiry whose idealizations, abstractions and deductions will eventually match in depth and subtlety those of the most advanced domains of modern science. *Generative grammar is turning into a natural science already, because of what it is now, not because of what it might one day turn into.* (Piatelli-Palmarini 1998)

In spite of its antiquity as an object of human enquiry, **linguistics as a science in the modern sense is very young**. It was only the last century [i.e., the twentieth century] that the study of language moved beyond observation and superficial description to attempts to explain why human language is the way it is. The primary impetus for **this dramatic and revolutionary shift was the work of Chomsky** (Chomsky 1957, 1965). (Peter Culicover and Andrzej Nowak 2003)

We are among those who are persuaded, on solid grounds we think, that in the past 50 years linguistics has progressively established itself as a **genuinely scientific discipline**. (Boeckx and Piatelli-Palmarini 2007)

When we read a few of these remarks from the nineteenth century, it is tempting to brush them off, but when one sees the same message published non-stop over 150 years, it forces us to pause and think about why scientists of the mind conclude, again and again, that they are the first to approach their problem scientifically.

Were they right or were they wrong? If later generations look back at the work of previous generations and doubt that what they found was the product of legitimate scientific practice, what is going on? Has the very meaning of the term *science* changed over time, or are the demands of scientific practice evolving in relation to how much has already been discovered? The answer to these latter two questions is undoubtedly yes, but we can also take away another entirely legitimate conclusion from this brief tour: the thoughtful actors in this story have always made special efforts to reflect on what science is, and how their discipline should develop in order to be taken seriously as a science by the neighboring scientific fields.

This, then, is a major theme that we will follow throughout this book: the best thinkers are constantly asking themselves, What does it mean to be a science and also to be interested in my questions? How should we be doing our work if we wish to be scientific?

The World of Ideas and the World of Social Relations

As linguists who came of age in the late 1960s, we ourselves saw and *felt* the two effects we mentioned above that were so common among the generative grammarians that we were proud to be a part of. We prided ourselves on our liberation from the shackles of behaviorism and other forms of empiricism, and we felt that generative grammar finally brought linguistics to the same playing field as other sciences. *Finally*, we thought, linguistics had developed formal theories that were worthy both of the complexity that emerges from a careful study of the data, and of the efforts of scholars who understood the power of formal mathematical models. We forgot that we were ourselves the children of a revolution which itself had been equally forgetful: the structuralist revolution which did its best to forget its past and reinvented the human and social sciences during the half century stretching from 1910 to 1960 by setting up the abstract notion of structure as the king in the kingdom of concepts that could explain everything.

We haven't forgotten the feeling, but we have come to realize that we are not the only ones to feel this way. We were preceded by a number of generations of researchers who felt just the way we did—and we have been followed by younger scholars who feel that linguistics is finally about to make it as a scientific field, for the very first time. We do appreciate the irony.[12] When we go back and read the early publications of such benighted predecessors as John B. Watson, the father of behaviorism, or the followers of Ernst Mach, the godfather of the Vienna logical positivists, or the structuralist linguists whose work forms the basis of our field today, we don't find stupid statements—we find people trying to cast off the chains of an official orthodoxy that they are certain is superannuated and standing in the way of scientific progress. And yet in the versions of intellectual history written by later generations of victors of the battles in the mind fields, the earlier approaches are described time and time again as so simple-minded that we can hardly take seriously anyone who went down that particular road. But we do know that those people, our intellectual ancestors, were no more stupid than we are today. Something must be wrong with the history books.

Maybe it's *not* simply the history books that are at fault. Maybe it's the simplistic conception of history that needs to be changed. There is nothing wrong with seeing history as a linear sequence of events, marked with dates and places—but that's only part of the picture. We cannot see the whole story unless we see the shifting tectonic plates of our history, composed of and populated by ideas and ideologies, including many global visions of the world that continued to have an impact on how scientists constructed and viewed the objects of their investigations.

Let's take an example. For the vast majority of contemporary linguists, regardless of the school to which they adhere, structuralist linguistics as it was practiced between 1925 and 1965 is as foreign as the blast of light that still reaches us from the Big Bang, a blast that is now reduced to a background buzz in the sky. Even contemporary views that recognize a debt to structuralism seem to view it as a dried-up well that was once the inspiration of a golden age.

In the history of science and that of ideas, the *thickness* of time is not uniform. On the one hand, 50 years separate us from the publication of *The Sound Pattern of English*, the manifesto of generative phonology published by Noam Chomsky and Morris Halle in 1968, and still it seems to be alive, living among its contemporaries in generative phonology. On the other hand, if 40 years separated Leonard Bloomfield's "Set of Postulates for the Science of Language" from Chomsky and Halle's opus, linguists in the 1970s could conceive of classical structuralism only as an obscure theory from an obscure time, a time that was almost literally prehistoric. It is as if space-time had been warped to such a degree that neither light nor information could reach us from that time. And this is all the more surprising given that any linguist over the age of 40, to say nothing of the founding fathers of our current schools, were trained in the methods and concepts of structuralism, whether we knew it or not. There isn't a student in linguistics from the over-40 generation who didn't burn the midnight oil trying to solve problems built from data torn from the pages of the *International Journal of American Linguistics*, to say nothing of problems photocopied out of the standard textbooks of the 1950s.[13] We know whereof we speak: we were there ourselves, we burned that midnight oil.

What is the cure for this selective amnesia that leaves us blind to our own origins? This is the work that we alluded to above under the rubric of Bourdieu's *anamnesis*, the first goal of any study of the history or the epistemology of a discipline. If it sounds suspiciously like psychotherapy, then so be it! We need to bring into the light of day the hidden linkages among ideas, sometimes denied because they show connections to ideas that seem embarrassing somehow. We need to bring out the underground ruptures that were never publicly acknowledged.

We are not willing to think about intellectuals as a spontaneous product of a virgin birth, or as creative powerhouses free of any and all external influences. We cannot understand theoretical frameworks without understanding the linkages and influences that helped to meld and form them. To speak today of intellectual *genealogies* is a bit loaded, as the word suggests Foucault's thoughts and his take on the history of thought, much of which finds no resonance in this account. But genealogy is important, both for understanding a patrimony passed down in ways both conscious and un-, and for trying to unravel the conflicts and tensions which sometimes are passed down more as that-about-which-nothing-should-be-said than as any sort of explicit inheritance. One of the themes that promises to teach us a lot about ourselves is the deathly silence that has for so long hovered over the question of how the work of Bloomfield, Sapir, and their students has been a fundamental component of all subsequent American linguistics, including the most dominant perspective, generative grammar.

We have learned that the value of studying academic genealogies was greater than we expected, and we will share with you quite a number of intellectual genealogies over the course of this book. To some degree, we were inspired by the work of Collins (1998), in which the study of individual and personal influence seems to shed considerable light on the way in which influence and authority is defined and aligned in academic fields.[14]

While Kuhn's *The Structure of Scientific Revolutions* (1962) plays a role directly in some of the literature that we will discuss, it is more generally viewed as part of a longer intellectual tradition that includes contributions by Pierre Duhem, Karl Popper, Imre Lakatos, Paul Feyerabend, Larry Laudan, and other people who have developed ways of understanding the history of science (or of sciences) and the way in which the nature of science as we know it includes alliances and conflicts. We are indebted to all of these authors for their insights, and we will refer to them at various points. Our primary goal is not to construct an overarching theory of science nor to align more with one of these scholars than another, but we are very much indebted to them in ways that will be clear throughout this book.

The history of the mind sciences is one of both rupture and continuities, and our principal task is to figure out how this can be so. A simple generalization can carry us quite a way. When we focus on the *ideas* in this story, what we see is a braid of ideas that interconnect and develop over time, and our story is one of continuity. When we focus on the *positions* taken by the individuals in the story, we find bold statements that separate rival camps, and we find ruptures of various sorts. Both of these perspectives are real, but neither of them, taken individually, is the whole story: this is found only in seeing both, together, at the same time.

We have, therefore, given ourselves total liberty to abstract away the human and social context when that is useful for our study of ideas, and *also* to ask how the intellectual positions of an individual or of a group are affected by the fact that such human agents are living in a world composed of human beings. The first is sometimes called *internal history*, the second *external history*; both are important for us.[15] At times we do the first, and follow the trajectory of an idea as it arises in one domain and evolves, perhaps touching down in two or three other domains. At other times, we examine the way in which real people interacted with other real people: even if they shared an interest and a passion for the study of the mind, they were all along flesh and blood human beings as well.

How do styles and forms of social interaction lead to direct and immediate effects on the growth and spread of ideas? Some ways are simply obvious. No one would deny the role that personal charisma can play in the spread of ideas. Some of the people we will discuss in this book were, or are, tremendously charismatic—for example, Franz Brentano, Edward Sapir, and Noam Chomsky. Others—such as Leonard Bloomfield—were anything but.

The complex relationship that exists between a dissertation advisor and his or her doctoral student is another social bond that will be part of the story we consider. We have provided quite a few genealogies that indicate the relationship between a thesis advisor and a student. And an equally important relationship, one that will play a major role in our discussion, will be that of *authority*, a complex notion that involves both people (who is the authority? in whose judgment is she the authority?) and fields (she is the authority when which questions are at stake?). As human beings, we all live in a complex patchwork of such *fields*: a Catholic may agree that in religious matters the pope is the final authority, but if the Catholic is a biologist, or for that matter a pharmacist called on to sell products her church does not approve of or may even condemn, she must come to a decision on how the forces and relationships in one field carry over into another. None of these ways of thinking should be taken as tools to oversimplify complex issues; none of them deny the fact that the scientific world enjoys greater autonomy in some respects than others, due precisely to the commitment to increase knowledge that lies at its heart.

We must not leave aside the very personal passion that a scientist has for *knowing*, which can be just as strong as any social ties with other human beings, or even stronger. We are very aware of this; we think it is well characterized by Augustine and his interpreter Pascal, who speak of *libido sciendi*, a human pleasure—"passion" might be a better word—that comes from snatching glimpses of truth. The seeker after truth is often willing to sacrifice a great deal if that sacrifice is the price of knowledge.[16] Such a seeker after truth also prizes the awareness that he or she is not alone in trying to pose questions to Mother Nature and in finding ways to quantify and calculate and specify explicit models of nature. A scientist discovers who he is—or rather, *that* he is a scientist—by recognizing that he finds pleasure and even joy in working, both alone and in teams, to better understand the natural world. We emphasize this point in order to underscore our view that we do not make science profane by considering its sociological aspect; that social side is one part, but only one part, of the larger picture.

Isaiah Berlin, the most profound raconteur one would ever hope to meet, wrote about his life in philosophy, and he put his finger on an interesting phenomenon that is not at all uncommon, and by its very nature involves the group within which one works—and in part, but only in part, its size. He wrote about what happens when one chooses an artificially small and personal group of associates to serve as one's intellectual cohort. Oxford University was his home for many decades.

> One of the shortcomings of these meetings is something that seems to me to apply to Oxford philosophy in general, at least in those days. We were excessively self-centered. The only persons whom we wished to convince were our own admired colleagues. There was no pressure upon us to publish. Consequently, when we succeeded in gaining from one of our philosophical peers acceptance or even understanding of some point which we regarded as original and important, whether rightly or, as was more often the case at any rate

> with me, in a state of happy delusion, this satisfied us completely, too completely. We felt no need to publish our ideas, for the only audience which was worth satisfying was the handful of our contemporaries who lived near us, and whom we met with agreeable regularity.

Berlin went on to say that the philosophers in that Oxford crowd did not feel that they had anything to learn from anyone outside the group. "This was vain and foolish and, I have no doubt, irritating to others." Of that, there is no doubt. This description is echoed by linguists in the early years of generative grammar, linguists who were not from MIT and whose remarks were not taken seriously by young generativists. Berlin ended with another telling observation: "But I suspect that those who have never been under the spell of this kind of illusion, even for a short while, have not known true intellectual happiness."[17]

Now, there is no way to tell whether Berlin's conjecture, offered in passing, is really *true*, but his point was this: For those seeking true intellectual happiness—what we referred to above as *libido sciendi*—the adventure must be done in a community, not as a solitary individual, and the optimal size of that community may be measured in scores or hundreds, but certainly not larger than that. And not only is the work done *within* that community, but it may well be that membership in this community goes hand in hand with an icy indifference to what is going on, intellectually speaking, outside of that community.

The significance of Isaiah Berlin's remark is that it reminds us of the importance of thinking about research in social terms—indeed, in sociological terms. Scientists quite rightly focus their attention on the subject of their science—whether that subject is language, rock formation, or mitochondria. But in *doing* science, each scientist is part of a social group—in fact, of many social groups, including the people from whom she learns directly (her teachers), indirectly (the authors of her textbooks and all the people who have established the field), and potentially (her cohort in grad school, her competitors, and so forth). We humans do very little that does not involve us *as members of social groups*—and that is our real advantage as a species.

Scientists spend *most* of their time thinking about science: that is their work. From time to time, they think about the nature of knowledge and the relationship between their theories and the world they study. We will have many occasions over the course of this book to hear the voice of scientists reflecting on the relationship between scientific theory and the reality that science aims to study. But rarely do scientists turn their attention to the more abstract question of the social structure of their activity.

Sociology is a field that could, in principle, be used to study the social structure of the scientists and their milieu.[18] Indeed, over the last several decades, sociologists such as David Bloor, Bruno Latour, and Steve Woolgar have done just that, developing a perspective that has been called the *strong program* in the sociology of science. We do not see our work as fitting within that perspective, but we do believe that it is of the greatest importance to include in our account of how science works an explicit understanding that scientists work in a complex world, one that is part of a larger human culture, situated in time and in space. We can learn from sociology how to ask questions that allow us to better understand how scientists accomplish what they do. We view the contributions of sociology to the questions we consider here to be entirely complementary to a study of the explicit logic of scientific research.

It is both helpful and important for us, as we organize our exploration of science and scientists, to draw a distinction between the pure world of ideas and of theory, and the social world in which scientists and disciplines exist and interact. Of course this is to some degree an artificial distinction: every paper is written by somebody who is a human being, every lecture is given by a person with strengths and foibles, and the audience in each case has a pretty good idea of who is writing and speaking—and this "pretty good idea" certainly has an effect on how the paper or lecture is received and understood. But the distinction is nonetheless important, and it allows us to view the developments in each of our sciences (and the activities undertaken by all of the scientists) in a way that provides new insights.

Our focus in this book is the nature of continuity and of rupture in the mind fields, and we are in a position now to observe that it is in the social world where rupture tends to be present and, indeed, to be a dominant characteristic, for there can be a clash between scientists over questions of authority even when there is little difference between their ideas. Of course there are important scientific debates over real scientific issues, and it can easily happen that these debates align with different scientific groups vying for scientific ascendency—and in a sense that is what we might hope to find. But that is not all we do find, a good deal of the time.

Here, then, is what we propose to show: as we look carefully at the development of linguistics, we find far more continuity in the world of ideas than the extant literature would have us believe. Ideas move from one discipline to another—from logic to linguistics, for example—and from one embattled subdomain to another—generative semantics to interpretive semantics, for example. Good ideas tend to flourish.

Things are much more complicated on the social level, where people are born, grow up, enter a discipline, and look for colleagues—compatriots—and a job. The world that they see before them is different from the world that their teachers saw in front of them a generation earlier. Those who persevere and remain active in their own discipline see it evolve, slowly or quickly depending on their own internal clock, and they do their best to help their students and those tendencies in their disciplines that they view as promising. And then they retire and leave, and all the while the process continues.

The social world of the scientist is built up out of relationships of communication, cooperation, and competition. No surprise there. Scientists communicate? Of course they do; they read and they publish papers, and they go to conferences. What more could we ask for? They cooperate: they share their results, and they go to extraordinary lengths to get their results out in front of their colleagues as fast as they possibly can. And they compete for such things as resources, and they compete for priority, and all of this competition makes the frontiers of knowledge move forward as quickly as possible.

As we say, that much is generally accepted, but there is more.

There is a far more important sense in which colleagues in a discipline support one other: they support each other's views as to what the important questions are that must be answered, and how those questions should be defined; they support each other's views as to how their discipline relates to neighboring disciplines.

They compete because life's resources are finite and limited; a scientist wants his work to be taught in courses being given by his colleagues at other institutions, and there is just a finite amount that can be showcased in a course. There is only so much space in a syllabus, and adding one new piece of work typically means throwing out something else: that's what it means for time to be finite. One could draw an analogy here to real-life commerce. We hear talk about "selling" a theory, but there is one enormous difference in the academic world (as Bourdieu[19] pointed out): by and large, the people to whom one wants to *sell* one's work are the very same ones who are out selling their own work. Much of science is a large souk, a bustling marketplace, where there are no customers for any merchant other than all the other merchants present that day. This simple fact has an enormous impact on the social structure of science.

Let's be clear: there is nothing wrong with this, and if a person knows how to do science well, then one of the consequences of that is simply that others will be interested in his work. There is no way to criticize a scientist for having made his work interesting: the closest one can get to that is the jealous response we sometimes hear that Professor So-and-So's presentations are *flashy* (which typically goes along with the follow-up that the material is not *deep*): *flashy* is a form of *interesting* that is not filling and does not last long, and one wonders afterward just what it was that seemed so appealing during that flashy presentation.

The issue of what questions are *interesting* and *important* is second only to the question of what is *true* (and sometimes may seem even more important). In some disciplines, external funding agencies can wield enormous influence in this regard, and they may be quite aware of the role they play in influencing what questions are viewed as interesting and important. If a federal agency decides to support documentation of endangered languages, then the importance of that field will, quite simply, rise. Much more often, however, it is the senior workers in a field, those in their mid-forties or older, who make the case, in public, as to what the important questions are that should be addressed, and these scientists work to create a *reputation* that will encourage others to take their suggestions seriously.

Over the next few pages, we will discuss three ideas from sociology that are useful for understanding the evolution of the mind sciences: *generations, authority*, and a fierce fighting word, *ideology*.

Generations

Generations play a large role in the story told in this book. Across cultures and across time, there are many different ways in which generations have been viewed. One view sees little but simple pairs of parent and child, as in the book of Genesis: "And Arphaxad begat Salah; and Salah begat Eber. And Joktan begat Almodad, and Sheleph, and Hazarmaveth. . . ." Such a view is useful when we are interested in keeping track of who is a descendant of whom. But there are other views, much richer in texture.

In Greek mythology, the most important organizing force in the pantheon of the gods and heroes is the generation: mother and father come together and engender a child—or, more often, a whole set of children. Children of the same parents (and even to a large extent, children of the same father) tend to form groups of solidarity. The Titans were 12 children of Gaia and Uranus. Just as important, there is more often than not a presumption of serious conflict between a father and his offspring.

We all know that Sigmund Freud alluded to the Greek legend of Oedipus, who killed his father, King Laius, and Freud took this as a way of talking about the challenges a young boy experiences as he grows up. But the powerful dynamics of generations play a much larger role than that particular one. Listen: Gaia was the very first solid thing that emerged after the aboriginal Chaos (whose name says it all), in Greek mythology. Gaia is the terra firma of our universe. After she came to be, she somehow managed to generate Uranus out of herself, without engaging in the usual procreative practices. Uranus was the sky, he was younger than Gaia, and he became her partner and her mate. The *very first* sexual union was thus that of Gaia and Uranus.

They were a fertile couple, but Uranus was a terrible father, and he refused to let any of his children emerge from Gaia—the first dozen of the offspring were the Titans, and the rest were equally awe-inspiring. The Titans *knew* their father would not let them become people in the world: they were in a much more dysfunctional family than Oedipus's (and that is already saying a lot). To bring this myth back to twenty-first-century academic life, these Titans were the role models for the graduate students whose academic father never wants to let them finish their degree and go out into the world.

Back to Greek mythology. Uranus forced his children to stay in the underworld, which caused Gaia great pain, as well as great grief. Gaia conferred with her children about what could be done, but it was only the youngest of the Titans, whose name was Cronos, who was willing to take on his father (everyone in this story, unlike Oedipus's, is immortal, even if they are vulnerable to attack). Cronos took a sickle and castrated his father Uranus. *This* symbolism requires no exegesis.

Life in dysfunctional families generally stays bad. Cronos himself was warned that he would someday have a son who would overpower him, and so *Cronos* in turn smothered his children. In his case, he did it not by keeping them inside their mother (as his father Uranus had done), but by swallowing them whole, which is a process in mythology that does not lead to imminent destruction—it is much like being swallowed by a whale in the Pinocchio story, or the story of Jonah in the Bible. Cronos's children were the role models for all of the graduate students whose urge to strike out on their own is met by their fierce teachers' will to keep them lashed tightly to the teachers' established truth.

Now, Rhea, who was both Cronos's sister and the mother of his children, saved her last child from being swallowed by Cronos. The last child was named Zeus, and she saved him by passing Cronos a surrogate child (always referred to as a "stone" in the myths) and secretly saving the real Zeus. When Zeus grew up, he managed to get his siblings out of his father (sources differ as to the method employed, though the use of an emetic remains the best hypothesis), and they all banded together to wage war on their hapless father Cronos and his siblings, the Titans. That 10-year war was the first great war in Greek mythology, and it pitted one generation against another and one set of siblings against another. Zeus's side (the younger generation) won the war, and their home, Mount Olympus, came to be known as the palace of the gods.

The power of these myths is that they bring together in one story the kinds of psychological and social forces that play major roles in how people act, individually and in groups. We certainly *could* understand intergenerational conflicts without bringing up Greek myths, but they do help us to focus on what makes us tick as the humans that we are, and they help us to understand, if only in a prescientific way, the kinds of forces that give rise to particular challenges that we will see in the chapters to follow, such as the rupture between the Neogrammarians and their teachers during an important moment in the development of modern linguistics.

But by no means is it *necessary* to embrace the metaphors of mythology to think seriously about generations. Sociologists since Karl Mannheim[20] have explored the consequences of knowing this very simple fact about humans: we are born, we mature, we age, and we leave the scene to be replaced by others. Given the kind of creatures that we are, we leave behind a record of what we have seen and what we have learned, but the next generation after ours never experiences the same things that we did. Each generation faces challenges (social, economics, political, and other sorts) that were never seen in quite the same way before, and each generation has just enough time and interest to learn what it can from what was left to it by the previous generation.[21] Science progresses fast *because* it gets rewritten by each generation: rethought, restructured, and rewritten. Some things get lost along the way, but hopefully not too much and hopefully nothing that we will regret having lost.

Or perhaps that is an oversimplification. Mannheim believed that it was just as important for society to forget as it was to remember, especially if the forgetting was a precondition for progress, or anything like it. He likened the lack of experience in the young to a lightening of the ballast in a ship: a lighter ship may be more agile, but then again, it may capsize in a storm. But on the whole, the inevitability of forgetting is, if not a good thing, a necessary thing, and a society (Mannheim wrote) composed of people who never died would have to come up with a new way to forget.

> Any two generations following one another always fight different opponents, both within and without. While the older people may still be combating something in themselves or in the external world in such fashion that all their feelings and efforts and even their concepts and categories of thought are determined by that adversary, for the younger people this adversary may be simply nonexistent: their primary orientation is an entirely different one. That historical development does not proceed in a straight line—a feature frequently observed particularly in the cultural sphere—is largely attributed to this shifting of the "polar" components of life, that is, to the fact that internal or external adversaries constantly disappear and are replaced by others.[22]

Let's illustrate this point with an example: Wilhelm Wundt's effect on the rise of psychology in the second half of the nineteenth century (see chapter 4). One of his students was Edward Titchener, who thought of himself as bringing Wundt's ideas to the United States—by way of Cornell University, in the event. Titchener, in turn, had a student named Edwin Boring, who became a successful and influential professor of psychology at Harvard University and whose writing on the history of psychology left a profound impact on how American students of psychology perceived the origins of their ideas. More recently, Kurt Danziger questioned Boring's efforts to go behind Titchener's own words to see what Wundt himself was arguing for. "Boring was himself," Danziger wrote, "deeply committed to the positivist philosophy of science whose influence on the early development of psychology is at issue here. But his is the commitment of the second generation: What had been for his teachers conclusions carefully arrived at and boldly asserted, have now become matters to be taken for granted, implicit certainties not open to debate or even worthy of mention." Danziger underscored the pernicious effect that this leads to: psychologists fail to see that many of their decisions about how to treat phenomena are the consequence of prior philosophical commitments, not realizing that there are indeed a range of philosophical positions that can legitimately be taken, all of which have an impact on the work in psychology. "This is a comforting attitude," Danziger wrote, "for those who have no wish to question fundamental assumptions, and that usually includes the conservative majority."[23] The more a philosophical view merges with the mainstream, the harder it can be to identify, even by those whose thinking is influenced by the idea all day long.

The effects on disciplinary knowledge that arise from the eternal shifting of generations are of two sorts. We have emphasized one kind, the more *epistemological* sort, which arises from the fact that the understanding of any one thing by a given generation will be different from the understanding of it by the preceding generation because the totality of what the newer generation has to learn has changed. The most striking instance of this occurs when a generation that struggled to learn something new and revolutionary passes the baton on to the generation that follows it, a generation which learned the revolutionary material in the classroom from textbooks.

The second kind of generational shift arises out of the fact that each generation begins young and then gets older and grows up, taking on greater personal and disciplinary responsibilities with each passing decade, expecting the generation of its teachers eventually to cede to it the positions of authority that once had been held by the older generation. This transfer of authority and influence is inevitable, but how smoothly and how graciously it occurs depends on many factors.

Let us look at an example of a shift in perspective that grows directly out of different generational perspectives. We will look at two remarks, made at different times by the same linguist, the first when he was a young man, and the second when he was an older and very distinguished figure (we'll let you know who he was after you've read what he wrote).

As a young man, he described what had happened when he sent a manifesto to an organization he belonged to. He thought the manifesto audacious, and he referred to the specific items as "theses":

> There were no substantial objections to the theses defended by [his group], and especially the resolutions about the tasks of [the larger organization] was accepted unanimously. If, however, [the manifesto] had been submitted to a secret ballot, it would have certainly provoked a few votes against it. Such was, at least, the impression gained from talks in the corridors. But, as a matter of fact, do the votes against mean much when they are devoid of any attempt towards argumentation? Such silent voices belong to those who realize that the recognition of the principles of . . . [linguistics] generates the necessity for fundamental changes in the field of synchrony, in linguistic history and geography,

> and in the description of literary languages, whereas such a thorough reorganization does not suit the adversaries' temperament.

This is a highly political view of the social structure that this young man was describing. He was just setting out on a career, in a period in which writing political manifestos was as natural as breathing the air. In the writer's fantasy, a vote was being taken, a secret vote, and there was a certain frisson that came from the thought that perhaps the theses would have been objected to if the pusillanimous scholars had let their true beliefs be known. And in that world of fantasy, those naysayers, those linguists who would have voted against, are not worthy of the privilege of having a vote: even if they had said no, they would have been the meaningless votes of the democracy in which everyone gets the same vote—just one—regardless of whether they really understand what they are voting on or not. This writer is a young man who is sure that he knows better.

Forty years later, this man, not so young any longer, has become the elder statesman of the field—it is Roman Jakobson, a major figure in twentieth-century linguistics. It is no longer of any value to think that the field of linguistics is riven by disagreement: what good is it to be a senior statesman if one isn't listened to? Now Jakobson preferred to see accord and unity, even when the rhetoric in the street seemed to say otherwise.

"Linguistic theory of our time seems to offer a stunning variety and disparity of clashing doctrines," Jakobson wrote in 1971.[24] But that is misleading, he suggested. Do we think we see "intensive contentions and tumultuous controversies"? That is mere appearance: do not be deceived. "A careful, unprejudiced examination of all these sectarian creeds and vehement polemics reveals an essentially monolithic whole behind the striking divergences in terms, slogans, and technical contrivances." That is quite interesting, if only because it invites us to face the question: when is it appropriate to tell two sides of an academic dispute to stop their disagreement, because the rest of the world sees them as arguing about how many angels can dance on the head of a pin? Over the course of this book, there are many occasions where the heated words and intemperate rejoinders seem, from our position today, quite out of proportion to what was at stake.

In 1971, Jakobson urged the younger linguists to see that "most of these allegedly irreconcilable contradictions appear to be confined to the surface of our science, whereas in its deep foundations the linguistics of the last decades exhibits an amazing uniformity." He wanted linguists to understand that when *he* was a young man, the field was rent asunder by real disagreements. Today, though, what linguists have in common "is particularly impressive in comparison with the substantially heterogeneous tenets that characterized some earlier epochs of this discipline, in particular, the nineteenth and the early years of the twentieth century." Jakobson urged linguists not to be led astray by terminology. "Most of the recent discord is based partly on dissimilarities in terminology and style of presentation and partly upon a different distribution of linguistic problems chosen and pointed out by single scholars or teams of inquirers as the most urgent and important." Be more open minded, Jakobson suggested, and recognize that what interests *you* need not set the limits to the questions the entire field is engaged in answering.

We do recognize that it is not possible to remove the role played by personality in matters of generational conflict. Consider the noted philosopher Ernst Cassirer (who is deeply connected to our skein of psychology and linguistics as well), who was born in the nineteenth century and chased to the United States by Hitler, like so many of his peers.[25] His view of generations was different, and he felt no attraction to the notion that "there is a deep and insurmountable gap between the generations; that every new generation must feel in its own way, think its own thoughts and speak its own language. I regard this as a misleading and dangerous dogma—and as a dogma that throughout my life I found constantly contradicted by my own personal experience."[26] His intellectual equilibrium was not matched by many others in this story.[27]

Then there is the question of age. Many people have pointed out, with varying degrees of graciousness, that the older one is, the harder it gets to change one's views about basic scientific questions, but even that observation (which is no doubt correct) stands in need of explanation. Is it to be explained by hormones and brain deterioration, or by rational risk-aversion, or by the possibility that the older scientists understand better than the younger ones do the range of good reasons why the current orthodoxy came to be dominant? Whichever account turns out to be correct (and all of those sound quite plausible) makes a difference for the conclusions that we draw from it.

Let us draw the tentative conclusion that some of the explanations for conflict and change may relate directly to a difference of generation. Still, that remark by itself leaves open a wide set of interpretations: the older generation may be irrationally clinging to a bygone tradition, the younger generation may be seeking something that is simply different from what came before, the older generation may be suffering from inadequate technical skills, the younger generation may be seeking job perks, or simply jobs. Technologies and dominant ideas may change, and a younger generation may be more willing or more capable of adapting and adopting them.

The generational character of a discipline is distinct from the effects growing out of the strong mentor/student relationship that invariably arises in the training of a young scholar; this latter lies properly in the domain of individual psychology, while the generational character of a discipline lies at the social level. This is a distinction that is useful, though we cannot pretend that it is always easy to draw neatly. Actors in our story make this point, in fact: we have already met Titchener, a psychologist important in the early years of American psychology, who had lost his father early in life; he remarked, "Until one is thoroughly settled for oneself, it must be good to have someone responsible in the prior generation upon whom one can lean." He wrote of William James, a bit older than himself and established as a psychologist, "James especially owed it to American psychology I think, to take some interest, and to deal out praise and blame."[28]

That is a very interesting and revealing remark, even if it is one with which we do not agree, and it is one that steps well outside the bounds of what can be justified on traditional scientific grounds. It is, at the end of the day, one of those things that one may say to oneself when one feels that the world is not quite fair and not quite the way one's parents had said it was going to be when we were grown up. We will see other cases where one generation disappoints another—sometimes the older one disappoints and sometimes the younger one.

Authority

A second aspect of science that arises because of its social character is the presence of *authorities*. There is no getting away from accepting the word of authorities, and what an authority declares is better protected from being falsified than is something that is declared by someone *not* viewed as an authority. What the great Charles Darwin wrote, with his characteristic modesty and his charm, is just what we all hope we can say:

> No doubt errors will have crept in, though I hope I have always been cautious in trusting to good authorities alone.[29]

Alexis de Tocqueville offered a number of astute observations regarding precisely this point: to accept an authority means to trust, and there is no option that avoids this:

> A man who should undertake to inquire into everything for himself could devote to each thing but little time and attention. His task would keep his mind in perpetual unrest, which would prevent him from penetrating to the depth of any truth or of making his mind adhere firmly to any conviction. His intellect would be at once independent and powerless. He must therefore make his choice from among the various objects of human belief and adopt many opinions without discussion in order to search the better into that smaller number which he sets apart for investigation. It is true that whoever receives an opinion on the word of another does so far enslave his mind, but it is a salutary servitude, which allows him to make a good use of freedom.[30]

Darwin and Tocqueville look at one side of the authority market, the demand side: we *need* authorities. As long as there is research to be done, there will be a demand for authority. The other side of the coin is the supply side (although the marketplace metaphor begins to feel a bit contrived): what the scientist wants above all else is to be the provider of authority to others, which is to say, to *be* the authority. The very word *authority* contains within it two important things: it is, first of all, relational. One can only be an authority for others, in the sense that one is never an "authority" to oneself (it is not even clear whether it makes sense to ask whether one is an authority in one's own eyes). If someone is an authority in an area, it is *to someone else* that he is an authority. Being an authority is by its very nature a relational, a social, phenomenon. And being an authority in science has much in common with the more general use of the term *authority*, as when we say that "he took his complaint to the appropriate authorities." The authorities have a certain power invested in them, and we expect them to exercise that power in a legitimate fashion, not overly influenced by self-interest.[31]

It seems to us that the natural history of science can only be understood if we look carefully and sensibly at both aspects of science. Each individual scientist works as hard as possible to move the accumulated wisdom of a discipline forward, and in doing so to establish himself as an authority, in some fashion, among those with whom he works, and among those with whom he communicates.

The notion that a scientist strives to be an authority is hardly surprising; when as professors we train beginning graduate students, we tell them that their work towards their doctorate will focus on their becoming the world's expert on some particular (and almost always) small domain. We expect them to *control* the literature in that area, and we hope that when they are done, no one else will be able to publish something on that topic without having to cite our student's (eventual) doctoral dissertation, or a journal article derived from it.

To a certain limited degree, the goal of achieving authority *may* under some circumstance act as a force binding larger groups and mitigating forces towards smaller groups, in the sense that the total amount of authority an individual reaps is heavily weighted by the size of the group in which that authority is recognized. Oversimplifying a bit, this is just to say that if one is invited to give a keynote address at a meeting of an association and derives from this invitation some measure of authority in his future interactions, the amount of authority (if we can speak of such a thing) is directly linked to how large the association is.

Group Identity

It does not take very much for a human act to become a social act: when the act involves language, all that is necessary is for the person to imagine that he is addressing someone else, or that he is speaking along with someone else—as a linguist would say, all that is necessary is for there to be a first-person plural, or a second person engaged in the conversation, real *or imagined*. That is a very low bar. Once those conditions are satisfied, the person begins to develop an understanding of thinking as a social act.

When an individual acts, he typically acts as a member of a group in which he views himself as participating, and as that participation grows, he adopts and develops an account of what that group is. And so we will speak of a person's *self-in-group identity*—or *group identity*, for short—and of the person's *group identity account*. Charles de Gaulle viewed himself as a Frenchman, and he had a historic account of what it meant to be a Frenchman, associated with many noteworthy moments, not the least of which was the French Revolution in 1789. A graduate student submitting a thesis proposal views herself as a graduate student in a particular discipline. As such groups arise, it is in the nature of human beings to develop accounts for themselves and for others as to just who they are. These accounts will often include a simplified story of a group's origins, its original aims, and its current aims. Often this story can play an additional role: helping to maintain group solidarity, or even sending an encouraging message to those who are outsiders. In this way, the story can begin to take on a function of *justification* of the project that formed the group.

As an example, let's look at part of the preface of an important work that we will discuss in chapter 7, when we turn to early twentieth-century philosophy. After stating what he intended to do in his book, the author (whose identity the reader will learn in due course) turned to the question of how his work related to the work of others, and he made this observation: "The basic orientation and the line of thought of this book are not properly an achievement of the author alone but belong to a certain scientific atmosphere which is neither created nor maintained by any single individual." Today we might paraphrase this by saying that knowledge is a social good that we share, rather than the possession of an individual, but when we say that, who is the *we* we have in mind when we say that *we* share it? We will come back to that. "The thoughts which I have written down here are supported by a group of active or receptive collaborators." We will look at the genealogical ancestry of this philosopher later, but for now we may observe that he was not referring to the larger movement his work was indebted to, and in particular he viewed himself as part of a much smaller group. He explained that members of the group had "in common especially a certain basic scientific orientation." In fact, this smaller group was as much as anything defined by what it had found, tested, and deemed to be no good at all in other philosophers' work. As for the group's own work, the fact that it rejected a traditional philosophy "is only a negative characteristic," he wrote. "The positive features are more important: it is not easy to describe them, but I shall try to give a loose characterization."

At this point, he began to describe specific characteristics of this group. "The new type of philosophy has arisen in close contact with the work of the special sciences, especially mathematics and physics. Consequently [members of the group] have taken the strict and responsible orientation of the scientific investigator as their guideline for philosophical work, while the attitude of the traditional philosopher is more like that of the poet." His group's members did science, and they did not associate with people who thought like poets. "This new attitude not only changes the style of thinking but also the type of problem that is posed."[32]

That is a very important point for those of us who are interested in the development of scientific groups; at the end of the day, nothing is more important than the characterization of what the *questions* are that we are interested in, and working on. We will see that in the area of the mind sciences, groups define strong principles that establish what these questions are for them. Those principles remain in a murky area that is neither quite a statement of fact, nor quite a statement of value; these are the principles that define what an interesting question is for people in the group. The group of our mystery speaker was the Vienna Circle; the author was Rudolf Carnap, writing in 1926, just as his fame and that of the Vienna Circle were about to expand. We will return to his views in chapter 7.

To sum up, then: for each social group that we belong to, we develop an account of that self-identity. This constitutes a story that we tell ourselves about who we are: about who we are as members of a certain group. In these stories it is convenient and useful to include what E. G. Boring called *Great Events* as well as *Great Men*. Boring was a psychologist, and he was talking to other psychologists, though he knew that they viewed him as a historian as well, and that he was therefore permitted to speak more freely than others might be. Nonetheless, he softened his point by using ironically Capitalized Nouns. He was discussing the fact that a discipline comes round to sharing stories of when its movements began and how they started, and anyone who makes an effort to understand the intellectual landscape will know that the Great Events are peaks in a chain of mountains: high and visible, but not isolated and not always the highest altogether. Boring said these Great Events helped "meet man's need to make history comfortable to understanding by personalizing it." He went on to observe that these stories often incorporated specific Great Events that were taken as moments when a movement started. His examples all came from the early decades of psychology, but we will see examples of this throughout the course of this book. Fechner thought his basic view of psychophysics first visited him in his bed on October 22, 1850. "People find pleasure in birthdays. To date the birth of a thought is to dignify it, and biographers pick up these anecdotes and embed them in history."[33]

Boring was all too aware of the inaccuracies that these comfortable stories might contain, despite the function they play. Should we try to get rid of the stories altogether? That was not possible. "The practical solution for all these predicaments seems to be to allow them, to use them, but to recognize them and every now and then to take measures to offset them."[34] This seems like a sensible piece of advice.

Twenty years earlier, Boring had been a bit less tolerant of the ways of the behaviorist movement: "A formal movement is thus a protest and the psychological reason for protest is, of course, insecurity. No established science feels insecure or protests, for, being secure, it turns to work without attention to itself." He was reflecting on the state of psychology in William James's day, which he thought was "insecure, self-conscious, protestant, and full of the business of founding itself";[35] it was *aggressive* and it exaggerated the importance of what it brought to the table.

Ideology

The term *ideology* ineluctably enters into the picture at this point, in light of the fact that one of its definitions is the *account that the members of a group construct to define who they are as a group*— what we referred to above as a group's account of its identity.[36] Still, that is not how the term has always been used, in serious literature and not just the media. We can recall the far more tendentious accusations of *ideology* found in discussions of the radical Left during the 1960s, including Lewis Feuer's intemperate *Ideology and the Ideologists*;[37] Feuer developed a Freudian view of 1960s rebels that has become part of the shared memory of that time. For his part, Feuer proposed three core items in *any* ideology of rebels: "an invariant mythological structure, an alternating set of philosophical tenets, and a historically determined chosen group."[38] The myth is some variant on the Moses story, a man leading his people to freedom. "What is distinctive in ideology," Feuer wrote, "is the drama it sets forth as the 'meaning' of the historical process, together with its assignment of the roles of leadership elite, chosen-class, and historical culmination."[39] In *The Conflict of Generations*, he wrote, "Student movements are the product of selfless, altruistic idealism combined with the resentment and aggression of one generation against another."[40]

The word *ideology* has, to be sure, an ingloriously checkered past, and it continues to be used today in everyday life in a casual and ill-defined sort of way, typically in a polemical context where a writer who is critical of somebody else's position calls it "ideological." "Free schools are a dangerous ideological experiment," thundered one British politician, concerned that the schools were teaching Islamic beliefs.[41] Another writer on the political scene describes as "ideologists" people who have a need to see more purpose in life than they find in staid bourgeois existence. A third commentator draws a parallel between militants who devoted their lives to the Communist Party and those who devote their lives to Islamic fundamentalism, declaring that such people have in common a need for an ideology (unlike the commentator). Yet another will characterize the "dominant ideology" of Western society as that of progress or reason.[42] In case after case, what is described as an ideology is a belief that is held to with great strength, but a belief which the writer is sure no one would take seriously if it were brought out into the open and debated rationally by people of good will. That very fact calls for an additional explanation of why *those* people over there do cling to that belief.

There is no getting away from the fact that in the broader world in which we live, the usage of the word *ideology* is both charged and tendentious. This fact is regrettable, because the *notion* of ideology, as developed by sociologists, *could* have been of use to us; it *could* have served to refer to the *group identity account*. We will have to be very careful as we consider the power and strength of adhering to groups, whether they are political or intellectual in their grounding.

The term *ideology* was coined by Destutt de Tracy at the end of the eighteenth century, and he used the term to describe the study of the sensory origins of ideas.[43] That usage did not last long. Napoleon, seeing in the Ideologists—a particular group of influential thinkers—critics of his political aims, began to use the word *idéologue* pejoratively to refer to a political actor whose abstractions are of dubious worth. Marx, and marxists after him, have often used the term as part of a way of arguing that most of the fundamental and tacit principles of a society emerge out of the economic relations found there, and these principles—the society's *ideology*—may be self-serving, oversimplified, and nearly invisible to those held in its sway.

There are essentially three ways in which we find tacit criticisms packaged under the rubric of ideology.

Ideology, in the first place, may be detected where there is an unhappy and unfortunate mixture of value and description. What passes itself off as description on the surface turns out to be heavily value-laden. Raymond Aron suggested, "Political ideologies always mix (with varying degrees of aptness) propositions of fact and judgments of values. They express a perspective on the world and an intention regarding the future. They do not fall neatly into the category of true and false statements." We will see an example of this in finer detail in volume 2, when we explore the influential work of Thomas Kuhn on scientific revolutions. One of his colleagues accused him of writing ideology disguised as history of science. Paul Feyerabend wrote to him,

> What you are writing is not just history. *It is ideology covered up as history.* Now please, do not misunderstand me. . . . [I do not] pretend that in history a nice distinction can be drawn between what is regarded as a factual report, and what is regarded as an interpretation according to some point of view. But points of view *can* be made explicit. . . . Nobody will think that the history of crime justifies crime, or shows that crime possesses an inherent "reason" or an inherent morality of its own. In the case of the sciences or of other disciplines [for] which we have respect the situation is much more difficult and the distinction cannot be drawn with equal ease. But in these cases it is of paramount importance *to make the reader realize that it still exists.* You have not done so. Quite on the contrary, you use a kind of double-talk where every assertion may be read in two ways, as the report of a historical fact, and as a methodological rule. You thereby take your readers in. [44]

A second criticism that is often brought under the general umbrella of *ideology* involves the accusation that one's opponent adheres inflexibly and intransigently to a belief that is much firmer when compared to other beliefs that one might maintain: a real *clinging* to a belief. People may be more resistant to *acknowledging* this belief, and they may find it harder to give up that belief in the face of what others see as reasons to abandon it. In short, they may be overly or irrationally committed to an idea—though not, needless to say, from their *own* point of view—and they may be unwilling or unable to formulate it explicitly. Lurking behind this view, more often than not, is the hope that the study of ideology can play the role of psychoanalysis, by curing and freeing the person who had been held under ideology's sway. Finally, charges of ideology are sometimes leveled when the accusation is really that one's own self-interest is at stake. A more complex version of this is that it is not so much one's own self-interest as it is the interests of those who already dominate (those who are "hegemonic," in Gramsci's usage). This sense of the word is often associated with a Marxist orientation. But it can be interpreted in various ways, and some of them are more appropriate for a discussion of science and of scientists (and their behavior as scientists). Researchers in a given field may all agree with one voice that tremendous advances have been made in the last ten years, say—and even if the person who repeats that sentiment out loud may not have any papers published that represent some part of that great leap forward, it is nonetheless true that he has a stake in the ongoing health and wealth of the discipline. That stake can be as simple as the belief that his work is part of a legitimate scientific enterprise (and not a waste of his time) or as complex as an effort to increase the money and jobs devoted to his style of research.

The first point (dressing up *shoulds* as *ares*) speaks directly to the *content* of the ideological belief; the second to the too-close-for-comfort relationship between the human believer and his belief; and the third, to the relationship among the objective economic conditions of the believer, his society, and the social role played by the belief.

Within all three uses of the term *ideology*—three uses which at times just barely contain the indignation of the accuser—there is one thing in common: they are ways in which an individual's thought process is deflected from the true path it would have taken if that person existed in a world where there were no friends, colleagues, competitors, self-delusion, idle curiosity, deadlines, mortgage payments, conferences, books, publishers, fatigue, tenure decisions, or time constraints. And because as scholars we often do our best to evaluate ideas abstracting away from those factors, we may naturally be led to the thought that when those factors do play a role in what we (or our colleagues) do, there is something deeply wrong. Raymond Boudon takes such a point of view; he uses the term *ideology* in his effort to better understand how it is that otherwise rational people can hold to a position that *seems* to rest on science, and yet does not, and which is nonetheless clung to with a force out of all proportion to what is rational.[45]

Jehovah's Problem and Noah's Solution

> Nothing is more usual and more natural for those, who pretend to discover anything new to the world in philosophy and the sciences, than to insinuate the praises of their own systems, by decrying all those, which have been advanced before them. —David Hume, *Treatise on Human Nature*, introduction

There is an odd and curious phenomenon that occurs and recurs in the history that we will tell. We call it "Jehovah's problem." You may not have realized that Jehovah had *any* sort of problem. Let's begin with a story that you know.

The reader will recall the state that Jehovah found Himself in, early in Genesis, just before the Flood.[46] He looked at the sorry mess that the human race had made for itself and for the rest of the world, and decided that He had had enough. He was going to eliminate it all, and start all over again, but do it right, the next time. After a bit of reflection, He realized that Noah was not at all half bad, and it would not be fair to eliminate him or his family. He would spare them, and the world would start all over again, but this time with just Noah and his closest kin. Noah built the ark; Jehovah sent the rain. Forty days later it was all over, and the only ones left were those who had made it onto Noah's Ark.

Noah was indeed a lucky man. He, and all of his descendants, did not have to contend with any competition from any of Noah's former friends, enemies, or teachers. They were all gone, all washed up and washed away. All of Noah's contemporaries, after the Flood, were highly beholden to him. The book of history was thenceforth rather short, too, because it consisted of everything that Noah wanted it to, and nothing else at all.

We will find many a mover and thinker in the mind sciences over the course of this book who felt himself to be both in Jehovah's shoes and in Noah's. This is someone who looks out on what he sees, who looks back on what he has been taught, and does not like it, not one bit. This is someone ready to chuck it all and start over: someone who would like to be able to call down 40 days of rain and a huge flood to wash away the competition, someone who is sure he could ride it out in an ark of his own design. Alas, no one can do that. Still, we find characters who do the best they can, characters who send forth the message that everything that is being done today is a worthless waste of time. They have a *new* story to tell, a new way to study the mind, and we can do it right this time.

We call this Jehovah's problem—and obviously, it is not a "problem" in the usual sense; it's more of a mind-set and a marketing strategy, and a particular interpretation of how one's own work relates to the preceding scholarship. But it is very common in the mind sciences, and coming to understand it, in all of its nuances, is one of the challenges that we will face. Most often, this mind-set goes hand in hand with the view that everything that has preceded has failed to be scientific, and now we can go forth and be scientific—a pattern we have already discussed briefly. We will see this in psychology, first when John Watson introduced behaviorism in 1913, and again when behaviorism was overthrown (note the metaphor!) by cognitivism in the 1950s. We see it in linguistics when Bloomfield declares (with his students' proud acclaim) that linguistics has finally become a science, in the 1920s, and again in the 1960s, when Chomsky declares (with his students' proud acclaim) that linguistics is finally a science.

In philosophy, we see this over and over again, in any number of different guises. The most famous philosopher who invited down upon himself a flood to wash away all assumptions and all former teachings was René Descartes, in the seventeenth century: he declared that he would doubt all things, wash away all certainty, and try to build up his beliefs and his knowledge from scratch. While some philosophers have been content to build upon the work of their predecessors, many have called for a complete washing away of what preceded them, on the grounds that it was all spoiled and rotten and worthless. The most ambitious of these was the logical positivists of the Vienna Circle, who urged a program that would relegate almost all of the work of their philosophical predecessors to the dustbin of history, where hopefully no one would ever read it again.[47]

Towards the end of the nineteenth century, the philosopher Josiah Royce (perhaps the last American philosopher whose thought could be said to be squarely and firmly tied into the European currents of philosophical thought) made a similar observation. He began by saying that scholars and thinkers all recognize their ties to earlier thinkers: "The time is long past when really intelligent thinkers sought to do anything outside of intimate relations to the history of thought," he wrote in 1892. But he thought about that again, realized that was not quite true, and continued, "It still happens, indeed that even in our day some lonesome student will occasionally publish a philosophical book that he regards as entirely revolutionary, as digging far beneath all that thought has ever yet accomplished, and as beginning quite afresh the labors of human reflection." He obviously had an example or two in mind, but he did not choose to share them with us. "Such men, when they appear nowadays, as once in a while they do appear, are anachronisms; and you will always find them either ignorant of the history of the very subject that they propose to revolutionize or incapable of reading this history intelligently." Yes, he clearly had some examples in mind. "What they give you is always an old doctrine, more or less distinguished in a poorly novel terminology, and much worse thought out than it has already been thought out, time after time." And having acknowledged that such people do exist, he went so far as to point a finger about modern liberalism: "It is one of the defects of the current liberalism in matters of opinion that it does encourage, only *too* often, this sort of thinking; and the sole corrective of the error is a certain amount of philosophical study of an historical sort before one begins to print one's speculation."[48]

It will be one of our main tasks in this book to document this pattern and to try to come to grips with what is wrong with it; what is, occasionally, right about it; and why this pattern is so common. It is not just a personal problem; we are not interested in a psychological analysis of anybody and certainly not of the people who have helped move these fields forward. The point is rather that we see a pattern, and not only that, we fail to see it reported in the literature as a generalization. When it is remarked upon, it is always as someone's personal failing—usually that of the would-be Jehovah (or Noah). But clearly there is more to it than that. The proof that there is more to it is simple: in the real world, there is no Flood. There is nothing that washes away the books and the publications of the earlier scholars. And yet John B. Watson was able to convince people not to read literature from before behaviorism, and B. F. Skinner's students were delighted to never have to read anything before Skinner. Chomsky's students did not have to read what had been published before 1957, and so it goes. Why did the world of scholars permit itself to become dumb? That is the question! Anyone can tell you not to read something. But what is it that makes you willing to follow that advice?

Credit Problem and Heroes

There is a problem encountered by the kind of approach we develop in this book that we should point out here: it does not provide any help in solving the problem of credit attribution. If anything, studying the scholars' work up close makes it all the harder to solve the problem of credit attribution. The more we learn about the evolution of the mind fields, the harder it may be to figure out who the *real* heroes are, and we find ourselves forced to question the reasonableness of asking that question.

In his brilliant book on Galileo, Kepler, and Newton, I. Bernard Cohen found himself trapped by the conflict of the two regimes of the world of ideas and the world of human beings. He surveyed the evolution of the ideas of motion, impetus, and inertia, and the development of these notions in the centuries before Galileo. One thing is perfectly clear: the world did not jump directly from Aristotle's view of motion to Galileo's, even if Galileo and more modern scholars would like to give that impression. Cohen wrote,

> Galileo's originality was therefore different from what he boastfully declared. No longer need we believe anything so absurd as that there had been no progress in understanding motion between the time of Aristotle and Galileo. And

> we may ignore the many accounts that make it appear that Galileo invented the modern science of motion in complete ignorance of any medieval or ancient predecessor.[49]

If you actually *read* the physics literature in the centuries preceding Galileo's work—that of Nicole Oresme, for example—you cannot fail to appreciate the continuous conceptual development during these centuries, and Cohen knew all of that material well: that was his discipline, after all. Cohen clearly sensed that there is a conflict at some level between demonstrating continuity in the development of ideas and the pointing out the brilliance and creativity of the work of such men as Galileo. And so he wrote,

> By making precise exactly how Galileo advanced beyond his predecessors, we may delineate more accurately his own heroic proportions.[50]

Heroic: that word says it all. When we focus on individuals and their life stories, we build heroes, and occasionally villains, and certainly buffoons. We explore the jealousy, we wonder at the rages, but the more we learn about the actual life of the ideas, the more we grow to distinguish the personal strengths and flaws from the advance of ideas.

Cohen was one of the greatest historians of science of the twentieth century, and we do take his perspective seriously, even if we do not agree with it. He insists on the importance of great leaps of individual minds:

> We do not fully understand why or under what conditions, a few hardy individuals are from time to time led to think in wholly new directions, but the fact is that they do.

Though he adds,

> New ideas are rarely creations unrelated to the general background of ideas.[51]

We would add: not only the general background of ideas but also the general sociohistorical background.

Here is another way in which the conflict between the regime of ideas and the regime of people has been treated. Claude Allegre, a well-known French scientist, described the origin of the notion of tectonic plates, first suggested by Antonio Snider-Pellegrini in 1868 and developed in the following years by others, including Elisée Reclus and Frank Taylor. But it was Alfred Wegener, writing in the second decade of the twentieth century, who is generally given credit for the idea. As Allegre notes,

> He defended his theory firmly but without excessive agressivity until his dying day. And so it is that he should be considered the father of the theory of shifting continents. As Georges Duby put it, in matters of reference and precedence, we must establish a simple rule, one which distinguishes clearly between an opinion which is simply one among many, expressed fleetingly, and a work that is built, argued for, and developed around a central idea. [In Duby's words:] "Reference to one is anecdotal, to the other is central and necessary." [52]

Allegre is trying to solve a problem that simply does not exist, which is to say, he is trying to resolve the conflict between the continuity that inheres in the world of ideas and the rupture that we insist must exist in the world of actors so that we can fairly and justly apportion credit for originality. Perhaps that is too crude a formulation. Of course there is *a* problem, a problem of credit assignment, because that is how our modern world today works: we expect there to be an answer to the question of who deserves the credit for the idea of continental drift, the idea that continents are floating on tectonic plates. But this credit problem is not one which aligns sharply with any significant, or even meaningful, question in the history of ideas. In the world of ideas, continuity is the dominant characteristic.

In the next chapter we will look at the rise of European linguistics in the nineteenth century, and in many of those accounts, the author feels the need to decide just who was responsible for the emergence of this new science. Two of the earliest candidates are William Jones and Friedrich Schlegel, but as one scholar notes, "despite the various claims that have been made in favor of Sir William Jones or Friedrich Schlegel in the history of linguistics, it is still generally held, and I believe with some justification, that Bopp's *Conjugationssystem* of 1816 constitutes the 'breakthrough' of the New Philology."[53]

Trying to determine who should get the credit for an intellectual advance is asking the wrong question. Sometimes it is an unavoidable question to ask in the heat of the moment, as when we make decisions about whom we should hire or who should receive an honorary doctoral degree, or a Nobel prize. But no such concerns drive us as we write this book. To adopt a metaphor dear to the heart of Americans, science is a team sport, and while we know that individuals will win prizes for outstanding performances that are statistically measured and individuals will be selected as Most Valuable Player at the end of each season, it is still the teams who play and win the games.

There is another reason to downplay the credit-attribution problem. Deciding who should get credit can drown out consideration of other questions that are also important. One such question concerns the natural passage of ideas from one of the mind sciences to another—from philosophy to psychology, from psychology to philosophy, and so on. The actors, the thinkers who are themselves engaged in this, are by no means the people who best understand how this passage happens. In a recent study of the origins of Karl Popper's influential position on the nature of modern science, Michel ter Hark argues convincingly that the important position Karl Popper published in the early 1930s involved ideas that Popper had not had in 1928, when he wrote his doctoral dissertation, and that the ideas were solidly rooted in his study of Otto Selz, a psychologist who had written his second (*Habilitation*) dissertation with Külpe just before World War I, and who by the early 1920s was professor of philosophy and psychology in Mannheim.

Ter Hark is well aware that his reader is likely to say, "who?" when he first encounters the name of Otto Selz. "Selz was, I think, the greatest scientist to emerge from the brief but extraordinarily creative phase of German psychology at the beginning of the twentieth century in Würzburg" and a member of a group that included Oswald Külpe and Karl Bühler. Now, it is true that neither Külpe nor Bühler is well known today, but anyone who has read a bit about the history of psychology will have encountered them, something that cannot be said of Otto Selz. It is clear that Ter Hark senses an injustice that began over 80 years ago. "Not to be credited for his scientific achievements seems to have been Otto Selz's destiny (and fear)," Ter Hark writes. And he proceeds to show in detail how Selz's ideas about creative thought were adapted and adopted by Popper. That Popper's work was an important example of the flow of ideas from psychology to philosophy is fine and admirable. But Ter Hark goes a step further and compares the intellectual influences that he has found with the stories that Karl Popper developed in his intellectual autobiography decades later. "Reading Otto Selz . . . brought about a significant change of perspective in this early psychology, one which would ultimately lead to his evolutionary stance in epistemology and philosophy of science. Because Popper never explains this formative role of Otto Selz in his published work, I even began to think of him as seriously distorting the historical record." Ter Hark ultimately set his goal to be to "reconstruct the immensely fruitful interaction that took place between psychology of thinking and epistemology"—between psychology and philosophy—and "simultaneously to give Otto Selz the credit that he especially deserves."[54]

Ter Hark may or may not have accomplished the task of getting Selz the credit that he deserved. But he comes very close to raising a question to which his work gives a partial answer, one that is more important for our task: how should we read accounts constructed by our mind scientists? How should we interpret their choice of what to talk about and what not to talk about? Their choice of whom to talk about and whom not to talk about? We might even say, their choice of what to remember and what to forget? Sometimes the answer to those questions is as simple as noting that the mind scientists are hoping to engage in the credit-assignment problem, or to engage in honorable or dishonorable efforts to influence future scholars engaging in credit assignment. When they do that, the stories they leave behind for us are not worth very much. But there may not be a better way, a *right* way, to carry out the credit-assignment problem. We are not convinced that there is.

Let's be clear on this, then: the distinction between the ideas and the intellectual positions taken by people that we will study is an artificial one, in the sense that one cannot exist without the other. There is no history of ideas to study if there are no scientists around to develop the ideas, and there are no scientists to make bold claims if there are no ideas. Nonetheless, the difference is both useful and important if we are to get a better understanding of much of what happens in the history of ideas, and in particular, if we are to understand how the history of the mind sciences could be simultaneously a story of rupture and of continuity.

If we could, we would simply dismiss the credit-assignment problem from all intellectual history: nothing so distorts the discussion of the development of our ideas than the passionate attachment to the assignment of personal credit. But we can't; there is some inevitable and unavoidable reason to take on the credit-assignment problem, as we will see. But the cost of going down that road is very high indeed.[55]

Mind and Materialism

One of the largest themes that will follow us throughout the book is the development of our understanding of mind, matter, and mechanism—and machines. Over the four or five centuries in which the Western scientific view has evolved, there has always been a sense of complementarity between mind and matter. For some, like Descartes, that complementarity is the reflection of a sharp division between the two, while for others, the separation has been less clear and more gradual. For almost everyone, the worlds of mind and matter differ at the very least by the ways that we describe and think of them, and the principles that we see guiding those two worlds. Over this time, our understanding of *both* mind and matter has changed considerably—indeed, radically.

A profound shift in Western thought occurred during the late sixteenth and seventeenth centuries in which a new picture of materiality emerged, one in which the most important aspects of what is real in the world we live in was directly tied to material shape, to location and movement, and to a new, measurable quantity called *mass*. This shift was deeply connected to the scientific advances that were made in the study of the movement of objects both in free flight and under the influence of gravity. Galileo, René Descartes, Isaac Newton, and others developed an understanding of the world according to which straight-line motion was a natural state for objects to remain in, and there was something about objects (not their size or shape, but something else) that determined both their resistance to change in speed and the degree to which gravity acted upon them. This is what Newton called *mass*. Since mass was revealed, in part, by the way it interacted with gravity, the mass of an object could be measured by setting it on a scale, to see by how much force it was pulled to the Earth.[56]

This was the first great scientific advance of the Western world, and it gave us a new sense of how the inorganic world fit together both beneath our feet and above our heads, both on the ground and in the heavens. But this scientific advance did *not* come with a mission to deny the reality of other aspects of the world, including most notably the spiritual side. Neither God nor the human mind was eliminated from the world views of Galileo, Descartes, or Newton. If the planets moved in paths that obeyed systems of quadratic equations in ways that people had never suspected, that was hardly a reason to doubt that a great mind lay behind the creation of this marvelous solar system that we live in. Yes, there was a revolution in how we viewed the physical universe; no, the revolution did *not* call for the deportation of God and spirit from the universe of the scientist. It was no accident that most of the greatest physicists were also great mathematicians: they were dazzled by the discovery that the language of nature, of God's creation, was mathematics.[57]

Our modern material view of the world was born in this period of 150 years—a view in which location, movement, and mass were central and essential properties, but several outstanding puzzles remained. The puzzles left little doubt that there was a great deal more about the universe than the distribution of matter in it. One mystery was why so many things retained fixed shapes. We call them solids, but why do some objects maintain a fixed shape as they move or rotate? Sticks, rocks, and bones (but not water or air) have a shape, a form, which means that the stuff inside them was bound together with a set of internal forces that remained to be explained. Whatever is responsible for holding things together is not matter itself. If there are atoms, what keeps the ones that are in solids in place? What keeps them from moving too far apart, or coming too close together? When two objects collide, why do they *collide*? Why do two solid physical objects refuse to mix and mingle, though two streams of water do? And how is it possible that things with the same size and shape can differ with regard to how much of this stuff called mass they are composed of? That is, why does a block of iron have more mass than a block of wood? Are there more tiny things jammed together inside a small piece of iron than there are in a small piece of wood? These were very basic questions about the fundamentals of the materialism that was emerging, and they had no obvious answers.

Behind these reflections was a hope cherished by our trio of scientists (Galileo, Descartes, and Newton) and those who came after them: they hoped that all interaction between things made of matter could be boiled down to two kinds of interactions. One was the local interaction between things that are colliding with one another, and the other was the non-local interaction that we call gravity, which mysteriously acts between massive objects over long distances.

This modern worldview began with an effort to carve out some aspects of the world we can understand, but succeeding generations wanted to explore the idea that *this material world is all there is*. Suppose we allow that there is matter that is revealed quantitatively by how much mass it has, that mass can somehow congeal into objects with shapes and sizes, that these objects can move in space, and that they interact with each other only when they collide with one another (and then there is gravity too). But suppose we say that that is all there is; there is nothing more. What then?

As we just noted, the mechanical view of the world that Galileo, Descartes, and Newton proposed did *not* require that there be nothing else; Descartes could not have been clearer on the subject, explaining that there is both mind and matter in the universe. He understood the limits of explanation coming from the study of mechanics: mechanics has nothing to tell us about the way people think or the way we use language. But others would follow who went to extremes, and of these the most famous was Descartes's fellow Frenchman La Mettrie who famously declared that man was a machine. La Mettrie was born a half century after Descartes died; from La Mettrie's point of view, he was *adopting* Descartes's idea and pushing it to its logical extreme. If Descartes had been there to disagree, he would have told La Mettrie that he himself had been *drawing a distinction* between mind and matter, that he could not have been clearer about this point, and that he was not trying to get rid of everything on the non-material side of that distinction. La Mettrie would have shaken his head, saying that he was just taking Descartes's ideas seriously. If he could have, Descartes would have told La Mettrie that what was important was not the mechanical side of the material world, but the overarching power of the rationalist point of view, capable both of informing us about how things work in the material world and of assuring us beyond any possible doubt that we ourselves exist as minds, and furthermore that God exists as well. Descartes was both a mechanist and a spiritualist. But it was La Mettrie's position that gained greater and greater traction.[58]

La Mettrie's position, the materialist position, was that once we understand how material objects interact (and we were very far from understanding that, but at least we had begun), we would find that all interactions other than gravity are local, and those interactions are strictly governed by the shape of objects, by their rigidity, their mass, and their motion.

And so materialism was born. It was a philosophy that was more smug than it had a right to be, because it declared that all that existed was material in space, yet there is a great deal that we do not know about material and that we do not know about space. But it was a very attractive philosophical position that will follow us throughout our story.

The biggest blow to materialism was the onward march of the scientific analysis of the material world, which never for a moment remained fixed and secure. Here are some of the things that science came up with that were serious challenges to early materialism: just as the amount of mass is conserved over time (matter can neither be created nor destroyed), so too energy is conserved. Like matter, energy can be neither created nor destroyed, but it can hop from one object to another during one of those local collisions. Heat is also an important part of the universe and cannot be reduced to matter; the laws that govern how objects can heat up and cool down differ from the laws of motion, and the laws of heat are what made possible the greatest inventions of the nineteenth century, starting with the steam engine. Gravity was not the only exception to the rule that all things interact only locally. There were also magnetism and electricity too, which came to be seen as part of a single invisible electromagnetic field that pervades the universe, allowing objects to interact at a distance as far as our eyes can tell.

The materialists continued to argue that man is a machine (or better yet, man is nothing but a machine). In this, they knew that they were waving their hands at any number of difficult questions that they were not prepared to answer, such as how it is that people can use language in a meaningful way. We will see three major themes in the battle (for that is what it is) between the materialists and all those who were not materialists.

In the first place, the non-materialists continued to devise better arguments that there were aspects of mind that were not explicable by known mechanist principles. Second, science itself gave up on the principles of mechanism (as we have just mentioned) to a degree unimaginable by someone like La Mettrie. The worldview of late twentieth-century physics is astonishingly different from Newton's understanding of the universe. And third, the very idea of machine and of mechanism was taken and adopted by the anti-materialists, as we will see in chapter 8, when mathematicians and logicians began to talk about "Turing machines," "things" that had all the trappings of machines and yet which could be defined outside the world of material objects.

The materialists continued to do their best to chip away at the challenges posed by the non-materialists. They did this by choosing various behaviors that revealed the presence of mind and spirit in the human, and then accounting for the behaviors in a way that was purely mechanical. Clever inventors would devote years to creating machines that could play chess; that would show that gears and wheels suffice to display intelligence, would it not? There was much discussion of *self-moving machines*, though this phrase did not carefully distinguish (as we would want it to do, today) between a machine that keeps on working without providing it with an external source of energy and a machine that controls its motion and movement in what appear to be intelligent ways.[59] Some inventors have come down to us as hoaxers: Johann Bessler is remembered as the man in the mid-1700s who claimed to have a working perpetual motion machine, and if we cannot prove that he was a fraud, we are certain that he was one nonetheless. But machines that controlled themselves? There was nothing fraudulent about that idea, and it became very important as soon as the steam engine was invented, at the end of the eighteenth century.

But while materialists (and agnostic engineers) continued to develop machines that could control themselves in significant ways, machines were always playing catch-up with humans. It was humans whose behavior defined what counted as intelligence, and it was for machines to show that they could do a few small things that could be seen as intelligent.

As we will learn in chapter 8, there was a time when that balance began to tip: it was the moment when Alan Turing invited the machine to move from the world of material into the world of ideas and mathematics, the non-material world. And now, in 2018, we are once again placed in a turbulent moment when the materialists and the non-materialists are at daggers drawn.

Conclusions

In the next three chapters, we will briefly cover the important currents of the nineteenth century that inform the development of linguistics, psychology, and some aspects of philosophy and logic. After that, we will consider more carefully five connected stories. The first chronicles the development of American psychology up through behaviorism, and the development of Gestalt psychology in Germany, followed by the transplantation of the Berlin Gestaltists to the United States. The second story is the rise of the linguistics of Edward Sapir and Leonard Bloomfield in the United States. In chapter 7, we look at a third development, involving two important philosophical movements in the early part of the twentieth century: the work of Edmund Husserl, and the development of the Vienna Circle of logical positivism. Chapter 8 explores some of the important developments of logic, and our understanding of mathematical logic, while chapter 9 explores the fifth and final story, the origins and the ideas of the European structuralists, focusing on Nikolai Trubetzkoy and Roman Jakobson.

CHAPTER TWO

The Nineteenth Century and Language

You must collect things for reasons you don't yet understand. —Daniel J. Boorstin

A conception can be understood only through its history. —Auguste Comte

Introduction: History, Typology, Structuralism

Some three-quarters of the way into the nineteenth century, the French historian Gabriel Monod offered what has become a famous observation about his century: "The history of languages, the history of literatures, the history of institutions, the history of philosophies, the history of religions, all of the studies which take man and the phenomena of the human spirit as their object have taken on a historical character. **Our century is the century of history**."[1]

As the nineteenth century began, its intellectual concerns were with history and with origins, and over the course of the century, these developed into a passion for typology, measurement, and classification, and then into a more abstract way of generalizing beyond typology. The notion of *time* was central to almost every question. And gradually another order of thought would emerge, one that we find by the end of the century, a concern with abstract structure, which had in its turn emerged from the study of taxonomies.[2]

Historical time *exploded* during the nineteenth century: the world went from being a few thousand years old to being much, much older. In large measure, this change occurred because the easy answers once provided by the Bible were no longer sufficient for everyone. Indeed, the sense that the Bible was not the final answer to many of these questions had begun with the Renaissance; the answers provided in the Bible were no longer consistent with what science was discovering about the physical world. This was a shock to Western sensibilities, though it was not the first of its sort: geographical space had exploded in a similar way during the sixteenth century. In the century that followed Christopher Columbus, as technology and commerce pushed explorers to map the globe, the world came to grips with the realization that while the Earth was finite, it had no edge, because it was a sphere.[3] Indeed, astronomical space had exploded during the seventeenth century, when we began to get a sense of how large the solar system is, by any terrestrial measure. But it was the nineteenth century that radically changed our notion of time, very much in line with the changes in our notion of space that had already begun.

There were many questions that interested thoughtful people: How old is the Earth, and how old the Sun? How did the different peoples come to populate the continents on the Earth? Is the history sketched in the Bible—a brief history of the Earth, a longer history of its people (longer, but still quite incomplete)—is this story right in its broad outlines, or is it so seriously off the mark as to be of no real interest any longer? How did so many different languages come into existence in our world? What was the first language, the language of Eden, and what would be the language of Heaven? Was the first language, the one that Adam and Eve spoke, perfect, and if so (as seemed likely to some), did a language like that express only perfectly clear and unambiguous messages? Why do descendants of the Romans speak languages that are similar, even if they no longer can be said to share a common language? Why are Africans dark-skinned, and northern Europeans fair-skinned? Why do we find some animals virtually everywhere, and some only on one continent? As these examples suggest, the question of the depth of time was deeply entwined with the discovery of the complexity of all of the stuff that there was on the Earth, all the teeming species of both organic and inorganic nature.

Trying to satisfy this grand curiosity demanded a tremendous effort to collect *things*. If you want to think dismissively, think butterfly collecting—is there anything we care less about today? Today butterfly collecting seems pointless because there is nothing to it but to pin a specimen in a book and set a label next to it.[4] But we have to remember that 200 years ago, it was not just butterflies that people were collecting, but species of every imaginable sort, from aardvarks to zebras, and not just mammals, but fossils and flowers, bugs and beetles, even soil and stone. Out of this grew a concern for finding the principles of organization that would account for the massive variety, and the hints of structure and organization that the expert could perceive. If the first four Aristotelian elements (earth, water, air, fire) and the fifth (ether, in the heavens) were no longer an acceptable catalog of the building blocks of the universe, what then would this system be replaced with? How many basic elements were there, in this sense, and why were there that many and no more? Why did different elements interact differently with others, and why did some weigh more? One of the greatest accomplishments of the century was Dmitri Mendeleev's periodic table of the elements, which imposed a simple ordering and organization on each and every known element, and this master stroke would feed the energies of scientists following on his trail. In chapter 9, for example, we will see how this model would have an impact on the work of Nikolai Trubetzkoy and Roman Jacobson. More clearly than anyone else at the time—and perhaps since—Mendeleev explained in a totally unexpected way the structure that lay behind a pattern of some 50 disparate objects that seemed indivisible. He proposed a principle of organization, the periodic table of the elements, that was *not* an analysis, in the core sense of the word: it did not *break down* the elements, but it showed that they reflected a common structure. This table would become the icon of nineteenth-century thinking, and we will return to this later in the chapter.

The understanding of history during the nineteenth century was part of the larger question of the meaning of time. The nature of the past is one that has in some respects, at least, been settled today, in the modern West, and science has had the last word. But our political spectrum today reflects a difference that divided Europeans at the beginning of the nineteenth century: Is the past the legacy of inequality and injustice that needs to be removed in order for society to achieve its potential, or is the past the sum total of the cultural achievements and the social practices that separate us from our prehistoric ancestors? Revolutionaries tend towards the first view, social conservatives towards the second, and the controversy continues today. This divide will appear again in our discussion of the Prague phonologists in chapter 9.

Not only was the past and its meaning for us rethought during the nineteenth century, so was the meaning of the future, and even that of the present. Taken together, this uncertainty, born of the need to reconsider the character of time, gave rise to a sort of anxiety that we will encounter over the course of this book. The central question for us is the sense and nature of change, especially at the level of society. Is it sensible to say that mankind is moving in a direction, or that it ever has, or that it ever will? The most basic question along these lines is that of *teleology*, a word that comes from the Greek *telos*, or goal: can human action—whether it be at the individual level, or that of a society—be understood in terms of goals? Most of us do not doubt that an individual's actions can only be understood in the context of the individual's own intentions, but does it make sense to say that a society changes in some fashion in order to better achieve a goal? Does it make sense to say that the natural world changes in such a fashion?

Over the course of the nineteenth century, we find three pairs of words in opposition to each other that were central to the way time was understood. The first pair, *static* and *dynamic*, was used in different ways at different times, sometimes clearly and sometimes obscurely. The second, *mechanical* and *teleological*, was almost always used in an obscure way, but those uses are no less important for all of their obscurity. Only the third, *synchronic* and *diachronic*, was used in a consistent and clear way: a synchronic analysis of a language looks at the condition of the languages at a given moment, while a diachronic account compares two chronologically distinct stages of a language, drawing conclusions from the way elements in the different stages of the language change in one way or another. These terms were brought to linguists' attention by Ferdinand de Saussure at the beginning of the twentieth century.

When, nearly 100 years earlier than Saussure, Auguste Comte used the pair of words *static/dynamic*, he had a similar distinction in mind. His *static* analysis was much like Saussure's later *synchronic* analysis, and his *dynamic* analysis was like Saussure's *diachronic* analysis. But later on, as we will see, when synchronic analyses came to be the dominant sort of analysis, linguists would find some aspects of synchronic accounts that were static and others that were dynamic, even though all of them were synchronic. We will have to figure out how this conceptual evolution occurred.

The opposition between *mechanical* approaches and *teleological* approaches will play an important role in this book, starting with our discussions of Charles Darwin's theory of evolution. The two terms were rarely defined clearly, and it will fall to us now to get as clear as we can what the issues were that were joined when there was controversy about mechanical analyses and teleology. Darwin's theory was viewed by most as a blow for mechanistic views, but that was only part of the story, and when we meet European structuralism, and Trubetzkoy and Jakobson in chapter 9, we will see them making the case for a teleological view of language change, which they see as being in keeping with the advanced scientific thought of the day, and explicitly rejecting Darwin's theory as they understood it.

Nations in Europe

To understand the development of the mind sciences in the twentieth century, we must first understand the broad political outline and the particular developments of the sciences during the nineteenth century. We will begin with a brief overview of the major political trends, as well as some of the broader themes of the period—the focus on time and history, on collecting and typologizing. By the end of chapter 5, we hope that it will be clear that all of the mind sciences are deeply rooted in discussions and controversies that were central to nineteenth-century thought.

The written form of language has long played an important role in the history of modern European languages and countries. During the Renaissance, fired by the views of the humanists and the Reformation, the modern languages spoken by real people took on a new importance. In northern Italy, Dante, and later Cardinal Pietro Bembo, accorded a new status to the modern forms of language, and they themselves wrote in the dialects of Tuscany and Florence. The Catholic scholar Erasmus translated the New Testament from the Greek. Protestants felt that it was essential to translate the Latin Bible into the modern vernaculars so that people had direct access to the holy word.[5]

And so, the Bible was translated into the languages of the people. Luther preached and wrote in German, as Calvin did in French, and Tyndale translated the Bible into English, as Nicolas Van Winghe did into Dutch.[6]

The immediate consequence was that the vernaculars of Western Europe took on the status of languages in the full sense of the term. The translation of the Bible established a norm, with consequences for spelling, for the lexicon, and for the syntax of each language.

It was during the nineteenth century that these linguistic issues had a direct impact on the political map, for the nineteenth century was the century of nation building, and by its end, the face of Western Europe had become one much closer to what we recognize today. The great British historian Eric Hobsbawm spoke of the *long nineteenth century*, starting with an age of revolution, which stretched from the great revolution of 1789 in France to the smaller revolutions across Europe in 1848, to the age of capital and on to the age of empire, which brought us to the brink of global war in 1914. Just listen to a few of the highlights of this part of the century. There was the Louisiana Purchase, by which the infant United States purchased a good running start on a coast-to-coast empire, buying out land—530 million acres—that Napoleon didn't need and didn't want, given his commitments in the Old World. There was Napoleon's march to establish an empire, which for a brief moment succeeded, with the military defeat of Prussia, Poland, Russia; Norway declared itself independent in 1814, but didn't quite make it till 1905. The Holy Roman Empire was dissolved; South America freed itself from Spain and Portugal; Greece became independent of the Ottoman Empire; Belgium became sovereign and independent, as did Colombia, Ecuador, Venezuela, and Panama, as well as Guatemala, El Salvador, Honduras, Nicaragua, and Costa Rica. Texas became independent, too, for a brief while.

It was easy to see the changes taking place on the map, and there was hardly a year when map publishers did not need to put out a new edition to keep up with reality. But there were also underlying themes and forces that had a lasting influence, many of which continue to have an influence on the political world. The nineteenth century saw the emergence of a strong pan-Slavist movement, for example, and it was one that would have direct consequences for the development of linguistics. The Czech language, a minority language within the Habsburg Empire, was displaced in many respects by German during the early part of the nineteenth century. This was also a period in which there was considerable migration from the countryside to the city, in part as a response of the abolition of serfdom. But efforts to develop and maintain a standard Czech language throughout this period were successful, and Czech was able to flourish as a language once Czechoslovakia was established as an independent country after World War I. Its first president was Tomáš Masaryk, who, as we will learn, had been a student both of Franz Brentano and of Wilhelm Wundt and had befriended Edmund Husserl before assuming the presidency, and afterward he provided important support for one of the most important centers of work on linguistics, the Prague Linguistic Circle (see chapter 9). In the figure of Masaryk, we see an incarnation of the close links between nationalist politics, philosophy, and linguistics.

Nationalisms in Europe

The relationship of language and nationhood has never been a simple one, and its importance has attracted the attention of political philosophers and other thinkers.[7] During the nineteenth century, several conceptions of what it is that makes a *people* took shape. These different approaches were rooted in disparate moments of Enlightenment thought, as well as French and German Romanticism.[8] We will focus on three views: political, ethnic, and class-based. In the first, the political view, the people is equivalent to the Greek *demos*, which is to say, the assembly of citizens, and it is this assembly that provides the legitimacy of political power.[9] This is the conception that lay at the core of the American and the French revolutions, and its sense is found at the beginning of the American Constitution:

> **We the People of the United States**, in Order to form a more perfect Union, establish Justice, insure domestic Tranquility, provide for the common defense, promote the general Welfare, and secure the Blessings of Liberty to ourselves and our Posterity, **do ordain and establish** this Constitution for the United States of America.

The second approach, the *ethnic* view, takes the people to be an *ethnos*, a group of individuals united by ties of blood and land. This conception was dominant in German Romanticism and was celebrated in the doctrine of "Blut und Boden": literally, blood and soil. Fichte (1808) believed that there was something that Germans and only Germans possessed—a belief rooted in an essentialist and linguistic conception of German identity:

> The first, original, and truly natural boundaries of states are beyond doubt their internal boundaries. Those who speak the same language are joined to each other by a multitude of invisible bonds by nature herself, long before any human art begins; they understand each other and have the power of continuing to make themselves understood more and more clearly; they belong together and are by nature one and an inseparable whole.[10]

The third vision sees the people as a *plebis*, the gathering of the poor among the lower class. This class-based view dominated the Marxist and socialist conception from the Paris Commune of 1871 until the rise of Leninism. This vision will play at best a marginal role in the questions of language to come.

Throughout the nineteenth century, the confrontation of these three conceptions gave rise to ceaseless changes in people's understandings of who they were and in the political struggles out of which the modern nations of Europe arose. The conflict between the first two views—the political and the ethnic—was stormy and sometimes violent during the period that we are considering, and the reader who has opened a newspaper recently will be aware that the last bullet has not yet been shot over this issue.

The ethnic view that emerged in German Romanticism offered a central role to *language*, and this conception was dominant in the European context. The popular movements of national identity in the Germanic countries inspired an internal critique of the rationalism of the Enlightenment led by philosophers such as Johann von Herder and Johann Wolfgang von Goethe and by the *Sturm und Drang* movement, which explored the cultural specificity of the German nation.[11] Herder himself saw language as a human creation and as the repository of the particular culture of each people—as that which made it unique, what he called its *Volksgeist.*

Herder's focus on the empirical study of each nationality was also the beginning of serious work on indigenous cultures and the science that would become the study of *folklore*, or traditional knowledge. This type of research in cultural ethnology would become especially important in the Germanic tradition, where traditional languages and practices, legends, myths, epics, and sagas were gathered and published. This work was carried out by linguists and philologists such as Rasmus Rask and Jakob and Wilhelm Grimm who specialized in Danish, German, and the other Germanic languages. The brothers Grimm would become known internationally for their traditional tales, of course: we will discuss their linguistic work later in the chapter.[12]

The communities that were formed by language and culture in the nineteenth century were also the communities that defined the political aspirations that led to the formation of the modern states of Europe. In many cases, the aspiration to the status of a *national language* entailed devising or recognizing something that *was* a language—but the status of *being a language* could not be reduced to being a recognized language of culture or a language with a written tradition. The national language in most cases was *popular*, not the language learned in school, but rather the everyday language that made people feel as though they had something in common. In a word, an oral language, a spoken language, or even what a linguist, or a sociolinguist, would later call a dialect.

Max Weinreich was a famous linguist in the first half of the twentieth century, a great specialist in Yiddish, and father of Uriel Weinreich, whom we will meet in volume 2. Weinreich told a story that is still told today, though more often than not without recalling that it came from Max Weinreich:

> A teacher at a Bronx high school once appeared among the auditors. He had come to America as a child and the entire time had never heard that Yiddish had a history and could also serve for higher matters. . . . Once after a lecture he approached me and asked, "What is the difference between a dialect and language?" I thought that the *maskilic* contempt had affected him, and tried to lead him to the right path, but he interrupted me: "I know that, but I will give you a better definition. A language is a dialect with an army and navy." From that very time I made sure to remember that I must convey this wonderful formulation of the social plight of Yiddish to a large audience.[13]

Deep Time

For someone who had traveled little and learned less, the Earth in 1800 seemed to be flat, even if this flatness is interrupted here and there by mountains and canyons. Circling the globe on a ship gives us a better sense of just how big the Earth really is, but even that experience will not give us a sense of how far away the Sun is, or the closest stars. As our ability to see and measure both land and sky has advanced, our imaginations have been forced to conceive of the universe we live in as immense, and almost indescribably larger than we had thought it was.[14]

There were three important sciences that informed the thinking of the nineteenth century as it recalibrated its sense of time: geology, zoology, and the study of human languages. In this section, we will look at geology and the impact of the Darwinian revolution on zoology, noting briefly the strong degree of interaction between the development of the ideas about the evolution of species and of languages in the middle of the century, and then we will return to linguistics in the next section.[15]

Geology

The Bible told a story of a great flood, from which only Noah's Ark saved humanity and the other surviving species on the planet. Could that story be substantiated by looking at the evidence before our eyes? Could the strata observed in rocks be linked to the Flood? These questions—as well as increasing industrial interest in ores and mining—led to careful geological surveys.

Later in this chapter we will encounter Ferdinand de Saussure, an important linguist in this century. One of Saussure's friends, Aimé Pictet, wrote this in his diary, after hearing Saussure speak excitedly about linguistics and Sanskrit during a visit:

> De Saussure begins to explain Sanskrit etymologies to Guillaume and me, how Sanskrit, Greek and Latin are sons of one and the same father, Indo-Germanic, and brothers of almost all the languages of Europe . . . the most comical thing was that I understood perfectly and that I began to admit, almost, that these philological studies might have a certain utility. In any event that of proving yet again that humanity is much older on this poor earth than it believes. It is curious how all the sciences, linguistics, geology, natural history, all arrive at this same result by a hundred different paths. And there is only this poor Genesis of Moses to battle against these conquests of the 19th century.[16]

Although thoughtful people have been interested in the nature of the Earth and what it contains, it was not until the end of the eighteenth century that geology reached a state of development where it was ready to become a science. There were many good and practical reasons why the study of the Earth was important, and of these the most pressing was the need to better understand where valuable resources inside the Earth were to be found and to be extracted—everything from gold and silver to coal and granite.

For the geologist, there was one major theme that began in the first half of the eighteenth century and developed throughout the nineteenth century: the discovery and identification of strata, or layers, of rock.[17] The central dogma at this time was that the Earth as we see it is covered by layers of distinct classes of materials, each formed on top of the previous one. Thus there was a temporal order to what we see on the Earth—the closer to the surface, the more recent. A number of conceptual breakthroughs were necessary for this central dogma to become dominant. Geologists had to discover the mechanisms that were responsible for why the strata that were observed were not always flat: the observed reality was not always as simple as the model predicted, and additional mechanisms needed to be formulated to allow for shifting up and down after the initial layers were created.

The second breakthrough that led to the central dogma of this early nineteenth century was considerably more breathtaking: it was the notion that the layers that we observe in one spot on the Earth could be aligned with the layers observed at other spots, often very distant from the first. The stratigraphy of the Earth was not merely local, but global as well. In fact, nineteenth-century geology centered around the development of an account of the structure of the Earth based on a single and global sequence, now called the *stratigraphical column*. The idea is simple and elegant: in many areas, it is easy to see that the ground is formed in layers of varying composition, and the hypothesis lying behind the stratigraphical column is that there is a single set of layers that accounts for the layers of the Earth everywhere. The surface of the Earth may correspond to different points on that column in different spots on Earth, but there was a single sequence found everywhere. Early in the nineteenth century, geological formations were relied upon to establish correspondences between spots at different locations on the Earth, but by 1835, fossils were being used for the same purpose. Long before this time, the notion of dry land being the result of sedimentation in long-gone lakes and oceans was understood, and this led naturally to the idea that the layers identified by the stratigraphical column corresponded to periods of time. But there was, during the nineteenth century, no way to associate any specific lengths of time. Moments in time could only be described in relative terms: before and after.

William Daniel Conybeare and William Phillips expressed this with great clarity in 1822. If geology is the study of the Earth's structure, then the crucial element is to ascertain "the order in which the materials constituting the *surface* of our planet (for beyond this observation cannot penetrate) are disposed." This declaration is by no means obvious; it amounts to a decision to treat data arrayed in three dimensions with a one dimensional model, a decision with momentous consequences for geology. "The superficial and hasty observer might suppose that these materials are scattered irregularly over the surface and thrown confusedly together, but a slight degree of attention will prove that such a conclusion would be entirely erroneous."[18]

They continued with a simple observation: if a traveler sets off from London and walks to an area where coal is to be found and keeps his eyes open, he will note that he passes through the same series of landscape, in the same order: first an area of clay and sand, next an area of chalk, followed by an area rich in calcareous freestone used in architecture. In the coal-rich areas, he will see (if he keeps both his eyes and his mind open) hills rich in the same sort of compact limestone, with gray and dark marble, and mines providing zinc and lead. A bit further on, he will find mountainous areas with slate, surrounding groups of granitic rock. This pattern, they wrote, was no accident:

> The intelligent enquirer, when he has once generalised these observations, can scarcely fail to conclude that such coincidences cannot be casual; but that they indicate a regular succession and order in the arrangement of the mineral masses constituting the Earth's surface; and he must at once perceive that, supposing such an order to exist, it must be of the highest importance to oeconomical as well as scientific objects, to trace and ascertain it.[19]

Geology was thus an outstanding exemplar of scientific enterprise during this century. Doing it right demanded enormous amounts of detailed observations, and a passion for seeking principles that brought regularities to light within the data. Geology was in a reconnaissance mode, as one historian has put it.[20] It has also used large amounts of observations to develop a picture of the past and the principles by which the past gave rise to the present.

This was heady stuff! William Whewell, president of the Geological Society, in England, waxed rapturous in 1833, just thinking about how great geology was:

> I confess indeed for my own part, I do not look to see the exertions of the present race of geologists surpassed by any who may succeed them. The great geological theorizers of the past belong to the Fabulous Period of the science; but I consider the eminent men by whom I am surrounded as the Heroic Age of Geology. They have slain its monsters and cleared its wildernesses; and founded here and there a great metropolis, the queen of future empires. They have exerted combinations of talents, which we cannot hope to see often again exhibited; especially when the condition of the science which produced them is changed. I consider that it is now the destiny of Geology to pass from the heroic to the Historical Period. She can no longer look for supernatural successes: but she is entering upon a career, I trust a long and prosperous one, in which she must carry her vigilance into every province of her territory, and extend her dominion over the earth, till it becomes, far more truly than any before, a universal empire.[21]

Alexander von Humboldt was a great German geographer and explorer, and brother to the linguist Wilhelm von Humboldt. Not long after Whewell made his comments, Humboldt connected geology and languages in his monumental *Kosmos*, published between 1845 and 1862.

> Languages compared together, and considered as objects of the natural history of the mind, and when separated into families according to the analogies existing in their internal structure, have become a rich source of historical knowledge; and this is probably one of the most brilliant results of modern study in the last sixty or seventy years. From the very fact of their being products of the intellectual force of mankind, they lead us, by means of the elements of their organism, into an obscure distance, unreached by traditionary records. The comparative study of languages shows us that races now separated by vast tracts of land are allied together, and have migrated from one common primitive seat; it indicates the course and direction of all migrations, and, in tracing the leading epochs of development, recognizes, by means of the more or less changed structure of the language, in the permanence of certain forms, or in the more or less advanced destruction of the formative system, *which* race has retained most nearly the language common to all who had migrated from the general seat of origin.[22]

Humboldt's remarks illustrate well how scientists at the time saw the history, geology, and historical linguistics engaged in a common enterprise at the cutting edge of knowledge.

Collecting and typologizing

During this period, each state, each king, each prince had his collection of plants, minerals, fossils, and animals mounted and stuffed, while passionate biologists, zoologists, geologists, and geographers joined the groups of men exploring the world, thanks in part to the developments in the technology of navigation that made such expeditions possible. Some of these voyages have become famous, of course: from those of James Cook and Jean-François de La Pérouse to those of Darwin and Alexander von Humboldt.[23]

One of the principal goals of the expeditions was to enrich these collections, and among the greatest scientists at the beginning of the nineteenth century were the curators of these great collections, which they discussed, presented, and analyzed in their theoretical writings.

Nineteenth-century science grew out of these collections and the commentaries that were written on them. In a very real sense, the collections *were* the research instruments of science in the nineteenth century. A collection would lead to a proposal for a system of classification, a taxonomy. And in this context, a science would begin as a systematic classification of what nature provides. Theoretical reflections on the principles that underlay classification led the further development of hierarchical organizations, which led to more abstract structures and systems of relations. These structures and relations would lead eventually to historical reconstructions and to the declaration of missing elements, places where an element should be within the structure even if it had not yet been observed.

The first great modern classification of forms of life had been proposed by Carl Linnaeus in the eighteenth century. His *Systema Naturae*,[24] which went through many editions during his lifetime, marked the beginning of a new scientific conception: that of *systematics*. The hierarchical taxonomic principles that served to classify the animal and vegetable world began in Linnaeus's scientific practice and evolved in effect into a theory—what Bourdieu called an *enacted theory*—and this systematic thinking would influence the entire nineteenth century, notably in Lyell's conception of geology, Darwin's of life, Mendeleev's of chemistry, and ultimately Saussure's reconstruction of the vowels of Indo-European. In a word, the greatest accomplishments of nineteenth-century science.[25]

Darwin and evolution

Perhaps the single greatest scientific discovery of the nineteenth century was the theory of evolution through natural selection, proposed and developed by Charles Darwin and by Alfred Russel Wallace. Darwin published *On the Origin of Species* in 1859, and the world has not been the same since. Building on the work of researchers before him, and the observations and lore of stockbreeders, as well as decades of his own observations and meditation, Darwin proposed a new account of why and how the world of living things had arisen. The world of living things had developed and evolved, with one species evolving out of another in a slow process of change effected within a community of all of the members of the species at any given time.

Darwin's central idea was that biological reproduction always involves variation, differences of sorts between the offspring and its progenitor.

(This variation is greater when sexual reproduction is involved, since the offspring is a never-before-seen amalgam of two parents' contributions.) The variations of the offspring will serve as a new basis from which the next generation is formed: each successive generation's genetic code (as we now call it) varies slightly from that of the one that preceded it. Nature—everything that surrounds the organism—would challenge every organism in its struggle to survive and prosper, and not just nature. Even the internal efficiency of an organism could contribute to its ability to function well. Those organisms that fared well would survive better, to the point of having better adapted offspring themselves, while those that were less fit than others were likely to fail to survive and reproduce. Thus Darwin proposed a model of evolution fueled by only two natural principles: the random variation generator that introduces variation in the recombination of the DNA with each generation and the natural selection that selects specimens best fitted to their ecological niche. With those two principles, he was able to justify and explain a general taxonomy of the living and make predictions in some cases of fossils that represented ancestral forms that had since disappeared.

Teleology

There were two ways in which Darwin's point could be seen in an even larger picture, and they relate to the opposition between mechanistic and teleological views that we discussed earlier. The interpretation that was mechanistic was viewed as a challenge to some conservative social views, while the weakly teleological interpretation view was that Darwin had shown exactly why we can be confident that the world is *progressing* (not unimportantly, this was the view that gave rise to social Darwinism).

The first interpretation, the mechanistic interpretation, emphasized the random character of the small changes that arise between generations. Some saw this as tantamount to the conclusion that there was no *finality* to the community of living organisms: there was no master plan of the living world that each organism was trying to satisfy. There was no *evolution towards something*, except in retrospect and in the eye of the beholder. Evolutionary changes took place in direct response to challenges and opportunities in the immediate world in which each organism found itself.

But there was another possible perspective that could be taken of Darwin's scientific views, even though this was not his own interpretation. His account could be joined with the belief that evolution in the animal world was *progressive*—species that died out died out for a reason, no? they were *losers*—and there will always be those who choose modern humans as the pinnacle of life's evolution. Darwin's account could be viewed as one that provides an explanation for progress, grounded in an explanation that the less adapted, the less fit, would fail to procreate, and this very failure was the underside of the larger picture of progress which the species as a whole was engaged in. This perspective, with all of its consequent social implications, came to be known as social Darwinism. At its heart lay the view that there is no better way to characterize large-scale change than as a fierce competition, and to succeed in that competition was quite obviously the goal of all living things.

But neither interpretation was of any comfort to people who believed that the world was evolving in a specific direction towards a particular state or in a particular direction, the view that we associate with teleologism. We are going to encounter three different interpretations of teleology, though, and we need to clearly distinguish among them. The first is the simplest: it is the belief that there exists a specification of where the world is heading, and that specification exists outside the universe as we know it—typically in the mind of a God who exists outside of the universe, and who can conceive of a universe that is different from the one in which we live. The traditional term for such a view is *transcendent*, in the sense that both God and His idea of the universe exist outside of the universe, and we can refer to this as transcendent teleology. The second is an *immanent* conception of the goal towards which the world is tending: immanent views of God's existence see Him as being one with the universe, not outside our universe: Spinoza is the ultimate spokesperson defending an immanent view of God. On this view, then, there are goals towards which the universe is heading, but they are of a piece with the universe that exists, just as God is (though there is no need to have an opinion about God in order to be a defender of immanent teleologism). The third interpretation of teleology is political. We will see this side of teleology emerge clearly in chapter 9, as we explore the meaning of Trubetzkoy's concept of Eurasianism.

To some, immanent teleologism seems impossible and contradictory, for how can the world be anything more than just what it is? And how can a goal towards which everything is moving not be part of what the universe already is? The immanent view of teleology is the most illusive, but also the most important in our story. It is a view that sets itself up as an alterative to a mechanistic view of the universe, but without casting about to find something exterior to the universe to explain what is going on within it. It is the belief that the seventeenth- and eighteenth-century Newtonian metaphor of the universe—billiard balls and complex systems of gears, as in a clock—are not sufficient to understand the world scientifically.

Two new ways of understanding physical systems that emerged in the nineteenth century would become extraordinarily important. One was the concept of a force field, such as an electromagnetic force field, while the second was the notion of entropy, which was crucial for understanding what heat is and how engines work: in a word, thermodynamics.

Before Michael Faraday and James Clerk Maxwell formulated the idea of electromagnetism as a force field, it was only natural to think of electrical (and magnetic) attraction and repulsion as forces that somehow related two objects that were at a distance from each another. Faraday rejected this and proposed in its stead a vision of a field of forces that is present everywhere and whose strength at any given point is created by charged objects. On this view, objects do not directly affect one another: each object affects the overall force field, which takes on specific values that can be calculated, and then the force field in turn acts on objects. This would be an important metaphor developed by Gestalt psychologists, as we will see in chapter 5.

Entropy was a very different sort of entity. During the nineteenth century, scientists studied the ways in which heat was created and how it flowed from one object to another. Heat had characteristics that it shared with other sorts of energy, which led scientists to conclude that in a closed system, energy is neither created nor destroyed but is, rather, conserved.[26] And yet in real systems, heat energy only flows from warmer to cooler objects, and never the reverse, an asymmetry in its style of conversion that seemed quite unlike energy in other domains. And entropy, a measurement of the disorder of a system, would always increase.

It followed from this that in at least one rather dismal sense, the universe did have a directionality to its evolution: it was always moving towards a state of greater entropy, and thus greater disorder, a thoroughly uninspiring conclusion. But a more appealing and elegant picture emerged as well: in predicting how an object will move in a force field and how a system of molecules will evolve in a larger thermodynamic system, the right way to understand the system is to conceive of it as a system in which the interaction of a large number of elements and forces seeks an equilibrium, a balancing of all the forces at every single place within the system.

Darwin and language

Darwin viewed language as an important factor in the development of the human brain—language did not simply reflect what was going on inside the brain, but was in addition a force for the development of the mind or brain.[27] One of the central challenges of an evolutionary account of the origins of the human race was to come up with an account of how the brain evolved to allow more complex thought, and Darwin sought an account that would involve language. Perhaps language came first, he thought, and the greater cognitive abilities that came with linguistic skills would in turn have an effect on the brain. What led Darwin to this perspective? Robert Richards argues that Darwin was influenced by Wilhelm von Humboldt, whom we will meet below, and by Hegel, both of whom were in turn greatly influenced by Herder. Hegel, of course, was the leading German philosopher in the generation following Immanuel Kant. Darwin wrote,

> If we possessed a perfect pedigree of mankind, a genealogical arrangement of the races of man would afford the best classification of the various languages now spoken throughout the world; and if all extinct languages, and all intermediate and slowly changing dialects, had to be included, such an arrangement would, I think, be the only possible one.[28]

By the time of Darwin's major publications, historical linguistics was an established discipline. Linguistics had taken the steps necessary for developing a rational taxonomy of languages whose groupings were historically meaningful and had gone a good deal of the way towards establishing the means to postulate a hypothetical series of stages in the history of the modern languages. This success at both a methodological and a conceptual level had enormous influence on Darwin's thinking, to the point where he devoted an entire chapter of his first major book to the parallels between the evolution of species and languages. He proposed 16 criteria in support of this parallel, noting the importance of the existence of variation and drift in language change as in the biological world. Over a century later, William Labov would take as the theme of his presidential address to the Linguistic Society of America the ways in which controversies in linguistics regarding lexical diffusion and regular sound change could be understood against the backdrop of parallel questions in the early work in biological evolution.[29] More recently, Labov has gone back to the 15 criteria that Darwin proposed, and showed that 14 of them had been verified in the linguistic context work in the twentieth century.

The fifteenth criterion concerns the question of whether adaptive changes are *improvements*. Linguists today do not characterize linguistic changes as improvements or as decay—and, as we shall see later in this chapter, the first linguistic scholars who felt themselves to be in opposition to their elders were the Neogrammarians, who rejected their teachers' willingness to characterize changes in languages' style of morphological analysis in a positive or negative sense.[30]

Darwin was in good company in looking for support of his view of evolution from the new and exciting work accomplished by linguists: in this, he was joined by the geologist Lyell, as well as by the biologists Asa Gray and Thomas Henry Huxley—friends of Darwin's—and Ernest Haeckel. And linguists, notably Hensleigh Wedgwood (Darwin's brother-in-law), Frederick William Farrar, and Max Müller, and August Schleicher, made a similar point about the bigger picture that linguistics was opening up.[31]

The periodic table

If exploration and classification were two of the essential passions of the nineteenth century, the quest for the basic building blocks of the physical universal was an important task as well. The Aristotelian world was built out of the elements of earth, water, air, and fire, plus a fifth element, ether, which could not be sampled here on Earth. The ancients did know something about the substances that later came to be understood to be "atomic": copper, silver, gold, iron, mercury, lead, tin, sulfur, and carbon; arsenic, antimony, and bismuth were added to the list later. But in the period beginning around 1735, some 40 such elements were discovered.

How did these elements differ? How should their properties be measured? How could they be organized so that they could be understood as more than just a hodgepodge of unrelated creations of God's imagination?

The story of how chemistry got it right—how it figured out what an element was, how to distinguish an atom from a molecule, how to understand what it meant for a molecule to have a certain weight, and how to count the relative contributions of different elements to a single molecule—is a great story, and greater still because it was carried out without scientists being absolutely certain that there were such things as atoms. It was not until the early twentieth century that the debate was finally settled about atoms and scientists came to agree that they do indeed exist.[32] But it was during the first three-quarters of the nineteenth century that chemists got clear on the central ideas of elements, compounds, and molecules.

The man who crystalized the notion of the periodic table of elements was Dmitri Mendeleev, who was born in Siberia in 1834 and became one of the most important Russian chemists of his day—and with the development of the periodic table, one of the most influential chemists of the nineteenth century. The periodic table was not only important for chemistry: its very form became the image of the best work that science could achieve.

Mendeleev published his work on the periodic table in 1869, at the age of 35, and as he noted explicitly, the work was made possible by the important empirical work that had been accomplished over the course of the 1860s. His work was a brilliant combination of careful study of numbers and observed properties, on the one hand, and the insistence on finding an overall schema that simplified everything, on the other. Looking back, he wrote about some of the basic principles that he had observed and taken very seriously: "The elements, if arranged according to their atomic weights, exhibit an evident periodicity of properties. . . . Elements which are similar as regards their chemical properties have atomic weights which are either of nearly the same value (e.g., platinum, iridium, osmium) or which **increase regularly** (e.g., potassium, rubidium, caesium)." "They increase regularly": this phrase hides the immensity of the creative leap required to recognize a pattern that no one had seen before. "We must expect the discovery of many yet unknown elements, for example, elements analogous to aluminium and silicon, whose atomic weight would be between 65 and 75," and "the atomic weight of an element may sometimes be amended by a knowledge of those of the contiguous elements. Thus, the atomic weight of tellurium must lie between 123 and 126, and cannot be 128."[33]

Some 20 years later, at the age of 55, Mendeleev gave the Faraday Lecture to the Fellows of the Chemical Society in England, and he looked back then at the results and impact of the approach he had developed. Chemistry, he said, had reached the ideal set by Bacon and Descartes—its results were submitted to the scrutiny of both experiment and reasoning:

> Willingly or not, in science we all must submit not to what seems to us attractive from one point of view or from another, but to what represents an agreement between theory and experiment; in other words, to demonstrated generalisation and to the approved experiment. . . . We still may hear the voices of its opponents; they enjoy perfect freedom, but vainly will their voices rise so long as they do not use the language of demonstrated facts.[34]

This was a lovely phrase: *an agreement between theory and experiment.* Together, jointly, they set the standard that scientists and their ideas must respect. Mendeleev explained why he was able to come up with his hypothesis. The first major factor was the development of detailed information about the atomic weights of each of the elements. Part of this development was empirical, in the laboratory sense of the word, and some of it was theoretical, in the sense that it was crucial to impose the condition that elements could contribute only in an integral (and not a fractional) way into the composition of molecules. All of this is very delicate, we know today, because the weight of an atom is simply related to the number of protons and neutrons in its nucleus, and since protons and neutrons weigh nearly the same amount, the weight will be nearly an integral multiple of the weight of a proton (or neutron). But the advancement through the periodic table, starting from the beginning, accords with the number of protons, regardless of the number of neutrons, and just why a collection of a certain number of protons wanted (or preferred?) a certain number of neutrons was a question wildly beyond the vision of any scientist in the nineteenth century. But that is not to criticize Mendeleev and his colleagues: their advance was tremendous, it was a necessary condition for moving ahead, and it required a great leap of abstraction, both quantitative and qualitative:

> The solution of the problem advanced but slowly, because the facts, and not the law, stood foremost in all attempts; and the law could not awaken a general interest so long as elements, having no apparent connection with each other, were included in the same octave.[35]

One of the most striking aspects of the development of the periodic table was the set of predictions that grew out of it regarding unknown elements that ought to exist, if the periodic table was a chart in which each natural position would be filled by a presence, by an element with its own weight and chemical properties; these predictions were subsequently proven correct.

The periodic table is a magnificent monument to the style of fundamental scientific work that characterized the nineteenth century. Every student of chemistry studies it today. Despite the fact that it has little to say about deeper scientific questions—why are the electron shells populated as they are? why do heavier elements need more neutrons in their nuclei to be stable?—one could hardly imagine someone arguing that the periodic table was outdated and needed to be replaced. Yes, theory goes much deeper now, but the periodic table of the elements is an extraordinary synthesis of careful, quantitative measurement and a thoughtful classification of chemical behavior. It is difficult to put ourselves in the mind-set of the nineteenth-century chemist, whose laboratory work gave him access to atomic *weight*, which is based on the number of protons and neutrons in the nucleus, but no access to atomic *number*, which is ultimately the number of protons in each nucleus. Since Mendeleev and his world had no concept of neutron or proton, he had to discreetly but adroitly find a way to fit elements with increasing—but irregularly increasing—atomic weight into a simple table, sometimes trusting in the elegance of nature to predict a place for a missing element (such as gallium) or even to invert the placement of elements (tellurium and iodine), and sometimes trusting in the existence of a structural simplicity in nature that would take many decades to establish empirically.

By its very nature, the periodic table makes an unlimited number of predictions, and we continue to discover (or create) new elements. As of this writing, mankind has reached element number 115. The periodic table provided a standard against which to propose and judge future models. The German psychologist Karl Bühler used the periodic table as a way to illustrate the importance of Trubetzkoy's ideas of phonology in the 1930s:

> Trubetzkoy's simple and lucid systematic idea is of great consequence in the theory of language. . . . Let us again recall Mendeleev's idea by way of comparison. The task there was to arrange the atomic weights of the chemical elements and it turned out that they form a discrete series according to a curious law of numbers. The theoretical reflections in chemistry began here and resulted in the well-known success in the analysis of the make-up of the chemical elements and finally of matter in general. Here we are considering vocalization in the word images of human languages; it turns out that it, too, displays a transparent arrangement if the aspect of diacrisis [phonological contrasts] is applied, but not otherwise. It turns out that the four dimensions of the vowel realm mentioned above successively become *diacritically relevant* [phonologically relevant]. Consequently, theoretical reflection will have to start at this point.[36]

The synoptic Bible

We have already mentioned that the study of the Bible—and, in particular, the story that it told—made it impossible for eighteenth-century scholars to construct models of language change that are believable today. But there is another side to studies of the Bible that may have helped early nineteenth-century students of language. Among students of Christian theology, a controversy raged regarding the relationship of the four books of the Gospel—Matthew, Mark, Luke, and John—because they each told more or less the same story, but not *exactly* the same story. Students of the Bible asked themselves whether there were earlier versions—which were perhaps no longer known to us—that could have served as sources for the texts that had come down to them in modern times.

This came to be known as the synoptic problem: what was the relationship among the Gospels? Through the late eighteenth and the nineteenth centuries, the problem was attacked by scholars of various sorts, especially in Germany. Gotthold Lessing and Johann von Herder, who were public intellectuals of their day (and were, in fact, friends), argued in favor of a reconstructed past of the Gospels in which their earliest form was oral, not written. Lessing was the first to posit (in 1778) an earlier gospel now lost; this hypothetical gospel has long been referred to as "Q" (for German *Quelle*).[37]

Let us take a brief look at the kind of problem these scholars were concerned with, because in retrospect, it is striking how similar the problems they approached were to those addressed by the scholars of historical and comparative linguistics in the coming years. The books of Matthew and Luke have a number of similarities: in some cases, the words themselves are quite similar. Matthew 6:24 and Luke 16:13, for example, are identical: "No man can serve two masters: for either he will hate the one, and love the other; or else he will hold to the one, and despise the other. Ye cannot serve God and mammon." More interesting, perhaps, is the relationship between the Sermon on the Plain that Jesus gives in Luke 6 and the Sermon on the Mount in Matthew and Luke, and in a much shorter form in Mark, where both the content and the grammar could be juxtaposed and differences identified.

The development of synoptic versions of the Bible was a major stimulus to the discussion of how differences among the Gospels could serve as the basis for a historical and causal account—*causal* in the sense that time came into the picture, and what was earlier was the cause of what came later—of the similarities and differences among the Gospels. The logic of this kind of intellectual puzzle is extraordinarily close to the problem that the historical linguists of the next generation were about to embark upon. In both cases, the heart of the matter is the observation of parallels followed by the creation of a hypothetical scenario that will provide a simple explanation for why things that are similar or identical are that way.

From a methodological point of view, and perhaps even an epistemological point of view, this was a remarkable step forward, employing a precise textual critique that looked a range of facts squarely in the face. Sentences in Mark, Luke, and Matthew were lined up one with another; differences were noted, similarities compared and evaluated. Arguments were made as to which came first, or at least which one had gone from an oral to a written tradition first.

Linguistics

> If a monument were to be erected to the admirable work done in the nineteenth century in language research, two words could not be left out of the inscription: *comparison* and *history*. —Karl Bühler, *Theory of Language*

Nineteenth-century linguistics engaged with the study of both living and dead languages in two ways: the first tried to uncover the ancestral relations between languages, which would show how one language changed and evolved over time into another, while the second was part of a more general effort to look into the functioning of languages and see how languages differed. This second trend focused on the morphology, or internal word structure, and morphology would continue to be the principal stage upon which these discussions would be played out over the course of the century.

Morphology is the word we use to describe the patterns of word-internal structure in language, but it would not be until halfway through the nineteenth century that the term would take on such a meaning in linguistics. Until then, linguists spoke generally of "word structure," and the rise of the term *morphology* in linguistics was encouraged by linguists (notably August Schleicher) who saw deep connections between biological and linguistic structures of various sorts. The word *morphology* had been coined at the end of the eighteenth century by Johann Wolfgang von Goethe, the poet and polymath who was deeply interested in the nature of biological species, their modifications, and their relations. For Goethe, *morphology* was the general study of form, which would include a typology of form, a set of criteria to establish similarity and differences among forms, and the ways in which forms could be composed with each other. This perspective would be applied to various sciences over the course of the nineteenth century, including chemistry, zoology, music, and linguistics.

Morphological analysis would rapidly become the central type of analysis done by linguists throughout the nineteenth century, and a number of influential suggestions would be made about which differences in morphology were important and which were not.

While we are on the subject of the changing use of technical terms, we should mention too that the term *inflection* is used differently today by linguists than it was in the early nineteenth century. The term then referred to internal changes in a root to mark grammatical features—as with *fall/fell* in English. Today, the term *inflectional morphology* has quite a different meaning; it refers to the ways in which the forms of a lexical paradigm differ from one another, depending on the verb's tense, mood, and subject. Inflectional morphology is distinguished today from derivational morphology, which is responsible for the analysis of the ways in which distinct words are related to one another (for example, a suffix such as the English suffix *-ize* can create a verb from a noun: *winter/winter-ize, scandal/scandal-ize*).

William Jones and the call of the Orient

As we noted just above, the greatest intellectual achievements in linguistics during the nineteenth century involved historical and comparative linguistics, the areas that ask questions about the historical relations among languages alive at a given moment and languages once spoken but now extinct.

In later years, the story would often be told, and then retold, of how comparative linguistics as we know it today began at the end of the eighteenth century with Sir William Jones, who put forward the view that a common language must have existed that gave rise historically to Greek, Latin, and Sanskrit, and likely to the Germanic and Celtic languages. He wrote, in 1788, a passage that has been cited in virtually every history of the development of modern linguistics:

> The Sanscrit language, whatever be its antiquity, is of a wonderful structure; more perfect than the Greek, more copious than the Latin, and more exquisitely refined than either, yet bearing to both of them a stronger affinity, both in the roots of verbs and the forms of grammar, than could possibly have been produced by accident; so strong indeed, that no philologer could examine them

> all three, without believing them to have sprung from some common source, which, perhaps, no longer exists.[38]

Most of the distinguished books in the standard linguistic canon point directly to Jones as the originator of notion that the Indo-European languages formed a family. Certainly it is true that in 1788, Jones was an Orientalist of considerable talent and renown, and he had already published a monumental grammar of Persian. He was a judge and a royal administrator, and was interested in the principles of traditional Hindu law.

And yet considerable work on the relationships among the languages of Europe had started in the early years of the Renaissance.[39] Current scholarship has shed considerable light on the richness and the complexity of the development in Europe of our understanding of the historical relationship of the European languages, and a number of scholars have argued that Jones's contribution has been exaggerated and the historical reality oversimplified. The current discussion seems to be structured in large measure to address the credit problem (how much intellectual credit should Jones receive?), a difficult question that we recused ourselves from in chapter 1, and yet it is a question that many find hard to simply let go of.

Renaissance scholars well before Jones had already established an understanding of the evolution of Latin into the Romance languages, and in the sixteenth century, Joseph Justus Scaliger had suggested groups corresponding to Germanic, Greek, Latin, and Slavic.[40] Seventeenth-century scholars such as Andreas Jäger and Georg Stiernhielm began to propose hypotheses linking the languages known to Western scholars, and a good deal of energy went into trying to determine which, if any, of these known languages was the first and original language. In the background of this discussion was the story of mankind as presented in the Judeo-Christian Bible, to be sure. The great philosopher Gottfried Leibniz spent an enormous amount of effort comparing the languages not just of Europe, but of Asia and Africa as well, and he proposed quite respectable hypotheses about the development of these languages. He put Arabic and Hebrew and other Semitic languages in a category he called Aramaic, and the European languages (other than Basque) into the Japhetic branch. As far as we can tell, his thinking was guided and constrained by his understanding of the history of mankind offered by the Christian Bible. It is striking, in retrospect, that the single most important intellectual shift that was necessary to arrive at the modern view was *drop-*

ping what had seemed like an essential constraint on possible theories, the Bible's rough account of the beginning of human history.

There was indeed an understanding common to a number of these works that there had once existed a "no longer spoken parent language which in turn produced the major linguistic groups of Asia and Europe."[41] A hundred years before Jones—in 1686—Andreas Jäger had written words not so very dissimilar:

> An ancient language, once spoken in the distant past in the area of the Caucasus mountains and spreading by waves of migration throughout Europe and Asia, had itself ceased to be spoken and had left no linguistic monuments behind, but had as a "mother" generated a host of "daughter languages," many of which in turn had become "mothers" to further "daughters." (For a language tends to develop dialects, and these dialects in the course of time become independent, mutually unintelligible languages.) Descendants of the ancestral languages include Persian, Greek, Italic (whence Latin and in time the modern Romance tongues), the Slavonic languages, Celtic, and finally Gothic and the other Germanic tongues.[42]

And the Jesuit Gaston-Laurent Coeurdoux was also responsible for an analysis prior to Jones's which postulated a family of languages which we would today call Indo-European.[43]

Even if the existence of a family of Indo-European languages and the postulation of an ancestral language that is no longer spoken were not totally new ideas, Jones's formulation had considerable impact, in no small part due to Jones's scholarly reputation—and, perhaps, the fact that the time was ripe for it. Jespersen gives Jones credit for stimulating interest in Sanskrit, but noted that Jones "did nothing to carry out in detail the comparison thus inaugurated, and it was reserved for younger men to follow up on the clue he had given."[44]

The cultural importance of Indo-European

Let us return now to the early nineteenth century. The dawning realization that there was an Indo-European family, with a considerable depth in time and cultural richness, served to provide a new answer to European issues of identity which became acute as a new set of nations would emerge over the course of the century. The Western tradition had long been ambivalent about the uncomfortable fact that it was to the Semitic sphere, not the European, that it reached when looking for its most important tradition of language, literature, and religion, one that reached back at least two thousand years. But with the emergence of the Indo-European family, a new family, the Aryan family, was now ready to appear with a rich and coherent tradition, not to mention a powerful set of myths. The word *Aryan* was coined by Schlegel in 1819 on the basis of the Sanskrit root *arya-*, "noble." In francophone and anglophone countries, one spoke of "Indo-European," while in Germany one spoke of "Indo-Germanic," a term created by Franz Bopp.[45]

The contemporary archeologist Jean-Paul Demoule put himself in the mind-set of a nineteenth-century European of this time:

> Had not the Europeans till then been indebted to the Jews, the image *par excellence* of the Other, in their accounts of their own origin?—surely this was scandalous. And so the European intellectual groped for the stuff from which a myth could be put together, a myth that would give Europeans indigenous roots; this quest was all the more intense in Germany, a country without a state, or a territory, or even a single language.[46]

Ernest Renan was a well-known writer in France in the mid-nineteenth century, an expert on the Middle East and especially Christianity—Renan's renown was based more than anything on his efforts to show that Jesus had somehow risen above his Semitic background to become Aryan. Unappealing as that hypothesis might appear today, it shows how work that began as strictly comparative linguistics opened the door for those, like Renan, who simply *needed* another account of who he and his countrymen were: they were rooted in a deep historical tradition that was, most importantly, *not Semitic*. It was Indo-European, Indo-Aryan—a fine alternative to the Semitic family, if one did not want to be culturally indebted to Jews and Arabs.

Renan was able then to divide his understanding of the Western world into two parts: "The most important conclusion that the historical and philological sciences have arrived at in the last fifty years has been to show in the general development of mankind two elements which (though they may have varied in their relative strengths over time) have served as the frame upon which the fabric of history has been woven." These two parts were the Semitic and the Indo-European.

> As far back as the 17th century and indeed almost as far back as the Middle Ages, it was recognized that the Hebrews, the Phoenicians, the Carthaginians,

> the Syrians, Babylonia (at least after a certain point in its history), the Arabs, the Abyssinians had all spoken languages that were of the same family. Eichhorn, in the last century, proposed the name "Semitic" for these languages, and inexact though it may be, may well continue to be used.[47]

But there was a whole new way to understand European history now, he wrote:

> In the first years of our present century an important discovery, both delicate and important, was made. Thanks to a knowledge of Sanskrit, due to the English scientists of Calcutta, the German philologists, notably Mr. Bopp, established solid principles with which one could establish that the ancient languages of Brahman India, the different dialects of Persian, Armenian, several dialects of the Caucasus, Latin and Greek and the languages derived from them, the Slavic, Germanic, and Celtic languages all form a vast system, profoundly different from the Semitic group, which was been termed Indo-Germanic or Indo-European.

In short, the discovery of the relatedness of Sanskrit and the vast majority of the languages of Europe was a matter of importance, value, and meaning, to thoughtful people—and in many cases, anxious people—everywhere in Europe. That work had an impact on the way in which Europe saw itself, and how it understood its history and its prehistory. This "return to India" passion inspired an Aryanism that was at first scientific but that evolved into a political and racial trend. The term *race* began as a descriptive term with no racist connotation, in our sense today, and we see that in the universalist humanism of Renan, for example. But it slowly took on the sense that we know today, which began with Arthur Gobineau and Georges Vacher de Lapouge and was developed by Houston Stewart Chamberlain in the larger context of a social Darwinism, which was, in turn, borrowed from Herbert Spencer and Ernst Haeckel.[48]

Aryanism and Indo-Germanism became the basis of a national movement, the *Volkisch* movement that arose at the end of the nineteenth century, incorporating racist elements in a quest to determine the identity of the German people. The movement was based on the anti-Semitic writings of Paul de Lagarde, who began his own career as a Semitic philologist, and who shifted his interest to Indo-Germanic and Sanskrit, influenced by Max Müller, a linguist we will meet shortly, and Herbert Spencer, the influential thinker who coined the phrase "survival of the fittest."[49] This nationalist movement was strongest in Germany, where the "return to India" movement brought cultural, ethnic, and mythological answers to the questions being posed by philosophers about their origin as a people.[50]

Within German Romanticism, Aryanism was tightly linked to organicism, the view of language as a living organism; it was Schleicher who most clearly expressed this very German position: "Languages," wrote Schleicher, "are natural organisms which, beyond human will and in accord with definite laws, are born, grow, develop, age, and die; languages thus manifest that series of phenomena which we normally view as aspects of *life*. Glottics, or the science of language, is hence a natural science."[51] This would become a hotly contested position later in the century, when linguistics moved further and further away from the biological metaphor. But linguists still refer to the smallest core of a word as its *root*.

Generations in linguistics

In the previous chapter, we discussed the important role that generations play in the development of the mind sciences, and nowhere is this clearer than in the development of linguistics in the nineteenth century. We are going to look now at the first three generations of linguists. The first generation was composed of the scholars who began the serious comparative study of the European languages after William Jones, bringing Sanskrit into that larger picture; the principal members of this group were Friedrich von Schlegel, Franz Bopp, Jacob Grimm, Wilhelm von Humboldt, and Rasmus Rask. It was only at the end of this generation that universities began to be a home for researchers: Bopp, as we will see, obtained a chair with the support of the influential Wilhelm von Humboldt. The first generation learned their linguistics by studying language directly.

The second generation included Georg Curtius, August Schleicher, William Dwight Whitney, and Max Müller, who deepened and broadened their teachers' work.[52]

The third generation brought the first instance of social rupture in linguistics, pitting the Neogrammarians in Leipzig against the image they had of their teachers. The Neogrammarians were the first generation of German linguists that was able to overcome the Romantic ideas of language decay and develop a modern account of sound change. These included such linguists as August Leskien, Berthold Delbrück, Hermann Osthoff, Karl Brugmann, and Hermann Paul. Heavily influenced by the

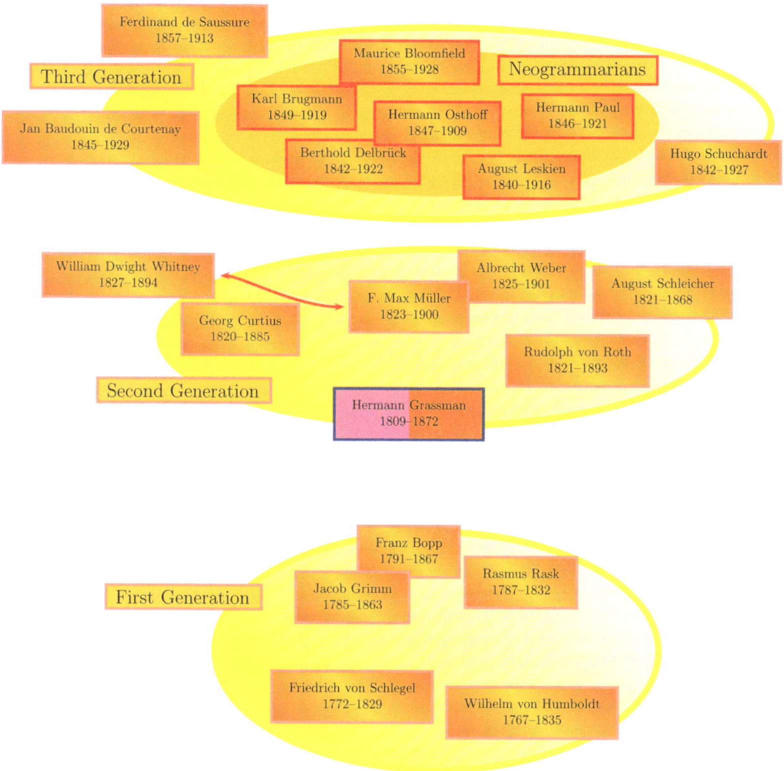

FIGURE 2.1. First three generations of linguists

Neogrammarians were two students who established schools outside of Germany: Ferdinand de Saussure, a Swiss francophone who studied briefly in Leipzig, and Jan Baudouin de Courtenay, a Pole who developed one of the first schools of phonology in Russia, after receiving his doctorate in Leipzig. Both Saussure and Baudouin viewed themselves as autodidacts, as scholars who had arrived at their insights without the help of the teachers who had lectured to them, as if they were unwilling or unable to incorporate anything worth keeping in their heads from the things that they heard in lectures.[53]

PROTO-INDO-EUROPEAN

We must briefly introduce some notions that will play an important role in the discussion of the first several generations of linguists in the nineteenth century, all of whom worked on the *reconstruction* of the early Indo-European language or some closely related issues. Once it was generally accepted that Jones's basic idea was correct, and that Sanskrit, Latin, Greek, and most of the other European languages descended historically from a single language of which no direct evidence exists today, it became perfectly clear that there was a great challenge to face: would it be possible to *reconstruct* the earlier language, to somehow spell out the words, and the roots, stems, and suffixes of this hitherto unknown language? To be sure, scholars had long been aware of the *fact* of language evolution: knowledge of Latin had not disappeared from Western Europe, and the common inheritance of the modern dialects of French, Italian, Spanish, and the other "Romance languages" was widely appreciated. But while there were a good number of evident commonalities among the Indo-European languages, the differences were far more numerous, and the challenge of *explaining* those differences was considerable. Ultimately the task for the linguist was to develop a set of hypotheses regarding what the word structure, and sentence structure, was in the most recent common ancestor of the Indo-European family, presented along with a set of hypotheses for how the 12 or so branches of Indo-European had changed as they went their individual ways—and this was precisely a matter of determining the *morphology* of the language in Goethe's sense. Behind this work was the belief that the reconstructed Indo-European language (or *Proto-Indo-European*) would have been spoken by a single group of people, in some particular geographical area whose location remained to be determined. That group of people, perhaps living somewhere in Central Asia, gradually migrated in various directions, beginning before any records that remain and continuing right up to the present.

TABLE 2.1. **Related Indo-European words**

Language	"father"	"mother"	"brother"	"one"	"two"	"three"	"nine"
Sanskrit	pitár	mātár-	bhrā́tar-	ḗka-	d(u)vā́	tráya-	náva-
Latin	pater	ma:ter	frāter	u:nus	duo	trēs	novem
Greek	paté:r	mā́tēr (Dor.)	phrā́te:r	oiné: ("ace")	dúo	treîs	enné(w)a
Old Irish	athir	māthir	brāthir	o:en	da:u, da:	trī	nōin-
Gothic	faðar		brōþar	áins		þriya	niun
Old Icelandic		mōðir					
Armenian	hayr	mayr	ełbayr				
Tocharian A	pācar	mācar	pracar			tri (fem.)	ñū
Old Church Slavonic		mati	bratrŭ		dŭva	trĭje	

An example or two might be helpful here. The Greek word for 'horse' was *hippoi*, and that word has been embedded in *hippopotamus* (from *horse of the river, hippo+potamos*), and the Latin word for 'horse' was *equī*. Sanskrit had *aśva*, and those are just three of the older forms that we know. From these and the general patterns we have uncovered, we can reconstruct the Indo-European word for 'horse' as **ekwo*, where we put the asterisk to indicate that this is our *hypothesis* of the form of the word at an earlier time. This example is a bit opaque; other examples of related words in the various Indo-European languages are more transparent.

First generation

FRIEDRICH VON SCHLEGEL

Friedrich von Schlegel holds an important place in the history of German culture: he was a major figure in the rise of the Romantic movement, which emphasized emotion over intellect, and beauty over rationality. In his late twenties, after studying ancient Greek, he moved to Paris, where he was able to study both Persian and Sanskrit, because of a rather unusual situation. A British citizen named Alexander Hamilton, born in India, had been a member of William Jones's Asiatic Society. After Jones's death, Hamilton moved to England from India, but he was working with Sanskrit archives in Paris when war broke out between France and England in 1803. Though technically an alien citizen of an enemy power, he was permitted to teach Sanskrit in the French capital, and he counted among his students Friedrich Schlegel and Franz Bopp, as well as a Frenchman named Antoine-Léonard de Chézy, who would later obtain the chair of Sanskrit in the Collège de France. Just below we will meet Franz Bopp, who studied Sanskrit in Paris with both Hamilton and Chézy.[54]

Schlegel's deepest interests lay in art and poetry, and by today's standards, he was much too quick, and too comfortable, in making decisions about what kind of word structure is degenerate, a style of expressing his ideas that a good many linguists continued to indulge in for another 50 years, until the time of the Neogrammarians. He was particularly unimpressed by languages in which it was easy to segment an inflected word into stem and affix, and he viewed this analytic simplicity as entirely artless.[55]

Schlegel proposed a first classification of ways in which words could be formed—of morphology, as we say today. When one looks at sets of grammatically related words (which might be as simple as *drink* and *drank*, or *jump* and *jumped*), we see that modifications may either be *inside* the

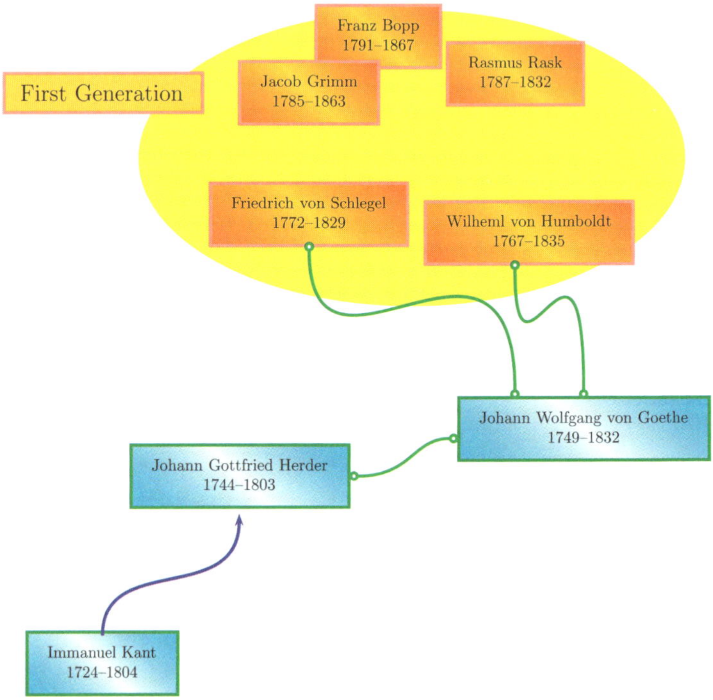

FIGURE 2.2. First generation of linguists

root (as in *drank*), which Schlegel called *flexional* modifications, or be formed by addition of material (such as *-ed*), which additions we call *affixes*. Ten years later, Schlegel's brother, August Wilhelm Schlegel, proposed a third category, that of languages that have neither flexional nor affixal morphology, which is possible if words remained isolated units, not forming paradigms or other sorts of sets of closely related words. Today these are called "isolating" languages, and Chinese illustrates this kind of language.

Schlegel saw something "organic"—which was, in his view, a very good thing—in a language with a flexional system, that is, one that makes internal changes to the root of a word (these changes can be to an internal vowel, or to a consonant). "In the Indian and Greek languages each root is actually that which bears signification, and thus seems like a living and productive germ, every modification of circumstance and degree being produced by internal changes." Now, reconstructed Proto-Indo-European contained many complex changes in the root that were occasioned by grammatical features, and this fact has been of great help in determining the relationship of the descendant languages. These complex changes were "woven into a fine artistic tissue," Schlegel wrote, and that has made reconstruction possible. When languages express their morphological features with distinct morphemes, each added almost one after another, it seems, there is no "bond of union," and their "roots present us with no living productive germ, but seem like an agglomeration of atoms, easily dispersed and scattered by every casual breath."[56]

Schlegel opposed the word *organic* to the word *mechanical* in his description of certain aspects of language, and famously wrote,

> The decisive point however which will clarify everything here is the inner structure of the languages or comparative grammar, which will give us quite new information about the genealogy of languages in a similar way as comparative anatomy has illuminated the higher natural history.[57]

We have already suggested that the astonishing growth of historical and comparative grammar at this time must be seen within the larger political and ideological context. If the "return to India" message and the development of orientalism as a science could be seen across Europe, in Paris as in London or Saint Petersburg, it produced an even greater effect in Germany, where, as we have observed, discussions were ongoing about the origin of language and ultimately about the nature of the German people, who were searching for a political unity as an *ethnos* as well as for cultural recognition. This accounts for the particular context in which German scholarly studies would from the start combine cultural studies with linguistic studies. Schlegel's account of the prehistory of Europe included a homeland in the East, where religion was originally discovered, in a place that was surely not too far from India. Though German speakers' ancestors came from this culture, Schlegel thought, Sanskrit was the oldest language, the closest language to this prehistoric common period. Egyptians and Hebrews were no doubt colonies of this original culture, which had somehow fallen apart for reasons that were no longer known, or even knowable. The ancestors of the Germans had traveled north from Central Asia, going through modern Ukraine or Russia up to Scandinavia.[58] This debate has by no means been settled today; there are still competing historical scenarios placing various prehistoric bands of Indo-European ancestors on different sides of the Black Sea, and while advances in DNA tracing and in archeology have clarified the situation, fundamental controversy about the historical character of the spread of the Indo-European languages remains alive today.[59] It seems likely that these questions will be settled quite soon.

FRANZ BOPP

Franz Bopp was the youngest member of this first generation of linguists, and also the first linguist to obtain a professorial position to *do* modern linguistics—in his case, developing Indo-European linguistics. As a student, he had been inspired by Schlegel's book to focus on Oriental studies, and he studied Sanskrit with Schlegel's tutor Chézy in Paris. He also interacted with Alexander Hamilton, as well as Wilhelm von Humboldt, who was a diplomat at the time in the Prussian embassy located in Paris. He was the first major linguist to develop an explicit account of the evolution of much of the morphological structure of Latin, Greek, Persian, and Sanskrit—inspired by the work of Schlegel in particular on Sanskrit, though Bopp was able to provide a clearer picture of the changes involved. One of his best-known discoveries illustrates well what nineteenth-century comparative linguists were looking for: not the evolution of words, but methods to use knots of complexity in separate languages in order to understand the historical relationship between the languages. Bopp did just this, showing that the Celtic languages (such as Welsh, Irish, and Breton) were part of the Indo-European family. It was well known at the time that in certain grammatical constructions, the initial consonant of a word could undergo systematic changes that were occasioned by the preceding word. In Irish, we see *a cara* 'her friend', but *a chara* 'his friend' (where the *ch* is a velar fricative), and *a gcara* 'their friend' (where *gc* represents the familiar sound /g/). The changes in the second word are determined by the word that precedes, and many other words of Irish show one of these three kinds of effects (called *mutations*). Bopp discovered that the kind of effect a word has in Irish can be determined by looking at the corresponding form in Sanskrit! Irish *a* 'her' corresponds to Sanskrit *asyās*, Irish *a* 'his' corresponds to Sanskrit *asya* 'his', and Irish *a* 'their' corresponds to Sanskrit *ēſām* 'their'. Irish words that correspond to vowel-final words in Sanskrit act like Irish 'her,' and those that end in a nasal consonant act like Irish 'their.'[60] What an astonishing discovery! Let us not lose sight of the fact that part of the beauty of this discovery derives from the fact that Bopp followed Schlegel's belief that studying the comparative linguistics of the grammatical system was more important than comparing the form of lexical roots and items.

TABLE 2.2. **Bopp's discovery**

Sanskrit		Celtic	
asyās	"her"	a cara	"her friend"
asya	"his"	a chara	"his friend"
ēʃām	"their"	a gcara	"their friend"

Bopp's work shifted the underlying assumptions about how European languages related to all of the rest of the world's languages. By identifying more rigorously what constituted the Indo-European languages, it became clearer how different the *non*-Indo-European languages were from the Indo-European ones. As we noted at the beginning of this chapter, this was the moment when the globe was becoming much *older* in people's minds, and the community of mankind was being rethought. Bopp's book on comparative grammar was published over a period of 20 years and had an enormous influence on the discipline; it was in its third edition at the time of his death.

Bopp saw morphology rather differently than Schlegel did, and their disagreement (which grew over time) reflects a difference of views that we still see today in some respects. Schlegel believed strongly that the governing principle of morphology in Sanskrit was what he called "inner change" in the roots, which would soon be called *Umlaut* and *Ablaut*, and which today might be analyzed as the effect of a rule on the phonological substance of a root. For Schlegel, any change that is not the addition of a prefix or a suffix (as these internal changes are not) is *organic*; the addition of an affix would be, in his view, *mechanical*. Most linguists today are in agreement that such a difference might well be analyzed in different theoretical ways, even if they would not agree that a language chooses just one sort of strategy or the other (many languages employ both). Few, if any, would defend the notion that one sort of language is better, or more organic, or glorious, than the other.[61] Bopp, on the other hand, was strongly inclined to divide words into morphemes, pieces that he could associate with grammatical function, in a fashion that some, but only some, linguists share today.[62] More than 100 years later, Charles Hockett would dub Bopp's theoretical preference as an *item and arrangement* model. Thus Bopp's conception of a perfect language shifted from that of Schlegel: for Bopp, perfection lay in the natural correspondence between the pieces of a word and the functions expressed by the word as a whole.

Bopp's work reads today like modern linguistics in certain ways. He notes, for example, the potential mood in Sanskrit is realized with a *ī* between the root and the suffix marking the subject; the *ī* becomes *y* before a vowel *adyàt, adìta* (the root is *ad* 'eat') and merges with a preceding *a* to become *e*, as in *bhavèt* 'he/she/it would be' instead of an expected (but incorrect) *bhavaìt*. Bopp notes the parallel situation in the Greek optative, where the *i* is "inserted just in the same way as *ì* is in Sanskrit, between the root and the personal termination."[63] And the same is found in Gothic, relating *Sokyam* 'we seek' and *Sokyaima* 'we may seek'. Or a few pages later, comparing a first-person singular suffix which is *m* in Sanskrit to a suffix which is *n* in Greek, he notes that the Greek *μ* is changed to *v* "conformably to the prevailing principle of the language, which does not permit the use of a final *μ*": as we might say today, word-final *m* becomes *n*—in any event, a hundred years later, Sapir will express such a generalization in exactly the same fashion as Bopp.[64] Bopp's discussion is engaged with talking about the language as it is at a given time and is not offered as an account of history. Bopp's analysis leaned heavily towards an abstract morphological analysis, in the sense that the internal pieces of the word—its analysis into morphemes—would each be assigned a grammatical function in the sentence in which it is found. He viewed the central relation in morphology to be binary, of which one item is basic, the other accessory (a "transparent agglutinative structure," as Verburg calls it): today we might say lexical and grammatical. His structural intuition is shared by many linguists today. By no means did his view in this regard convince his colleagues.

Bopp's introduction to his most famous work (1856) contained a passage that expressed well Bopp's perception that he was doing something new by studying languages "for their own sake":

> As in this work the languages it embraces are treated for their own sakes, i.e., as objects and not means of knowledge, and as I aim rather at giving a physiology of them than an introduction to their practical use, it has been in my power to omit many particulars which contribute nothing to the character of the whole; and I have gained thereby more space for the discussion of matters more important, and more intimately incorporated with the vital spirit of the language.[65]

This phrase would be echoed decades later in Ferdinand de Saussure's work.

Verburg argued that Bopp's analysis was essentially synchronic in outlook: "Bopp compares forms of language which he takes to be static . . . the diachronic process with its transitions do not interest him as such. Especially the typically Romantic ethos with regard to history is wholly absent in him. . . . Everywhere Bopp doggedly insists on the reduction to affixing, e.g., infixing, on the elementary addition of formerly independent linguistic elements (except for reduplication), in which the primitive element, the root, bears the [fundamental concept] and the affixed (or infixed) elements bear the [relational concept]."[66]

From Humboldt's point of view, Bopp's work, and more generally historical and comparative grammar, was strategically important for the construction of a German identity, and he was able to easily persuade the kaiser to create an academic chair for Bopp in Berlin.

WILHELM VON HUMBOLDT

Wilhelm von Humboldt's placement in the first generation of linguists seems odd if one simply looks at the year in which he was born: he was in fact 20 years older than Rask and Bopp, but his own career as a linguist did not begin in earnest until he retired from a career as a Prussian civil servant and diplomat in 1820, at the age of 53. Not that he did not work on language and languages while he was a diplomat: he certainly did. When he was Prussian ambassador to England, he took private lessons on Sanskrit from Bopp, who was doing research in England at the time. And Humboldt did much the same when he was in Paris.

Humboldt articulated a view according to which the ability of humans to learn languages, with all their richness and depth, was surely based on a deep-seated human capacity. "Language could not have been discovered," he wrote, "if its prototype, its *Typus*, had not already been present in the human mind."[67] What a human ear hears is not a stimulus, like one that an animal might respond to; it is a "an articulated sound designating an idea." And he believed that language could not have been slowly developed by humans: "its discovery could only happen all at once. The human being is only a human being because of language: in order to discover language he must already be a human being." Of course, languages are learned, but that learning is possible because inherent in humans is a modality that allows them to be creatures who speak. He saw that in babies: "Anyone who observes infants carefully will agree that their way of imitating the sounds of those who are around them is less a matter of learning than of guessing and creating."[68]

Humboldt developed the morphological typology that the Schlegels had begun, proposing that many North American languages should be classified as *incorporating*, allowing for the occurrence of more than one stem in a word[69]—indeed, his typology remains widely used today.[70] If we look at what Humboldt wrote with our present-day eyes, we can interpret his discussion charitably, and we can see that he was sensitive to at least three important characteristics of some, but not all, morphological systems. Some languages present internal changes to stems for grammatical reasons—*internal* in a left-to-right sense, just as we see in English in the changes that relate the three words *sing, sang, sung*. Languages that express grammatical information in this way were called *inflecting* during this period, as we have noted. The principal alternative to this kind of morphology was the affixal morphology, whereby a word was composed of a stem, with one or more affixes (i.e., prefixes or suffixes). There was a tremendous suspicion and bias in the minds of linguists such as Humboldt against affixation and in favor of inflection, and the reasons that Humboldt gave in support of his view, from today's perspective, seem amateurish at best.[71] Still the considerable influence of his point of view on Darwin was significant, and it allowed Darwin to talk about the parallels between species and languages.

There were two more distinctions brought out: in some languages, there were changes in the sounds that appeared at the end of a stem or the beginning of a suffix, typically changes that made the union of the stem plus the suffix seem better glued together—for example, a stem-final vowel might be elided before a suffix beginning with a vowel (as we do in English, when we combine *buddha* and *-ism* to form *Buddhism*). And finally, there were languages in which each affix could be associated with a single grammatical category and those in which that was not the case. Indo-European languages typically fell into the second category, in that verbal suffixes more often than not indicate both person and number, and nominal (and adjectival) suffixes often mark some combination of number, case, and gender.

During the nineteenth century, the notion of progress was visible everywhere, a fact that made it difficult to think about morphology and change without bringing the notions of progress, and later fitness, into the picture. Germany was home to a great deal of the early research on Indo-European, and the influence of Romanticism on it is indelible in the work of Bopp, Schlegel, and other early linguists, such as Jacob Grimm. From Romanticism came the image that language is organic, and that the course of an organism over time is not merely change, but a movement along a dimension that can be labeled as positive development and perfection at one end and decay and degeneration at the other.[72]

Organicism took on another form in the context of the comparative analysis of morphological systems that Humboldt undertook. Humboldt made an Aristotelian distinction between language as *ergon*—something produced or made—and *energeia*, which is an activity or a virtual dynamic, something more likely to be conceived of as a vital power, hence an organism. It was notably the distinction among the three types of morphology developed by Schlegel and Humboldt that was conducive to a typological classification of languages and an account by which each typological class could be graded along a cline of adaptive evolution. Inflectional languages were the most evolved, while isolating languages were the least, with agglutinative languages falling in between.[73]

THE NEW UNIVERSITY SYSTEM

The rapid development of a new system of German universities which was focused on research over the nineteenth century played a key role in the developments of all of the disciplines that we discuss in this book; we will just briefly point out some of the crucial elements of this history.[74] Before the remarkable revolution in the German university system at the beginning of the nineteenth century, universities had come out of the Middle Ages as an embattled institution with a history of close connection to the Church. One reformist model for education that was discussed in the eighteenth century was the complete replacement of universities by state-funded and -directed professional schools, with scientific research academies operating in a different social sphere. Major parts of this proposal were adopted in France in 1793, as the 24 universities in France were abolished wholesale, at a time when universities across Europe were struggling with unsustainably low enrollments. In France, a highly centralized system of higher education was then established, which was maintained essentially through the nineteenth century and which served to educate those entering the state civil service,[75] while in Germany a radically new system was established that would eventually take root throughout the world, one in which the research function of the professoriate was at the heart of the university's raison d'être, and the life of the people at the university included active seminars and dissertations with novel ideas. The central belief was that the highest order of education would create people who could themselves be intellectually creative, and the best way to accomplish this was to organically link the teaching and the research functions of the university faculty.

This process began in 1810 with the establishment of the Königliche Friedrich-Wilhelms-Universität—today, Humboldt University in Berlin—by Wilhelm von Humboldt, whose work after his retirement in 1820 we have just looked at. In 1810, Humboldt headed the section responsible for education within the ministry of the interior, and he was able to develop in practice many of the radical ideas proposed by the philosophers Johann Gottlieb Fichte and Friedrich Schleiermacher in their discussions of the goals of education. The university at Halle, the country's premier university, had been closed by Napoleon in 1806 (the reader will recall Napoleon's march across much of Western Europe), and the king of Prussia wanted to establish a new university. Humboldt proposed that henceforth, teaching would be valued and viewed as an integral part of doing research. *Creating* knowledge would be a central mission of professors, a radically novel idea. Humboldt himself noted that henceforth, science would be viewed as "a problem that has not yet been completely solved and thus remains always in a state of research. . . . The relationship between teacher and student is therefore utterly different from what it was in the past." In the past, the teacher was there for the sake of the student, but in the new system, "both are there for the sake of *Wissenschaft*," that is, for the sake of science.[76] (This was the model that would come to the United States later in the century: first to Johns Hopkins, then to Clark University and the University of Chicago.[77]) This conception of the university shared an important feature of its medieval predecessor, which was its autonomy with respect to civil authority. Structurally, the German system was conducive to competition among universities, in part because Germany was not united politically.

As universities were founded, linguistics was viewed by many as a field that was young and up-and-coming, and also one that should be supported by professorships at such places as Berlin, Bonn, Leipzig, Munich, and Breslau (the reader who thinks that linguistics departments are new inventions should think again). Schlegel was named professor of Sanskrit in Bonn in 1818, and Bopp in Berlin in the same year; in both cases, Humboldt played a significant role in the appointment process.

It is important to recognize the impact on the development of the sciences by the change in the very conception of what universities are for, and how they were to be organized. Scientific research, both theoretical and applied, became an important driving force behind economic development and social progress. Rivalries among the major states of Europe—England, France, Prussia, Austro-Hungary, Russia—for political and economic superiority included competition for the best scientific laboratories as well, and universities played a significant role in this area. Germany's victory over France in the Franco-Prussian War of 1870 was viewed by the French intelligentsia not so much as a defeat of France but as a victory of German science.[78] For many intellectuals, linguistics and especially historical and comparative grammar was a German science—an opinion that would play a role in some of the polemics that would ensue in the years to come.

RASMUS RASK

Rasmus Kristian Rask was a Danish linguist who was particularly interested in classification and the typological relatedness of languages based on their morphology. He saw in his own work a descent from Linnaeus's systematic classification. From the beginning of the century he worked on Old Norse, and he compared the morphology of contemporary Danish with that of its ancestor in the etymological grammar he wrote, in Danish, in 1810.[79] This was the first work to explicitly compare two distinct states of a language, and it proposes a systematic analysis. What has come to be known as Grimm's Law, describing the major features of the changes of consonants in the Germanic branch of Indo-European, was first discovered by Rask; we will return to this shortly.

JACOB GRIMM

Much of the world remembers the brothers Grimm for their fairy tales; Jacob Grimm, the older brother, was a professor of Germanic linguistics at Göttingen and Berlin. His careful and precise analyses have left their mark on the names given to many classic phenomena, such as strong and weak verbs, umlaut, and ablaut. As we have noted, the Germanists' world was focused on the oral language and on the diversity that lies in the range of dialects found in real life, and Grimm's *Deutsche Grammatik*, published in 1819, was an important moment in the comparativist movement.[80] The second edition, which came out in 1822, included a systematization of Rask's work, and introduced his *Lautverschiebung*, known to posterity as Grimm's Law (a name later coined by Max Müller). Grimm devoted nearly 600 pages to the study of phonetic changes and the laws that organized them, thereby changing the very face of linguistics: his work for the first time justified using the term *law* in the same sentence as *language*.

No longer were linguists studying words; they were rather studying *sounds* (still often called *letters*, it is true). Linguists were studying the system of sounds in a language: that was an enormous step forward.

Grimm embraced the judgmentalism of the first two generations of linguists, as a sentence like this illustrates well:

> Since the High German of the 13th century shows nobler, purer forms than the language of the present day, and those of the 8th and 9th centuries are purer still than those of the 13th, and finally since the Gothic of the 4th and 5th centuries shows even more complex forms, so it follows that the language spoken by the German people in the first century will have surpassed even Gothic.[81]

Second generation

AUGUST SCHLEICHER

August Schleicher was a major figure in the midcentury study of Proto-Indo-European, in the generation following Bopp and Schlegel, the second generation of modern linguists. His vision of language was deeply influenced both by Hegel and by biological thought and metaphor, and he emphasized the way in which the reconstruction of Indo-European was the reconstruction of a linguistic genealogy: he wrote of the *family tree* of languages—his term, *Stammbaum theory*, "family-tree theory," is still widely used today, and he used the same graphical notation for languages that people used for family genealogies. He was also the first to attempt to reconstruct the phonological form of a hypothetical proto-word.[82]

Schleicher shared with his times the need to view change as either development or decay, but his was the last generation in which this kind of value-laden judgment would be viewed as natural for the linguist. For someone who studies linguistics today, the very idea of using linguistics to decide which languages are better and which are worse in some absolute sense is very odd.[83]

Schleicher also retained the Humboldtian distinction that we discussed above regarding the types of morphology found in languages: isolating, agglutinating, inflectional. In keeping with his concern with the concept of development, he saw not just a difference among these types, but a direction of positive change (away from isolating, and towards inflectional, systems). The question of *teleology* arises here, which is to say, the notion that change is always towards a goal that is defined, somehow or other, before the change in question takes place.

But the evidence does not point to such a directionality in the evolution of inflectional languages, even if Schleicher perceived that the inflectional system constituted the highest level of development. How can this apparent contradiction be maintained? Here the influence of Hegel was crucial. Hegel distinguished development *before* history and *during* history. Schleicher believed that the development of language was prehistoric, in Hegel's sense, and once the Hegelian spirit entered history, language was no longer on its developmental track: it had actually entered into the period of *decay*.

Schleicher read Darwin's *Origin of Species* a few years after it came out, and he immediately saw the closeness of the fit between his thought and Darwin's; we noted earlier that Darwin himself drew a parallel between languages and species and was deeply influenced by historical linguistics. Schleicher published a book in 1863 exploring the relationship between work in historical linguistics and Darwinian explanation, and he had no difficulty in accepting the notion that languages were "organisms of nature."[84] Schleicher went so far as to write that languages "have never been directed by the will of man; they rose, and developed themselves according to definite laws; they grew old, and died out."[85] For other linguists, as we shall see, this was going too far.

MAX MÜLLER

There were two other important linguists of the second generation: Max Müller, a German linguist who spent his career in England, and William Dwight Whitney, the first great American linguist, who would have an impact on linguists to come, notably Saussure and Bloomfield. Müller and Whitney became embroiled in the first public linguistic controversy.

Max Müller was a German-born linguist who spent 50 years in England, where he was widely recognized as a great scholar of Sanskrit and a popularizer of the new science of linguistics, and today he is recognized as one of the most important contributors to the founding of the study of religions. Müller proposed the term *Turanian* to refer to what he viewed as a large family of languages spoken on the land mass to the east of Europe and spreading over much of Asia. He also spoke and wrote elegantly about what linguistics had accomplished and what it had yet to do. Let us take a brief look at one lecture, much of which is modern and insightful and at the same time is thoroughly grounded in a mind-set of development and decay, making some of it hard to take seriously today.[86]

"What has the Science of Language, as yet, really accomplished? It has achieved much, considering that real work only began about fifty years ago; it has achieved little, if we look at what still remains to be done," he wrote. "The first discovery was that languages admit of classification," he continued, and certainly that was an important step forward; he had in mind, first of all, the genetic classifications that linguists had been exploring since the end of the eighteenth century, and also the classifications regarding morphology and word order that we discussed briefly above. Of the linguistic families, there were two that had been studied in detail, the Aryan—that is, the Indo-European—and the Semitic. Others have been studied to some degree, including the Ural-Altaic, the Indo-Chinese, and the Dravidian, but Müller argued this was really not enough, because

> considered by themselves, and placed in their proper place in the vast realm of human speech, they describe but a very small segment of the entire circle . . . nearly all the discoveries that have been made as to the laws of language, the process of composition, derivation, and inflection, have been gained by Aryan and Semitic scholars.

There are so many more languages to be studied:

> We must not shut our eyes to the fact that our field of observation has been thus far extremely limited, and that we should act in defiance of the simplest rules of sound induction, were we to generalize on such scanty evidence.

Now his discussion became, all of a sudden, something that seems like it came from a mind-set we no longer share. Müller wrote that since we know the Indo-European and Semitic families of languages better, let us see what they can show us—a not unreasonable suggestion:

> All we know of them is their period of decay, not their period of growth, their descending, not their ascending career, their Being, as we say in German, not their Becoming.

It is taken for granted, and without argument, that languages first went through a period of growth and development whose details are hidden from us in the mists of time, and that period was followed by a period of decay (the only period that you and we have ever known). This was how the world was seen and how it was felt.

> Even in the earliest literary documents both the Aryan and Semitic speech appear before us as fixed and petrified. They had left for ever that stage during which language grows and expands until it is arrested in its exuberant fertility by means of religious or political concentration, by means of oral tradition, or finally by means of a written literature.

In Müller's view, languages *grew* and developed for a period, an exuberant period, and then that development was cut short. Why? It was cut short by becoming a written language, even though "writing, or what in early times takes the place of writing, oral tradition, is something merely accidental." That is to say, do not suppose or speculate that something significant has happened to a language when its speakers begin to write it down, or develop an oral tradition. "It represents a foreign influence which, in natural history, can only be compared to the influence exercised by domestication on plants and animals." Language would be *more truly language*—those are his words—if there were neither oral nor written literature. And so if we take our study of the ancient written languages of the Semitic family or the Aryan family as our guides, they will never be anything but *monstra*, "unnatural, exceptional formations which can never disclose to us the real character of language left to itself to follow out its own laws without let or hindrance."

Müller then sketched the Humboldtian view, which he thoroughly endorsed, that languages may begin in a period of word isolation, in which sentences are strings of indivisible words, after which some of the words may lose semantic robustness and become regular affixes on a neighboring word, which is then a stem. After that, the stem-affix unity may become more organically composed, and we then have an inflecting language, such as the languages of Indo-European and Semitic. So far, this is still widely agreed to today. But Müller went on to say that many of the languages of the world were "arrested in their growth during their earlier stages, and had remained on the surface [note the geological metaphor] in this primitive state, exposed only to the decomposing influence of atmospheric action [geology again!]." Chinese is a language that was arrested in its development very early on, Müller continued, which makes it both instructive and fascinating. Chinese is like someone "groping his way, and so delighted with his first successful grasps that he repeats them again and again. It is child's play, if you like, but it displays, like all child's play, that wisdom and strength which is perfect in the mouth of babes and sucklings." Müller then went on to describe very plausibly how languages could evolve from Humboldt's isolating type to his agglutinative type, and then on to an inflectional system—again, plausible by today's understanding, but cast within a worldview in which this morphological and syntactic change is part and parcel of a picture of decay, not just of language but decay in the modes of thinking that are due to that linguistic decay.

Müller emphasized in the strongest possible terms that languages could and did evolve through these three stages, and he could not understand why so many of his colleagues in the discipline seemed to think that "by some inexplicable grammatical instinct, or by some kind of inherent necessity, languages were from the beginning created as isolating, or agglutinative, or inflectional, and must remain so to the end." "But when we analyse each language more carefully we find there is none exclusively isolating, or exclusively agglutinative, or exclusively inflectional": each language can retain parts of the its older inheritance, which serve like fossils to show how the language used to be. But again the metaphor of decay is never far away: "Unless Sanskrit and Hebrew had passed through the agglutinative stratum, nay unless, at some time or other, they had been no better than Chinese" (imagine that!) their present state would be inexplicable.

WILLIAM DWIGHT WHITNEY

The first great American linguist was William Dwight Whitney. He was born in Northampton, Massachusetts, and early on studied ornithology and botany; as a young man, he worked with his brother Josiah on a geological survey of Lake Superior. But in the end it was Sanskrit that captivated Whitney's imagination, and he went to Yale to pursue this interest. Alas, Yale was not the place to study Sanskrit—Germany was—and so he set off for Berlin. His first Sanskrit teacher was Franz Bopp. "I don't find lectures at the University of Berlin so great shakes as I had supposed," Whitney wrote to his brother. Whitney was appalled at how much he was being charged. "I have had the last experience with the old gentleman, and never really want to see his face again. . . . Bopp's lecture this afternoon, stupid enough as usual."[87] He stopped attending Bopp's lectures and went on to study with two other scholars, both of them former students of Bopp: Rudolph von Roth and Albrecht Weber. Then he returned to Yale in 1854, where he taught Sanskrit—and French and German too, since demand for Sanskrit was limited.

Whitney published two books on general linguistics, the first in 1867, and the second in 1875. The latter, *Life and Growth of Language*, was a much greater success, and French and German translations appeared almost immediately.[88]

In 1875, he was back to Leipzig, where he began his *Sanskrit Grammar*.[89] The next year, a young Swiss linguist by the name of Ferdinand de Saussure arrived in Leipzig to continue his studies of Sanskrit, which culminated with his thesis on "The Primitive System of Vowels in the Indo-European Languages."

A good deal of Whitney's work is remarkably contemporary in tone. And his work in general linguistics (which is to say, that which is apart from his technical work on Sanskrit) emphasized a distinctively American practical sense, a realization that while language is everywhere—in his words—the product of the conscious action of human beings, it is not what "men have voluntarily or intentionally placed there."[90] In fact, when we read Whitney's observations on language and its scientific study, it sounds very familiar. His 1867 book, *Language and the Study of Language: 12 Lectures on the Principles of Linguistic Science*, in some respects seems more up-to-date than books written in the 1920s or 1930s.

> Each separate item in the production or modification of language is a satisfaction of the need of the moment; it is prompted by the exigencies of the particular case; it is brought forth for the practical end of convenient communication, and with no ulterior air or object whatsoever; it is accepted by the community only because it supplies a perceived want, and answers an acknowledged purpose in the uses of social intercourse. . . . A language is, in very truth, a grand system, of a highly complicated and symmetrical structure; it is fitly comparable with an organized body; but this is not because any human has planned such a structure and skillfully worked it out. Each single part is conscious and intentional; the whole is instinctive and natural.[91]

Whitney drew heavily upon analogies (though these always remained analogies, whose purpose was for clarification and for instruction) with two of the most important scientific movements of his day: Darwinian evolution and geology.

> There is a yet closer parallelism between the life of language and that of the animal kingdom in general. The speech of each person is, as it were, an individual of a species, with its general inherited conformity to the specific type, but also with its individual peculiarities, its tendency to variation and the formation of a new species. The dialects, languages, groups, families, stocks,

> set up by the linguistic student, correspond with the varieties, species, general, and so on, of the zoölogist. And the questions which the students of nature are so excitedly discussing at the present day—the nature of specific distinctions, the derivations of species by individual variation and natural selection, the unity of origin of animal life—all are closely akin with those which the linguistic student has constant occasion to treat. We need not here dwell further upon the comparison: it is so naturally suggested, and so fruitful of interesting and instructive analogies, that it has been repeatedly drawn out and employed, by students both of nature and of language.[92]

One of Whitney's concerns was with the autonomy of linguistics as a discipline and as a profession. He pointed out that "physical science on the one side, and psychology on the other, are striving to take possession of linguistic science, which in truth belongs to neither."[93]

As we will see below, Edward Sapir would say precisely the same thing in the late 1920s. Whitney certainly had no confidence that an intellectual alliance with psychology was in the interest of linguists. He recognized first that psychology as a discipline was making advances:

> For . . . besides the "recognition of the creative power of man," we have in this century the advantage of a rational psychology, which strives to discover a mechanism in the movements of consciousness, laws in mental life, and so on; since all the creations of man will be found not less subject to the dominion of rational laws than are the productions of nature.[94]

That sounds very modern: psychology has the right to expect to find laws just like the natural sciences do. Whitney went on to say, then, that understanding the mind better will doubtless help us to understand language better. This is followed then by a big *but*:

> Now we also, on our part, expect decided advantage to the study of language, as of every other human production, from an improved comprehension of the operations of the human mind, as of all the other determining conditions of a difficult problem. But whether the advance of psychology is or is not to bring about a revolution in the science of language, is a question depending on the manner and degree in which language is a "mental production."

Is a linguistic production just a psychological act? No, says Whitney; it would be a serious error to think so:

> It is very possible here to fall into the serious error of looking upon words and phrases as an immediate emanation of the mind, and so of settling the laws of mental action, and out of them evolving the events of language-history. The soul of man and its power and operation are, after all, the mystery of mysteries to us; the phenomena of language are one of its external manifestations, and comparatively a simple matter; the light which these shall cast upon the soul must probably be greater than that which they shall receive from our comprehension of the soul.

We would not put it this way today, but it is not hard to see what he is getting at. Trying to explain language through psychology amounts to explaining something we understand somewhat by something we understand much less well.

> If the linguistic student, in his devotion to psychology, shall invert this relation, he is very likely to add one more to the already numerous instances in which metaphysics has shown its inaptitude for dealing with facts of observation and induction. Only the result can decide, and that we will proceed to test.

We can agree with Whitney or not, but his point is clear enough: to look for an explanation of linguistic generalizations in psychology assumes that the principles of psychology are better understood than those of linguistics, and he was clearly dubious about any assumption of that sort.

> Whether, in the first place, men be willing to allow to the study the name of a science or not, is a matter of the smallest moment. It has its own character, its own sphere, its own importance of bearing on other departments of knowledge. If there are those whose definition of a science excludes it, let it be so; the point is one on which no student of language need insist.
>
> What he does need to insist upon is that the character of his department of study be not misrepresented, in order to arrogate to it a kind and degree of consequence to which it is not entitled—by declaring it, for example, a physical or natural science, in these days when the physical sciences are filling men's minds with wonder at their achievements, and almost presuming to claim the title of science as belonging to themselves alone. [He argues that linguistics is a "historical or moral science."] . . . Not one item of any existing tongue is ever uttered except by the will of the utterer; not one is produced, not one that has been produced or acquired is changed, except by causes residing in the human will, consisting in human needs and preferences and economies.[95]

Let's take a look at how Whitney introduced the idea of a set of sounds in a language. He called it our "spoken alphabet" and wrote that it was "an orderly system of sounds," with its "lines and degrees of relationship that bind its members together, and help to determine their transitions."[96] He introduced the various parts of the articulatory apparatus—the lungs, the larynx, the pharynx, the parts of the mouth. In short, voice was "the audible result of a column of air emitted by the lungs, impressed with sonancy and variety of pitch by the larynx, and individualized by the mouth-organs."

He then presented the sounds of the language as forming a continuum, from the widest open /a/ to the most closed stop. If the friction of the breath "as driven out through the aperture, forms the conspicuous element," then it is called a *fricative consonant*. The main plan of the inventory of sounds is thus along one axis stretching from most open to most closed. "But there

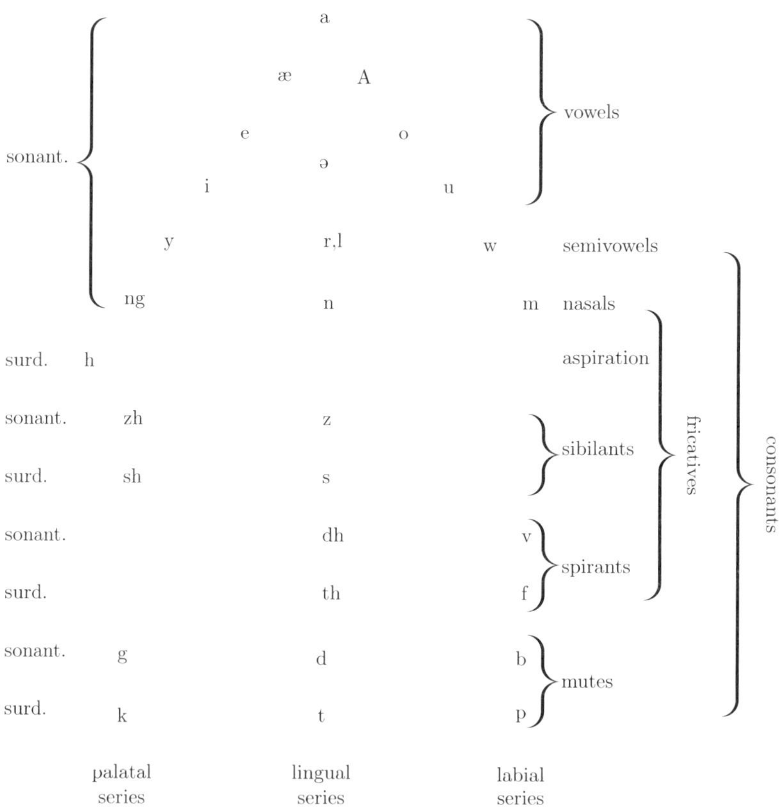

FIGURE 2.3. Whitney's analysis of sonority

are other lines and degrees of relationship in it." The most important involves a position in the mouth—which today we generally call a *point of articulation*, but which Whitney called a *position*, or a *position of mute-closure*. There were three important positions: "one in the front, made by lip against lip, the labial closure, giving *p*; one in the back of the mouth, made against the soft palate by the rear upper surface of the tongue, the palatal (or guttural) closure, giving *k*; and one intermediate between the other two, made by the point or front of the tongue against the roof of the mouth near the front teeth, the lingual (or dental) closure, giving *t*." He put this together graphically in a simple chart, reproduced here.

All of this sounds quite modern and up-to-date, but there is more, and we wish to emphasize this fact because what he wrote next is often taken to be an insight that was not made explicit until the work of the Prague Circle in the 1930s. "Along with *k,t,p*, in the first place, go their nearest kindred, *g,d,b*," Whitney wrote, and he referred to *g* as the sonant, or vocal, "counterpart" of *k*, and so on. The two sound-counterparts are identical *except* that the "sonant utterance begins in [pa] just when the contact is broken, and in [ba] just before." The two sounds share something in common, and they are distinguished from each other by the difference in the temporal alignment of the laryngeal voicing with respect to the oral release.

"But there is a third product of the same three positions of mute-closure," he continued: there are also the *nasals*, each of which shares the position with a pair of stops. And then he discussed the other sounds of English and some familiar European languages. "The scheme thus drawn up and described may be taken as a general model, on the plan of which the spoken alphabet of any language may best be arranged in order to the determination of its internal relations and to its comparison with other alphabets."[97]

He emphasized the continuity that exists between the consonant and the vowel system, even if "their distinction is of the highest importance in phonetics," because they are "only poles, as it were, in one continuous unitary series, and with a doubtful or neutral territory between them: they are simply the opener and closer sounds of the alphabetic system."

Whitney drew careful analogies between the work that he and his colleagues were doing in linguistics with the work that other scientists were doing, particularly in geology, the queen of the physical sciences at the time, a status that Whitney took advantage of:

> Once more, a noteworthy and often-remarked similarity exists between the facts and methods of geology and those of linguistic study. The science

> of language is, as it were, the geology of the most modern period, the Age of Man, having for its task to construct the history of development of the earth and its inhabitants from the time when the proper geological record remains silent; when man, no longer a mere animal, begins by the aid of language to bear witness respecting his own progress and that of the world about him. The remains of ancient speech are like strata deposited in bygone ages, telling of the forms of life then existing, and of the circumstances which determined or affected them; while words are as rolled pebbles, relics of yet more ancient formations, or as fossils, whose grade indicates the progress of organic life, and whose resemblances and relations show the correspondence or sequence of the different strata; while, everywhere, extensive denudation has marred the completeness of the record, and rendered impossible a detailed exhibition of the whole course of development.[98]

But one of the deepest questions regarding language, like every aspect of human culture, is how it can be analyzed both as the intentional result of many individual acts, and also as a system shared by a community that each individual confronts when he or she learns it, with blessed little power to make changes with it. Whitney wrote,

> While . . . we are . . . forced to the acknowledgment that everything in human speech is a product of the conscious action of human beings, we should be leaving out of sight a matter of essential consequence in linguistic investigation if we failed to notice that what the linguistic student seeks in language is not what men have voluntarily or intentionally placed there. As we have already seen, each separate item in the production or modification of language is a satisfaction of the need of the moment; it is prompted by the exigencies of the particular case; it is brought forth for the practical end of convenient communication, and with no ulterior aim or object whatsoever; it is accepted by the community only because it supplies a perceived want, and answers an acknowledged purpose in the uses of social intercourse. The language-makers are quite heedless of its position and value as part of a system, or as a record with historical content, nor do they analyze and set before their consciousness the mental tendencies which it gratifies. A language is, in very truth, a grand system, of a highly complicated and symmetrical structure; it is fitly comparable with an organized body; but this is not because any human mind has planned such a structure and skilfully worked it out. Each single part is conscious and intentional; the whole is instinctive and natural. The unity and symmetry of the system is the unconscious product of the efforts of the human mind, grappling with the facts of the

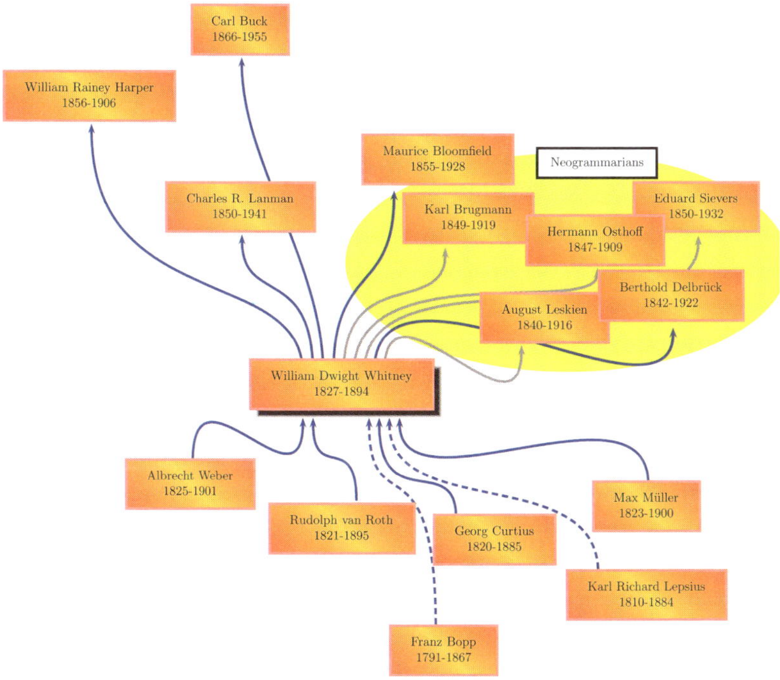

FIGURE 2.4. William Dwight Whitney

> world without and the world within itself, and recording each separate result in speech. Herein is a real language fundamentally different from the elaborate and philosophical structures with which ingenious men have sometimes thought to replace them. These are indeed artful devices, in which the character and bearing of each part is painfully weighed and determined in advance: compared with them, language is a real growth; and human thought will as readily exchange its natural covering for one of them as the growing crustacean will give up its shell for a casing of silver, wrought by the most skilful hands. Their symmetry is that of a mathematical figure, carefully laid out, and drawn to rule and line; in language, the human mind, tethered by its limited capacities in the midst of creation, reaches out as far as it can in every direction and makes its mark, and is surprised at the end to find the result a circle.[99]

Whitney concluded that language is an *institution*, a social institution. His views left a strong impression on all of the linguists who followed him and on none of them more than Ferdinand de Saussure.

Third generation: Neogrammarians

Among the members of the generation of German students of linguistics in the 1870s—the *third* academic generation of German linguists—there were many who felt that they were part of a revolutionary vanguard that eventually came to be known as the *Junggrammatiker*, a name that carries an ironic, even self-deprecatory overtone that the translation *Neogrammarian*, unfortunately, does not. But as *Neogrammarians* they are known in the anglophone and francophone world, and this is how we shall refer to them.[100] The origins of Neogrammarians can be traced to both developments in the understanding of linguistic evolution and to issues of authority within the German profession of linguistics. In the 1860s and 1870s, there were exciting discoveries—notably those of Hermann Grassmann and Karl Verner—that made sense of a large number of apparent counter-examples to the sound correspondences that had been discovered in the preceding 50 years, most especially in connection with Grimm's Law regarding the evolution of consonants. Grassmann's discovery made sense out of the distribution of voiced aspirates in Greek and Sanskrit, and Verner's made sense out of exceptions to Grimm's principles in the German languages, on the basis of the placement of stress. The Neogrammarians were confident that these breakthroughs were not isolated but rather showed the value of continued search for deeper explanations and the importance of not letting counter-examples be dismissed lightly. This confidence played an important role in the creation of the Neogrammarians as a self-defined group.

This growing confidence led them to the conclusion that linguists were discovering generalizations that deserved to be called *laws*. This is not to say that such a term had not been used by earlier workers; Bopp wrote of "phonetic laws" and "sound laws" in the 1820s, and for Schleicher, *Lautgesetz* was an important concept as well.

Just as important, the young linguists rejected the deeply held ideas about the development and the decay of languages. The young linguists chose what was known as *uniformitarianism*,[101] first identified in the study of geology: the notion that the most general laws of change have been fixed over time. Languages in general are not getting better or getting worse; they are just changing. They are evolving.

In 1878, Hermann Osthoff and Karl Brugmann published a statement that is a perfect example of an account of group identity, the kind of text which we discussed in chapter 1: a description of who they were, this new group of linguists studying the reconstruction of Proto-Indo-European.[102]

The manifesto began by explaining that something new happened with the publication of Wilhelm Scherer's book on the history of German (Scherer was at that point a German philologist, whose career would turn in the following years towards literature). Now a new method arose, and with it a group of people had arisen consciously aligned with this new method. Here is how they put it: "Since the appearance of Scherer's book . . . and principally through the impulses that went out from this book, the physiognomy of comparative linguistics has changed considerably. A method of research has been instituted since then and is winning more and more supporters; it differs in essential respects from the method by which comparative linguistics proceeded in the first half-century of its existence."[103]

In this manifesto we find a constant mixture of statements of fact and statements that emphasize (or exaggerate) differences of position that relate to being in a group and being outside of a group. Here is an example: Osthoff and Karl Brugmann pointed out that human speech has a mental and a physical side. Articulatory phonetics has focused on the physical side of speech. "This science is several decades old, and the older linguistics, since about the 1850s, has also profited by its results; for this we must give it great credit." In that sentence alone, there are two striking expressions that emphasize Osthoff and Brugmann's distancing: one is the phrase "the older linguistics," and the other is the simple matter of giving someone credit, which is something you normally do to someone else, not to yourself. They went on to say that there were many aspects of common sound changes for which an articulatory account is not at all sufficient. "The first outlines of this science were drawn by Steinthal in the essay 'Assimilation und Attraction, psychologisch beleuchtet,' which up to now has received little notice from linguistic science and articulatory phonetics." This remark, too, is an example of a flourish, a trope common to manifestos: they identify an important piece of work that has come out but received (they believe) far less attention than it has deserved.

The manifesto continued:

> The older comparative linguistics, while it readily accepted and utilized the teachings of articulatory phonetics, hardly concerned itself at all with the psychological aspect of the speech process, and as a consequence it fell into numerous errors. Only in very recent times is one becoming more aware of that neglect. Fortunately the movement starting with Scherer's efforts, the "Neogrammarian" movement, has already done away with some of the fundamental errors which dominated the entire older linguistics.

Empirical psychology—what was then called the "new psychology"—was just coming into its own in the 1860s in Germany. That this should have an effect on how young linguists were thinking about their field is hardly surprising: indeed, it is gratifying and speaks to the influence across disciplinary boundaries that rarely fails to appear eventually. But Osthoff and Brugmann did not describe this as an opportunity for their field to move forward so much as an opportunity to identify errors that the preceding generation had fallen into. This metaphor, by which errors are like ditches, and one falls into them by not paying attention to where one's feet are, is quite the opposite of the fallibilism generally espoused today, which recognizes that everything we believe now will at some point be found to be replaceable by a view that is more accurate. Strictly taken, fallibilism, our own philosophy of scientific modesty, is comfortable with the view that an "entire older linguistics" could be beset by "fundamental errors," as the Neogrammarians wrote about their teachers, but it is certainly salutary to bear in mind that the earlier generation had rid the world of another and different set of fundamental errors, just as the next generation of linguists coming up at the end of the nineteenth century would do as well.

> That this, however, was not the right way to arrive at correct guiding principles for the investigation of form change and form innovation in our Indo-European languages is so very obvious that one must be surprised at how many have not yet become clear about it. Is not, after all, the authenticity, the scientific probability, of the original Indo-European forms, which are of course all purely hypothetical creations, totally dependent on whether they agree in general with the proper conception of the development of linguistic forms and on whether they are constructed according to correct methodological principles? Thus the investigators went around in the most obvious circles, and even today many still do so, without knowing it or wanting to admit it.[104]

This passage illustrates another characteristic of manifestos: they not only make assertions and argue for them, they provide some instruction on how to respond to hearing differences of opinion on matters of concern: in this case, one might be "surprised"' to note how many of the preceding generation have not come to agree with them—yet.

> Some linguists, particularly a few most directly concerned passed over the question and, abruptly rejecting it, remained satisfied with the old way. No

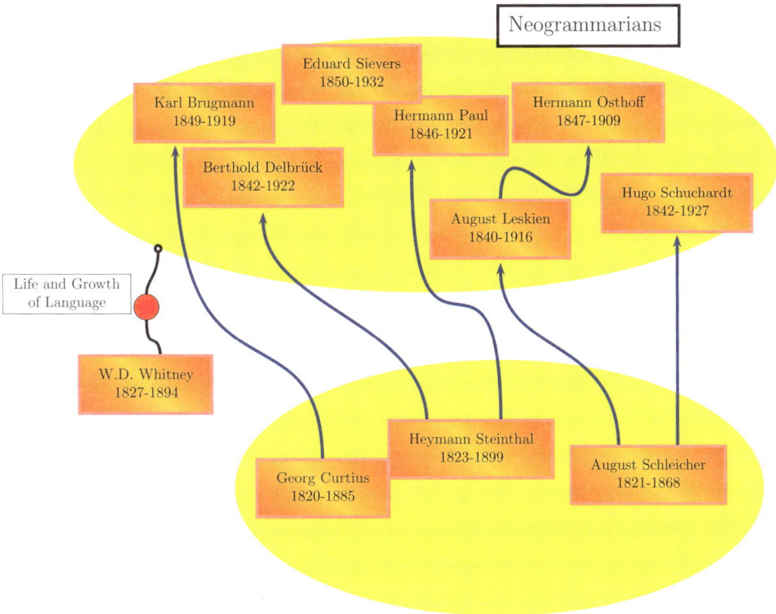

FIGURE 2.5. Neogrammarians

> wonder. When serious attempts at upset are directed against a procedure that one is used to and with which one feels comfortable, one is always more readily stimulated to ward off the disturbance than to undertake a thorough revision and possible alteration of the accustomed procedure. But with others, especially younger scholars, the seed scattered by Scherer fell on fruitful ground.[105]

The Neogrammarians declared that their positive focus was on two principles that were central to language change: first, laws of sound change applied without exception, and second, analogy played a central role in the evolution of paradigms. Exceptionless sound change was based on shifting pronunciations in the direction of greater articulatory ease, while the effects of analogy would increase the simplicity of the correlation between form and meaning. But much more important was the view that language change was *not* a process of decay, and that a language at every moment is the product of the creative, psychological forces of the individuals who learn it. This recognition marked the beginning in Europe of the modern conception of historical linguistics.

Paul Kiparsky[106] has suggested that the new shift in perspective in the 1860s and 1870s was that a new perspective came into the picture of what constituted an explanation. No longer was data from a later stage of a language useful only for reconstructing an earlier stage; now the discipline could demand an explanation of the later forms on the basis of the earlier forms. There were no mechanical procedures, so to speak, which would allow data from a later stage to automatically reconstruct the earlier forms—that would remain the challenge for the creative linguist.[107]

But if those reconstructed forms were accurate, then it should be possible to provide a small set of relatively simple rules of sound change that would explain the later stage of the language, given the earlier stage.

> Soon after the appearance of Whitney's major work [1875] a movement began in Germany among the Indo-Europeanists which rejected a number of widely used methodological errors, which developed out of errors that Whitney had uncovered. And if suggestions about this movement support came from elsewhere, it is those from Whitney that were perhaps the most important and enduring.[108]

There was intense reaction to some of the Neogrammarians' perspectives by other linguists, none sharper than those of Hugo Schuchardt, notably in a long paper on sound laws entitled "On Sounds Laws: Against the Neogrammarians." He wrote, "The positing of the Neogrammarian principle does not mean for me a revolution in the history of linguistics with which the science began to progress with greater rapidity and assurance."

Georg Curtius, too, was puzzled by the Neogrammarians' fervor. "For sixty years," he wrote, "Indo-Germanic linguistics has been developing smoothly and without significant inner contradictions," and now this? Whence came this "need for a thorough divergence from views previously held in many areas"? Curtius's own work emphasized his indebtedness to Wilhelm von Humboldt, Max Müller, and William Dwight Whitney.[109] Hermann Collitz was a little bit younger than the Neogrammarians—he was five years younger than Sievers, six years younger than Delbrück—and he was approached by Sievers to see whether he would join forces with them. The necrology of Collitz noted that Collitz "considered many of the attacks against the elders as totally unjustified" and refused to join them. He later came to the United States, and he was the first president of the Linguistic Society of America.[110]

In response to Curtius, the Neogrammarian Hermann Paul wrote a review of Schuchardt's book.[111] He starts by quoting from Schuchardt, who had written, "The only position of the so-called Neogrammarian school that they can claim as their exclusive property is the exceptionless application of sound change." Paul was not happy with this. "Against his claim with which the writer begins I must lodge a decisive protest. First of all, what gives him the right to speak of a Neogrammarian 'school'?" This is a striking question: what, indeed? Who has the authority to determine whether certain authors form a bloc, a school—and what difference does it make, anyway? The answer to Paul's question was the fact that Osthoff and Brugmann published their manifesto. It was an *act*, a declaration that was a major step towards the formation of the Neogrammarians as a self-conscious group.

Paul referred to a recent review by Osthoff and Brugmann written by Bezzenberger.

> This work is referred to as "a link in a chain of publications of personally closely allied authors." It continues, "the young researchers who have joined with Leskien, working as a closed party." I am sorry that I did not immediately protest this remark, since it has given rise to misunderstanding. There is a misunderstanding on Bezzenberger's side which he would not have made if he had read my own words with an unbiased attention and not been blinded from the outset.

He continued:

> The truth is that over the last decade various ideas that had been seen obscurely have become clearer and their larger role better understood. But that does not mean there is a sense of solidarity to the point of mutual responsibility. There is, then, no *school* of Neogrammarians—though there is a Neogrammarian "direction."

Olga Amsterdamska[112] has argued that the fervor and self-conscious rise of the Neogrammarians can only be understood in historical and sociological terms, and we find her case compelling. The Neogrammarian movement arose at a time when the German university system was growing rapidly, not least in linguistics, philology, and psychology. The number of university faculty in this area as well as the number of regularly published journals were rising at a breakthtaking speed.

The heart of the Neogrammarian movement lay in Leipzig. Brugmann spent most of his career there, first as an assistant at the Russian Institute in the mid-1870s, and then as a lecturer at the University of Leipzig, eventually taking the chair of Sanskrit and comparative linguistics when Georg Curtius retired. Other Neogrammarians were students at Leipzig during the 1870s, such as Wilhelm Braune, Adolf Noreen, Hermann Osthoff, Ferdinand de Saussure, and Maurice Bloomfield; August Leskien was a professor there after 1870. We will return to Leipzig during this period again: it was the spot where the New Psychology of Wilhelm Wundt would develop, when Wundt arrived there in 1875.

The Neogrammarians have left a legacy for modern linguistics, one that puts a high priority on careful historical derivations of observed forms and that refuses cultural evaluations of a language, or a family of languages, or its speakers.

Jan Baudouin de Courtenay, Ferdinand de Saussure, and Maurice Bloomfield

From a cultural point of view, the Neogrammarians were insiders: they were Germans, defending a new German linguistics, largely against an older German generation. There were, however, three other linguists of an age not too different from that of the Neogrammarians, but who came from different cultural backgrounds and whose professional trajectories were quite different from those of the Neogrammarians. The first was Jan Baudouin de Courtenay, a Polish linguist (born at a time when Poland was under the sway of the Russian Empire); the second was Ferdinand de Saussure, a Swiss linguist; the third was Maurice Bloomfield, an American who had studied with Whitney and who came to Leipzig to study the latest linguistics (he had been born in Austria, so the German language was not an issue). They all spent time studying with German linguists during the 1870s. But Baudouin and Saussure never felt like participants in a great scientific venture with the more senior scholars they met in Germany, and both left Germany feeling that they were autodidacts, that they were the proper students of no one, and that they had to fight to have their views taken seriously.[113]

Baudouin, Saussure, and Bloomfield illustrate three distinct strategies that an outsider can take in the face of a dominant and external imperium. Baudouin, from Poland, was a Slavist, and he spent most of his

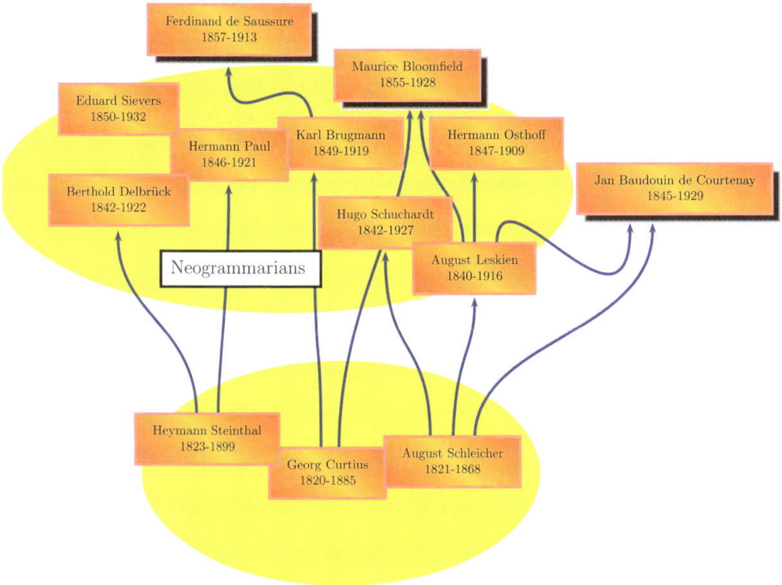

FIGURE 2.6. Baudouin de Courtenay, Saussure, and Bloomfield

professional life in Russia; his was a career that represents the Russian strategy of radical autonomy.

Maurice Bloomfield illustrates the policy taken up more generally by American science in the nineteenth century: it undertook a tacit licensing agreement with Germany, and brought to the New World a simulacrum of German universities and German sciences. It hired scientists trained in Germany, and the network of those scientists increasingly covered the field of American academia. This strategy was abandoned by the 1930s, as the United States began its climb to leadership of the international academic community, a position that would become clear in the postwar world.

But France's strategy towards its traditional enemy was different. France suffered a stinging defeat at Prussian hands in 1870, and its response was to strengthen its traditional forces. German scientists had once, not that long ago, been educated in France: it was time again to form a professoriate that could compete with the German model. For linguistics, that meant developing a French school, and that was precisely what Saussure was engaged in during the second period of his career, the French period.

Let us look first at Baudouin de Courtenay and Saussure, and then return to Maurice Bloomfield.

Saussure was 12 years younger than Baudouin, but each influenced the other; the book that Saussure published when he was 21 years old deeply marked Baudouin's understanding of the sound systems early on, and Baudouin's writings were cited with approval by Saussure in his lectures: in 1908 he referred to Baudouin, and to Baudouin's brilliant student Mikołaj Kruszewski, as "closer than anyone to a theoretical view of language."[114] When we turn to Baudouin and to Saussure, we encounter for the first time the explicit notion of the *phoneme*, one of the most important contributions of linguistics to modern thought, and a notion that we will see develop over the course of this book. Like any important concept, its precise definition changes over time, and it is often used in different ways by different linguists. But the core meaning is straightforward enough: the sounds of an utterance, in a spoken language, can be represented as a sequence of symbols, regardless of whether the language has a traditional writing system or not. As we learn more about carefully observing sounds, and with the aid of modern technology, we may find that the number of different symbols we employ to produce a very fine-grained description of the utterance will grow. Phonemic analysis is a way of thinking about the system of symbols used to represent utterances that does not allow the set of symbols to grow unmanageably: phonemic analysis allows for the introduction of a symbol (to represent a consonant or a vowel, typically) *only* if that symbol is necessary in order to represent a meaningful difference in the language. In short, a phoneme is a member of the inventory of sounds used by a language to mark differences between words. There will be much more to say about the phoneme as we proceed.

JAN BAUDOUIN DE COURTENAY

Jan Baudouin de Courtenay was born in Poland in 1845, in an old aristocratic family of French origin which could trace its origins to the times of the Crusades. He always considered himself Polish, though he lived and worked in many Central European countries over the course of his career. He studied linguistics in the late 1860s with Schleicher, Leskien, Brugmann, and Delbrück in Leipzig, and, in Russia, with Izmail Sreznevskij, but like so many scholars discussed in this book, he considered himself self-taught, an academic orphan who was the student of no one. His first major academic position was in Kazan, where he taught from 1875 to 1883, and it was there that he mentored Mikołaj Kruszewski, who in turn would leave a heavy mark on Baudouin's thought. It was during this period of his life that he created what has come to be known as the Kazan school. Later, in the years nearer to the end of his life (1900–1918), he established one of the two important schools of Russian linguistics in the early twentieth century, the St. Petersburg school.

Baudouin may have felt that his work was produced off the beaten path, and it may have taken a while for the linguistic world to recognize the validity of his work, but his effort to develop a theory of phonology and morphophonology produced important results which we view as the first theory of phonology. For linguistic scholars, his most accessible work was the 130-page monograph entitled *Versuch einer Theorie phonetischer Alternationen*, published in 1895. Some of the terms that are essential to the theory of phonology are due to Baudouin and his school: these include *morpheme*, *alternation*, and *distinctive feature*. (The linguist Meillet found *morpheme* to be "un joli mot," and so the term made its way into his translation of Brugmann; the rest is diffusion and history.) It is by no means easy to separate the contributions of Baudouin and Kruszewski; below we will speak of the views of the Kazan school and of Baudouin, but Kruszewski was deeply involved in the initial development of these views.[115]

Eight years after Kruszewski's death at the age of 35, Baudouin discussed how he viewed his student's contribution to the new ideas in linguistics that he was exploring himself. The remarks were not generous. Perhaps that is made up for by Baudouin's acknowledgment of his student's contributions; perhaps not. Baudouin noted that "Kruszewski developed the 'theory' of alternations more 'philosophically,' more comprehensively, and more precisely than I myself have done." And yet, "it cannot be denied, however, that Kruszewski merely gave another, finer form to what he had learned from someone else"—and that someone else was, of course, Baudouin. And Kruszewski "left many things unnoticed"; imagine that! And when Baudouin brought up one of Kruszewski's "most original ideas," he added that this idea was "nothing new for me personally, since I had been developing it for several years in my lectures." When we look at Saussure's experiences in Leipzig, we will see the inverted case, the view of a student who goes to a great deal of trouble to document his personal development of an idea despite his teachers' parallel efforts.[116]

The word *phoneme* had just been created, but because of its central role in the development of phonological theory, the precise definition and

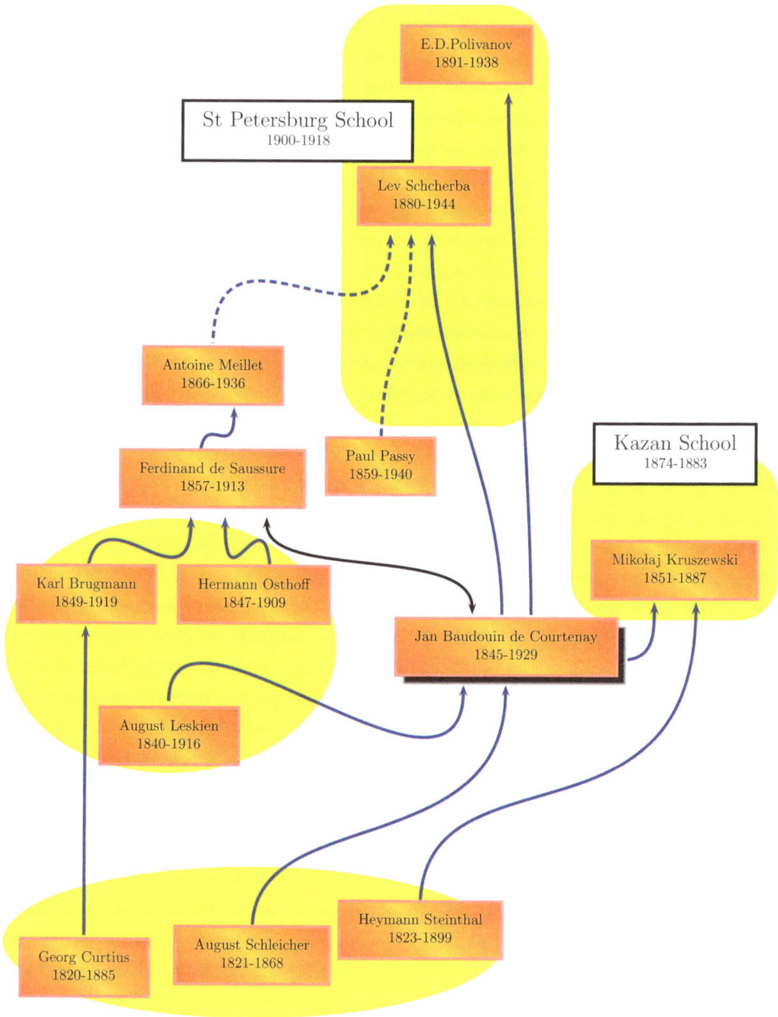

FIGURE 2.7. Jan Baudouin de Courtenay

usage assigned to it has shifted and fluctuated over time, never so much as in its early years. He offered a definition: a "unitary concept belonging to the sphere of phonetics which exists in the mind thanks to a psychological fusion of the impressions resulting from the pronunciation of one and the same sound: it is the psychological equivalent of a speech sound."

Baudouin expressed the feeling that he shared with so many others in this book, the sense that he was likely to be forgotten:

> My attempt at presenting a theory of alternations will perhaps receive no recognition. It cannot, however, be denied that the concept of "alternation" and "alternants" is relevant to an enormous mass of linguistic facts, for there is probably no sound in any language which is completely isolated and does not alternate with another sound, just as there is no word to which the study of phonetic alternations cannot be applied.

More than any linguistic school before or since, the phonological analyses of the Kazan school focused on *pairs* of words that were similar but not identical and on words that were related to each other *etymologically*, that is, in the history of the language. The focus on pairs, or more generally, sets of words that are related in the history of a language is, for the most part, a way of talking about words that are related either by inflectional morphology (which is to say, the various members of a paradigm) or derivational morphology (related nouns and verbs, for example). In English, this would *not* include the set of words *trip*, *train*, *travel*, even though they contain a common initial sequence *tr* and a shared meaning, nor would it include the pair *male*, *female*, because their etymologies are unrelated to each other (they come from French *mâle*, *femelle*, and *mâle* has nothing to do with *femelle*). But there are many families of words that are related in this way. For example, *f* and *v* are paired with each other in words such as *life/lives*, *wife/wives*, *leaf/leaves*, and a long *i* and a short *i* are related to each other in *divine/divinity*, *dine/dinner*, *line/linear*, and *ride/ridden*. Baudouin used the word *alternation*: in English, *f* and *v* alternate in *life/lives*, and so on.

Here is the central point: Baudouin proposed important links among etymologies, diachronic evolutions, and synchronic phonological processes. As we watch the evolution of thought about phonology from Baudouin down to our time, the varying ways in which these links were understood lie at the heart of the discussion. It is at the center of Leonard Bloomfield's "Menominee Morphophonemics,"[117] which we will discuss in chapter 6, and Morris Halle's "Phonology in Generative Grammar."[118] It represents a central part of the effort of linguists to come to grips with the question of how diachrony, the evolution of the system, is embedded within synchrony, the system as viewed at a moment in time. Baudouin was quite explicit on the importance of distinguishing what he called *static* and *dynamic* aspects of language—what today we call *synchronic* and *diachronic* aspects.

An essential part of the Kazan school's contribution was the recognition that in languages, alternations need to be divided into two sorts.

One sort, which Baudouin called *anthropophonics* (today known as *phonological*), has to do with the sound context in which the sounds in question appear; the other sort, which he called *psychophonetic*, relates to grammar and will be discussed in a moment. American English has two pronunciations of /t/, one of which is a flap, as in words such as *Italy*, and one of which is not a flap but rather is aspirated, as in the word *Italian*. Which of these two pronunciations is used in a given word is entirely determined by the nature of the sounds appearing around it, and this is an example of what Baudouin called an *anthropophonic* alternation, and more specifically, a *divergence*: "this kind of contemporary, live split of a psychologically homogeneous phoneme into two or more phonemes we shall call *divergence*."[119] And if the parallelism that forms the alternation is completely general, leaving no word an exception to it, then he called it a *purely anthropophonic phonetic* divergence.

Typically there is something about the immediate context that either enhances a sound or suppresses it: a consonant may well be enhanced if it is followed by a vowel, or suppressed in some way if it is followed by a consonant, for example. Such an environment may be said to "cause" the divergence, but the causality is not simple and elementary. In some cases, the causes might be general, applying to all languages at all times, while in other cases, the cause might be a "transient temporary force which acts only in a particular period of a given language." And just as important, these forces may operate synchronically, or not—and if not, then they "belong to the past of a given language; and finally, the effects being described may be either obligatory or optional."

The effects of divergences may go unnoticed, Baudouin underscored: "Since the properties of the phonemes [*x*] and [*y*] are not individual, psychological properties that are stored in the brain center, but conditioned variants which depend on the phonetic environment . . . rather than on psychological factors, the anthropophonic variants of the dependent phoneme may go unnoticed."[120] For Baudouin, there are psychological properties that are stored in the brain center, which are the foundation for a number of distinct realizations of the stored element whose differences are logically dependent on surrounding sounds. This is the heart of pure phonology.

The other sort of alternation was what Baudouin called a *psychophonetic* alternation or "correlation," and the two examples from English given earlier (*dine/dinner, life/lives*) are psychophonetic alternations; an even better example from English is *sing/song*, because the difference there is linked to nothing other than a grammatical (that is, psychological) condition. Sound differences of this sort are not the logical effect of a surrounding sound; they are the logical effect of something that lies within the grammatical system. The term *correlation* was adopted by Roman Jakobson in the 1920s, as we will see below in chapter 9, where we discuss other antecedents of the term *correlation*.

Baudouin strikingly emphasized that this shifting of one phoneme to another is an option for morphology "in the same way as it utilizes mobile *word-forming morphemes* (affixes), i.e., prefixes, suffixes, desinences, etc. . . . And like the suffixes or prefixes, the correlatives may serve to distinguish morphological categories."

Baudouin emphasized that the vowel gradation widely discussed in the Indo-European literature was "closely connected with a correlative alternation of phonemes in which one member of the alternating pair has a zero-phoneme, i.e., lacks any phoneme," which he wrote as "x||∅," and just as with divergences, correlations may be productive, or their productivity may be waning. (It would be a mistake to ask what the *causes* are for the correlations of a language: they are what they are, and they exist because learners learn them from their environment. Baudouin's view was that if we wish to learn more about them, then we may undertake historical and comparative linguistics, but that is something outside of our study of the language itself.) Furthermore, we do not expect the sounds related by correlations to be phonetically similar.

It is important to bear in mind that the theoretical work involved in defining and justifying the phoneme was the development of a practical effort to develop writing systems, both for languages that had not yet been provided with an orthography and for languages whose orthographies were sadly out of date, at least as some people saw it—like English or French. And we can learn a good deal about how sophisticated people thought about sound systems if we pay careful attention to a discussion of the orthography proposed for a language such as Zulu, a Bantu language of South Africa, that did not have an orthography in the mid-nineteenth century. A series of three papers was published in the *Journal of the American Oriental Society* in the early 1850s about a proposed orthographic system for Zulu. Lewis Grout, an American missionary in South Africa, had already published a lengthy paper on Zulu morphology in the first volume of the *Journal of the American Oriental Society*, in 1849, and he followed up with a paper on phonology and orthography in 1853, to which two responses were published. The first was by Josiah

W. Gibbs, professor at Yale and one of the three editors of the journal; it was followed a bit later by a paper by C. A. Holmboe, a Norwegian linguist from Oslo. We can see how they thought about symbols and sounds from their discussions.

Grout's account gives the impression that he has studied Zulu and thought a good deal about it, but he has not quite made the transition from thinking about Zulu from an anglophone's point of view to that of a linguist (or a zuluphone). He worried about phrase-final vowels, whose vowel quality was difficult to determine, but noted that in many cases, he could add the locative suffix *-ni* and the uncertainty vanished. If a word with the locative ends in *ini*, then the original word ended in *i*, if it ended in *eni*, then the original ended in *e*, and so on. But this sophisticated linguistic thinking does not help him decide just how many distinct vowels there need to be in the orthography: he hears a range of vowels between *e* and *i*, for example, and between other pairs of vowels for which he proposed to assign basic symbols. Those intermediate vowels do not need their own symbols: "those which are really intermediate between two different vowels may doubtless be gathered under one or the other fundamental and distinct vowel cognate, without inconvenience or violence to the language."[121] And Gibbs agrees with him entirely: "There are also in this language many intermediate gradations of sound between *p* and *b*, and between *f* and *v*. But these, as is suggested by Mr. Grout, it is neither necessary nor expedient to indicate. . . . These different grades will naturally arrange themselves under the two extreme sounds."[122] Today we would not agree with that at all.

But the most interesting part of the account is the treatment of clicks. The proposal that Grout made is the one that has been adopted today, using the letters *q*, *x*, and *c* for what Grout called the palatal, the lateral, and the dental clicks. Gibbs did not like that—he thought that it would lead to confusion—and suggested that a rational system with new geometric symbols would be best: he suggested an empty box □ as a basic symbol for clicks, with or without a horizontal or a vertical line to give the different points of articulation (and other elaborations to indicate other complexities, such as nasality, by adding diagonal lines at one corner or another), such as: 

From our point of view today, the most interesting comment was made by Holmboc, in response to Gibbs's proposal to add a set of new symbols for the various clicks. He was skeptical that Zulu really needed four or more symbols for the clicks; he suspected that the different pronunciations were predictable from context.

> As to the clicks, Messrs. Grout and Gibbs propose four varieties of form for each of them. I do not doubt the existence of that number of perceptible varieties of sound, but I doubt the necessity of making use of so many signs in order to express them all. It is the case in every language that a letter exhibits certain differences of pronunciation, caused by its contact with other letters, but it may nevertheless always be written with the same sign, without leading to confusion. If the varieties of the clicks have this origin, it is superfluous to multiply signs for them. If, on the other hand, there are in these languages words, or forms of words, which would be confounded, if the signs of the clicks were not varied, I acknowledge that it would be necessary to vary them.[123]

This was a straightforward statement of the phonemic principle as it would be applied to writing systems. It was clear that the principle was implicitly understood by a good number of people thinking about language at the time.

Baudouin himself seemed to hedge on the question of whether it is reasonable to think of alternations as being *changes* of some sort. For the most part, he was against taking the idea seriously that there really was a change; the ambivalence he expressed would be echoed 60 years later by Zellig Harris. At times, he was clear in his statement that there were no changes here: "there are neither phonetic changes nor phonetic laws, and there never can be," because change takes place in the world of space and time, and the differences that we note with alternations are embedded in utterances that are separated in time and space. What we note is "simply the phonetic difference between etymologically related morphemes." Still, he allowed that these could be called "phonetic change" or "transformation."[124]

Baudouin saw himself as breaking away from the linguistic scene of his contemporaries, and he made this clear in the inaugural lecture that he gave in 1870, when he was 25 years old. He explained that he saw three approaches to linguistics in the world about him: a descriptive approach, an aprioristic approach, and a "truly scientific, *historical*, genetic approach." The descriptive approach was flat, devoid of significant consequences, and boring; the "speculative, philosophizing, aprioristic and childish approach" is pursued by people who recognize the inadequacy of the descriptive approach but do not have the slightest idea how to do linguistics other than by making up hypotheses and then forcing the data to fit the hypotheses, an approach which in the final analysis "is decidedly harmful"; while the scientific approach embodies a set of inductive methods. It views language as "the sum total of actual phenomena, of actual facts," and as an inductive science, it limits itself to two things. The first is comparing phenomena to each other, and the second is to "establish forces and laws . . . that connect the phenomena and present them as a chain of cause and effect." The first activity lends order to a large collection of facts, and the second allows for the crucial element of deduction at the heart of science. Baudouin took special aim at linguists who chose to ignore data: "all facts have equal right and can be viewed only as more or less significant; by no means can some be deliberately ignored, and it is simply ridiculous to sneer at facts. All that exists is reasonable, natural, and **lawful**; that is the watchword of any science."[125]

Baudouin argued strongly against any confusion of linguistics with its intellectual neighbors. He wrote:

> I must caution against the very confusion of linguistics and philology. Philology, as it has developed historically and as it is usually presented by its practitioners, is a conglomerate of knowledge, of detailed information about a variety of matters, and not a science in the strict sense of the word; on the other hand, linguistics is a monolithic and well-defined discipline. . . . In origin, linguistics is everywhere in the debt of those philologists who first studied language for special purposes, that is, as a means of investigating other aspects of the intellectual life of a nation, but then discovered the pleasure of studying language for its own sake and created the science of grammar. . . . In his discussions of language, the philologist asks: what is its parentage? Does it boast ancient written records? Does it have a history of several dozens of generations who used it for literary purposes?[126]

But linguistics is something else:

> In fact, however, the study of modern languages, accessible to us in all their facts, is far and away more important. My statement may strike some people as eccentric, but the natural scientist will understand me immediately. The study of paleontology presupposes the study of zoology, botany, etc., and not the reverse. As the name itself indicates, linguistics is the scientific investigation of language, or human speech, in all its diversity. Like other phenomena, linguistic phenomena give a first impression of chaos, disorder, confusion. But the human mind has

> an innate ability to shed light upon seeming chaos and to find harmony, order, system, and causal relationships in it. Linguistics is the goal-directed activity of the human mind to find order in the phenomena of language.

Baudouin made much the same point with regard to psychology, from the perspective of linguistics:

> Two elements are inseparably linked in language: a physical and a psychological one. . . . The forces and laws and the life of a language in general are based on processes which are of concern to physiology (to anatomy and acoustics) and to psychology. But the same physiological and psychological categories make up a rigidly defined subject which is investigated by the historically developed science of linguistics. Most of the questions raised by the linguist are never broached by the physiologist or psychologist; consequently, linguistics must be regarded as an independent science, not to be confused with either physiology or psychology.[127]

We offer this brief discussion of Baudouin's perspective as an example of what we referred to in chapter 1 as the internal history of a discipline, one in which we are able to treat his views as relevant to questions that are still active and in play on the contemporary scene.

FERDINAND DE SAUSSURE

For the twenty-first-century reader, Ferdinand de Saussure is a character from the late nineteenth and early twentieth centuries who is an amalgam of contradictions, and of all the contradictions, the deepest is this: today he is revered as the great and famous linguist who set linguistics on its course as a modern discipline. He is the founding father of modern linguistics, and perhaps the founder of structuralism more generally, but here is the irony: he himself had no idea that this cloak would ever fall on his shoulders. He died a relatively young man—he was 55 years old—and all indications are that he viewed his own attempts to pursue a scholarly career as failures that ultimately led him nowhere. Yes, he finished his career as a professor, but not in an intellectually vibrant city (he was professor in Geneva, not Paris or Leipzig), and he had very few students during his Geneva years. He had written a brilliant essay as a very young man, but it was a vanity publication, and his father had covered the costs to have it published, while Saussure's quite famous teachers in Leipzig—Brugmann and Osthoff—were always rather dubious about Saussure's claim to owe none of his ideas to those of the professors whose courses he took. Over the thirty-some years that followed the publication of the book of his youth, he had a publishing record that could be called modest, if we were to be unusually generous, and disappointing, if we were to peer inside Saussure's head to try to see his own view of himself.[128]

And we do have the opportunity to see a bit of what went on inside his head. Although he published little, he was a man who *wrote* a great deal, and with the publication of many of his incomplete writings, we understand better what his point of view was. In addition, he left behind a number of manuscripts that described his own view of his life, and how his work related to that of others; most importantly, he wrote a number of drafts in the fall of 1903 of an account of his life, and how things had happened about him and how he had responded to what life had provided.

In the entire history of linguistics, there is no more striking example than Saussure of a brilliant and creative individual who insists that he is a free-floating intellect who has learned little or nothing from those around him, and whose life is made utterly miserable by his efforts to persuade himself of his own intellectual autonomy. His legacy to us consists of work that he did as a young man on the analysis of the vowel system of Indo-European, and the published form of lectures that he gave on the general theory of linguistics—though these lectures were assembled and published posthumously by two linguists from his department, Charles Bally and Albert Sechehaye.

We will examine four periods, and three cities, in Saussure's life. He grew up in Geneva, where his gifts were recognized early. He was next the brilliant student in Leipzig, then the young genius in Paris, and finally the professor once again in Geneva.

Saussure was born in 1857 to a family of Calvinist aristocrats of French origin.[129] His family tree included the famous Horace-Benedict de Saussure, alpinist and, along with Jacques Balmat, the first to climb Mont Blanc in 1787. Saussure was thus the scion of a distinguished family of naturalists for whom a break with traditional taxonomies and classical systems of classification was almost second nature. His childhood and early education was spent in an environment where "the highest intellectual culture had long been a tradition."[130]

These things, then—the collection of natural data, of description, and ultimately the elaboration of an explanatory system of classification—formed a zeitgeist, a climate of thought and action, that was singularly rooted in the Saussure family, giving each member of the family (but none as much as Ferdinand) what De Mauro calls a scientific *forma mentis*.[131]

We see just this scientific culture in an extraordinary observation that Saussure made later in his life, in connection with his work on Sanskrit, in a letter he wrote to Meillet (January 4, 1894):

> The beginning of my paper on intonation is about to appear. . . . But I am quite disgusted by all of this, and by the constant difficulty of writing even ten lines which do not violate common sense when it comes to facts about language. Preoccupied most of all, and for so long, about the logical classification of these facts, of the classification of the points of view from which we look at them, I see more and more both the immensity of the work that would be necessary to show the linguist what he does; in reducing each operation to its predicted category; and at the same time the great vanity of what we can ultimately do in linguistics.[132]

That was Saussure, out of breath, but what a statement.

Saussure's education in modern linguistics began in the fall of 1875, when he was 18 years old and a student at the University of Geneva. There he attended lectures by Louis Morel, who had just returned from Leipzig after studying with Curtius. During that freshman year, Saussure carefully studied the work of Bopp, Schleicher, and Curtius.[133]

Saussure was also reading the recent literature by French Indo-Europeanists, and devouring the substance of work in the field at an astonishing rate; here is something he wrote in the second half of that year:

> What opinion should I have of Bréal? I haven't read much by him; but here for example is an article of his on Latin *aut* . . . where in the end he does nothing more than repeat what Bopp said, and it cannot be unwitting since he is the translator of Bopp's *Comparative Grammar*. . . . In the same article he says in a note: *we* derive *indiges* from *indu* and *ga* "bring into the world," an explanation given long hence by Curtius and cited as well in the *Comparative Grammar* of Bopp translated by Bréal.[134]

In his recent biography of Saussure, John Joseph discussed these diary entries, noting that the hubris in these entries is "obvious," and indeed it is, and it is matched by any number of remarks that Saussure made at the time.

Saussure left the University of Geneva after that first year. Geneva had been his home, and it was time for him to go elsewhere. He went to the University of Leipzig, where the local linguistic world had just seen the establishment of a new school, the Neogrammarian movement, and where prominent linguists of the preceding generation—notably Georg Curtius—had a lot of difficulty taking the movement seriously as something that was really *new*. But at 19, Saussure was simply too young and immature (the two are not the same thing) to understand how the young teachers he would meet there viewed themselves. From their *own* point of view, Brugmann and Osthoff were still young scholars trying to push their new ideas and defending them from the criticisms of their own teachers. After all, Karl Brugmann was only 27 years old; Hermann Osthoff was 29. They were not prepared to be thought of as the old guys. But when Saussure arrived in Leipzig, he saw Brugmann, Leskien, and Osthoff as the older generation, and he was perhaps ready to be unimpressed with what he found. For their part, the Neogrammarians were not ready to be treated as part of the linguistic Establishment; they were still young and still revolutionaries at heart.[135]

One of the central problems that the Leipzig Neogrammarians were working on was the reconstruction of the vowel system of Proto-Indo-European: how many vowels were there, what were they, and what were their pronunciations? This too was the question that drove Saussure. Perhaps he was already passionately engaged in that question before he got there, or perhaps not; it is difficult to be sure. Either way, Saussure was hoping to go *mano a mano* with Brugmann and Osthoff and to be the first to figure out what the vowels were in Proto-Indo-European.

But things did not work out the way Saussure had hoped. On the positive side, he threw himself into his studies and worked very hard, and he became friends with some of his teachers, such as Brugmann (who was not so much older than him, anyway),[136] and rubbed shoulders with all of the active Indo-Europeanists, so many of whom came to town for professional reasons. But the culture and the language, both of which were foreign, put him off, and he did not feel like one of the students.[137] Worse yet, he felt he had his own ideas about the Indo-European vowel system, and he wanted to make a big splash and be recognized as the person who solved the difficult analytic problem of Indo-European structure. By the end of that first year, he had been studying linguistics in a serious way for no more than two years, though he had been thinking about it on his own for at least a year or two more. But Brugmann and Osthoff had quite a few more years on him, and they had (or so he felt) already made their reputations.

This left Saussure in a difficult situation. He wanted to keep up with the competition, which meant going to Brugmann's and Osthoff's lectures, but he did not want to be labeled just as one of their students, and he wanted the scholarly world to recognize *him* as the original thinker.

The incident that seems to have been really traumatic for Saussure, one that hurt him and defined the difficulties that he had with the German linguists and their lack of recognition for his genius, was the case of the sonant nasal. Even 35 years later, he came back to it in his personal diary with a sense of bitterness, and a sense that something had happened that he could not step back from. Saussure wrote a number of descriptions of his life, but none of them was authorized for publication during his lifetime, so we certainly have to take them with a grain of salt, and it is up to the reader to decide how serious Saussure was. The story, in any event, goes back to something Saussure had noticed while he was still in high school.

> The moment I saw the form *tetákhatai*, my attention, in general extremely distracted, as was natural during this year of covering old ground, was suddenly seized in an extraordinary manner, for I made this mental analogy which is still vivid in my mind even now: *legómetha: légontai*, consequently *tetágmetha: tetákhNtai*, and consequently *N=a*. I left the Collège wondering how n could have become a, and trying to make the sounds in a way that would answer the question. I conceived, in repeating these experiments, that it really was possible to pass from *tetákhntai* to *tetákhatai*, but naturally without marking this n with any special sign (such as ṇ or something else) even in my mind. Its characteristic was for me (which is physiologically right) its position between two consonants, which caused it to give rise to a Greek *a*, but it was an *n* like any other.[138]

Three years later, Saussure was a student in Leipzig. Brugmann (not thinking of Saussure, obviously) published a paper on the sonant nasal that was received as a great advance by the Neogrammarian community. Thirty-five years later, Saussure wrote down his recollections of finding this out.

> I went to [M. Hubschmann's] house, to introduce myself to him. He was the first German professor that I would meet, and I was right away delighted by his good mood in receiving me. He began to speak to me almost immediately of Indo-European linguistics, and he asked me if I had read the article, that had come out over vacation, by Brugmann on the sonant nasal. I hadn't even

> *heard* of Brugmann, which was pardonable at that point, especially for me, and so M. Hubschmann explained to me that there had been a flurry of excitement over the past several weeks about the question as to whether there were certain Greek *α*s that derived from *ν*, or whether certain *ν*s hadn't produced *α*. I couldn't believe my ears, since in my first meeting with a German scientist, he describes as a great scientific conquest something that I had believed for three and a half years to be just a piece of elementary truth that I did not even dare to mention because it was probably too well known. So I said, a bit self-consciously, that this did not seem all that extraordinary or new. Then Hubschmann insisted on the importance that Germanists were placing on this.[139]

The upshot was that Saussure explained to Brugmann that he did not want to sit in on his lectures for fear of committing unintended plagiarism of sorts, and then he sat down at the end of his first year of classes in Leipzig, in the summer of 1877, and worked non-stop to produce a 200-page masterpiece on the treatment of the vowel system of Proto-Indo-European. This is the *Mémoire on the Primitive System of Vowels in the Indo-European Languages*, an utterly brilliant work that was instantly recognizable as a work of genius by anyone keeping up with linguistics at that moment.

In 1878, Saussure left Leipzig, and the whole group of German Neogrammarians, as quickly as possible, but his reputation as a brilliant young man was unfolding before him. After a brief period in Berlin, where he met Whitney, and in Paris and Geneva, he worked on his thesis. He went back to Leipzig in 1880 to defend it, and then went back to Paris, crowned in the glory that shone from his thesis, entitled *De l'emploi du génitif absolu en sanscrit*. The scientific atmosphere he found in Paris was very different from that he had lived with in Leipzig. France had been thoroughly defeated by Prussia in 1870; it had lost two whole *départements*, and such important cities as Strasbourg and Colmar. Revenge was the watchword of the moment. The reasons for the defeat were not obscure: it was not so much that France had been beaten as it was that the administration, the modern university, and the science of Germany had won the war. The recipe for revenge was to redesign the fusty French universities and to match—no, to surpass—the German standards of science. In leading German universities, such as Leipzig, Berlin, and Jena, the jewel of the sciences was comparative linguistics. What had to be done, therefore, was to build in France a school of linguistics that could challenge, or even replace, the Neogrammarians. The Napoleonic university, the old Sorbonne, was hardly the place where new ideas were going to be received and developed. As is often the case in France (though not in the United States, which has maintained the German model), innovation was to come from the periphery of the classical university system: the Ecole Normale Superieure, the Ecole Pratique des Hautes Etudes (EPHE), and the Collège de France. These were three of the top institutions in France, and comparative linguistics was a known quantity at all three. Michel Bréal, professor at the Collège de France and man of the hour, recognized the brilliance of Saussure's *Mémoire* and recruited him.

And so Saussure would become the founder and leader of the French school of linguistics. As Emile Benveniste, one of the great linguists of this movement, wrote, "Ferdinand de Saussure's career began in Paris, with the teaching of comparative grammar at the [EPHE] from 1881 to 1891, from the age of 24 to 34, which had a decisive importance for the development of French linguistics."[140]

As he began at the EPHE, Saussure was 24 years old; his students were 20 years old; his prestige was immense and based on his two books, the *Mémoire* and his thesis. He had everything going for him: his competence in Germanic languages and Indo-European more generally, his thorough knowledge of the Neogrammarians and their work, his culture and independence of spirit, his new ideas that arguably beat the Germans at their own game—all of this was the mark of the young genius, the charismatic leader of a new school. Maurice Grammont, who would be his student, wrote, "His teaching at the [EPHE] gave birth to a veritable school, the French school of linguistics."[141] The young Saussure was everywhere on the Parisian scene, at the EPHE and at the Linguistic Society of Paris; he published and gave courses and lectures, and the public came to hear him. Benveniste remembered, later on:

> The luster that F. de Saussure shone on our school [the EPHE] is not only the reflection of the posthumous glory that today shines on his name. From the first day those who listened to him—confirming the judgment of those senior to him, Bréal, Havet—had the revelation of his mastery. The initiated knew that three years before, at the age of 21, he had written this *Mémoire on the Primitive System of Vowels in the Indo-European Languages* which gave new life to the methods and perspectives of comparative grammar, and that his thesis on the genitive absolutive in Sanskrit was the work of a consummate Indianist. The brilliant intuitions on the one hand, and the extreme analytic rigor on the other, and also his charm and distinction, this alliance of such great gifts had seduced Bréal, who, we know, wanted to keep Saussure on a permanent basis.[142]

Saussure was professionally active throughout the ten years of his career in Paris. He published notes and papers in the journal that was maintained by the Linguistic Society of Paris, and he moved from being an active member to one of the inner circle. Around 1894, he worked on a book on phonology, though he never finished it, and the manuscript was many years later acquired by the Houghton Library at Harvard (on the advice of Roman Jakobson) and published in 1995.[143]

Bréal continued to support Saussure, and Saussure was awarded the highest distinction that France could offer, the Legion of Honor, and Bréal also proposed in 1888 that Saussure should take his place at the Collège de France when he stepped down. But that would have required Saussure to accept French nationality, and to give up his Swiss citizenship as well, and Saussure declined. He was not satisfied with his status in France—neither from a financial point of view, nor from a professional one. And he decided, eventually, to go back to Switzerland.

In 1891, Saussure resigned his position in Paris and returned to Geneva, and the following year he was appointed professor at the University of Geneva. He taught comparative grammar of the Germanic languages, Greek, Latin, and Sanskrit, and while he published little, his research continued, as we know now from the archives of his own notebooks. From 1894 to 1897, he worked intensively on the theory of the syllable, writing much of a book on the subject.[144]

In 1907, the University of Geneva created for Saussure a chair of general linguistics, and he gave a course three years in a row for a small group of students and younger colleagues on what he called linguistic "metaphysics," reflections on the foundations of linguistics, with the goal of creating a program of research and a coherent method for this science. This material was to become a book, one that would be a worthy successor to William Dwight Whitney's great book, the book which had captivated Saussure for decades. But Saussure fell ill, and he died on February 22, 1914, before he was able to put any of this new material into publishable form.

But the work did not stop with Saussure's death. Two years later, two of his students and younger colleagues, Charles Bally and Albert Sechehaye, published *Cours de linguistique générale*, with Saussure's name as the author on the title page, and the book entered into history. It has been published, translated, and commented on around the world, and it is certainly the best known and most widely distributed book in all of linguistics. In the 1920s, Roman Jakobson referred to it as a crucial moment in the awakening of structuralism, and newer readings, first by Benveniste and then by Lévi-Strauss, Lacan, and many others, paved the way for a new wave of structuralism in the human and social sciences in the 1960s.

The *Cours* is famous for its *dichotomies*: synchronic vs. diachronic, syntagmatic vs. paradigmatic, relations *in praesentia* vs. relations *in absentia*, opposition vs. difference, signified vs. signifying, and the biggest of them all: *langue* vs. *parole*, which is roughly the distinction between *language* and *speech*. We cannot possibly review all of these here, but we cannot simply leave them be either. Let's consider just the last one, the crucial opposition between language and speech, which has been picked up by grammarians from Benveniste to Chomsky in order to underscore the primacy of grammar over the speech act, and to justify the position that the science of language is the science of grammar. Speech is an activity undertaken by individuals at given times and given places, for specific ends and purposes, often composed of sentences that are meaningful and yet spoken for the very first time in history. The activities we call speech also include repetitions of phrases and sentences that have been said many, many times. Language is the name we give to the system that unites all of these acts of speech, and which we see as part of the skill and knowledge of a fluent speaker of a particular language.[145]

It would be wrong to ascribe to Saussure the view that language, in his sense, had a priority in any sense over speech (though it is not impossible to draw this conclusion from some of the modern accounts of Saussure's views), and we know that Saussure planned to offer a course on *parole* which his demise prevented. Saussure saw himself as following Whitney, for whom language presented a "double essence," with language and speech indissolubly linked. There is a duality between the study of language as *langue* and language as *parole*, and neither can be studied without taking the other into account. This is apparent in one of the uncompleted manuscripts of Saussure, which has been published only recently,[146] in which Saussure wrote,

> Only linguistics, I dare say, is vast. There are, notably, two parts to it: one is closer to language, the passive warehouse, the other is closer to speech, the active force and the real source of the phenomena which appear little by little in the other half of the system.
>
> In summary:
>
> i. *Not* what is individual, but rather what is established by social usage, thereby fulfilling the conditions whereby something is linguistic;

ii. *Not* necessarily what is written, but preferably what is spoken;
iii. *Not* with a normative goal or in order to give the rules of proper expression;
iv. And finally, with the goal of generalizing observations, to arrive at a theory that is applicable to languages.[147]

If we read his students' notes, and the fragments left behind, Saussure's position is clear: "No doubt, language itself emerges from speech in a certain sense; we need the speech of thousands of individuals in order to establish the agreement from which language will emerge. Language is not the initial phenomenon." Saussure fully understood the paradox that every linguist must face, which is that without speech, there is no language, but the deepest beauty and structure is found in language, not in speech. This is a paradox of which linguists today are not always mindful, because the computational metaphor for language has made it easy to think of grammar as a thing that exists in time and in space (in someone's brain, in particular)—a story that will be central to our concerns in volume 2. That was not an option for Saussure in his day. He was deeply sensitive to the fact that understanding a language means discovering complex structures: he was, indeed, one of the great masters of discovering such structure.

Very close to the end of the *Cours* the reader finds a sentence that has become famous: "We must study language in itself and for itself."[148] This is a striking turn of phrase, the sort we might expect to find in a manifesto, but contemporary scholars of Saussure's *oeuvre* have expressed serious reservations about whether these are Saussure's words or someone else's. The sentence itself is highly reminiscent of a similar statement by Bopp, as we saw earlier in this chapter. Scholars familiar with all of Saussure's work have questioned whether this sentence faithfully reflects his intent.[149] Tracing rupture and continuity in texts can be treacherous.

SAUSSURE'S *MÉMOIRE*

Let's return to Saussure's first work, his *Mémoire*, as he called it: it is brilliant, and you can tell from the first three pages that there is a transcendent intelligence at work. Antoine Meillet would one day call it "the most beautiful book of comparative grammar ever written."[150] There are two characteristics of the German school of nineteenth-century linguistics that Saussure eschewed: paragraphs of endless data whose significance is left to the reader to figure out, and, at least in some of the earlier writing, a tendency from time to time to get lost in flights of fantasy. Saussure's *Mémoire* is neither of those: it is sharp and to the point.

Saussure's *Mémoire* begins with some comments that were certain to leave an impression of audacity, pretention, and even cockiness: the third sentence reads, "no other subject is more controversial; opinions are divided virtually beyond limit, and the various authors have rarely applied their ideas in a purely rigorous way." This young man would explain to his teachers what *rigorous* really meant! And he then spells out the gist of the analyses given by Bopp, by Curtius, by Fisk, by Schleicher, by Amelung—and then by Brugmann, all in an introduction of five pages, before he gets down to business. And when he does, it is much neater and cleaner than in the earlier books he has cited. One of the striking differences is Saussure's book makes it very clear that he is concerned with the evolution of particular *phonemes*, while the earlier linguistic tradition focused on words and morphemes, leaving it to the reader to translate this into a statement about phonemes, if the reader is so inclined. Contrast this with the way Brugmann wrote, for example, when he introduced how he would be employing the notion of *ablaut*, which refers to the kind of vowel changes we still note today in such triples as *sing-sang-sung*. Here is what he wrote; this passage is clear if you already know what he means to say, but otherwise you are lost.

> The following e.g. stand in ablaut relation to one another: Lat. *datus: dō-num; s-iē-s* (O.Lat for *sīs*): *s-ī-mus;* ἄγ-ω: στρατ-ᾱγ-ός (Ion. Att. στρατ-ηγ-ός); λείπ-ειν : λοιπ-ός : λιπ-εῖν; ἄγ-ε-τε: ἄγ-ο-μεν; με-τέρ-ες: με-τρ-ός: με-τρά-σι; Goth. aúh-sin : aúhs-an-s : aúhs-n-ē(loc. sg., nom. pl., gen. pl of *aúhsa* 'ox'). Skr. *pác-āmi* 'I cook': *pak-tás* 'cooked' (difference of accentuation).[151]

Saussure's style of presentation, on the other hand, is crystal clear: take a look at table 2-3, for example, exactly as it is found on the printed page.[152] It is perfectly clear that three sets of sounds are being proposed, and how each member of one set corresponds to a member in the other. The rest of the explanation is found in the text, just above the chart. The table that Saussure presented was neat and clean, and full of meaning arising out of its structure, just like the periodic table of the elements, or a synoptic edition of the New Testament.

Saussure's central point concerns the deletion of the vowel of a syllable in Indo-European, what might cause that deletion, and what might prevent

TABLE 2.3. **Root vocalism in Indo-European**

Full root	a_1	a_1i	a_1u	a_1n	a_1m	a_1r	a_1A	a_1ọ̬
	a_2	a_2i	a_2u	a_2n	a_2m	a_2r	a_2A	a_2ọ̬
Reduced root	–	–*i*	–*u*	–*n̥*	–*m̥*	–*r̥*	–A	–ọ̬

it—and what the consequences are when the deletion occurs. Here is a remarkable passage:

> The root *wak* is in Sanskrit *vac* and appears as *uc-más* in the plural. . . . What is this phenomenon? A weakening of the root, no doubt; only it is essential to make it clear that the word *weakening* never means anything other than *deletion of the a*. It would be too vague [*c'est laisser trop de latitude que de dire*] to say, as M. Brugmann does . . . "vowel deletion under the influence of accentuation." . . . Among other examples given we find indo-eur. *sunsú* 'bru, daughter-in-law' for *sunusá*. . . . When in such a word a *u* drops . . . , it would be an absolutely abnormal case, and one which we could not take to be parallel and which would in fact be in contradiction with the law of the deletion of the *a*, because a corollary of the law of *a*-deletion is precisely that the *coefficients* of *a* must remain. Let us avoid using the word *samprasāraṇa*: this term, it is true, simply denotes the shift of a semi-vowel to a vowel; but in reality it is equivalent in any linguistic work to: shrinking of the syllables *ya, wa, ra (ye,we; yo wo)* to *i,u,*. In the mind of someone who uses the word *samprasāraṇa*, there is inevitably the idea of a special action of *y,w,r* on the vowel that follows, and of an absorbing force that the phonemes somehow enjoy. If that is the meaning that we give to *samprasāraṇa*, it is essential that we say that the weakenings we are looking at have nothing to do with *samprasāraṇa*. The *a* drops: that's all [La tombe, voilà tout]. And it is not by means of several distinct phenomena, but by one simple phenomenon, that *pa-pt-ús* derives from *pat*, and *s-mási* from *as*, *rih-mási* from *raigh*. . . .[153]

The nature of the intellectual problem that was the determination of the vowel system of Indo-European had become more abstract in the literature leading up to the moment when the young Saussure wrote his *Mémoire*. There were two sides to this problem: one involved comparative linguistics, and the other involved internal reconstruction. The comparative task was this: if we compare the roots in two different languages, and divide them up on the basis of what vowel they contain, then when those sets align well across pairs of languages, the vowels of both sets must be descendants of the same vowel in the ancestral language (though the actual pronunciation of the vowel may have changed in one, or in both, of the daughter languages). The internal reconstruction task is rather different, and it was especially difficult for Indo-European, because in essentially all of the descendant languages, there are sets of related words in which the vowel of the root has been modified in a regular pattern: the stem of a verb has one vowel in one tense, and a different vowel in a different tense. But the differences were regular and appeared to follow a pattern, if it could only be found. Typically these patterns strongly suggest that one member of the set has a strengthened vowel (one which is stressed, or longer), while the other has a weakened vowel (unstressed or deleted). Saussure aimed to solve both problems with a single proposal for early Indo-European.

The linguist who wanted to solve this task had to be ready to explore an abstract analysis and to imagine hypotheses in which it did not even matter what the *sound* was that was posited—a symbol would do—and would serve as the basis for a regularity in the evolution from an earlier stage to a later stage. Saussure's teacher Brugmann made this point clearly when he wrote about two vowels which he called a_1 and a_2 (even if the reader might mentally take them as *e* and *o*, respectively): writing the two vowels that way made the point that it was not the phonetic value that was important, but rather the role that the symbol played in the larger system.

This was the beginning of what would come to be called linguistic structuralism, the moment when what mattered in solving the problem was coming up with a hypothesis that could be expressed in terms of symbols, and which would be rooted—but only after all the symbolic computations had been accomplished—in audible vowels and consonants.

When we read the *Mémoire* today, not just with the eyes of a historian, but with those of a linguist, those of a phonologist, we cannot help but be struck by the absolute modernity of the reasoning presented there. We have already met this feeling of amazement when we read other nineteenth-century linguists, whether it is Humboldt, Baudouin, or other giants of the field. We indicated at the beginning of this book that this is an essay on continuity (and, yes, rupture), and this is a fine moment to acknowledge and to appreciate the way in which we can read Saussure as a contemporary of ours. We know something about what would happen to his ideas, something that he had no way to know during his lifetime, and we can place him in a genealogical tree to see his ancestors and his descendants, but

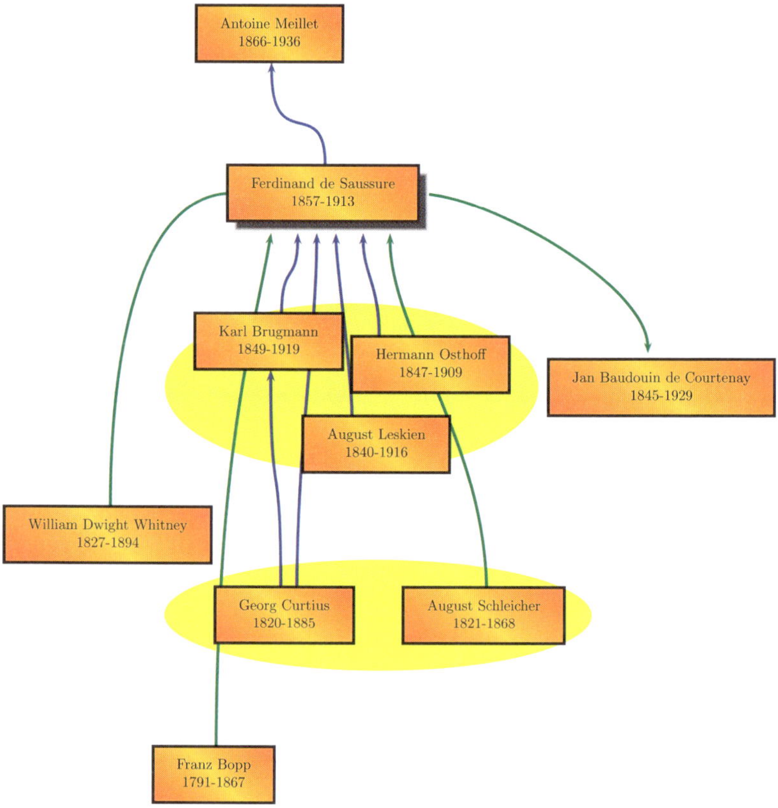

FIGURE 2.8. Ferdinand de Saussure

what is most important is that he can actually speak to us, in our disciplinary language, and with arguments that speak to us today. This is what we mean when we insist that we can read the great and important texts of the discipline, those of Sapir, Bloomfield, or Harris—we mean that we can read them as our contemporaries. It takes a little bit of work at first, but it can be done, and the pleasure of reading a brilliant mind has few equals. Saussure's theory of the syllable, for example, his analysis of the phonemic system of Indo-European, and his critiques of the Neogrammarians are all pieces of work from which the phonologist who is working right now can learn and to which he or she can respond.

Let's turn now to a quite different career, that of the American linguist Maurice Bloomfield.

MAURICE BLOOMFIELD

Maurice Bloomfield was born in Austrian Silesia in 1855, and he came to the United States as a boy of 12, just after the American Civil War. With him had come his parents, of course, as well as his sister Fannie and his brother Sigmund (Sigmund would later be the father of the important American linguist Leonard Bloomfield, whose work we will explore at length below). Maurice Bloomfield excelled at the University of Chicago as an undergraduate, and after obtaining a master's degree from Furman University in South Carolina, he went to Yale University, where he spent a year studying under William Dwight Whitney.

Whitney, in turn, proposed that he should go to the new Johns Hopkins University to study with Charles R. Lanman, who had been Whitney's student at Yale and who had just started teaching at Hopkins. Whitney thought very highly of Lanman; after Lanman had gotten his degree from Yale, he had gone to Germany, just as any young American linguist needed to do, and he had studied with Whitney's teachers in Berlin and Tübingen.[154] After study in Germany, Lanman was ready to take an academic position in the United States, and he started his academic career at Hopkins.

So Whitney sent Maurice Bloomfield to Johns Hopkins, and he received his PhD in 1879. What Bloomfield did next was inevitable: he went to Germany, just as his teacher, and his teacher's teacher, had done. In Berlin, he studied with Albrecht Weber, and in Leipzig with the most important Indo-Europeanists: Brugmann, Leskien, Curtius. In Leipzig, he met Saussure, and he would remain in contact with him over the years that followed.

A passionate Neogrammarian, Maurice Bloomfield published a paper called "On the Probability of the Existence of Phonetic Law" in 1884, in the *American Journal of Philology*, the principal American journal of linguistics at the time. It was an flattering defense of the Neogrammarian principle of the exceptionlessness of phonetic change. The main problem, though, was that many a Neogrammarian could read that paper and say that with friends like this, who needs enemies? Bloomfield's defense of the exceptionlessness brings him to a final moment of fervor in which he declares that *even if the Neogrammarian principle is wrong, it is right from a larger perspective!*—because it made linguists do better linguistics.

Schuchardt's long response was published the following year, and towards the end he wrote, "In light of what we have just seen, the thesis of the exceptionlessness of phonetic laws can neither be proven through deductive reasoning, nor inductive reasoning; whoever holds to it must recognize it as a dogma . . . and that is precisely what [Maurice] Bloomfield did. . . . Bloomfield is quite open about it . . . [saying] that if the thesis of exceptionlessness of sound laws turns out eventually to have been false, that will have no impact on its value as method, because it has proven itself by its fruits." Schuchardt is, quite rightly, aghast at this admission. "Connecting correct results from possibly false premises is contrary to scientific thinking. One cannot simply identify a scientific procedure with a scientific theorem; but many linguists would no doubt be—more or less consciously—in agreement with Bloomfield."[155]

Schuchardt scratched his head when he thought about the alternatives that young Bloomfield, and also young Kruszewski, seemed to have left him. He felt that he was being told that if he did not accept that sound laws had no exceptions, then he was saying that language was just chaotic. "[Bloomfield's] fundamental error, which he shares with others, is a deep one, and it resides in fact in the hypothesis that there exist really, or at least that one could imagine, some domain in which no laws whatsoever would hold."[156] But that is simply not the case, Schuchardt said. In fact, when we look at the world, we see that there is a hierarchy of contexts in which regularity reigns, ranging from "the caprice of a game of chance all the way to the firmly established rules of the mechanical universe." It is not that sound changes *must* have exceptions, but rather that incomplete or sporadic sound changes can certainly be found—and that is what the Neogrammarians try to rule out.

When Bloomfield returned to the United States, his PhD advisor, Charles Lanman, had just been recruited by Harvard, and Hopkins offered Lanman's position to Bloomfield, which of course he accepted. He would have a major impact on the early development of linguistics in the United States, not the least through his nephew Leonard, and years later, in 1926, he was chosen to be the second president of the Linguistic Society of America.

Bloomfield would maintain close professional connections with German linguists and philologists for the rest of his life, and he was especially close, in an epistolary fashion, with Max Müller. His greatest recognition came for his work on myths and the Hindu religion,[157] but he also contributed significantly to comparative grammar,[158] and in Vedic studies, his work is still considered a classic.[159]

Franklin Edgerton wrote:

> His achievements are probably greatest in the field of Indology, where he stands unquestioned as the first of all Atharvanists and in the foremost rank of interpreters of the Vedas in general. To it he has bequeathed a gigantic tool, the Vedic Concordance, unmatched in its usefulness to future investigators, save for the Petersburg Lexicon alone. It is a monument to his insight into the needs of Vedic study, his patient industry, his willingness for self-sacrifice, and his talent for organization. . . . It is as a linguist that he belongs to us. In this field his studies fell at a peculiarly fortunate time—towards the close of the seventies of the last century, the decade of the great discoveries that led to the rebirth of Indo-European grammar and of general linguistics. It was he, more than any other, who brought the seeds of these new sciences to this country and worked for their dissemination. The bibliography, especially for the earlier years, bristles with discussions of special problems in Indo-European grammar that have made his name familiar to the readers of its handbooks.[160]

Grammars, dialects, languages: Retrospect and prospect

When we look at linguistics throughout the nineteenth century from the perspective of an internal history of continuities and of ruptures, we see it as the culture from which our modern discipline has grown. Intellectuals and thinkers of staggering talent created what all hoped would be a new science, but without any certainty that their program would bear fruit in the long run. We have emphasized the fact that the central concern of the century was with time and history, and that focus encouraged a view of language that was more Aristotelian and less Platonic than ever before in the West. In this context, the Aristotelian turn meant the recognition that oral language, uttered at a moment located in time and space, was a central concern, and that variation and dialect formation were an integral part of language. This was not merely an idle observation; variation and dialect formation were parts of a raging sea of nationalism and cultural identity, and linguists were caught up in this vortex as much as anyone else.

The Aristotelian turn was not everywhere, to be sure: if the study of the evolution of Indo-European was central to much of linguistics, then its object was surely in time and space, and yet most linguists were willing to leave wide open the question as to whether the early stages of Indo-European were spoken in Anatolia, or north of the Black Sea, or somewhere else entirely. There was an intense insistence, before the Neogrammarians, that some forms of language (like languages in which stem-internal changes of vowels were more important than adding suffixes) were *better* than others, and this insistence was not Aristotelian. But Whitney, and many linguists influenced by him, such as Saussure, were willing to see language as anchored in practice and performed by individuals in a community. The importance of Darwinian thinking after 1859 cannot be overestimated: while no one was certain whether Darwinian competition had a counterpart in the world of languages and dialects, Darwin's ideas served as proof of concept that out of variation within communities could emerge differences which after the fact seemed, and indeed were, qualitative changes. We will take a brief look at how this Aristotelian turn played out in the study of the modern dialects and languages.

While historical and comparative studies of language were new,[161] the systematic study of the classical languages of Europe was not. Dictionaries (which went hand in hand with normalizations of spelling, in many cases), pedagogical grammars and textbooks, and other sorts of grammars existed for a number of languages. In several cases, the development of this scholarly capital was instrumental in raising a dialect to the status of a language. At the center of this grammar-oriented view was the role played by normativity: there is a right way (and hence, a wrong way) to write things in this language, and we will explain to you what that right way is.

But the study of *dialects* was something quite new, and with it came a different point of view, one which was essentially descriptive, and one which put much less emphasis on the importance of a written form of the language (though it happened often enough that the scholar who worked on a dialect, often his own, wanted to see an orthography developed for it), to the point where the oral form of language would be even more important than the written. Linguists even today will sometimes be uncomfortable when they need to analyze a grammatical structure which is perfectly normal in the written language, but very rare in spoken language.[162]

DIALECTOLOGY

The case of Latin, and the Romance languages that descended from it, was particularly clear: what had begun as dialectal variation within the Empire (or occasionally, perhaps, a process of creolization) became distinct languages, each varying in a different direction, each with its own speed of change. Viewed in retrospect from modern times, it was hard to decide when these systems should be called dialects and when they should be called new languages. Whatever they were, they were the intermediate stages, the links in the chain of the genealogical trees that linguists were constructing: how could they *not* be of great interest to a linguist of this era? Grimm, for example, built his famous grammar of German as a grammar of dialects of German. Later on, Schuchardt turned phenomena of language mixing and creolization into a central question of linguistics, and he would become interested in lingua francas, pidgins, and creoles of all sorts. Studying dialects involves a serious elaboration and organization of data and observations: it is a descriptive account of how people speak. The nineteenth century saw the first linguistic surveys, first by questionnaire sent through the mail, and later by fieldwork and direct descriptions. All of Europe was slowly covered by linguistic atlases showing the reach of dialectal characteristics of all sorts: lexical, phonological, morphological.

EXPERIMENTAL PHONETICS

Spoken language thus became the center of interest for a good number of linguists, and they established systems of transcriptions and detailed phonetic alphabets in order to document variation and change. This development paralleled the development of a new science, phonetics, the detailed study of the sounds of language.

Detailed studies of articulatory movement in speech had emerged since the Renaissance, and this material was well known to teachers of diction and of singing and to doctors dealing with language-related issues. People working with the deaf, such as the Abbé de l'Épee in Paris, had developed detailed notations for diagnosis and training, but the greatest advances came from progress in the scientific understanding of acoustics.

In that development, Hermann von Helmholtz played a central role. Helmholtz began his career as a physiologist and eventually became the most famous German scientist of his age. His work on acoustics, on the perception of tone and the decomposition of sounds into formants, had enormous consequences. His coupled resonators were capable of producing vocalic sounds, and the extraction of these formants made it possible to visualize the components of linguistic sound. His presentation of the physics behind speech was clear and detailed; he explained that "the vowels of speech are in reality tones produced by membranous tongues (the vocal cords), with a resonance chamber (the mouth) capable of altering in length, width, and pitch of resonance, and hence capable also of reinforcing at different times different partials [overtones, harmonics] of the compound tone to which it is applied," which is exactly what we teach students today. He noted, "we find pretty uniformly that the first six to eight partials [overtones, harmonics] are clearly perceptible, but with very different degrees of force according to the different forms of the cavity of the mouth, sometimes screaming loudly into the ear, at others scarcely audible."[163] He noted that the loudest partial corresponded to the height of the vowel and noted that the back vowels seemed to have a discernable second partial. His technology was not as advanced as ours today, but he took some of the most important steps in setting the stage for our understanding today of the acoustics of vowel production.

Helmholtz's influence on researchers at the end of the century was considerable, including Jean-Pierre Rousselot, abbey and dialectologist, who was deeply involved in the description and transcription of linguistic sounds and whose work moved into acoustics. Rousselot built a series of devices that were able to trace the formant structure of vowels, and he is often viewed as the true founder of experimental phonetics, based both on his writing[164] and the phonetics laboratory he created at the Collège de France in 1897.

At the end of the nineteenth century, the pace of technological advances accelerated: Charles Cros, and later Alexander Graham Bell, developed systems for the creation and transmission of artificial sound, eventually leading to the patent of the modern telephone in 1876. Even more importantly, technologies for recording the human voice were developed as well. With the ability to record and play back sound, to transmit it over distances, and to analyze its spectral properties, the reinvigorated field of acoustic phonetics began to change the way that speech would be studied by linguists at the end of the nineteenth century. New instruments and new descriptive tools arose hand in hand and had a profound impact on linguistic studies, nowhere more than in the field of dialectology.

In England, Henry Sweet was a Germanist, with a special interest in Old English and Old Norsk. He developed practical tools for teaching living languages, and his work contributed to the founding of phonetics as an autonomous field of linguistic study.[165]

In Germany, Edward Sievers, who with Hermann Paul directed the Neogrammarians' journal, *Beiträge zur Geschichte der Sprache und Literatur*, specialized in articulatory phonetics, and his work *Grundzüge der Phonetik* constituted a fundamental reference for a whole generation of linguists in Europe and in Russia. He emphasized the notion of a scale of sonority, a notion that has remained very important to accounts of syllabification up to the present. In Denmark, at the end of the nineteenth century, Otto Jespersen contributed to the development of phonetics; along with Sweet and Sievers, Jespersen was involved with the use of phonetics in the teaching of living languages in the classroom, especially English and German. Indeed, throughout the second half of the nineteenth century, the teaching of living foreign languages underwent tremendous growth throughout the educational systems of the countries we have looked at, and there was a good deal of controversy over the role of oral usage, phonology, and morphology in the classroom.

CHAPTER THREE

Philosophy and Logic in the Nineteenth Century

Philosophy

Over the course of this book, we will see that the connections between philosophy and linguistics—as with the connections between philosophy and psychology—are deep and present everywhere. The reasons for this are not hard to find: two of the central concerns of philosophy are the nature of mind and the nature of scientific knowledge, and both are at the heart of what psychologists and linguists grapple with in virtually everything that they do.

In this chapter, we will give an overview of several of the philosophers whose impact on twentieth-century developments was profound and offer a sketch of how their work fit together to form a larger structure. We have no choice but to begin with Immanuel Kant, whose work at the end of the eighteenth century defined the questions for the following generation of philosophers, and more generally thinkers of all stripes. And we must give an introduction to three important figures, for the most part not well known today, who cast a long shadow over the twentieth century: Auguste Comte, Ernst Mach, and Franz Brentano. Their obscurity today is in no way deserved.

Immanuel Kant

Immanuel Kant's inquiry into the nature of rational knowledge began with a familiar pair of positions: first, that philosophical discussions wander where they do not seem to have a right to go, a conclusion that we can draw from the simple fact that discussions among philosophers are uncertain and fraught with controversy; and second, that modern science, of the sort established by Galileo and by Newton, is a great example of the kind of knowledge that we should always be looking for.[1]

The rationalists and the empiricists of the seventeenth and eighteenth centuries, philosophers such as Descartes and Leibniz, and Hume and Locke, were two parties addressing one and the same set of questions, expressing two variant points of view but in agreement that understanding the success of modern science was a central philosophical question—and that science provided the best clue to help us understand the central philosophical question of knowledge, which is to say, how it is that we can come to understand the universe. Kant hoped to overcome the apparent conflict between the rationalist and the empiricist point of view, since both sides had strong points and both had vulnerable points, and ultimately Kant sought to reconcile these two embattled alternatives.[2]

Kant's central point regarding knowledge was that it is a synthesis that brings together two different things: what the senses provide, and the active mechanism which registers that. If you wish, that active mechanism might initially be blank and homogeneous before the world has its way with it, but how it responds to the senses is its own nature. The mind may be a tabula rasa, a blank slate of some sort. But of *what* sort? A blackboard and a whiteboard are both fine models of a tabula rasa, and so is a piece of glass, or an unexposed piece of X-ray film or photography paper, but they all respond differently to light and to scratches. Which one is most like the true tabula rasa, the tabula rasa of the human mind?

Kant argued that there were some basic facts about our thoughts that reflected the active mental organizing principle, and of these, the most important were space, time, and causality. For space and time, he used the term *Anschauung*, which unfortunately is translated into English and French as *intuition* whenever Kant is being discussed, though *sense-perception* is a better translation. Actually, he distinguished between empirical *Anschauungen* (the suffix *-en* marks the plural) and pure *Anschauungen*, and space and time were the distinctive members of that second group; he also called space and time the *forms* of *Anschauung*.[3]

Kant's argument against Lockean empiricism was more sophisticated than Descartes's or Leibniz's had been. Descartes's typical example of an idea that had not come through the senses was a *triangle*, while Leibniz's was the unstated premise in a logical argument that we accept as compelling even when the premise is left unstated. In both cases, the nature of that which is innate is very much like that which comes through the senses. Kant's sense of what is innate is a good deal more abstract than what Descartes or Leibniz had pointed to: for Kant, the crucial points were the aspects of the scaffold of thought without which no thinking about reality is possible and which is yet not contained inside a sense-perception, and his prime examples are space and time.

It is no exaggeration to say that Kant's proposal is the clearest antecedent of the modern, cognitive view of the mind as an active agent of thought. There is merit in viewing Leibniz as an important antecedent, certainly, and the modern cognitive view could not have become what it is without the invention and development of the modern computer. All that may be true: and yet, Kant's determination to make it clear that our thought is not possible without an active thinker—and not just a recumbent onlooker but an active participant—is the most important moment in the rise of the modern, cognitive view of the mind.

To think about thinking, Kant needed a general framework for understanding thoughts, and for better or for worse, his logical analysis of the content of a thought fell clearly in the classical tradition which sees a universal division of thoughts into subject and predicate. It is hard to overemphasize how important this logical starting point was, and it will not be until we encounter Gottlob Frege below that we will see a clear rejection of this view. The subject/predicate split lies at the heart of a critical distinction for Kant, the difference between an *analytic* and a *synthetic* judgment: a judgment is analytic if the subject is contained within the predicate, and synthetic if it is not.

Kant gave an example of what he had in mind. "All bodies are extended" is an analytic truth that we know with certainty, while "All bodies are heavy" is a synthetic truth; heaviness is nowhere to be found in the concept lying behind "all bodies."

What was not obvious at the time was how much effort would be spent over the next century in coming to grips with understanding what the phrase "is contained within" actually means. This would come to be known as analyzing the logic of parts and wholes: what does it mean for one thing to be part of another? Franz Brentano, a philosopher we will meet shortly, brought out how important and difficult a question this is, and his students (and then his students' students) continued to develop the question in various contexts. There would be consequences of this work all over the intellectual map: when Georg Cantor established set theory at the end of the nineteenth century, one of the first steps was to distinguish between "being a member of a set" and "being a subset of a set," a distinction which was far from obvious at the time but one which flowed from the effort to understand the logic of parts and wholes. Getting clear on this Cantorian distinction also set the stage for the famous antinomy, the apparent contradiction at the heart of set theory, that Bertrand Russell would discover at the end of the century, the diabolical set of all sets that do not contain themselves.[4]

Any linguist would find it striking that there is a clear line of intellectual descent from Kant's use of the German word *Merkmal* (a word that *could* be translated into English as "characteristic" or "trait") all the way down to Trubetzkoy's use of the very same term in a way that is now translated into English as "feature," in the context of distinctive features in linguistics and most especially in phonology. For Kant, a feature (using that word now to translate his word *Merkmal*) was one of the components of a concept, and thus to understand a concept, one must understand its features, and how its features are put together to form a concept. This understanding would pass down through the nineteenth century and the work of Friedrich Trendelenburg, Rudolf Hermann Lotze, and Edmund Husserl—all people whose work would rarely be thought of as significant for modern linguistics, though indeed we begin now to see that it is. One of the major goals of philosophy, for Kant and later philosophers inspired by him, was to work out and better understand synthetic *a priori* truths: truths that were not dependent on any particular experience, but which expressed statements in which the predicate asserted something that went beyond the essential meaning of the subject. This goal was adopted by many in philosophy in the following years of the nineteenth century, and each could be identified, therefore, as some sort of *neo-Kantian* approach to philosophy. A century later, as we will see, Frege and Russell would present arguments that the true logical form of many sentences was not, at its highest level, *subject/predicate*; the highest level was a statement of existence or some other sort of quantification. But all this was a long way off when Kant was alive.

The rationalist tradition of Descartes and Leibniz emphasized the existence and the importance of certain innate ideas, and in a new way, Kant insisted on the prior existence of certain intuitions that make thought possible. These innatist tendencies of Descartes, Leibniz, and Kant are quite different from nativist views espoused today. There were already in the nineteenth century thinkers who were quite nativist, sometimes under the influence of Darwin, but the innate dispositions that were studied by such writers as Herbert Spencer should not be confused with rationalist or Kantian innatism. As Max Müller wrote,

> The prehistoric genesis of these congenital dispositions or inherited necessities of thought, as suggested by Mr. Herbert Spencer, seems to me a mere shifting back of the difficulty which other philosophers meet boldly face to face, and whether right or wrong, it would not help in solving the problem which Kant is dealing with.[5]

And he cited T. H. Green approvingly, who wrote,

> People who think that the development of habits through hereditary transmission will account for the necessity of necessary truth, show that they do not know what is meant by such necessity.[6]

Space, time, causality, and *quantity*: as far as what humans can know, these were inescapable categories that we use and that make our kind of thought possible. After Kant, philosophers would no longer call them innate, but they were brought by the active mind to the process of encountering the world.

Over the course of the nineteenth century and then into the twentieth, Kant's assumptions and his framework were taken apart and put back together in a vast number of ways. *Space* was the first to be subjected to serious and intense scrutiny, beginning not long after Kant's system was published.

Until the time of Descartes, geometry was considered the heart and soul of mathematics, and the axiomatic method established by Euclid was held up as one of the greatest achievements of the human mind: a method for achieving certainty from a small set of transparent axioms and methods of inference. But careful reflection over the centuries had led geometers to question whether Euclid's fifth postulate was indeed as evident as some had once taken it to be. Could it be proven? The fifth postulate—also known as the parallel postulate—is a bit convoluted. This is how it reads, in its original form:

> If a line segment intersects two straight lines, forming two interior angles on the same side that add up to less than two right angles, then the two lines, if extended indefinitely, meet on that side on which the angles sum to less than two right angles.

In the first three decades of the nineteenth century, opinion began to turn against what came to be known as the parallel postulate, or the fifth postulate. There had been many attempts made to *prove* it, which is to say, to show that it should not be taken as a postulate, but rather as a consequence of the other postulates. But could sense be made of space if this postulate were not assumed, if space somehow violated the postulate? Surely Kant's space, as a necessary condition for our perception, required going along with this postulate. And surely the world that we see around us does indeed satisfy this postulate. No?

Carl Friedrich Gauss was the greatest mathematician of the nineteenth century. He wrote in a letter in 1817,

> I am becoming more and more convinced that the necessity of our geometry cannot be proved. . . . Perhaps only in another life will we attain another insight into the nature of space, which is unattainable to us now. Until then we must not place geometry with arithmetic, which is purely a priori, but rather in the same rank as mechanics.[7]

and, 12 years later,

> My conviction that we cannot base geometry completely a priori has, if anything, become even stronger . . . it is my inner conviction that the study of space occupies a quite different place in our a priori knowledge than the study of quantity . . . we must humbly admit that if Number is the pure product of our mind, Space has a reality outside of our minds and we cannot completely prescribe its laws a priori.[8]

In 1829, Nikolai Lobachevsky—and independently in 1832, János Bolyai—established the consistency of a non-Euclidean geometry. What then was the status of our familiar Euclidean geometry? Is it still an essential part of our knowledge, and do we now need to put the word "knowledge" in scare quotes and recognize the irony of speaking of knowledge which may not even be true? There was no immediate satisfactory answer, none at all.

We will return to this theme in chapter 8, when it will be seen as one of the steps that led to foundational crises in mathematics.

German philosophy (and that of much of Western Europe) in the century after Kant divided into several traditions. One rejected the optimism of the Enlightenment that is found throughout Kant and instead glorified a view of life as subjective activity and action—Nietzsche is a leading figure in this tradition—while another saw itself as correcting the errors and blind spots of Kant. In this second tradition, we can detect three distinct niches: positivism, neo-Kantianism, and the semantic (or linguistic) tradition. We will look first at two important figures in the positivist tradition, Auguste Comte and Ernst Mach, and then at Franz Brentano, an early figure in the semantic tradition who was also influenced by positivism.[9]

Positivism and the anti-metaphysical backlash

Auguste Comte, a French philosopher well outside of the academic system of his day, was the most influential of all of the major figures in the backlash against the philosophical systems of Germanic philosophy. He proposed what he called a *positive* philosophy, and what he championed has come to be known as positivism (though that term is most commonly used today in a larger sense, one which includes many positions that Comte would disown if he were still alive). Comte wrote in a spare and precise style, quite modern and untraditional. Though his work is not widely read today, he articulated a view that not only was very appealing to his contemporaries, it is also one that continues to appeal to us today in many respects. His conclusions may strike us as extreme, but that is hardly surprising: when you are the leading figure of a movement that is responding to a powerful force, you are likely to adopt an extreme position. More important, positivist sensibilities have played a major role in many of the discussions of language and the mind, and the views of psychologists and linguists in the periods that we will explore. What we identify as concrete analyses in contemporary linguistics are often motivated by just these sensibilities.[10]

Comte emphasized that our only true and reliable knowledge is of "phenomena," and he said further that we are able to notice just two basic kinds of relations of phenomena—the similarity between two phenomena and their regular succession in time and space. We fall into error if we think we can draw grander conclusions. In the final state of knowledge, towards which we are all evolving, "the human spirit gives up looking for the origin and the destination of the universe, and knowing the inner causes of phenomena."[11]

Like all movements that attract a large following, positivism offered a liberation from the heavy chains of the past. Comte was a very appealing writer, as we noted; he wrote simply, with conviction, and with erudition as well, and he left the reader with the sense that the truth was really quite simple and straightforward, once you looked at things the right way.

In the development of human intelligence in all its various spheres of activity, from the dawn of time until the present, Comte believed he had discerned a fundamental law which governed all of this dynamic, and he said that he could demonstrate the basis of this law both by following its inner logic and by looking at the history of human thought. In every domain of human endeavor, knowledge passes through three stages of development: the theological stage, the metaphysical (or abstract) stage, and the scientific (or positivist) stage. This development was natural, in the sense that it was the result of human nature, the nature of the human mind. The three stages were fundamentally incompatible—or perhaps a better way to put it would be that they are logically incompatible—nonetheless, in the evolution of thought, sometimes earlier forms of thinking survive as holdovers of the past for a while, as intellectual atavisms.

The logic of the situation, Comte believed, was that the original, theological point of view was necessary as a starting point for humankind, and that the final, positivist point of view was the end point of the evolution of human thought, and that the metaphysical intermediate stage was a necessary transition period between the first and the final. The theological stage of development saw the explanation of everything in the world as grounded in the intentions of individuals, but the individuals need not be just the visible human beings on Earth: they may be all-powerful deities that we cannot see. Still, in this primitive perspective, it is what the gods (or an individual God) *wants* that is the reason for why things are as they are.

The next stage, the metaphysical, is one in which much the same kind of explanation is sought by humans, but instead of attributing the cause to an unobserved deity, we attribute it to something else unseen and unseeable—what Comte sometimes calls "personified abstractions."

In the third stage, the scientific stage, we move past searching for absolutes and rest content with understanding the inviolable rules that govern relations of succession in time and of similarity of occurrences. Actually, there was an additional point that would come to be quite important later on: not only was the goal of a positivist science to find invariable laws about what can be observed, it was also necessary to find the smallest number possible of laws that accomplish this.[12]

SCIENTISM

Positivism, then, was deeply committed to the view that to advance knowledge was to do more and better science, and the view that the most important way to improve philosophy was to better understand how philosophy can be accomplished in a world in which science is the most important way we attain knowledge. Part of the message that philosophy ought to learn, while learning from science, is that a certain humility (built on a deep respect for the fact that brand new discoveries can dethrone what had once seemed like an eternal verity) is always in order. The flip side to that is that if we are interested in understanding how actions can be understood as steps towards the accomplishment of some future goal (which is always going to be true when we study the mind), traditional science is generally not able to offer helpful advice. Most scientific models take it for granted that when science provides an answer to the question of what caused an event, the answer will be based on events that are close to one another in space and time: think of how we explain the recoil of one billiard ball bouncing off of another. That way of thinking is not likely to provide us with much insight into how to develop a theory of goal-oriented activities.

But by the same token, the actual statements that physicists and chemists were making during this early period of positivism were changing, not because of the influence of philosophy but because of the advances in scientific understanding. The notions of *energy* and of *entropy* were cutting-edge conceptual tools in the physical sciences. The idea that there was both a convertibility from one form of energy to another, and that there was nonetheless an overall perfect maintenance of the total quantity of energy, was a very fruitful one, but it also suggested that energy was a kind of *thing*, and not merely a way of describing the activities of things.[13] Over the course of the century, the understanding was growing that there were different *kinds* of energy, and that while one kind of energy (like heat, for example) could arise (heat can be created by rubbing two sticks of wood against each other), the heat did not arise out of nothing; it was a transformation of another kind of energy. All of this would eventually become formulated as a law of conservation of energy. Just as importantly, mid-nineteenth-century science was coming to grips with the notion that the *entropy* of a closed physical system would always increase, a conclusion that seemed at times dangerously close to saying that the evolution of a physical system was *towards* a state—of maximum entropy, in particular.[14]

PSYCHOLOGY

Auguste Comte was quite skeptical regarding the prospects for a positivistic discipline of psychology. He argued that introspection simply could not provide a reliable, or useful, kind of observation from which a science of psychology could arise: the best way to study the mind is to study other people from the outside. "As regards . . . observing intellectual phenomena while they are unfurling, this is obviously impossible. The thinking individual would not be able to divide itself in two, one part reasoning while the other watched the reasoning. The observed organ and the observing organ being, here, identical, how could the observation take place?"[15]

Comte's influence and renown was greatly enhanced by a book that John Stuart Mill wrote about him and his positivism in 1865, just shortly before Comte's death. Mill wrote that Comte "rejects totally, as an invalid process, psychological observation properly so called, or in other words, internal consciousness, at least as regards our intellectual operations. He gives no place in his series of the science of Psychology, and always speaks of it with contempt. The study of mental phaenomena, or, as he expresses it, of moral and intellectual functions, has a place in his scheme, under the head of Biology, but only as a branch of physiology. Our knowledge of the human mind must, he thinks, be acquired by observing other people." Mill found this view difficult to take seriously—how can we understand someone else if we do not start with some understanding of ourselves? Comte was convinced "we can learn very little about the feelings, and nothing at all about the intellect, by self-observation." Comte preferred phrenology—and Mill was aghast at that choice. Was this what it took to make psychology a positive science, Mill asked. "The condition of mental science would be sad indeed if this were its best chance of being positive,"[16] and despite what Comte had hoped, serious studies showed little hope for the future of phrenology.

What was phrenology? It was the creation of Franz Gall, a German physiologist about a generation older than Comte whose work concerned what we today call "localization" in the brain. It gained prominence at the end of the eighteenth and the beginning of the nineteenth centuries, when Gall proposed the creation of a sort of cartography of the human brain, where areas of the brain corresponded to recognizable abilities or motivations. He catalogued 27 different zones, including friendship, memory of things, sense of language, mechanics and architecture, and religion.[17]

Gall proposed a "theory of faculties," and the view that each human personality can be characterized by the way in which, and the degree to which, certain faculties dominate others in individuals. The areas corresponding to the strongest faculties in an individual are more developed, and Gall could therefore read the dominant characteristics right from the scalp: this was the famous theory of bumps, his phrenology. Phrenology would be totally abandoned by the beginning of the twentieth century, after a long run of great popularity during the nineteenth, when it offered what was so greatly sought by anthropometry: measurements, classifications, and typologies. We will return to this in connection with the work of Paul Broca, at a time later in the century when measurements became the central concern of the age, and when measurable differences across the range of human beings became an equally hot topic.[18]

DISCOVERY AND JUSTIFICATION

Comte urged his readers to distinguish between two styles of understanding the development of scientific thought. John Stuart Mill put it this way:

> The philosophy of Science consists of two principal parts; the methods of investigation, and the requisites of proof. The one points out the roads by which the human intellect arrives at conclusions, the other the mode of testing their evidence. The former if complete would be an Organon of Discovery, the latter of Proof. It is to the first of these that M. Comte principally confines himself, and he treats it with a degree of perfection hitherto unrivalled. . . . We are taught the right way of searching for results, but when a result has been reached, how shall we know that it is true?[19]

Mill wrote this, about Comte's notion of philosophy of a science, as it differs from the science itself: it is "the science itself, considered not as to its results, the truths which it ascertains, but as to the processes by which the mind attains them, the marks by which it recognizes them, and the co-ordinating and methodizing of them with a view to the greatest clearness of conception and the fullest and readiest availability for us: in one word, the logic of the science."[20] This conception would be adopted with fervor by the logical positivists (see chapter 7) as they discussed the difference that they saw between "the context of discovery" and "the context of justification," precisely Comte's distinction.

Comte wrote with great insight about the advantages and disadvantages of presenting a science in a historical fashion.

> All science can be presented in two essentially distinct ways, and any other manner of exposition can only be some sort of combination of the two: a **historical** approach, and a **dogmatic** approach. With the former, we present knowledge in stages, in the same order that the human mind actually encountered them, and adopting, as much as possible, the same paths to arrive there. With the latter, we present a system of ideas as it could be understood today by a single mind which from an appropriate intellectual position and with sufficient knowledge could rebuild all of science. . . . The human mind constantly tends to substitute the second, the dogmatic order, for the first, the historic order; the dogmatic order is the only one that satisfies the final state of our intelligence.[21]

Comte's simple account of the history of ideas contained within it an invitation at particular moments to sweep clean the past—in the transition from the metaphysical perspective to the scientific perspective, the leap forward consists in large part in letting go of concepts that had once been satisfying and even comforting. But those comforting metaphysical visions would have to be dropped, as all of the vestiges of metaphysics were rooted out, exposed, expunged, eliminated.

SYNCHRONY AND DIACHRONY

There is a distinction which became essential to the study of language, late in the nineteenth century, which can be traced back to Comte, who first clearly marked the distinction between analysis which is *synchronic* and analysis which is *diachronic*. While those terms are due to Saussure, who emphasized the relevance of the distinction to the analysis of language, the more general distinction had been clearly developed by Comte decades earlier. Comte emphasized that a positivist philosophy required the development in parallel of both a *static* and a *dynamic* analysis. In sociology, always a central concern of his, a radical distinction needed to be borne in mind, one which kept distinct the static conditions which are vital to the existence of a society at a given moment in time, on the one hand, and the dynamic principles reflected by the changes over time that a society undergoes.

WILLIAM DWIGHT WHITNEY

We have already met the American linguist William Dwight Whitney. He too was obviously influenced by Comte's perspective.[22] Whitney found the three stages in the development of linguistics. The first was the theological stage, which can be found in the writings of those who see language as God's divine creation and His gift to man. More interestingly, perhaps, Whitney perceived a second view which would in a few decades be labeled as *behaviorism*, and he saw it as emanating out of this theological phase as well: a doctrine in which "speech is a direct product of the physical constitution of its speakers, a kind of secretion of organs provided for that purpose . . . its varieties represent differences of animal organization. Both these alike cut off all possibility of a real science of language."[23] Whitney saw the metaphysical stage of linguistic thought wherever people saw language as having an existence independent of the people who spoke and understood it, as being endowed with properties and being affected by forces. For Whitney, the third stage was the positive stage, which would be achieved when linguists "keep themselves strictly upon the basis of observed fact and legitimate induction . . . and not to cover up ignorance and obscurity of thought with sounding and philosophical phraseology."[24]

THE POSTHUMOUS EMBARRASSMENT OF SCIENCE'S DIRECTION

When we look at the history of science as it evolved after Comte, there are quite a few cases that turn out to be embarrassments for his position—for example, the treatment of gravity and of heat, two of the most important areas of physics in the hundred years following Comte.[25] Newton's theory of gravitation, he said, was a great success because its predictions were accurate, and it could equally handle astronomical phenomena and the fall of an apple to Earth. "But as to what this attraction or weight is in itself, or what are its causes, these are questions that we consider to be unsolvable and which are no longer in the domain of positive philosophy." Einstein would show this to be wrong; by looking more deeply for an account of a world in which things are not quite what they seem to be (an account in which the fabric of space and time is intertwined, and deeply affected by the presence of matter within), physics made great strides. Comte's second case turned out to have the same ironic fate. Comte cited Fourier's work on the nature of heat, and said that it was greatly to its credit that it took no stance on the controversy between those who believed that heat was a substance called *caloric* and those who took heat to be the vibrations of a universal *ether.* At the turn of the next century, the analysis of heat as the vibration of molecules would turn out to be one of the great achievements of all of science and would show that science could, indeed, discover the truth about domains that once seemed totally out of scientific reach.

Thus for Comte, the fundamental challenges of a healthy science were to develop methods of classification and methods of relating categories, including ways to develop *hierarchies* of classes, perhaps the very best way of relating categories, just as botanists and zoologists were engaged in doing during Comte's lifetime.

PHYSICALISM

If we were to identify the single most important position associated with positivism, it would be this: that all that is truly real are the objects that are identifiable as perduring things in space, over a certain and determinate period of time. That is all there is to reality.

This leaves no room in reality for sentences, ideas, intents, or meanings, or really anything whose principles of operation are not those of the physical, causal sort. That is a serious problem for most people who want to deal with the nature of the mind: they must either give up on the hope that the things that they care about are actually real, or else they must find some secondary way for thoughts and the like to be real. Some people, most notably those in the behaviorist movement in the twentieth century, would be willing to eliminate ideas and intents from their understanding of reality, but others would not.

We will turn to Wilhelm Wundt in chapter 4. An important innovator in nineteenth-century psychology, he saw psychology as rooted both in natural science and in human science:

> The question, whether it is possible to indicate principles of psychic causality, which are equivalent to those physical principles expressed in causal equations, includes the question about the legitimacy of psychology as an independent discipline, which in turn includes the further questions about the significance of the *Geisteswissenschaften* in general. If there is nothing but physical causality, then the fate of all these disciplines is sealed.[26]

This is perhaps the most important point of all that we will address in this book: all accounts of the mind that do not *deny* its existence include some account of causality that is not physical causality. And we will be able to identify, in each approach to the mind, *where* that kind of causality is thought to exist. Precisely this point will come back early in chapter

4, when we look at the positivist split that occurred in a later generation of psychologists with Wundt's view of psychology.

Ernst Mach

Like Comte, Ernst Mach's name is not well known today, outside of its use in expressions about the speed of an airplane: an airplane traveling at Mach 3 is flying three times the speed of sound. But Mach was one of the most influential thinkers in the second half of the nineteenth century among philosophers, scientists, and intellectuals generally. Einstein would later say that Mach's perspective had helped him greatly in coming to grips with the nature of space and time. Mach was an empiricist, an intellectual descendant of George Berkeley, John Locke, David Hume, and Auguste Comte, and he viewed science as a marvelous means for summarizing an enormous amount of observational data. As an empiricist, he believed that all knowledge is founded on immediate sensation.[27]

Mach was born in 1838 in Moravia (then part of Austria, now part of the Czech Republic), but he grew up and was educated in Vienna. (We will see throughout this book how strong the axis was linking Berlin, Prague, and Vienna, and it is helpful to bear in mind that these three cities lie nearly on a north-south line defining the central axis of Europe.) Mach became professor in Graz in 1864, and after 12 years, moved to Prague, where he worked and taught for 19 years. In 1895, he moved back to Vienna (supported in this move by Franz Brentano) to take a chair as professor, but three years later he suffered a stroke. He retired in 1901, when his chair was taken by the physicist Ludwig Boltzmann.[28]

Mach's influence was great in physics, psychology, and the philosophy of science. He became interested in psychology early in his career: psychology was being done by physicists, after all, people such as von Helmholtz and Fechner, whose work we will look at in the next chapter.[29]

Mach saw clearly how important it is to learn the logic of a science through its historical development, as we in fact are trying to do here, the alternative to what Comte called a dogmatic approach:

> The historical study of the evolution of a science is absolutely necessary, for without it the laws which it has acquired through its arduous labor may well turn into a system of half-understood precepts, or worse, a system of preconceived ideas. The historical approach not only helps us to understand what our present understanding is, it also opens up in us new possibilities by showing

> us that what exists is to a large extent conventional and fortuitous. By taking a historical perspective in which different conceptual avenues converge, we can also see ourselves better and see paths as yet undiscovered.[30]

Mach waged intellectual war on the notions of absolute space and absolute time. He worked hard to force physics to reject the assumption that Newton had made, that space and time were absolutes—that they really exist, even if all we ever see and touch are the things that are *in* them, whatever that means. What would it mean if space were not something absolute? It would mean that space everywhere was a framework that allows us to compare two lengths that are put next to one another, like a ruler and something whose length we want to measure, such as a pencil. By lining them up, and assuming that the ruler can be trusted not to change sizes, we can determine that the pencil is six inches long. If we can also move the same ruler around to measure other things, then we have the beginning of a good sense of space that is purely relational. There are no absolutes in space, but there are local comparisons of length (or distance) that can be sensibly made.

If we accept Mach's reasoning up to this point, there is a second problem we encounter. Mach's principle so far tells us that all motion in a straight line is relative; that is, we cannot say what an object's velocity is in any absolute sense, but only relative to some arbitrarily chosen frame of reference—and that seems to be right, as far as the universe is concerned. But the same reasoning suggests that there is no absolute sense in which something could be said to be spinning around. That, however, does *not* appear to be true: the whole big universe out there does provide a backdrop against which we can define which objects are spinning around others, and which objects are not spinning at all. So Mach raised the question: can we say that somehow it is the distant stars that are fundamentally responsible for our ability (right here on Earth) to know what counts as spinning around and what does not? We noted just above that Einstein, whose work would settle this question in the second decade of the twentieth century, pointed to Mach as the figure whose thoroughgoing skepticism about the Newtonian theory of space created the intellectual room for him to rethink the nature of space and time.

Mach was one of the most important figures in the rise of an anti-metaphysical movement in Austria. By "anti-metaphysical," we mean a great skepticism with regard to the reality and the existence of unseen objects. Objects may be unseen because our senses are ill-equipped to perceive them directly (think of radio waves), or because something more fundamental makes them invisible (like neutrinos), or some religious factors make them invisible (like angels), or because they exist in a world to which there is no sensory connection, like mathematical objects (whole numbers, fractions, complex numbers). These are quite different sorts of invisible things, to be sure, and philosophers can easily disagree as to which of them are real despite our inability to see them.

It is difficult for today's reader to get a sense of what it means to be on one point or another of this spectrum regarding what should count as real. Chomsky's style of theorizing, for example, is one in which there is a low bar for positing the reality of theoretical objects, such as cases and traces, while most psychologists, we daresay, share a different sense of what may be considered to be real, demanding results caused in a laboratory setting before agreeing that something is real. It is always difficult for people who are at distant positions along this axis to understand why the other person believes what he does.

Mach was thus the poster child of skepticism directed at unobservable entities postulated by his scientific colleagues. This skepticism served him in good stead, but only to a certain degree. His skepticism about absolute space and time helped Einstein, in a sense, deconstruct and reconstruct our understanding of how the two relate to one another in his special theory, but Einstein's *general theory* of relativity brought us back to a picture of space-time which had real properties of its own—in particular, a complex sort of curvature which lay at the heart of his general theory. Mach was also skeptical about the existence of atoms, but in this he was simply wrong: he was one of the very last skeptics who doubted the existence of atoms. Sometimes unobserved entities, like phlogiston and like ether (the hypothetical object eagerly sought during the nineteenth century as the substance whose vibrations would be what we call "light"), turn out to be myths, but others, like atoms, turn out to be quite real.[31]

Mach worried about the fact that if we can appeal to nothing but sensation in doing science, we limit ourselves greatly when we want to provide explanations. As an intellectual heir to the British sensualists, he was convinced that it was sensation that was real, but he was perfectly well aware that there are good reasons to believe that there are complex entities that were more salient, more real from a subjective point of view: a musical chord or melody, an equilateral triangle, a face, a rhythmic beat all have more salience than a single flash of the color red on the retina. In 1865, he concluded that there must be some additional "presentations" that are shared by all of the perceptions of the melody (or the triangle, and so on). In the case of visual figures, it is the *shape* that we humans grasp so readily, and the German word for this is *Gestalt*: we grasp a Gestalt. (Later we will see Gestalt psychology coming into existence, with a recognition of its early roots in Mach's ideas.) Mach appealed to a second-order sort of presentation, one that was about the relation between sensory presentations, and he called these new ones *Muskelempfindungen*, but it is *these* presentations which are the important ones, and the study of these more complex shapes would eventually become extraordinarily important. The right way to understand the principles of combinations of sensations into Gestalts would eventually include Mach's other idea, the one that placed a great emphasis on the simplicity of the description of the data.[32]

Mach also influenced the discussion of the notions of cause and effect throughout the natural and social sciences.[33] The positivist view held that causation was largely illusory, and Mach argued that in the core physical sciences, a mathematical formulation does not attribute the role of cause to one part of it and effect to another; the scientist who formulates a mathematical expression has little or no need to engage in discussions of what causes what. "We call cause an event to which another (the effect) is constantly bound."[34] That, of course, is pure Hume. "In the more highly developed natural sciences the concepts of cause and effect are becoming more limited and more seldom used. . . . As soon as we succeed in characterizing the elements of events by means of measurable quantities . . . the mutual dependence of elements is much more completely and precisely represented by the concept of a function than by those of cause and effect, themselves poorly delimited."[35] Once you have written an equation that describes what you will observe, Mach would have said, you are not improving your account by giving names to unobserved things, like forces.[36] Mach's impact was wide, as we have noted. Another person whose intellectual direction was influenced by Mach was Jacques Loeb, a German physiologist who helped develop the skeptical worldview that grew into American behaviorism: Loeb immigrated to the United States and was an influential teacher of John B. Watson, the godfather of behaviorism, as we will see in chapter 5.

Franz Brentano

Brentano is the third principal character we need to meet in our perspective on European philosophy in the middle of the nineteenth century. He may not be widely remembered in this day and age—his name is certainly not as familiar as Kant's or Hegel's—but his influence on the ways in which we think about the mind today is considerable, and it runs through the separate strands of philosophy and psychology, and even of linguistics. There may be no one who plays a more important role in the development of the mind fields over the course of this book than Brentano. It is essential to our task to see how this impact was felt—and the fact that his sway is so poorly remembered today is a fact of capital importance to our broader questions of intellectual rupture and continuity.[37] Brentano's students testified to the power of his intellect and its effects on their own development, but he published relatively little during his lifetime, and with the passing of his students by the end of the 1930s, awareness of his significance faded, though important studies of his work continue to this day.[38]

Brentano's influence is evident in the intellectual range and power of the work that his students engaged in, over a range of disciplines and professions. In a long teaching career, Brentano had quite a large number of students who became both famous and influential, especially in Central and Eastern Europe (thus Vienna, Prague, and Moscow). These included such psychologists as Sigmund Freud, Carl Stumpf, Christian von Ehrenfels, and Karl Bühler; such philosophers as Edmund Husserl, Anton Marty, and Alexius Meinong; the logician Kazimierz Twardowski; and the unclassifiable inventor Nikola Tesla, the man who invented alternating current. Husserl the student is now much better known than Brentano the teacher; Husserl is widely viewed as one of the very most important and influential European philosophers of the first half of the twentieth century. Tomáš Masaryk, the first president of Czechoslovakia, was also a student of Brentano: before becoming a politician, he was a philosopher, earning his PhD in 1879. To Masaryk we will return as well.

Brentano was a scholar of Aristotle, and his entire career was devoted to bringing Aristotle's questions to life in a way that made sense to a nineteenth-century sensibility. This was an enormous undertaking, but several of these questions are directly relevant to our subject, such as the nature of mind, of certainty, and the categories of logic. Before we explore Brentano's account of these questions, it will be helpful to place Brentano in the larger history of the period.

Franz Brentano was born in Germany in 1838 in a small town in western Germany near Koblenz, and he first studied philosophy at Tübingen

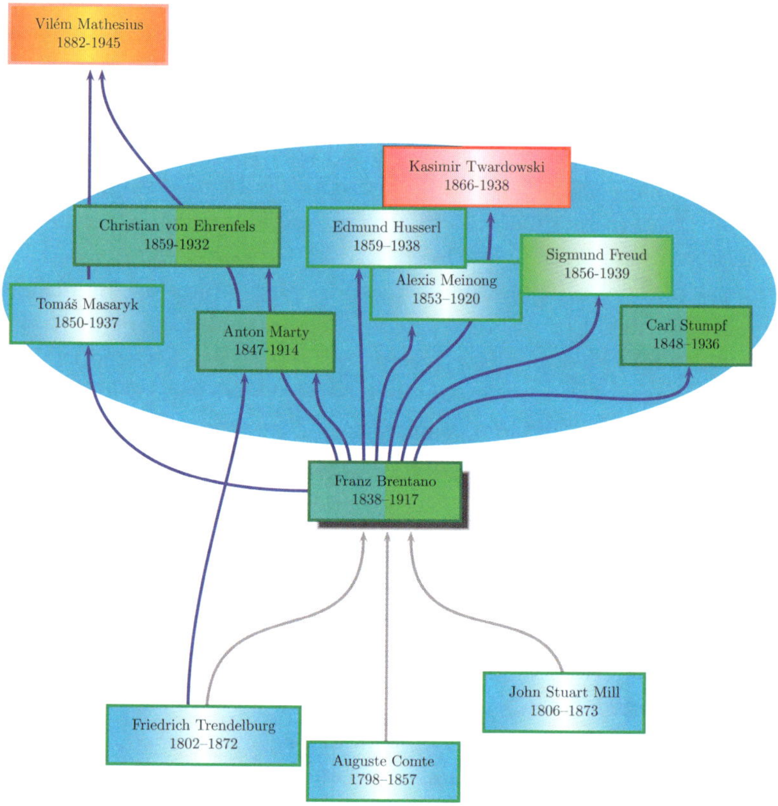

FIGURE 3.1. Franz Brentano

and then theology, preparing to enter the priesthood in 1864. In his late twenties, he began to lecture at the University of Würzburg, where his students included Carl Stumpf, who would become a leading psychologist of his generation, and Anton Marty. Stumpf wrote that he "had never met an academic, neither in my student days nor since I have been a professor, who dedicated himself to such an extent, both verbally and in writing, to his task as a teacher," and "the friendly relations with his students, based upon an equally absolute devotion to the highest purposes, was one of the strongest needs of his life."[39] Brentano had a difficult academic career. He became full professor of philosophy at the University of Vienna, but was forced by the Austrian government to leave his chair because he gave up the priesthood and married. He remained an unsalaried lecturer (*Privatdozent*) at the university in Vienna until 1895, when he left Austria following the death of his wife. He spent a number of productive years after that in Florence, as his sight gradually deteriorated, but when World War I broke out, he moved to neutral Switzerland, where he died in 1917.

There are several reasons why Brentano plays an important role in our story: the first is his influence on the development of philosophy, largely through the many students who studied with him and were influenced by his thought at different moments of his life. The second reason involves the development of psychology. At the height of his academic activity, the period around 1890, it was widely thought that there were two quite different poles emerging about how psychology should be done—one with Brentano at its center, developing an "act psychology," and another with Wilhelm Wundt at its center, focusing on experimental psychology. We will turn to Wundt in chapter 4 and come back again to the question of how Brentano's ideas influenced the mind sciences in the late nineteenth century.[40] But we should emphasize now that Brentano viewed his own work as developing a scientific form of philosophy that included psychological questions. At one point, he wrote to Carl Stumpf, one of his first students, "I am at the moment wholly a metaphysician. I must confess that, after having been exclusively a psychologist for a few years, the change makes me happy."[41] We have to some degree separated philosophers from psychologists in this chapter and the next, and Brentano's case is one which makes clear how artificial that distinction can be.

Brentano drew a distinction between what he called *genetic* psychology and what he called *empirical* or *descriptive* psychology. Brentano's genetic psychology is much closer to what we think of as psychology today, certainly to the sort of psychology done in laboratories.[42] Genetic psychology seeks answers that involve the notion of causality and particular events in space and time: they "specify the conditions with which the individual phenomena are bound up causally."[43] Descriptive psychology, on the other hand, was exact and exceptionless, and we each have a privileged special relation with the principles of descriptive psychology. The modern linguist can think of this sort of psychology as including a speaker's intuitions about their own native language: in a sentence such as *The last king of France was bald*, a speaker may know that the sentence is composed of two parts, *The last king of France* and *was bald*, and that this is a grammatical sentence. Here is a sharper example to illustrate the difference between the domains of these two kinds of analysis. Imagine hearing someone say, "That's no way to . . . to . . .

to help a friend." The object of genetic psychology is a linguistic act that contains three successive occurrences of the word "to," and we might develop an account of when it is that a person hesitates in this way. The study of empirical psychology is a sentence, "That's no way to treat a friend," and how a speaker or a hearer analyzes and understands it. It was this side of Brentano's thought that led to phenomenology. If Brentano had heard of Saussure's distinction between language and speech, he would have put Saussure's language on the side of his empirical psychology.

Brentano recognized that a considerable part of the challenge in developing a descriptive psychology was understanding the relationship between wholes and parts. We have a term today, "mereology," for the study of this relationship; the word is due to Leśniewski, who was a student of Kasimir Twardowski, himself a student of Brentano's. At the most inclusive level, understanding the nature of individual consciousness requires understanding how it is that *all* of what we experience is ours, and forms a whole that we perceive as ours, as all ours. At lower and smaller levels too, we understand the objects of our thought in terms of their component properties and of how those component properties are integrated into a whole. The component properties may be bound together in quite different ways: a statue may have a head, body, and limbs, just as book has a jacket, a binding, and 236 numbered pages. A consonant such as *p* may be produced with the lips and involve the complete stoppage of air and a ceasing of the vibration of the vocal cords. The analysis of the sound into subgestures seems different from the analysis of a statue. The number 15 can be analyzed as 5 times 3, or as 12 plus 3. A mathematician may find 5 times 3 a much more interesting analysis, because he knows that the structure that multiplication imposes on whole numbers is much richer and more interesting than that imposed by addition, just as a linguist knows that the analysis of our earlier sentence into [[The last king [of France]] [was bald]] is richer and more interesting than the analysis [[The last king] [of France was bald]]. He wrote something that would make great sense to a phonologist:

> If someone believes in atoms he believes in particles [*Körperchen*] that cannot be dissolved into smaller bodies, but even in the case of such particles he may speak of halves, quarters, etc.: parts which, although not really separable, are yet distinguishable. We can call these latter distinctive [*distinktionelle*] parts. In human consciousness, too, there are also, apart from separable parts, mere

distinctive parts. And, in so far as the distinguishing continues further than real separability, one might speak of parts (or elements) of elements.[44]

Brentano argued by example that descriptive psychology had the responsibility to show the logical structure of the properties of various sorts of sensory perception. If we consider the visual field, for example, we find that objects have a position in our visual field, a shape, a size, and a color which itself can be analyzed in terms of brightness and saturation (and the list does not end there). This analysis is philosophical and psychological, and necessary (rather than contingent). Virtually all of Brentano's students, and most of his students' students, accepted this task as a major goal of psychology and philosophy, and this became a major component of phenomenology (as developed by Husserl and others), of Gestalt psychology (as developed by Köhler and others), and of phonology (by Jakobson and Trubetzkoy).

Brentano responded to Comte's concern about the reliability and trustworthiness of introspection in the development of a scientific psychology. Brentano proposed that a distinction needed to be drawn between *inner observation* (*innere Beobachtung*) and *inner perception* (*innere Wahrnehmung*), and this response would later be adopted by Wundt as well. The distinction was between an *observation* of an inner state or process, which requires a split of sorts in the consciousness of the subject, a distancing of the cognizer from the process observed, and an inner perception in which one is aware of where one is, mentally speaking, without an effort to provide an observer seated in a different mental location.[45]

Franz Brentano, the man, is also someone who can teach us something about the nature of rupture and continuity in the academic world. He was a charismatic teacher, and throughout his life inspired students to come work with him and to take him as a role model for their own lives. His lectures—whose importance cannot be overstated—were powerful, logical, and clear. Husserl, one of his most famous students, spoke with some emotion of how he saw Brentano when he listened to his lectures: "in every feature, in every movement, in his soulful, introspective eyes, filled with determination, in his whole manner, was expressed the consciousness of a great mission."[46] Is there a better description of charisma? And at the same time, Husserl recalled Brentano's language as "the language of dispassionate scientific discourse, though it did have a certain elevated and artistic style through which Brentano could express himself in a completely appropriate and natural way."[47] Brentano left an impression: "he stood before his young students like a seer of eternal truths and the prophet of an other-worldly realm." Even after Brentano's death, Husserl recalled the force of *attraction* to his former teacher: he wrote, "In spite of all my prejudices, I could not resist the power of his personality for long. I was soon fascinated and then overcome by the unique clarity and dialectical acuity of his explanations." Brentano was, Husserl noted, convinced of the truth of his philosophy. "In fact, his self-confidence was complete. The inner certainty that he was moving in the right direction and was founding a purely scientific philosophy never wavered," and developing his philosophy was "something he felt himself called to do, both from within and from above. I would like to call this absolutely doubt-free conviction of his mission the ultimate fact of his life. Without it one cannot understand nor rightly judge Brentano's personality."[48]

Still, Brentano was "very touchy about any deviation from his firmly held convictions," and he became "excited when he encountered criticisms of them, adhered rather rigidly to the already well defined formulations and aporetic proofs, and held out victoriously, thanks to his masterly dialectic, which, however, could leave the objector dissatisfied if he had based his argument on opposing original intuitions." And "no-one took it harder when his own firmly entrenched convictions were attacked."

And so the connections that form tight bonds, both personal and intellectual, at the beginning of a student's career evolve into forces that lend themselves to rupture. Husserl, again, explained straightforwardly how this happened in his relationship with his teacher: "At the beginning I was his enthusiastic pupil, and I never ceased to have the highest regard for him as a teacher; still, it was not to be that I should remain a member of his school." Husserl knew that he was going to move out and become an independent thinker. "I knew, however, how much it agitated him when people went their own way, even if they used his ideas as a starting point." *Even if?* Surely Husserl knew perfectly well that that was the worst possible case, from Brentano's point of view. "He could often be unjust in such situations; this is what happened to me, and it was painful."[49]

As we listen to Husserl's words about his relationship with his teacher Brentano, it may be hard for us not to think that Husserl should have had more forbearance in his interaction with Brentano, since, after all, it was Husserl who would become far more famous a century later. But of course Husserl had no way of knowing that. Like all of us, he was swimming in uncertain waters. And Husserl knew that he could not provide an argument for his point of view that Brentano would find persuasive. Husserl could give in to his teacher's criticisms, or he could set out on his own, even while he knew that Brentano had better arguments than he did, for the moment. He was obviously talking about himself when he wrote, "the person who is driven from within by unclarified and yet overpowering motives of thought, or who seeks to give expression to intuitions which are as yet conceptually incomprehensible and do not conform to the received theories, is not inclined to reveal his thoughts to someone who is convinced that his theories are right—and certainly not to a master logician like Brentano." We are left to conclude that Husserl tried, early on, and failed to engage Brentano in a conversation in which Husserl's ideas were something other than a heresy. He was not able to meet his teacher's standards for logical persuasion. "One's own lack of clarity is painful enough," Husserl went on. But he could neither convince Brentano that something was wrong in his teachings, nor persuade him that Husserl's alternatives made sense. "One finds oneself in the unfortunate position of neither being able to produce clear refutations nor being able to set forth anything sufficiently clear and definite." An unfortunate position, indeed: to be struck dumb in the presence of one's teacher. "My development was like that," wrote Husserl, "and this was the reason for a certain remoteness, although not a personal estrangement, from my teacher, which made close intellectual contact so difficult later on. Never, I must freely admit, was this his fault. He repeatedly made efforts to re-establish scientific relations. He must have felt that my great respect for him had never lessened during these decades. On the contrary, it has only increased."[50]

But then many years went by with each man going his own way. Towards the end of Brentano's life, while he was living in Florence, Husserl went to visit him there. Brentano was blind at that point, unable to read and able to write only if someone took dictation. His hair had turned gray, and his eyes had lost the gleam that had once captivated his students. Husserl could see that his former teacher was chafing under the conditions he had to live in, with rarely a colleague to speak to about philosophy. Husserl could listen, though.

> Once more I felt like a shy beginner before this towering, powerful intellect. I preferred to listen rather than speak myself. And how great, how beautifully and firmly articulate, was the speech that poured out.
>
> Once, however, he himself wanted to listen, and without ever interrupting me with objections, he let me speak about the significance of the phenomenological

> method of investigation and my old fight against psychologism. We did not reach any agreement.
>
> And perhaps some of the fault lies with me. I was handicapped by the inner conviction that he, having become firmly entrenched in his way of looking at things, and having established a firm system of concepts and arguments, was no longer flexible enough to be able to understand the necessity of the changes in his basic intuitions which had been so compelling to me.[51]

Husserl never lost his love for Brentano the teacher. In that final meeting in Florence, Husserl found that Brentano had "a slight aura of transfiguration, as though he no longer belonged entirely to this world and as though he already half lived in that higher world he believed in so firmly." The world would soon lose a brilliant thinker and teacher. Husserl ended his note with these words: "This is how he lives on in my memory—as a figure from a higher world."[52]

Of Brentano's students who became philosophers, Husserl was the best known, and in chapter 7, we will look in greater detail at his ideas and his influence. Among Brentano's students who became psychologists, one of the most famous was Carl Stumpf, whose work we will consider in the next chapter. Stumpf's recollections of Brentano the teacher share something of the heartfelt emotions of Husserl's: "I wish to express the love and gratitude which I owe to my great teacher," he wrote, "which I will retain until the day I die. The close relationships he established with his students and which he was so eager to maintain played a more important part in his inner life than is the case with many other thinkers."[53] Stumpf spoke of the metamorphosis that Brentano produced in him: he had started at the university expecting to study law, but after some weeks that resolution weakened. "Before Christmas I sought him out to inform him of my intention of choosing philosophy and theology as my life's work. I even wanted to follow him into the priesthood, so much of an example had he set for me."[54] From that day on, Brentano spent many hours walking and talking with Stumpf. As Stumpf's professional stature grew, and the two no longer lived in the same city, they naturally grew apart intellectually, and Stumpf bore some of the burden of having been Brentano's first student; he wrote about "a certain touchiness on Brentano's part toward dissension that he thought to be unfounded," echoing Husserl's sense as well. If Brentano "encountered basic intuitions in his students' publications which were considerably different from his own, and which were not thoroughly justified and defended on the spot, he was inclined to consider them at first as unmotivated, arbitrary statements even though they may have been subject to several years' thorough study or may have matured imperceptibly without one's having been expressly aware of it. Occasional ill-feelings were unavoidable in the face of this."[55]

Brentano's philosophical ideas were central in the development of the third tradition that filled out the nineteenth century's response to Kant. The first, as we have seen, was the neo-Kantian tradition that shared with Kant the belief that philosophy could and should develop a deeper understanding of *a priori* truths about the world and our place in it; in various ways, history and culture were integrated into these neo-Kantian perspectives in ways that would not have pleased Kant himself. The second tradition was positivism: though Kant shared with positivism a deep respect for the work of contemporary science, positivists were gardeners of a sort, intent on clearing out the unwanted shrubbery of philosophical discourse, keeping only the part that passed the inspection of the observational sciences. Brentano was instrumental in the development of the third tradition, the linguistic, or semantical, tradition. This development looked to language and its analysis to provide a different kind of explanation for the statements that Kant had taken to be both synthetic and known *a priori*. It is to our language we must turn, held the third tradition, if we want to understand our statements and our beliefs.[56]

Brentano's intellect and personality left a deep mark on Tomáš Masaryk, sparking an interest in English empiricism, French positivism, and Aristotle.[57] After defending his dissertation with Brentano, who had encouraged him to study Auguste Comte, Masaryk moved to Leipzig, where he studied with Wilhelm Wundt and others. Leipzig in the 1870s: this was a crucial nexus, and we will find all of the mind sciences deeply influenced by what went on there and then. We have already been there, to the largest university in Germany at the time: this is the place where the Neogrammarians burst on the scene. And Masaryk met Edmund Husserl there in Leipzig; Husserl was nine years Masaryk's junior, but the two were both Moravians, which seemed to matter to them and give them a reason to bond. The two of them attended Wundt's lectures together and participated together in the Academic Philosophical Association, a group which was heavily influenced by the philosopher Richard Avenarius. Masaryk then moved back to Vienna to work again with Brentano.[58] Masaryk became chair of the department of philosophy at the Czech university in Prague when it was founded, in 1882—the university in Prague, called the Charles-Ferdinand University, was divided into a Czech-speaking and a

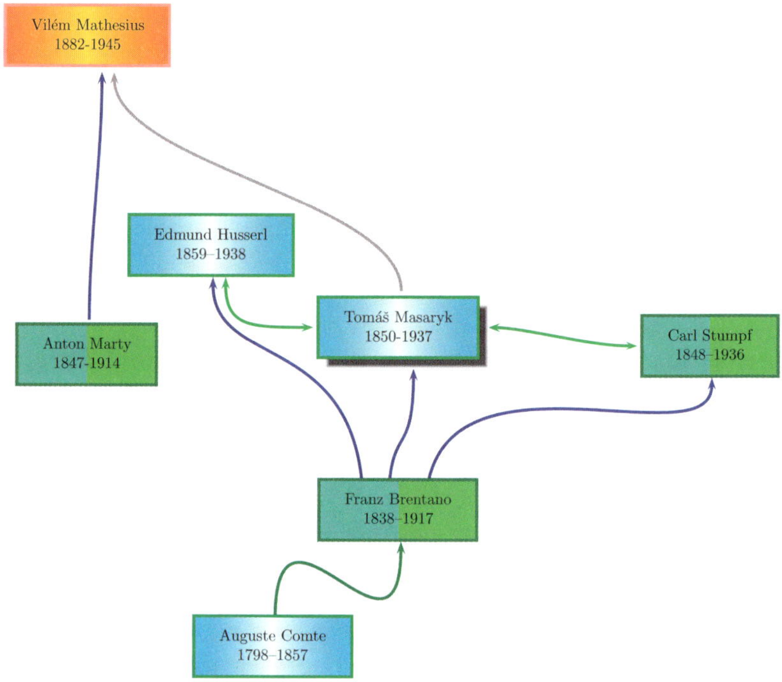

FIGURE 3.2. Tomáš Masaryk

German-speaking university that year. There he became friends with two Brentano students: Carl Stumpf and Anton Marty, who was also lecturing in Prague.[59] Jakobson would later note the important impact that Marty had on Masaryk. As we will see below as well, Masaryk was the teacher of Vilém Mathesius, founder of the Prague Linguistic Circle. Still later, during the years leading up to the First World War, Masaryk spent a great deal of his time on party and national politics, and he became president of Czechoslovakia on October 28, 1918, at the end of the war, when the Austro-Hungarian Empire was dissolved, and an independent Czechoslovakia established.[60]

In his scientific work, Masaryk emphasized the importance of maintaining a distinction between static and dynamic analysis, two complementary perspectives that, as we have seen, Auguste Comte had emphasized earlier in the century, and which are essentially no different from what Saussure would later call synchronic and diachronic analysis.

Christian von Ehrenfels

Christian von Ehrenfels was an aristocratic Austrian who studied with Brentano in Vienna, and with Brentano's student Alexius Meinong both in Vienna and in Graz. After 1896, Ehrenfels was professor of philosophy at Charles University, the German university of Prague.[61]

His most famous work was *Über Gestaltqualitäten* (*On Gestalt Qualities*), published in 1890, in which he developed the major Brentanian theme of the logical relationship of parts and wholes—and he developed the idea of a shape, or *Gestalt*, that Ernst Mach had begun to write about in 1865. As we noted just above, Mach talked about a *Gestalt* as the thing which is important but which is not the sensory perception itself. Ehrenfels came back to that, with a nod of the head to Mach, and made the study of these Gestalts the centerpiece of his paper, which would ultimately become one of the most influential papers in the entire history of psychology. Now, we often say in casual speech that the whole is not the sum of its parts, but the task was to say *exactly* what the difference was between the whole and the collection of its parts. Ehrenfels took the example of a melody as the perfect example of a whole that is so much more than its parts (though a word or a sentence is a good example too). A melody is easily recognized as the same even if it is raised or lowered by a musical interval. What is it that is the melody, then, if all of the component notes have changed? It is something relational that ties the parts together.[62]

Ehrenfels's article begins, right from the first paragraph, with the recognition that the starting point of his work lay in Mach's *Analysis of Sensation*. Ehrenfels wrote to a friend, "I sent Mach 'Gestalt Qualities' and he replied in a friendly manner that he had already given the main thoughts in 1865 in volume 46 of Fichte's journal, and had expressed them in a more psychological way."[63] This seems like an unusually gracious recognition of intellectual continuity, but it also seems that it was, alas, dismissed by Mach with a toss of a hand.

> The theory of gestalt qualities began with the attempt to answer a question: What is melody. First and most obvious answer: the sum of the individual notes which make up the melody. But opposed to this is the fact that the same melody may be made up of quite different groups of notes. . . . Mach, who was struck by this fact, drew from it the conclusion that the essence of melody must reside in a sum of special sensations which as note sensations (*Tonempfindungen*) accompany the notes.[64]

Mach's great success was in perceiving that his focus on the individual sensations, which he was certain formed the basis of our perception, was inadequate to provide an account of our perceptions, and he, perhaps grudgingly, acknowledged that more was needed. Ehrenfels recognized that what was there was something else, something past the sensations. Instead of a melody, let us take as our example a sequence of integers: 7, 9, 11, 13, 15. We can remember them and recite them, and in all likelihood you will recognize that there was a difference of 2 between each of the successive numbers. From Ehrenfels's point of view, we remember not just those five numbers, but a sixth also: the number 2. But it is not until we come to Ehrenfels's student Max Wertheimer (and the psychologists who got on board with Wertheimer's program) that something much closer to the real story would come about: the Gestalt was based on an active principle that binds all the component sensations by an activity that is capable of generating them.

Ehrenfels was the first teacher of Max Wertheimer, who would go on to be a graduate student in Berlin with Stumpf, and then in Würzburg with Oswald Külpe, where he finished his degree; Wertheimer would later develop Gestalt psychology, which will be one of the principal themes of chapter 5. Gestalt psychology would succeed better than any preceding school of psychology in making explicit what the active principles are that dynamically organize the perceived world. Ehrenfels provided an important step forward in emphasizing the logical gap between the perception of the parts and the perception of the whole, which can subjectively be far more important than the parts.

Logic: Boole, Frege, Russell

The middle of the nineteenth century saw a new trajectory in the study of logic with the work of George Boole, Gottlob Frege, and Bertrand Russell. Before we turn to what they proposed, we need to stop a moment and reflect upon what logic is, and what it is intended to accomplish.[65]

Today, we might say that the goal of logic to make precise what counts as valid inference, from one thought to another, or one sentence to another, or one proposition to another. But in the longer tradition of logic, this is just a part of what is covered by the term *logic*. Logic is the discipline more generally *that studies what makes thought possible*. There are at least three areas that need to be covered by this: first, the nature of valid inference; second, the categories of thoughts; and third, the modes of existence. By "categories," we mean the *sorts* of things that we need in order to think, such as objects, properties, and relations. By "modes of existence," we mean the various ways in which statements can be used in their descriptive function, of which the most everyday ways are related to time: a statement that *the front door was locked at 9 p.m. last night* does not say anything about the state of the lock right now. Modes of existence can also be nontemporal: *cars stop at red lights* is true in a fashion that is not rendered false by noting that this car or that one went through a red light. Statements can be about a more abstract reality than what is observed in space and time—though positivists will urge us not to be deceived about that way of speaking, to be sure. In short, a mode of existence could be one which specifies socially accepted actions (like stopping at red lights), or rules of a game (like moving a bishop in chess only on the diagonal).

Thinking about categories in this sense is one of the oldest traditions in philosophy, rooted in ideas that Aristotle first laid out in such works as *Categories* and *Analytics*, and that others famously treated later, like Kant, in *Critique of Pure Reason*, and Husserl, as we will see.

George Boole

George Boole was the author of *An Investigation of the Laws of Thought*, which he published in 1854, at the age of 39. He was born in the north of England to a family of modest means but one which surrounded him from an early age with opportunities to learn languages, science, and mathematics. For a number of years, he was a teacher and educator, but in 1849, when he was 34, he assumed a position at Queen's College Cork, in Ireland, where he spent the last 15 years of his life.

He spent his thirties exploring the development of the laws of thought (which is to say, logic) through the intellectual tools of mathematics. The laws that he referred in the title of his book were divided into those of logic and those of probability, and for both it was his intent to develop a calculus. He was at pains to note the genealogy of this work: on the logic side, he traced it back to Aristotle, and to many others since: Abelard, Ramus, Descartes, Bacon, and Locke. Probability, he noted, finds its roots in the work of Pascal and Laplace, and while its origins lay in the study of gambling and insurance, its importance for understanding the laws of thought was far greater. For this was his goal: to understand the very laws of thought:

> To unfold the secret laws and relations of those high faculties of thought by which all beyond the merely perceptive knowledge of the world and of ourselves is attained or matured, is an object which does not stand in need of commendation to a rational mind.[66]

"To unfold the secret laws"! This was a book that promised to do a great deal, systematically investigating traditional mathematics, and a new sort of mathematics based on the old one but which aimed to develop an *algebra*, a formal system which resembled arithmetic and which would model thought and logic. Traditional algebra uses variables to stand for measurements of quantity; this innovation was made by François Viète in the late sixteenth century and quickly became standard practice. Algebra became a language and a symbolic system with a grammar; understanding that grammar soon became a standard part of being an educated person. What Boole did was to say that much the same language could be retained, with some significant modifications, to be sure, but still and all, a new dialect of algebra could be developed that said reasonable things about logic and inference.

Boole divided propositions into two sorts, primary and secondary. Primary propositions are about objects, or at least they are in simple cases: "all dogs are mammals" is a primary proposition, as is "Socrates is mortal." Secondary propositions are formally complex, and made up of propositional parts: "if you leave, you will miss the ice cream" is a secondary proposition.

Boole began his discussion with classes of objects, and used lowercase letters, such as x and y, to represent them. "If the name is 'men,' for instance, let x represent 'all men,' or the class of 'men.'" And adjectives worked with the same kind of algebra as nouns: "if an adjective, as 'good,' is employed as a term of description, let us represent by a letter, as y, all things to which the description 'good' is applicable, i.e., 'all good things,' or the class of 'good things.'" And then the combination xy is said to refer to the objects that satisfy both x and y's description: xy stands for "good men." It follows immediately, Boole wrote, that $xy = yx$. The phrase *navigable rivers that are estuaries*, with x, y, and z defined in the appropriate way, could be represented as xyz, or zyx, or any linear combination of the three variables. The commutativity (and he used that term) of this operation is, he said, a law of thought "and not, properly, a law of things." And what is more, as a law of thought, "it is actually developed in a law of Language." Boole noted that poetic diction allows inverted orders of nouns and adjectives, as in "the rising world of *waters dark and deep*." We are tempted to say that Boole was approaching dangerously close to the point of doing bad linguistics. He proposed that these inverted forms were not simply the fruits of poetry: they were "sanctioned by the intimate laws of thought," and it was just "convenience" (whatever that might mean) that leads us to rarely exercise our noun-adjective word order more often. But he also noted that if $x=y$, then $xy=x$, from which it follows that $xx=x$, for any x. This leads him to some bad linguistics again: "to say, 'good, good,' in relation to any subject, though a cumbrous and useless pleonasms, is the same as to say 'good.'" Well, no: we can say, "there's good, and good"—just as we say "there's *good*, and then there's *good*," and the two *good*s are taken to mean quite different *good*s: different amounts of *good*, perhaps. In any event, Boole went on to write that classes of objects, so long as they are disjoint and share no elements in common, can be combined with "and" and "or," as when we say "trees and minerals," or "men and women."[67]

There are further signs, those which express *relations*, and Boole argued that all verbs would be represented by such signs. *But*, he wrote, we really only need one such symbol, because all verbs can be paraphrased with *is* or *are*: *Caesar conquered the Gauls* can be rephrased as *Caesar is he who conquered the Gauls,* and we might as well just use the symbol "=" to mark *is* and *are*.

Boole turned then to secondary propositions. The clearest examples of secondary propositions are of the form *either X is true or Y is true*, or *if X is true, then Y is true*. Boole argued that these secondary propositions are fundamentally all about time: the second example can be construed as saying that *at all times at which X is true, Y is true*.[68] In the end, Boole gave every indication that he was quite comfortable with the Kantian category of time as a basis for the meaning of secondary propositions. He was much less certain that there is a corresponding Kantian intuition of space that is the basis for primary propositions, but he gave the very strong impression that something *like* that must be correct.

Boole discussed the logical distinction between *intension* and *extension*, a difference that had long been made in the literature on logic, and one that would play an important role in the decades to come. Frege, whose work we will turn to shortly, chose the German words *Sinn* and

Bedeutung, and others since have used yet other terms. Husserl, at the turn of the next century, would develop a vaster analysis of the ways in which intensions can interact and relate to each other. In phonology, as we will see, the development of an intensional logic of classes was the great contribution of Trubetzkoy, which today we call *features* in English. (We have noted that Trubetzkoy, writing in German, used the word *Merkmal*, a term already used by Ehrenfels, and in the context of intensional logic.)

We probably could not do better than Boole in providing an account of the distinction:

> According to a recognised division the concept of a class of things may be considered 1st with reference to its extension as a whole made up of parts, 2ndly with reference to its intension as formed by the union or combination of qualities common to all the individuals which it comprehends.[69]

And so, he wrote, "minerals" is a concept, and it can be viewed as a "class of things including gold, silver, iron, aluminium, etc."—and this is viewing it in extension. The concept "minerals" can be viewed in intension, and so involves "the qualities of ductility, fisibility, a peculiar lustre etc. common to all the individuals of the class." Extension is particularly apt for understanding the operation represented by the English word *and*, wrote Boole, and he called this *addition*; then he described subtraction, "that operation by which from the concepts of a whole and the concept of one of its parts we form the concept of the other part, as when from the concept 'stars' and the concept 'planets' we form the concept 'stars which are not planets' or 'stars except planets'"—as the word *except* can be used to indicate.

But there is much more to say about "operations founded on intension," such as *composition*, whereby the concept "white men" is formed by composing the concept "men" and the concept "white things." Boole called the opposite of this *abstraction*, and noted that there are some subtleties and complexities here. Suppose we take the concept of "white flowers," and that of "flowers." Can we reconstruct that of "white things"? Yes, we could, and having done so, composition could bring us from "white things" and "flowers" to "white flowers." But: "this concept 'white things' is not the only one from which by composition with that of flowers the concept 'white flowers' may be formed." The process does not yield a unique concept. If instead of "white things" we had taken the class consisting of "white things" and also ("and," now, in the extensional sense:

we are making a bigger class of things) "red leaves" (that was Boole's example), we will also arrive at "white flowers" if we compose it with "white things." Here is how Boole put it:

> To the concept "white things" we might add the concept of any class of things possessing neither whiteness nor the floral character or of any indefinite portion of such a class of things and still the resulting concept would be one which by composition with [that] of "flower" would generate the concept "white flowers." Thus if from that class of things which consists of "white things together with red leaves" we mentally select those individuals which answer to the description "flower" we arrive at the concept "white flowers."[70]

The upshot is this: abstraction may well be understood as the inverse of composition, but what we get out of it is a whole class of concepts—just as in the particular example Boole gave, we can derive not simply "white things," but an infinite number of variations on that.

We have noted that Boole was quite aware that it was difficult to know just where the study of thought ended and the study of language began. Early on, he observed,

> That Language is an instrument of human reason, and not merely a medium for the expression of thought, is a truth generally admitted. It is proposed in this chapter to inquire what it is that renders Language thus subservient to the most important of our intellectual faculties. In the various steps of this inquiry we shall be led to consider the constitution of Language, considered as a system adapted to an end or purpose; to investigate its elements; [and] to seek to determine their mutual relation and dependence. . . .[71]

He then noted that there was no great loss suffered if it turns out that the conclusions he draws about thought are more properly said to be about language; the conclusions will still be valid (if slightly misplaced). But he made then an interesting point: the conclusions he expects to draw will hold, he thought, for all speakers of all languages, which suggests that what is important is thought, not language:

> Nor could we easily conceive, that the unnumbered tongues and dialects of the earth should have preserved through a long succession of ages so much that is common and universal, were we not assured of the existence of some deep foundation of their agreement in the laws of the mind itself.

Boole certainly expected to have significant things to say about grammar and the logical structure of language:

> The substantive, the adjective, and the verb, together with the particles *and*, *except*, we have already considered. The pronoun may be regarded as a particular form of the substantive or the adjective. The adverb modifies the meaning of the verb, but does not affect its nature. Prepositions contribute to the expression of circumstance or relation, and thus tend to give precision and detail to the meaning of the literal symbols. The conjunctions *if*, *either*, *or*, are used chiefly in the expression of relation among propositions, and it will hereafter be shown that the same relations can be completely expressed by elementary symbols analogous in interpretation, and identical in form and law with the symbols whose use and meaning have been explained in this Chapter. As to any remaining elements of speech, it will, upon examination, be found that they are used either to give a more definite significance to the terms of discourse, and thus enter into the interpretation of the literal symbols already considered, or to express some emotion or state of feeling accompanying the utterance of a proposition, and thus do not belong to the province of the understanding, with which alone our present concern lies. Experience of its use will testify to the sufficiency of the classification which has been adopted.

Gottlob Frege

Gottlob Frege was born in 1848, in what would soon be northern Germany. After studying philosophy and mathematics (as well as other topics), he began lecturing at the University of Jena in 1874, and he would remain there until 1918. During his lifetime, he felt that his fame was limited and that his work little commented upon. His work was reviewed by his peers in the serious journals, but reviewers saw nothing revolutionary in it. Frege was not given to explicitly criticizing views that preceded his, and he was taken to task for not appreciating Boole's system. In the decades since, the world has come to his door: his lifework has had enormous influence.

During his lifetime, Frege's work influenced Bertrand Russell greatly, and Whitehead called Frege the greatest logician of the nineteenth century at the 1936 meeting where the Association for Symbolic Logic was established, a decade after Frege's death.[72] And sitting in on his lectures in Jena in 1914 was Rudolf Carnap, who was to be an enormously influential in the years to come, and it was through Carnap, in part, that Frege's ideas were developed.[73]

By the time Carnap attended his weekly one-hour lecture, Frege was no longer a young man: he was 62, and Carnap thought that he "looked old beyond his years." He was retiring and shy, faced the blackboard rather than the class (which had just a small number of students), and filled the board with his strange formalisms. Never would a student raise a question, either in class or afterwards, and Frege's lectures were just that: there was no discussion, simply lecture by the professor.

Carnap's friend Wilhelm Flitner, whom Carnap had roped into taking Frege's classes with him, noted that Frege would walk on the street with his eyes cast down to the ground and one hand on his back. Flitner spoke only a few insignificant words to him over the several years he took courses with him; Carnap, he said, never did even that.[74]

Looking back, Carnap saw mostly sadness. "It was obvious," he wrote later, "that Frege was deeply disappointed and sometimes bitter about this dead silence. No publishing house was willing to bring out his main work, the two volumes of *Gesetze der Arithmetik*; he had it printed at his own expense. In addition, there was the disappointment over Russell's discovery of the famous antinomy which occurs both in Frege's system and in Cantor's set theory."[75] Many years after Frege had passed away, Bertrand Russell wrote an enormously successful history of Western philosophy, and he wrote into it an important role for Frege. He noted that Frege "remained wholly without recognition until I drew attention to him in 1903,"[76] when Russell published his *Principles of Mathematics*.[77]

It was in 1879, when he was not quite 30 years old, that Frege published a revolutionary work, his *Begriffsschrift* (or "concept-notation") that attempted to go well beyond what George Boole had accomplished. He was not at all satisfied with what logic had produced before him and was willing to put forward a set of new ideas to shake the tree of logic.

> If we view Boole's formal system as a whole, we see that it consists of dressing up abstract logic in the clothing of algebraic symbols; it is not appropriate for the expression of content and that was not its aim, in any event. But that is precisely my aim. I want to blend the several signs that I have introduced with the mathematical signs to arrive at a single formal system. *The symbols that already exist would correspond roughly to the stems of words, while the signs I introduce may be compared to suffixes and grammatical elements which establish the logical relationships between the contents of the stems.*[78]

LOGICAL FORM

While Frege's actual formal system—his formalism, let us say—has not stood the test of time, much of his conceptual project has. Frege's goal was to establish that there was a *logical form* to a sentence, one which is far from obvious, and that project has had an enormous impact on how language has been analyzed in the years since. The three most important ideas that he developed were these: first, that the fundamental structure of a sentence should be understood as a predicate and a set of arguments, just as a function in mathematics is. The arguments of a predicate are typically nouns or noun phrases. Second, it is possible for the meaning of a sentence to bind together two (or even more) arguments from a logical point of view. And third, quantifiers such as *all* and *every* have to be understood not as playing a role parallel to that of words such as *this* or *red* in the logical form of a sentence, but rather as *operators*, which take a whole sentence as its scope, "binding" some of the variables within. We will take a look at each of those in a bit more detail.

Frege was at pains to show that it is not sufficient to take the logical form of a sentence to be the attribution of a predicate to a subject—and he insisted that the logician and the philosopher must uncover the logical form of a sentence formulated in a natural language. We need to understand what the view was that he was trying to move beyond. Till then, philosophers had largely been in agreement with traditional grammars: they emphasized the division of a sentence into subject and predicate. A sentence like *Socrates is bald* consists of a subject and a predicate which expresses something said about the subject. The same was taken to be true of a sentence like *The Yankees beat the Mets*: it is formed of a subject *the Yankees* and a predicate *beat the Mets*. At the same time, the sentence *The Mets were beaten by the Yankees* is formed of a subject *the Mets* and a predicate *were beaten by the Yankees*. This was essential to Kant's notion of *analyticity*, which pertained to sentences in which the predicate is contained in the subject. Frege said that from the point of view of the meaning of the sentences, the active and the passive sentences were the same, and a better way to specify what they have in common is to think of a mathematical function of two variables, as when we define a mathematical function $f(x,y)=x^2+2y$, with the understanding that in the case of the baseball sentence, our function has as its arguments not two numbers, but two baseball teams.[79] A more perspicuous representation would then be: *beat (the Yankees, the Mets)*, a representation that is valid regardless of whether the *the Yankees* is chosen as the subject, in the active voice, or *the Mets* is chosen, in the passive voice. What Frege actually wrote is a bit more interesting: "In the mind of the speaker the subject is ordinarily the main argument; the next in importance often appears as object." Semantically driven grammatical theories still appeal to such statements. "Through the choice between [grammatical] forms, such as active-passive, or between words, such as 'heavier'-'lighter' and 'give'-'receive', ordinary language is free to allow this or that component of the sentence to appear as main argument at will, a freedom that, however, is restricted by the scarcity of words."[80]

The binding of variables is something that is perfectly clear in our mathematical notation, but which needed to be made explicit when Frege tried to lay bare the logical form lying behind a sentence. From a mathematical point of view, if we define a function $f(x, y)$ as x^2+2y, then we know that $f(2, 1)=4+2=6$, $f(1, 2)=1+4=5$, and $f(2, 2)=4+4=8$. The notation makes it perfectly clear that x and y are conceptually independent, and we can choose any value we want for either variable; those values can be different, or they can be equal. If we wish to, we can bind those two variables to create a function that has but a single variable; the natural way to express this is to say that we define a new function $g(x)=f(x, x)$. The two variables in the function $f(x, y)$ are now bound, and we can also say that $g(x)=x^2+2x$.

Something very similar is true in language, but in a less obvious way. Let us consider a function in language that has two arguments and try to find one in which the two arguments can perfectly well be different, or perfectly well be the same. We can say *John saw Kim* or *John saw himself*, just as we can say *Paul killed Kim* or *Paul killed himself*.[81] Frege proposed that a predicate can be understood as having a single variable which binds two argument positions in a function, as we might express with an expression like *for someone to kill himself*. A similar predicate in which two distinct variables bind the two argument positions could be expressed as *for someone to kill someone (else)*.[82]

The grammar of natural language may not make explicit where all of the variables in a sentence are. In the sentence *John tried to open the safe*, there are two logical positions bound to the same variable: the subject of *try* and the subject of *open*. And the discrepancy can be even greater. Consider the statement *no integer is greater than its square*, and compare it to *my suitcase is larger than its handle*. The sentence about my suitcase is much more straightforward: we can identify what object we are talking about by the description I give of it, and we can measure its size, and do the same thing with its handle, and the sentence makes the assertion that the first is larger than the second. But the sentence *no integer is larger than its square* is not about an object (or a set) "no integer"; what is really involved is a property that we could describe as "being greater than one's square," or the property that x would have just in case $x > x^2$. The sentence has the logical structure of denying that there are any integers that have that property: there are no integers x for which $x > x^2$. That is quite different from the suitcase sentence, which is about one very specific suitcase.

It is easy to lose sight of the fact that Frege's accomplishment lay not only in the solution of a set of problems, but also in the *selection* of the problems that he solved. Consider three words that can be used to conjoin two sentences: *and*, *or*, and *but*. Of the three, *and* is the easiest to deal with; few will object if we analyze *S and T* as being true just in case both *S* and *T* are individually true. *Or* is more problematic, because in some cases, it seems that a sentence which is formed by connecting two sentences by *or* is intended to be false if both of the connected sentences are true; this is the kind of *or* we call an *exclusive*, as in *You can learn to put your things away when you are done with them, or you can live like a slob*, while in most cases, a sentence built up from *or* connecting two sentences is true even if both sentences are true (*You left the hall light on last night, or someone slipped into the house while we were asleep*); Frege decided that he would ignore the first meaning and focus on the second. And the word *but*? Frege wrote, "The distinction between 'and' and 'but' is of the kind that is not expressed in the present ideography. The speaker uses 'but' when he wants to hint that what follows is different from what one might at first expect."[83]

From our vantage point today, we can see places where Frege's insights were more fragile than they appeared to be. For example, there are many examples in natural language when it is not clear how many arguments appear in a given sentence. The case of *The Yankees beat the Mets* is a clear one, containing two noun phrases that each play a similar role in the action described. But what about *John drinks a little bit too much*? There is no apparent direct object, though we know perfectly well that there is an implied object: John is drinking too many glasses of liquids that have too much alcohol in them. That paraphrase reassures us that there is another usage of *drink* in which an overt noun phrase is present, suggesting that the logical form of our first sentence has two arguments—one for the drinker, and one for what is being drunk. But we may be fooling ourselves. *John drinks a little bit too much* seems quite parallel to *he sleeps too much and he works too much*, and neither of *those* sentences has easy and plausible paraphrases with objects. Even a simple sentence like *John has a car*, which is quite plausibly viewed as describing a relation between two things, John and his car, is not so simple. We cannot say *Which car does John have?*, though we can say *What kind of car does John have?*, suggesting that *John has a car* describes a relation between John and a kind of object, or perhaps a kind of car—in any event, not a particular car; *John has a car* does not have the same logical structure as *John bought a car.* All of this is simply to say: Frege's decisions about what aspects of natural language he would clean up and deal with in his logical form were decisions, not discoveries, and he opted for the analysis of types of language usage which are particularly prominent in discussions of mathematics, as others after him chose to look at aspects of language which were prominent in philosophy. If Frege were reading this, he would probably protest that he never claimed that he had provided a tool that solved all problems. He was quite clear about the benefits that accrue to incremental improvements in our analysis of language. We too should bear in mind that Frege would not have claimed that he had discovered the true logical form of sentences; he had taken some steps towards uncovering that logical form, but many others would remain to be taken. Furthermore, it would be a serious mistake to come away from our discussion thinking that Frege hoped that we could learn about logic by studying language *per se*. He made his view on this perfectly clear in a letter he wrote to Edmund Husserl:

> It cannot be the object of logic to investigate language and to determine what lies in linguistic gaps. Someone who wants to learn logic from language is like an adult who wants to learn thinking from a child.[84]

MENTAL ACTS AND THEIR OBJECTS

Although readers of the published literature would not encounter the term *psychologism* until later in the century, the groundwork that made the term not only possible but necessary arose during the period that followed Kant and Hegel.[85] Like all of the interesting terms that we care about in this book, the word *psychologism* has been used in a raft of ways, and we cannot expect too much consistency of usage. But this we *can* say:

psychologism refers to a perspective that sees the major grounding of studies of mind in the empirical—read *laboratory*, for the most part—study of human faculties in particular situations. The basic argument in favor of psychologism? Everything we can learn about how people reason will inform us about the nature and possibilities of reasoning, and if laboratory science can teach us about atoms and molecules, surely it can teach us about how we reason. The basic argument against psychologism? People reason well and reason poorly, depending on who they are, what context they are in, and how much they know. Studying what they do and how they reason in context combines in an unholy mess what constitutes good reasoning and what bad reasoning—and bad reasoning is really not what we are interested in.

Bear in mind that the term *psychologism* is almost always used as an accusation, with the sure sense that no one (except perhaps a psychologist!) would chose to pursue psychologism in their account of human knowledge. In one way or another, the charge of psychologism comes down to this: there are deep things to explain about the mind, and psychologism is the easiest way to misunderstand how deep those things are, and to fail to give an appropriately deep explanation.

Frege's perspective on mental acts (like judging a brownie to be tasty or an argument to be convincing) stands out against the dominant Lockean tradition: Frege always emphasized the importance of distinguishing psychological acts from the content of those acts. My act of judgment is different from yours (for example, it typically occurs at a different time), but the content of the judgments may be exactly the same. We may feel exactly the same way about a given brownie, or a published paper. Frege was thus drawn to the conclusion that to understand a sentence required an understanding that there was something about it (and its meaning) that went well beyond anything that was psychological and subjective, that there was something interpersonal, and perhaps even objective, about it. These other things are propositions (though Frege used the word *Denken*, normally translated as *thoughts*).

Bertrand Russell and his antinomy

Bertrand Russell was a philosopher who would have tremendous influence in several areas, most notably the development of philosophy in the first half of the twentieth century, and we will return to his work in chapter 8. But we must point out here the importance of a discovery that he made, finding a deep flaw in Frege's system of logic in June 1901, when Russell was 29 years old. He realized that it was not possible to allow notions like "belonging to a set" to be used uncritically and without reservation, at pain of finding oneself in self-contradiction. In talking about sets, it seemed natural both to say that one set is a subset of another—the even numbers are a subset, for example, of the whole numbers—and also, in other cases, that one set is a *member*, not a subset, of another set. For example, we could define a set of all the real numbers between 0 and 1, another of all the real numbers between 1 and 2, and so on, up to those between 9 and 10—and then define a set containing exactly those 10 sets. The larger set has 10 members; each element in that larger set has an infinite number of elements.

But if we allow ourselves to talk about one set being a member of another set, Russell realized that without some other restriction, we will arrive at a contradiction, because nothing will stop us from asking questions about whether a particular set is a member of itself. Suppose we define a set to be *normal* if it is not a member of itself, and just about any set one has ever thought about is normal, in this sense. The set of integers is not a member of itself, for example; it is normal, because it contains integers (all the integers, in fact), but no sets of integers.

But what about the set of all normal sets, *N*—is it normal? We cannot answer the question without falling into a contradiction, which is tantamount to saying that even *asking* the question lands us in a contradiction. If it is normal, then it is not a member of itself, by the definition we set up for *normal*. But by the very definition of how we define sets (we say something is a member of a set if it satisfied the property we use to define that set), that set *N is* a member of itself, because it is a member of the set of all normal sets. In just the same way, we land in a contradiction if we start by assuming that *N* is *not* normal. If it is not normal, then by the definition of *normal*, it is a member of itself. But if *N* is not normal, that also means that *N* is not a member of *N*. And we arrive at the same contradiction.[86]

Frege's effort to provide a solid logical foundation for set theory, and arithmetic, came to a standstill once Frege realized the import of Russell's contradiction. Russell sent Frege a letter in 1902, outlining the nature of the problem, ending with the statement, "in your works I find the best I know of our time, and therefore I have permitted myself to express my deep respect for you." Many years later, when asked if the entirety of his letter might be published, Russell wrote,

> As I think about acts of integrity and grace, I realize that there is nothing in my knowledge to compare with Frege's dedication to truth. His entire life's work was on the verge of completion, much of his work had been ignored to the benefit of men infinitely less capable, his second volume was about to be published, and upon finding out that his fundamental assumption was in error, he responded with intellectual pleasure clearly submerging any feelings of personal disappointment.[87]

These efforts by Boole and Frege were two, but only two, of the most important steps to invigorate the design of logical systems in the nineteenth century. We see a great deal more effort made to be conscious and aware of the characteristics of the language used in the logical analysis and an equally great awareness of the subtleties and the sometimes treacherous features of natural language. *Treacherous*? That would only be the logician's perspective, the logician who has certain expectations of how a communicative system ought to work and who feels a bit let down when natural language does not live up to those expectations.

The systems that the logician explored, and the new systems that he designed, were things that existed outside of time and space, and they were in that sense abstract. At the same time, psychologists were exploring the abilities as well as the limits and frailties of the ways in which human beings acted and reasoned in real life and in the laboratory. This was a time of tremendous growth in the field of psychology. And it is to this activity that we will now turn.

CHAPTER FOUR

The Mind Has a Body

Psychology and Intelligent Machines in the Nineteenth Century

If you ask a psychologist who knows a bit about the history of her discipline when it all began, chances are good that she will say that it all started during the second half of the nineteenth century, and that psychology emerged out of two fields: philosophy and, especially in Germany, physiology.[1]

There is much truth to that. Nowadays, the opening of Wilhelm Wundt's experimental laboratory is often taken as an icon symbolizing this shift, and the year 1879, when the lab opened its doors at the University of Leipzig, is taken as a symbolic starting moment for psychology. But there was no particular event in 1879 that would have made someone look up and say that a new discipline had just begun. Quite the contrary: we have already observed that Whitney, quite a few years earlier, discussed the relationship of linguistics and psychology.

The tacit agreement that psychology was born when Wundt's lab was opened is no doubt a reflection of the belief that the methods—in this instance, the laboratory methods—are what comes closest to the heart of psychology as a discipline. We will look at the steps that the field of psychology took over the last decades of the nineteenth century, in Germany, in the United States, and in France. Despite the creation of much that was new, despite the development of new methods and new questions, its elder statesmen continued to view the field as still bound to the mother discipline of philosophy. But the younger psychologists were ready to jettison that older tradition.

Germany, the Homeland of Psychology in the Nineteenth Century

The first chapter in the book of modern psychology was written by Ernst Heinrich Weber and Gustav Fechner at the University of Leipzig, several decades before Wundt opened his laboratory there. Weber was a professor of anatomy and physiology at Leipzig, where he had started teaching in 1818, and over the course of his career he developed many of the familiar and basic experimental techniques learned today by budding psychologists, like determining how great a sensation has to be to be noticeable to a human being. For many sensations, it was possible to develop a theory of *thresholds*. For example, if we want to know how close together two pin pricks can be felt, but felt as happening at different places, we can perform experiments on human beings, and determine what physical distances on one's finger, one's leg, or one's lips are needed for two sharp, simultaneous sensations to be perceived as two sensations rather than just one. Weber became famous for a series of experiments in which he found that subjects who were asked whether two objects were the same in weight would display a threshold of weight difference that was not absolute, but was rather more complex: if the *proportion* of the weight of the heavier weight to the lighter weight were greater than a specific threshold, then the weight difference would be perceived. Today, this notion of *just noticeable difference* is a standard concept in experimental psychology. Weber's discovery was one that fit perfectly with the orientation of the age, with its focus on measurement, and it demonstrated for the first time that an unexpected but quantitative law could be established that extended to both sides of the body/mind divide.

Gustav Fechner's book *Psychophysics* was published in 1860, and for some that was a momentous and earth-shaking event. He advanced Weber's discoveries to a mathematical level that seemed astonishing to scientists who had had no inkling that something having to do with the mind could be aptly characterized in quantitative and algebraic terms. Ernst Mach spoke of the liberation he felt it granted him from "the greatest intellectual discomfort of my life."[2] Fechner proposed that the strength of the subjective sensation is proportional to the logarithm of the intensity of the stimulus. He made the argument step by mathematical step: he denoted the stimulus by β and its increment by $d\beta$, the sensation γ and its increment $d\gamma$, with the terms $d\beta$ and $d\gamma$ "each to be considered as referring to an arbitrary unit of their own nature."[3] He then gave Weber's Law[4] as this: $d\gamma = K\frac{d\beta}{\beta}$, and then derives: $\gamma = x \log \frac{\beta}{b}$, where b is the threshold of perception, and x is a constant whose value is determined by the nature of the units used (and the base of the logarithm chosen). This was as stunning a result as was Kepler's discovery that the planets move in elliptical orbits, in the sense that it showed that once again, a new domain was open to quantitative analysis and mathematics.

Wilhelm Wundt

Wilhelm Wundt is the defining figure in the history of modern psychology. Like Ernst Heinrich Weber, he was trained as a medical doctor and turned his attention towards the question of applying the methods of experimental science to the exploration of human thought and behavior. His book *Physiological Psychology*, published in 1874, had a good deal of influence, and in 1875, he moved to the University of Leipzig, where he would soon establish a psychology laboratory. As we have seen, Leipzig was where both Fechner and Weber had been professors; they had retired by the time Wundt arrived, but they were supportive of his orientation. And Leipzig was the home of the Neogrammarians, whose controversies were at their peak when Wundt joined the faculty. In the years that followed Wundt's arrival, he trained an enormous number of academic psychologists who took his perspectives to the four corners of the globe.[5]

One important aspect of his modernity lies in the simple fact that his laboratory ran experiments, and it did so at a wholesale pace, with a cadre of graduate students and assistants: he had 186 students, of whom 116 earned their degrees in psychology. Wundt had learned how to manage a group like this as a student of physiology. An experiment was designed to allow the experimenter to control the stimulus presented to the subject, and ideally the subject was constrained to a response that could easily be characterized. In between the stimulus and the response might be a brief period of time during which the subject, who had been suitably prepared to introspect, would report what he observed. Laboratory methods were appropriate for individual psychology, but this was only half of Wundt's world. He also saw another side to modern psychology, which he called *Völkerpsychologie*, which has been translated in various ways into English: there is the flat-footed "folk psychology," and "ethnic psychology," and also the more apt "collective psychology." This was the discipline that

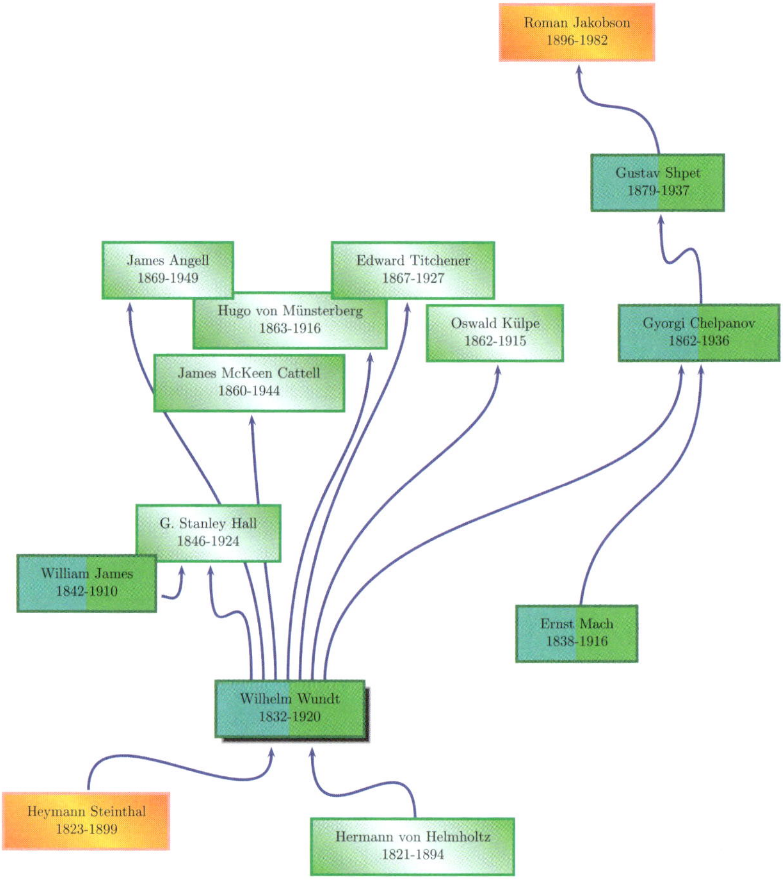

FIGURE 4.1. Wilhelm Wundt

would study language, customs, myth, and other aspects of mind that are socially based.[6] As we noted earlier, Wundt was adopting an idea that came from Herder. Over the last 20 years of his life, he published a ten-volume treatise entitled *Völkerpsychologie*, beginning with a treatment of the nature of language. That very substance and devotion makes clear how important the social was to Wundt's view of the mind: the individual mind is an abstraction that we make in a world where all minds are above all part of a community.[7]

Wundt saw himself not only as following in the tradition of philosophy, but also as rejecting important psychological traditions before him, most strenuously the *associationism* that followed so easily from empiricism, especially anglophone empiricism, such as that of John Locke. The issue that separated Wundt from associationism is how the interaction of ideas, in the broadest sense, should be understood: the extreme associationist position treats ideas (and perceptions, a kind of idea) as *thing*-like in certain regards. Ideas exist somehow in space and time (or so the associationists believed), since we can see two ideas appear together frequently, and these ideas then take on some sort of connection (think: glue) between them. In this picture, there is no active backdrop playing a role: the activity is playing on a stage on which appear only two disconnected ideas, so to speak, and if we extend the metaphor, the activity is spontaneous, without a script or a director. The opposite view, which Wundt embraced, is *active*: at the very least, ideas are attended to (and Wundt would have used the word *apperceived*, but that is a word that carries no meaning today), and they are juxtaposed by the active mind. The world of billiard balls (explained by Newton's laws of motion) is the inspiration for the associationists, who thereby saw unity in the physical and psychic world, but Wundt was thoroughly opposed to that, and he emphasized the active side of mind.

Wundt defended a position that he called "voluntarism," though the connotations of that word are treacherous: he would emphasize that what is psychologically real are events, or occurrences, that occur in time; these *events* are not ideas that may leave and then come back to consciousness. If we set foot in the metaphor of thought as a river, then for Wundt the water swirling at our feet (and which has flowed downstream already in the moment that we notice it) represents the mental occurrences, and nothing will ever bring them back to us again. On this point, Wundt is on the side that others will call *psychologism*, which we discussed in the last chapter. Thoughts are not *impressions*; they are activities. Wundt believed that others had fallen into the fallacy of incorporating into their view of ideas certain conclusions that they had reached about the physical world: that objects perdure, most of all. An idea of this particular book, for example, has none of the durability that the book itself possesses. (Brentano is surely an excellent example of someone who has fallen into what Wundt took to be a fallacy.) It is certainly difficult to avoid the urge to understand the mind in terms of something not at all mind-like, especially when that other thing is tangible, and we see this happen in other contexts in this book.

Perhaps the single most important distinction separating camps of scientists in Germany who studied human beings in one fashion or another was the question of whether all explanations could be reduced to statements about causal relationships among events in space and time. It may seem to us highly reasonable to accept the idea that science can study causal relationships in the physical world! The issue, however, was not whether such causal relationships exist, but whether there were any relationships that were studied by psychology that obeyed laws that were not fundamentally physical laws—and the most striking case was the case of one belief serving as a reasonable basis for believing another statement.

There is a second issue here whose importance is often underplayed. It is easy to accept the affirmative side of the positivist creed: that all elements of reality can be localized in space and in time. But the flip side is that nothing else really exists. And the dynamics—that is, the change or evolution over time—of a real system is bound by the laws discovered by the basic physical sciences, which essentially means physics, chemistry, and almost certainly physiology and biology more generally. A positivist psychologist can only propose the existence of some *thing* if he can explain it in terms allowed by those positivistically more basic sciences.[8]

And *ideas* may be difficult to define in those terms, but far more difficult to define in physical terms are the relations that we know (or *think* we know) must exist *between* ideas, such as the relation of *implying*, or *serving as a grounds for believing*.

In Germany, the controversy was placed in the context of the relationship between natural and social sciences. Natural sciences could be defined as those which without undue controversy fall under the positivist's principles; the objects of study exist in space and time, and they interact in ways ultimately governed by physical laws, all of which reduce to physics. But the human sciences are different; the Germans refer to these as *Geisteswissenschaften*. *Wissenschaften* simply means *sciences*, while *Geist* can be translated as *mind*, *spirit*, or even *soul*. Kusch[9] emphasizes that Wundt noted four distinctions between causality in the physical and the psychological realms: causality was imputed by a theory in the physical world but was consciously experienced in psychology; the psychological notion of causality is "everywhere shot through with value determinations," unlike the situation in the physical realm; the psychological notion cannot be understood without reference to the notion of *purpose*; and causation in the physical world assumes a substantial foundation—in the classical sense of a perduring *substance*—that it does not in the world of psychology.[10] And it should be noted that the non-substantiality of the psychological world was related to the rejection of a role for the unconscious:

"no underlying substrate was assumed, and all processes and events were 'depictive' (*anschaulich*) and thus observable."[11] As we will see later on, the difference between these two styles of analysis would give rise to the Praguean distinction between phonetics (a study in the physical world) and phonology (a study in the psychological world).

Kurt Danziger has argued that Wundt "is a singularly inappropriate figure to choose as the originator of the modern psychologist's professional identity."[12] Certainly we do not find Wundt making a clear call for the separation of psychology from philosophy (quite the opposite!) or offering a definition of research questions that are independent of speculative philosophy. In a letter to Adolf Meyer in 1918, Edward Titchener, one of those who helped establish Wundt's reputation, wrote that "Wundt is . . . the very first large figure in the history of thought who is *temperamentally* psychological. . . . I believe that Wundt's generalizations are mostly wrong; I do not at the moment recall any one of the larger ones that I accept today—though I have in my time swallowed most of them; but I still affirm that Wundt's *instinct* is psychological, even where it leads him astray. . . . That is his importance: not the special things he teaches."[13]

There was a tension in nineteenth-century psychology concerning both the significance and the usefulness of introspection as a source of data for the psychologist; it was a tension that divided roughly along the lines of the British versus the Germans. For the British (not just the English: we include the Scotsmen), introspection remained a privileged source of information about the mind; the mind was, for them as for Descartes, a transparency visible to the user. On the continent, that was no longer the dominant view and had not been since the time of Leibniz. Kant had made it especially clear that what is phenomenally accessible to the end user is only the phenomenal self, which is to say, only the *appearances* of psychological reality.[14]

Wundt discussed two processes, both of which have been translated into the English word *introspection*. The first is *Selbstbeobachtung* "self-observation," which he dismissed; the second is *innere Wahrnehmung* "internal perception," which he held to be important—indeed, foundational. (Recall our discussions of introspection in the last chapter, in connection with Comte and Brentano.) Internal perception was immediate and entirely independent of memory, and the use of practiced observers was encouraged because they were in the habit of responding without reflecting on the answer before responding. Wundtian psychology has achieved a certain degree of suspicion in anglophone countries because of the distorted image of Wundt's work brought by Wundt's student Titchener, as we will see below.

Wundt's view about the inner, subjective experience of the mind was that it was *not distinct* from the outer, objective experience of the mind: if there are "psychical" and "physical" objects, they are "not different objects at all, but one and the same content of experience." There are two different ways of looking at human experience, though:

> This content is examined in the one case, that is, in the natural sciences, after abstracting from the subject. In the other case, that is, in psychology, it is examined in its immediate character and its complete relation to the subject. All metaphysical hypotheses as to the relation of psychical and physical objects are, when viewed from this position, attempts to solve a problem which never would have existed if the case had been correctly stated.[15]

This view embraces the *duality of human experience* and rejects dichotomizing, either/or *dualisms*. Wundt's view was that the same experience can be viewed and analyzed in more than one way, and the psychological point of view and the physicist's point of view are about as distant and complementary as two views can be. But both are among the many ways that humans can talk about and appreciate their experience, and their experiences. Wundt's view was a version of what has been called *psychophysical parallelism*, a view that Wundt could point to in Fechner's work, and one which we can still see as a dominant view today.[16]

Oswald Külpe

Oswald Külpe was a student of Wundt's who took an independent stance on several of the most important questions in psychology of his time. Külpe was born in 1862 and studied psychology during the 1880s: first at Leipzig, where he studied with Wundt, then at Berlin and Göttingen, after which he went back to Leipzig to earn a doctorate with Wundt. He continued to work with Wundt until 1894, when he was appointed professor at Würzburg. He developed a strong laboratory there, perhaps second in Germany only to Wundt's, and he directed the research of many of the top psychologists of the next generation, include Max Wertheimer and Kurt Koffka, leaders of the Gestalt psychology movement.

Külpe was much more sanguine than Wundt about the notion that experimental methods could be used to study human intelligence, and his Würzburg psychologists were among the first to use subjects' own reports of how their mental process had proceeded as they tried to solve a difficult task given to them. Külpe himself developed a series of experiments challenging the traditional view that sensation preceded an individual's conscious intent, and that sensation arose independently of that intent as well. He presented images briefly, after asking the subjects to identify the color, or the shape, or perhaps what letters were present. When he then asked them about *other* characteristics that he had not prepared the subjects to attend to, what he found was that subjects did

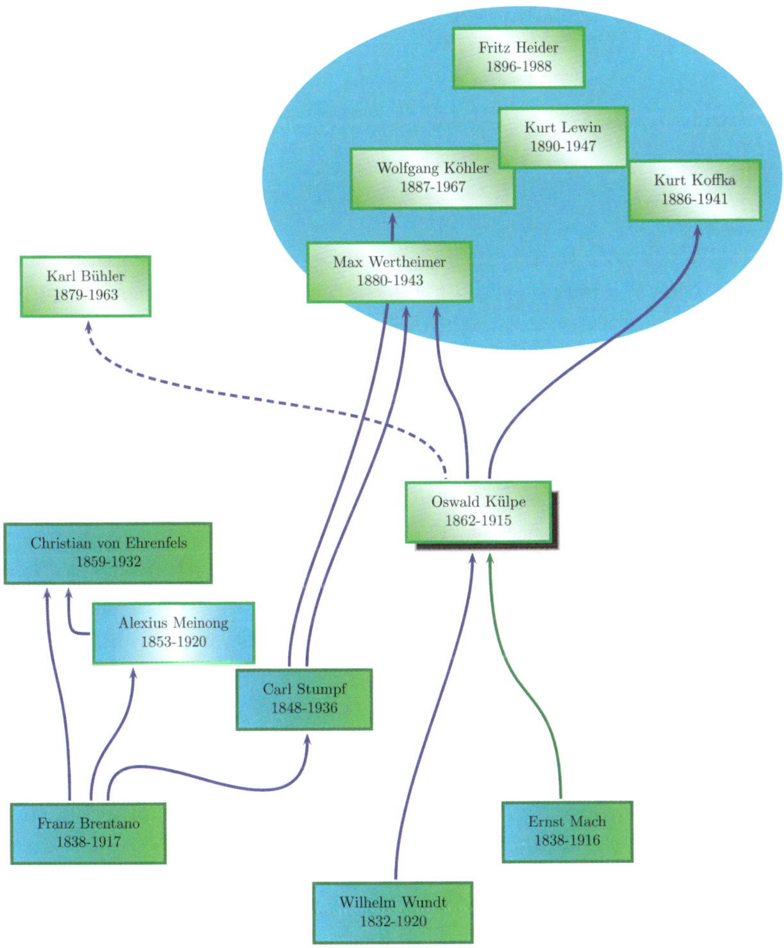

FIGURE 4.2. Oswald Külpe

poorly, and the more aspects of an image the subjects were asked to pay attention to, the worse they did on reporting the *other* aspects. In short, even basic and elemental aspects of sensation had the character of an act performed by the subject. This view of sensation would be central to the development of Gestalt psychology, as we will see in the next chapter. Külpe was the teacher both of Max Wertheimer, who founded the Berlin school of Gestalt psychology, and of Karl Bühler, a Gestalt psychologist of a different school, working in Vienna, who will also be a significant actor in what follows. Wundt, on the other hand, saw Külpe's position as a setback for psychology and would have no part of an attempt to reduce (as he saw it) psychology to physiology. This is an issue that remains with us today.[17]

In Külpe's view, then, sensation was an activity, not a passive process, and higher-level cognition was not an amalgam of sensations either.

Carl Stumpf

Carl Stumpf was a student of Franz Brentano (and also of Hermann Lotze, who officially directed his dissertation), and like so many of his time, he trained first as a philosopher, and then left his greatest mark on psychology. His career was strikingly successful, as the German-language universities vied with each other to make offers to persuade Stumpf to come and settle down. He taught first at Würzburg (filling the position that Brentano had just left) and then at Prague, where he did important work on the psychology of musical perception and interacted with colleagues Anton Marty, Ernst Mach, and Ewald Hering; and then he moved to Halle (where he was a colleague of Georg Cantor) and then to Munich. Finally, in 1894, at the age of 46, he accepted an offer to move to the University of Berlin, to teach both philosophy and psychology, and he spent the rest of his career there, developing the Psychological Institute and training some of the most important psychologists of the next generation, such as Max Wertheimer, Kurt Koffka, Wolfgang Köhler, and Kurt Lewin, as we will see in the next chapter, and they would be the stars of the Berlin school of Gestalt psychology. Stumpf would pass on the directorship of the Institute of Psychology to Wolfgang Köhler in 1922, and he served as Husserl's dissertation advisor.[18]

Stumpf's work on phonetics, the sounds of language, was extremely influential, and would leave an important mark on the theory of phonology. His work was both a continuation of Helmholtz's earlier work, and

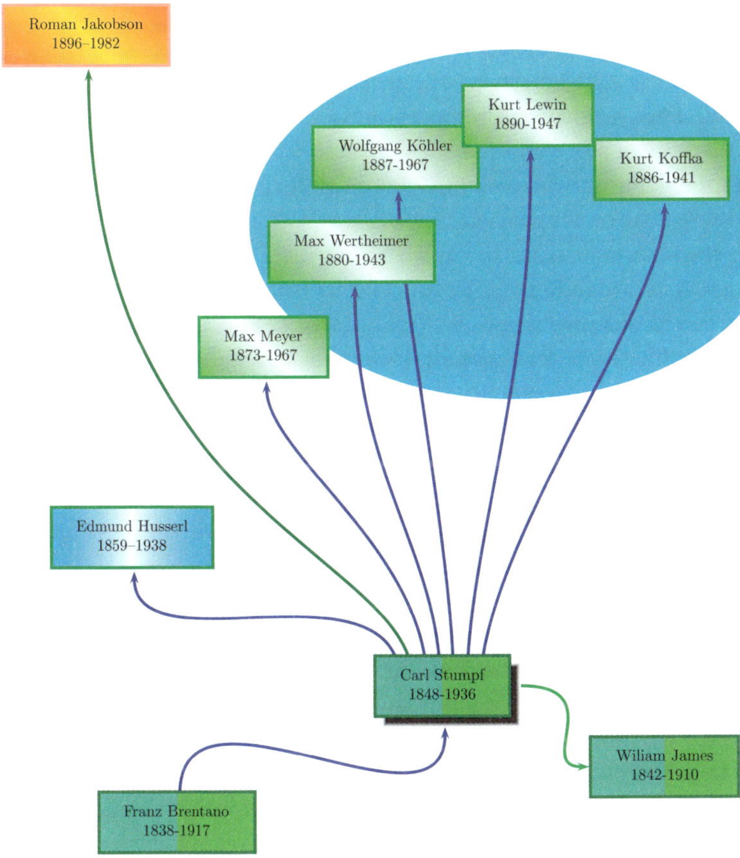

FIGURE 4.3. Carl Stumpf

also a strong criticism of it; it was published in 1926 in *Die Sprachlaute*. Stumpf's analysis of the phonetic characteristics of vowels and consonants would be adopted and developed by Roman Jakobson in Prague and later with Morris Halle in Cambridge, Massachusetts.

Max Meyer was one of Stumpf's less famous students. He was not a great success academically but he published a significant book on behaviorism two years before the more famous publication by John B. Watson on the subject. This would be of no great consequence to us but for another aspect of his academic career. He had one, just one, doctoral student in his career at the University of Missouri, and that one student was Albert Paul Weiss. Weiss was a behaviorist who spent most of his academic career at the Ohio State University; he had a weak heart, and died at the age of 51. But Weiss was a close friend of Leonard Bloomfield, the linguist, and during the few years they shared together at Ohio State University, Weiss's behaviorist orientation would have an enormous impact on Bloomfield, as we will see in chapter 6.

There is another important line of influence of Wundt's work, one which flowed to Russia and then beyond. This stream of influence involves a number of important psychologists whose names are rarely encountered today but who play an important role in the work of Trubetzkoy and Jakobson, both in Russia and elsewhere. Of these students of Wundt, the most important were Georgi Chelbanov, who became professor of psychology at the University of Moscow in 1907, and his student Gustav Shpet. We will return to this stream of influence in chapter 9, when we discuss Trubetzkoy and Jakobson.

Psychology Comes to the New World

During the 1880s and 1890s, psychology came to the United States, packed in the trunks and suitcases of the handful of young Americans who traveled to Europe—mainly to Germany—to study what was happening over there. These were largely students of philosophy who were not satisfied with what was being taught in philosophy courses—young Americans such as William James, G. Stanley Hall, and James McKeen Cattell; there was also Edward Titchener, an Englishman who came to the United States after studying in Germany. Of these four, all studied in Europe (which was the normal case for any American seeking an advanced education), and all but James studied with Wilhelm Wundt.

William James

William James's greatest contribution to psychology was *The Principles of Psychology*, published in 1890. It was a massive book, 12 years in the writing; it was so long that James was obliged to produce a shorter version for the slow of reading. When the book was finally finished, much of James's interest in psychology was finished along with it. But it was such a magnificent creation that James remained till the end of his life many psychologists' conception of the true American psychologist, the leader of his discipline and the authority of his field. John Dewey, foremost

American philosopher of his time, wrote of James, "By common consent he was far and away the greatest of American psychologists—it was a case of James first and no second. Were it not for the unreasoned admiration of men and things German, there would be no question, I think, that he was the greatest psychologist of his time in any country—perhaps of any time."[19]

James was 48 at the time *The Principles of Psychology* was published, and professor of philosophy and psychology at Harvard. His origins were unusual: his grandfather had been an immigrant, but one who made a considerable fortune in upstate New York. James's father suffered from being unappreciated by his own very successful father, and this sense of incompleteness was passed down, as can happen in families, to William. The family fortune allowed for private tutoring and travel in Europe—this was the first generation of wealthy Americans, midcentury, who could spend considerable lengths of time in Europe, learning the languages and the culture of the Old World, and James drank deeply of the art he found in Europe, and mastered both French and German—an important step, as we have noted, for a scientist-in-training. James's siblings were just as extraordinary as he was. His younger brother Henry would become one of the most famous men of letters of his generation, and the diary of his sister Alice would contribute significantly to scholars' understanding of her day. William's interests ranged widely, and scarcely would remain fixed; they changed, they evolved. As a young man, he was very interested in the arts, and in his twenties he spent a good deal of time studying art in Europe. He came back to Boston, earned a degree in medicine, and was hired to teach anatomy and physiology at Harvard University in 1873. His interests moved towards psychology at that point—he was greatly impressed by Wundt's book (1874) on physiological psychology, and he wrote a review of it—but James had not studied the subject formally, and learned by teaching, beginning in 1876. Two years later, he signed a book contract and started writing his book; its publication in 1890 was an iconic moment in early American psychology.

He was ready then to say to himself that he had had enough. *Basta.* He worked hard to persuade Hugo Münsterberg to come to Harvard and take over the psychology lab, and eventually the department. Münsterberg was a generation younger than James, was German, and had earned his doctorate with Wundt in 1885, so he was the spitting image of a successful academic psychologist. It was not an easy task to persuade Münsterberg that coming to the United States did not constitute academic suicide, and the process of persuasion took quite a few years. In the end, Harvard won out over Freiburg in Münsterberg's decisions, and this was a beginning of the tipping towards the New World's academic leadership. But as a step it was hardly noticed at the time.

When he was able to devote less time to psychology, thanks to Münsterberg's presence and energy, James was able to think more about philosophy, developing the distinctively American school of pragmatism. He had been friends with Charles Sanders Peirce in college at Harvard, and James interacted with the luminaries of American philosophy in the years that followed his engagement with psychology. He developed during this time his own style of pragmatism, distinct from what Peirce presented and published. Like many of his generation, James became seriously involved in the study of parapsychology and life after death. There hardly seemed to be a limit to his interests.[20] If James had come to psychology from anatomy, he had never been a great lover of laboratory experiments, certainly not as Wundt and his colleagues conceived of them, and he was happy to leave all that to Münsterberg. There is a wonderful passage in his book which reveals much of his opinion about a range of things: he wrote that "psychology is passing into a less simple phase. Within a few years what one may call a microscopic psychology has arisen in Germany, carried on by experimental methods, asking of course every moment for introspective data, but eliminating their uncertainty by operating on a large scale and taking statistical means."[21] This sounds very reasonable, does it not? James plows on: "This method taxes patience to the utmost, and could hardly have arisen in a country whose natives could be *bored*." We might like to think that this was just good, clean fun, but there was more to it than that. *Could not be bored*? James had continued, "Such Germans as Weber, Fechner, Vierordt, and Wundt obviously cannot; and their success has brought into the field an array of younger experimental psychologists, bent on studying the *elements* of the mental life." Maybe Germans could not be bored, but they *could* read English. Wundt's laboratory did not appreciate these remarks. Charles Judd, an American working there, recalled that "especially was there a very pronounced antipathy to James. James had done what was thought to be quite out of order; not only had he criticized Wundt but in some cases—as, for example in discussing the innervation theory—he had allowed his criticism to take the form of witty sarcasm. This was far too much. Not only that, but he indulged in that remark about patient laboratory work in a land where they did not know what it means to be bored. As a result diplomatic relations were promptly suspended."[22] We noted just above that James seemed to have had it with psychology by the time his book came out. Hear him again, as he wrote, "it would be terrible if even such a dear old man as [Fechner] could saddle our Science forever with his patient whimsies, and, in a world so full of more nutritious objects of attention, compel all future students to plough through the difficulties, not only of his own works, but of the still drier ones written in his refutation."[23]

There is something in William James's style of writing, most of the time, which marks the beginning of a down-to-earth, no-nonsense American style of writing, a style that he shared with John Dewey, and that we will see more and more (we will shortly see it in John B. Watson, for example). That style certainly left open the possibility that one might bring up German psychology and say that it couldn't have arisen in a country where people could get bored. In another paper, James skewers "scholastic philosophy," saying that it is "common sense grown pedantic."[24] *That* is everything that Americans don't want and don't like. There is a culture clash here, a we-are-not-Europeans kind of identity, and it is one that grows and develops over the course of the following century. We will see a replay of this later on, when a kind of folksy and at times self-deprecating sense of humor runs full tilt into a no-nonsense East Coast earnestness. Self-deprecating? That was a style that William James could adopt, and frankly it would be *unimaginable* for a German professor to give a presidential address to his society and make a remark like the one that James did: "You will agree with me that I have brought no new insight to the subject, and that I have only gossiped to while away this unlucky presidential hour to which the constellations doomed me at my birth. But since gossip we have had to have, let me make the hour more gossipy still by saying a final word about the position taken up in my own *Principles of Psychology*" and off he went again.[25]

The Principles of Psychology begins with the sentence, "Psychology is the Science of Mental Life, both of its phenomena and of their conditions."[26] There is a lot to unpack in that sentence. For James, the central fact of psychology was the flow of consciousness, of awareness, and how that flow was accessible to our inner awareness, and how it was related to our place as biological organisms in the larger world we live in. To James we owe the expression of how the baby must encounter the world, "as one great blooming, buzzing confusion."

There is no better way to get a sense of James's way of thinking than to listen to him; and he wrote many sensible things about the ways in which a person is constantly engaged in social acts. Here is a typical example of how William James reminds us of who we are:

> Our social self-seeking, in turn, is carried on directly through our amativeness and friendliness, our desire to please and attract notice and admiration, our emulation and jealousy, our love of glory, influence, and power, and indirectly through whichever of the material self-seeking impulses prove serviceable as means to social ends. That the direct social self-seeking impulses are probably pure instincts is easily seen. The noteworthy thing about the desire to be "recognized" by others is that its strength has so little to do with the worth of the recognition computed in sensational or rational terms.[27]

We all have enough insight into ourselves, you can almost hear him say. Surely we can all see ourselves well enough to see how predictable we are, even if we are not entirely rational about it. What follows might almost come from a novel: "We are crazy to get a visiting-list which shall be large, to be able to say when any one is mentioned, 'Oh! I know him well,' and to be bowed to in the street by half the people we meet."

A moment later, James remembered a passage from William Makepeace Thackeray's *Book of Snobs*: "Thackeray somewhere asks his readers to confess whether it would not give each of them an exquisite pleasure to be met walking down Pall Mall with a duke on either arm. But in default of dukes and envious salutations almost anything will do for some of us; and there is a whole race of beings to-day whose passion is to keep their names in the newspapers, no matter under what heading." If the rhythm were a bit different, we might almost be reading Alexander Pope.

For James, psychology was not the study of the individual limited to the interactions she might have in our psychology laboratory. It was the study of people, a study that all of us have been engaged in since our youngest years, and we have all reached some of the same conclusions:

> Not only the people but the places and things I know enlarge my Self in a sort of metaphoric social way. "Ça me connaît," as the French workman says of the implement he can use well. So that it comes about that persons for whose opinion we care nothing are nevertheless persons whose notice we woo; and that many a man truly great, many a woman truly fastidious in most respects, will take a deal of trouble to dazzle some insignificant cad whose whole personality they heartily despise.

G. Stanley Hall

G. Stanley Hall was the other leading psychologist in the United States in James's day. Recent scholarship has not been kind to Hall's reputation.[28] He worked hard to develop the authority of the leading psychologist in the United States, a status that would place him, had he achieved it, above William James, but despite a range of professional accomplishments, the American world of psychology did not come to view him in that light.

Hall arrived at Harvard University to do doctoral work in psychology in 1876, where he worked with William James, who was only two years his senior. The two developed a warm friendship which "would eventually prove to cover a good deal of competitive hostility."[29] Hall was awarded his PhD just two years later (which was unremarkable at the time)—and it was the first PhD awarded by Harvard's department of philosophy, as well as the first awarded in the United States in the field of psychology.

Hall then went to Germany, and spent the first year in Berlin. While there he worked closely with Hugo Kronecker at the Physiology Institute. He wrote back to James at one point, "I have stood in much the same terms of intimacy and recipiency [to Kronecker] as last year to you, and to [him] I am likely to owe a scarcely smaller debt of gratitude."[30] The next year, in 1879, he went to Leipzig and took advantage of the opportunity to attend Wundt's lectures.

He was not impressed. "Wundt is more and more exasperating," he wrote to James.

> He seems to me a grand importer of English ideas . . . and an exporter of the generalized commonplaces of German physiology . . . inexact . . . and as a man who has done more speculation and less valuable observing than any man I know who has had his career.[31]

It's not clear exactly what that means, but it's not good (and sounds a bit like the disparaging remarks that Whitney made about his German professors). "His experiments, which I attend," Hall went on, "I think utterly unreliable and defective in method." Hall returned to the United States, and tried to get a position at Johns Hopkins University, which, like Harvard, awarded the PhD degree. James wrote a strong letter of recommendation for him, describing him as "a more learned man than I can ever hope to become . . . I feel his exceptional merits, moral as well as intellectual, so strongly that I cannot bear to think of his being any longer with a place commensurate with them."[32] High praise, but Hopkins was not ready to hire Hall at that moment.

Within a few years, things changed. In 1884, Hall was appointed professor of psychology at Johns Hopkins (over Charles Sanders Peirce, in the event),[33] and in 1889, he became the first president of Clark University, in Worcester, Massachusetts. For this achievement, Hall has long been recognized as one of the early academic leaders who created in the United States a German-style university with a heavy emphasis on graduate studies and on research. Clark opened its doors in 1887 with a world-class faculty (which included Franz Boas). Sokal noted, "By 1892, however, Hall's chronic secretiveness and dishonesty had alienated most of his colleagues, and by the end of the year most of them had left, primarily to go to the University of Chicago."[34]

Edward Titchener

Edward Titchener studied with Wundt in Leipzig in 1890, just after getting his degree from Oxford. He became a professor of psychology at Cornell University in upstate New York two years later, where he continued Wundt's goal of studying the character of the content of consciousness; his approach was known both as *structuralism* (he coined the term)[35] and as *introspective psychology*.[36]

In coming to Cornell when he did, Titchener was in a position to develop an influential group of researchers trained in his ways of doing psychology, and he did precisely that over the next several decades, supervising the writing of 56 PhD dissertations in psychology.

Titchener viewed himself as bringing Wundt's work to the United States. One of this students recalled that Titchener was "a brilliant young man who would give us the latest news from Leipzig rather than one to be heard for his own sake."[37]

By all accounts, his was a dominant—we could say *domineering*—personality. Two of his students who had married each other decided that they would "accept 'insults' and arbitrary control from Titchener in order to retain the stimulus and charm of his sometimes paternal and sometimes patronizing friendship." One of them wrote, "I never broke with the master and I still feel the credit balance remained on my side." This was Edwin Boring, who would become a highly influential professor of psychology at Harvard. Still, years later he recalled the "personalized magnetism" that Titchener generated, as well as the "real kindness [Titchener demonstrated] to those disciples who avoided transgression."[38]

Eventually, like so many, Titchener began to refer to what younger people did as "fads." In 1907, he wrote to Robert Yerkes, who was perhaps the first leading American figure in animal learning:

> Animal behavior is like the functional standpoint—extremely in fashion. There is a new field opened for mediocrity: That is the real secret. As soon as a novel standpoint is announced or a new region or work opened, the rank and file rush in, because anything that they say or do will for the time pass muster: And it is contrariwise, deucedly difficult to get a hearing where the bulk of past work is great and the methods established. Let us keep our heads: that is the important thing. Work on animal behavior is decidedly important, and the functional standpoint is decidedly worth thinking through—my favorite expression, you see! But they are not the be-all and end-all of "scientific" psychology.[39]

Towards the end of his life, Titchener offered an overview of the kinds of confrontation that modern psychology had encountered when it was first being developed by Wundt. Titchener's goal was to show to those most active in psychology that they were likely to be misled about the growth of psychology because of the generational effect: *his* teachers' generation was responding to the *previous* generation, but they did not make it their business to say it out loud. Wundt was responding to Herbart in everything he wrote, but how could someone reading Wundt *today* have even the slightest inkling of it, since Wundt left that fact out?

Wundt's psychology "took shape against a background of physics and of a physiology informed by physics. Fechner was a physicist; Helmholtz was physicist and physiologist; Weber and Hering and Wundt were physiologists." The principal drive, Titchener said, was to establish a psychology which can be housed (as he put it) under one roof with physics and biology, without which, "science can have no stable meaning." His picture is clear: biology is based on physics, and psychology is based on biology. But the academic environment was hostile, and there were three forces holding psychology back. The first was the influence of Herbartian psychology; the second was what he called "empirical psychology," and the third was quite simply philosophy.[40]

Herbartian psychology? A modern psychologist who could identify who Herbart was and what he did would be a rara avis indeed. Nor was

Herbart a household name at the time, either: Titchener wrote, "To most of us, I suppose, Herbart is little more than an historical name; to Wundt in the seventies Herbart was an all-pervading institutional opponent." Titchener listed nearly a dozen Herbartian psychologists who were prominent before Wundt, and then wrote, "Bonitz and Exner; Waitz was at Marburg; Stoy was at Jena. To us, truly, these men are just a list of names and dates; to a nascent experimental psychology they represented a highly formidable opposition." Then Titchener emphasized what he suspected the current reader was in no position to perceive: that "all through the second part of the *Physiological Psychology* Wundt has Herbart steadily in mind; that the improvised doctrine of apperception is meant as a counterblast to Herbart; that the whole of Wundt's psychology beyond the chapter on perception is shaped with polemical reference to Herbart."

Titchener was telling his reader that when they read Wundt, they may have heard and understood the *answer*, but they did not understand the question: they did not even realize that there was a question that had flown over their heads. He went on:

> That, you see, is the sort of trick that history plays upon us; a good many of the later Wundtians have been anti-Herbartians without realising it. So that the effect of Herbartianism upon experimental psychology was in reality twofold, internal and external. The Herbartians were in possession; they must be dislodged, superseded, discredited, if experimental psychology was to grow to power; that was the external side of things. Internally, meantime, the very fact of having to combat a well-rounded and critically tested system tinged and moulded the doctrines of experimental psychology itself.

Titchener hit the nail clean on the head: this is a tremendously important phenomenon, and one that is not always easy to note in real time. Titchener could hope that his students would understand him when he wrote that, but *we* will never know if they did. Bjork notes that Titchener "was conscious of engaging in 'a battle between the principles of science, as they were built up by the men of the nineteenth-century, Darwin included . . . and the principles of teleology, which have been adopted . . . by the biologizing psychologists.' And he told Adolf Meyer something he might have said to James: "I shall use every weapon at my disposal to resist your revolutionary aims: I have not the faintest intention of allowing myself to be killed."[41]

Let us take a look at what was on Titchener's mind as the nineteenth century came to a close. William Caldwell, a philosopher at Northwestern University, published a critique of some of Titchener's work in the *Psychology Review*. Caldwell's concern was that Titchener's sole focus was on the consciousness of sensation and affection, and he argued that this leaves out some very important things: *will* or *intention*, first of all, as well as the integrative act of consciousness, the pulling together of all the sensations and affects into a perceived unity. Titchener, he wrote, "explains the highest psychical formations from their lowest elements; it is equally important for psychology to explain the lowest elements from the point of view of the highest psychical formation, such as control and conduct and self-affirmation." The critical point to emphasize, he wrote, was the "active, unifying, synthetic self."[42]

Titchener's response began with an off-hand remark in a footnote that is nonetheless worth mentioning. If Caldwell had presented 12 numbered arguments against Titchener, Titchener notes that at least three of them "rest upon technical errors. . . . Such lapses are hardly to be avoided by anyone who travels out of his own special field into that of another discipline; they do not at all impair the value of Professor Caldwell's contentions regarded as a whole."[43] Here is a chide, gentle or not, that philosophers do not have all the chops necessary to do psychology, or even to understand it in a sound, professional way.

When a biologist studies an organism, Titchener wrote, he may study form, or structure, and this is anatomy or "morphology"; he may study function (physiology); or he may study development (ontogeny). He may do any of these not only from the point of view of an individual, but he may also do this at the level of life as a whole. Its parts are its species, its physiology is its ecology, and the historical version of ontogeny is phylogeny, the development of species over time.

The psychologist's task runs parallel to that of the biologist, Titchener explained. The study of form is the attempt to "ravel out the elemental processes from the tangle of consciousness." But just as dissecting a frog does not teach us the functions of its organs, discovering the elements of consciousness is not the same thing as discovering the functions of these elements.

Yes, there is function, he wrote. In the biological world, this includes digestion, locomotion, secretion, excretion, and so on. In the psychical world, this includes memory, recognition, imagination, conception, judgment, attention, apperception, and volition—to mention just a few. If we want to understand such things, we have to engage in *functional psychology*, and that is simply not what the experimental psychologist, engaged in Titchenerian psychology, is doing: he is doing *structural psychology*, akin to the task of the student of biological anatomy—not to that of the student of physiology.

"It cannot be said that this functional psychology . . . has been worked out either with as much patient enthusiasm or with as much scientific accuracy as has the psychology of mind structure," Titchener observed. But the main interest has been in morphological analysis, not in function. And why is that? "The reasons [are] not far to seek: We must remember that experimental psychology arose by way of reaction against the faculty psychology of the last century. This was a metaphysical, not a scientific, psychology." If we focus on function before we get structure clearly worked out, we run the great risk of allowing for pseudo-explanations that things are as they are just so that they can do what they in fact do. If we did allow that kind of account, then psychology would just become philosophy again (a slight dig at the philosophers). "In a word, the historical conditions of psychology rendered it inevitable that, when the time came for the transformation from philosophy to science, problems should be formulated, explicitly or implicitly, as static rather than dynamic, structural rather than functional."

Now that is an interesting remark that we should pick up on: Titchener was associating *dynamic* with *functional*, and also with less scientific. This is a linkage that we will see again later, most notably in discussions of linguistics in the second quarter of the twentieth century.

E. G. Boring, in his own memoirs, remembered studying psychology at Cornell, from 1910 to 1918, which "revolved around and was kept in its orbit by the personality of E. B. Titchener. What a man! To me he has always seemed the nearest approach to genius of anyone with whom I have been closely associated." This comment came after a long career in which Boring had been a professor at Harvard for decades. "I used to watch my conversations with him, hoping I might gain an insight into why his thinking was so much better than mine. I decided presently that his superiority lay in his easy command of memory traces, his ready entertainment of novel relationships, his equally ready abandonment of unprofitable hypotheses, and his avidity in the pursuit of goals." And he goes on in the same vein for quite a while. "Seldom did he distinguish between his wisdom and his convictions and he never hid either."[44]

Recent scholarship has been highly critical of Titchener's presentation of Wundt to American scholars. Kurt Danziger drew scholars' attention to the fact that "Titchener practically made a career out of interpreting Wundt in his own highly idiosyncratic fashion," and many Americans got their first and only taste of Wundt from reading Boring's classic history of psychology—and "it is apparent that Boring took his admired teacher, E. B. Titchener, as a guide in these matters."[45]

Danziger echoes a point we raised in chapter 1: we must not lose sight of the fact that questions, once settled, become *no longer understood*, and this forgetting becomes deeper as more generations pass. Danziger wrote,

> Boring was himself deeply committed to the positivist philosophy of science whose influence on the early development of psychology is at issue here. But his is the commitment of the second generation: What had been for his teachers conclusions carefully arrived at and boldly asserted, have now become matters to be taken for granted, implicit certainties not open to debate or even worthy of mention. For the historiography of psychology the major consequence of this stance is that the dependence of psychological theory and method on prior philosophical commitments is lost from view.

Danziger is being perhaps too modest, or in any event underselling his point: the consequence that he identifies is matters not merely for the historiography of psychology; the historians of psychology can take care of themselves. The more important consequence is the loss of consciousness among the research psychologists themselves. He continues:

> Because only one kind of philosophy of science is regarded as legitimate (or even conceivable), differences on scientific issues are not seen as the consequence of philosophical differences. This is a comforting attitude for those who have no wish to question fundamental assumptions, and that usually includes the conservative majority. When a particular philosophical commitment becomes characteristic of the mainstream of development in a certain field, it is usually so much taken for granted that it is not even identified by its practitioners.

Functionalism, John Dewey, and the University of Chicago

The University of Chicago was founded in 1891 with money provided by John D. Rockefeller and a vision that came from William Rainey Harper, a teacher of Hebrew, a man of boundless energies, and a scholar with a PhD in linguistics, a former student of William Dwight Whitney's at Yale. One of Chicago's early priorities was to establish an important psychology department, and Harper brought John Dewey in from the University of Michigan in 1894.

Under the leadership of Dewey and of James Rowland Angell, a school of psychology arose which came to be known as *functionalism*, a perspective interested not so much in the structure of consciousness as in the way the mind functions to meet goals in a given context or environment. Early in his career, John Dewey was deeply involved in both disciplines, psychology and philosophy (in this respect following the path of William James), but pragmatism, as a disciplinary label, largely kept itself to philosophy. Was functionalism, in psychology and as developed at the University of Chicago, just the interpretation of Deweyan pragmatism in a psychological context? This, it turns out, was a touchy issue.[46]

John Dewey was the American intellectual par excellence. He was born in Burlington, Vermont: could there be a more American place to be born, in the year of our Lord 1859? This was the same year in which Darwin's first, great book was published, and it would influence Dewey intellectually when he first encountered it at the University of Vermont. Dewey taught school for two years after college, then headed for graduate study at Johns Hopkins University, where he studied philosophy with Charles Sanders Peirce and G. S. Morris, and experimental psychology with G. Stanley Hall. He spent ten years as a faculty member at the University of Michigan, and then in 1894, he moved to the University of Chicago, which had only been in existence for just two years, where he would be actively involved in both questions of philosophy and psychology. Dewey also set up the Laboratory School of the University of Chicago, because his pragmatism was a philosophy that needed to work in the wild if it was to work anywhere at all. In 1904, he moved to Columbia University.

Early on, before there was something that would be called functionalist psychology, Dewey cast his lot with the new kind of psychologist, the one who no longer spent his hours introspecting but rather running subjects in experiments. Already by 1884, he felt that there had been a revolution in psychology:

> What can be meant, then, by saying that the rise of this physiological psychology has produced a revolution in psychology? This: that it has given a new instrument, introduced a new method, that of experiment, which has supple-

> mented and corrected the old method of introspection. Psychical facts still remain psychical, and are to be explained through psychical conditions; but our means of ascertaining what these facts are and how they are conditioned have been indefinitely widened.[47]

What was characteristic of the new methods was the use of experiments under controlled conditions, and the commitment to quantitative measurement, both of which were inconsistent with the older style of introspection.

Dewey also made some remarks in this early paper on the character of the relationship between those who embrace a new revolution in science and those who have come before. It is so unusual and so mature that it bears both citation and reflection. We have no need to castigate those who came before us, Dewey wrote:

> There is no need to cast stones at those who, having a work to do, did that work well and departed. With Sir William Hamilton and J. Stuart Mill the school passed away. It is true that many psychologists still use their language and follow their respective fashions. Their influence, no doubt, is yet everywhere felt. But changed conditions are upon us, and thought, no more than revolution, goes backward. Psychology can live no better in the past than physiology or physics; but there is no more need for us to revile Hume and Reid for not giving birth to a full and complete science, than there is for complaining that Newton did not anticipate the physical knowledge of to-day, or Harvey the physiological.[48]

Yesterday's psychologists did the work that needed to be done when they did it, and we can be grateful to them for doing it. And their work was not just solving scientific problems: it was identifying the problems, it was asking the questions for the first time.

> The history of all science demonstrates that much of its progress consists in bringing to light problems. Lack of consciousness of problems, even more than lack of ability to solve them, is the characteristic of the non-scientific mind. Problems cannot be solved till they are seen and stated, and the work of the earlier psychologists consisted largely in this sort of work. Further, they were filled with the Zeitgeist of their age, the age of the eighteenth century and the *Aufklärung*, which found nothing difficult, which hated mystery and complexity, which believed with all its heart in principles, the simpler and more abstract the better, and which had the passion of completion. By this spirit,

> the psychologists as well as the other thinkers of the day were mastered, and under its influence they thought and wrote.[49]

And then Dewey called upon all of his contemporaries to set down their arms that had been raised against scientists of the past, and to live in the present:

> Thus their work was conditioned by the nature of science itself, and by the age in which they lived. This work they did, and left to us a heritage of problems, of terminology, and of principles which we are to solve, reject, or employ as best we may. And the best we can do is to thank them, and then go about our own work; the worst is to make them the dividing lines of schools, or settle in hostile camps according to their banners. We are not called upon to defend them, for their work is in the past; we are not called upon to attack them, for our work is in the future.

It is hard to imagine an expression of a nobler attitude towards one's discipline.

James R. Angell

Angell was 10 years younger than Dewey. He had been a student of Dewey's as an undergraduate, and then master's student, at the University of Michigan (where Angell's father was president). Angell went to graduate school at Harvard, where he was influenced by William James as well. Angell seems to have had an amiable soul; towards the end of his life, he wrote that he thought highly of his Harvard teachers, William James, Josiah Royce, and George Herbert Palmer: "all three remained my warm friends as long as they lived and two of them, James and Palmer, I am sure considerably overestimated my abilities."[50] Together, Dewey and Angell established functional psychology at the University of Chicago, and after Dewey left for Columbia, it was Angell who remained at Chicago as the doyen of the functional psychologists there.

Functionalism took its cue first from Darwin: it put the emphasis on the primordial fact that all people (and all animals, for that matter) have evolved in a biosphere in which individuals survive if they are able to formulate goals—finding food, avoiding predators—and establish behaviors that allow them to achieve those goals. *The central fact is that people act*, and the goal of psychology should be to understand how people carry out activities in order to achieve the ends that they have defined; consciousness may play a role in this, but it may not, and it was therefore a part of the story, but not the central part.

James Rowland Angell explained in a lecture in 1906 how functionalism differed from earlier views. First he pointed out how functionalism distinguished itself from structuralism.

> [Functionalism involves] the effort to discern and portray the typical operations of consciousness under actual life conditions, as over against the attempt to analyze and describe its elementary and complex contents. The structural psychology of sensation, e.g., undertakes to determine the number and character of the various unanalyzable sensory materials, such as the varieties of color, tone, taste, etc. The functional psychology of sensation would on the other hand find its appropriate sphere of interest in the determination of the character of the various sense activities as differing in their modus operandi from one another and from other mental processes such as judging, conceiving, willing and the like.[51]

The structuralist might sit in a quiet room and introspect; the functionalist would rather know how a person accomplished a task in an environment that resembled a real-life situation. This difference regarding method would emerge again and again; in late twentieth-century linguistics, linguists differed on whether linguists' judgments about specific sentences, out of context, were of any value: some thought these judgments were the rock-bottom empirical basis of the science, while others thought they were poor substitutes for evidence about how sentences were "really" used, in real contexts by real people.

> [Functionalism's] fundamental intellectual prepossessions are often revealed by the classifications of mental process adopted from time to time. Witness the Aristotelian bipartite division of intellect and will and the modern tripartite division of mental activities. What are cognition, feeling and will but three basally distinct modes of mental action? To be sure this classification has often carried with it the assertion, or at least the implication, that these fundamental attributes of mental life were based upon the presence in the mind of corresponding and ultimately distinct mental elements.

One of the consequences of the Darwinian revolution was the realization that one of the means for better understanding people was by studying the behavior of animals, in the hope that such studies, conducted under controlled laboratory conditions, would reveal general laws about learning and behavior that were true across all mammals, or perhaps even across all animals. Psychologists were now encouraged to open laboratories to the study of a range of animals: rats, pigeons, and dogs, for example. Angell remarked on

> the rejuvenation of interest in the quasi-biological field which we designate animal psychology. This movement is surely among the most pregnant with which we meet in our own generation. Its problems are in no sense of the merely theoretical and speculative kind, although, like all scientific endeavor, it possesses an intellectual and methodological background on which such problems loom large. But the frontier upon which it is pushing forward its explorations is a region of definite, concrete fact, tangled and confused and often most difficult of access, but nevertheless a region of fact, accessible like all other facts to persistent and intelligent interrogation.[52]

Angell will come back into our story twice more—first, when his student, John B. Watson, rebels against Angell's functionalism and establishes a new framework that he calls *behaviorism*, and then later when he becomes president of Yale University and succeeds in hiring Edward Sapir, the superstar professor of linguistics, away from the University of Chicago.

Psychology in France

A good deal of our attention has been focused on actors and events in Germany and in the United States, but what was happening in France throughout this period can give us a sense of some general patterns as well. France was closer to Germany in every sense than the United States was, yet its culture and tradition were quite different from that of Germany.

In the second half of the nineteenth century, France enthusiastically embraced the methods and goals of experimental psychology that were being developed at a brisker rate in Germany. When we juxtapose the developments in neighboring and competing countries, as we do throughout this book, we find three strategies (in a broad sense of the term).[53]

The first strategy was to develop a distinct approach, well informed about the progress being realized elsewhere but conducted according to the logic and the dynamic that is specific to a particular country. This is basically what we find in Russia during this period, and to a large extent, what we see in Great Britain.

The second strategy was to import and license a foreign trademark on a massive scale, so to speak, which would eventually lead to a transfer of the requisite competencies and ultimately an autonomy which could lead to a transfer of leadership. This is the model that we see very strikingly throughout this book, as the United States engages in a transfer of knowledge and competence from Germany, supplemented by the equally important flight of intellectuals during the Nazi period to the United States.

The third strategy was straightforward competition. This was France's way of dealing with Germany's domination in linguistics: the French linguistic scene got behind the creation of a Paris school that could stand up to the comparative linguists in Germany, and to the young Neogrammarians in particular. But that strategy demanded a leader who had the intellectual stature of a young Saussure. It was less clear in psychology that France could come up with a leader in psychology who had the breadth, the depth, and the energy of a Wilhelm Wundt. There were a few possible candidates who we will mention shortly, but for reasons both personal and structural, an equally powerful French school did not arise.

We should not lose sight of the fact that France had another more structural disadvantage when compared to Germany: France was still suffering from the organization of the university that Napoleon had instituted, one which kept professional training apart from research. The importance of research seminars and laboratories of the sort that were developed in Germany but not in France cannot be overestimated. The sheer number of graduate students, of dissertations, and of ongoing research projects in the German world of psychology not only led to its domination as a model, but also helped the field to become distinct from that of philosophy and from the work done in medical schools.

What France did have during the nineteenth century was a national educational system that put a great deal of effort into formulating a standardized curriculum for philosophy taught to all students in the country during their final year of high school (*lycée*); this education set the stage for the training of advanced students who would then, after earning their advanced degrees, go on to become high school teachers. A small number of these 600 or so teachers each year might hope to move up the academic ladder to a university appointment after a decade or so, but the professional goal of a university student studying philosophy was typically to become a high school teacher and to teach philosophy as a capstone course in the final year of the high school curriculum.[54]

This powerful machine was in large measure the lifework of one man, Victor Cousin, about whom a historian has recently written, "So indelibly did that pedagogical experiment bear the stamp of one Victor Cousin, philosopher and educational administrator, and of the innovations of the early 1830s by which he raised philosophy from its 'hitherto rather humble role in our system of education,' [as Emile Durkheim later put it] that Cousin's predecessors had been all but obliterated from the collective memory."[55] Cousin instituted a composite philosophy in the early 1830s for a country that was on the rebound from a massive social and political revolution in 1789, followed by a Napoleonic empire, and then 15 years of uncertain monarchy. Cousin's philosophy allowed room for an important role to be played by sensation and experience—sensation and experience were often emphasized in philosophies that were sympathetic to social and political revolution—but also for reason, will, and introspection, three faculties that tended to feel more comfortable in the eyes of those who were not so revolutionary.

This "eclectic philosophy" (as it called itself) was the state-sanctioned philosophy, and thus nearly the official philosophy of the nation. It was also devoted to the cultivation both of "spiritualism" (what we would prefer to call today "mentalism") and of science in the study of the mind. This emphasis would naturally lead to the presence of an important component of scientific psychology in the French curriculum even at a time when psychology in Germany had barely begun to emerge.[56]

Théodule Ribot

In some respects, Théodule Ribot was the closest thing that France came to producing a Wundt of its own, and he is often considered the father of psychology in France. Trained as a philosopher, Ribot taught philosophy at the *lycée* level for a number of years before moving up to the higher academic world in Paris. He not only followed closely the developments in psychology in Great Britain and in Germany, he published translations and overviews of that work into French for his fellow francophones to read.[57]

In 1870, Ribot published an influential book on English psychology in which he joined his strong criticism of French spiritual eclecticism with a defense of the possibility of a scientific approach to psychology, which

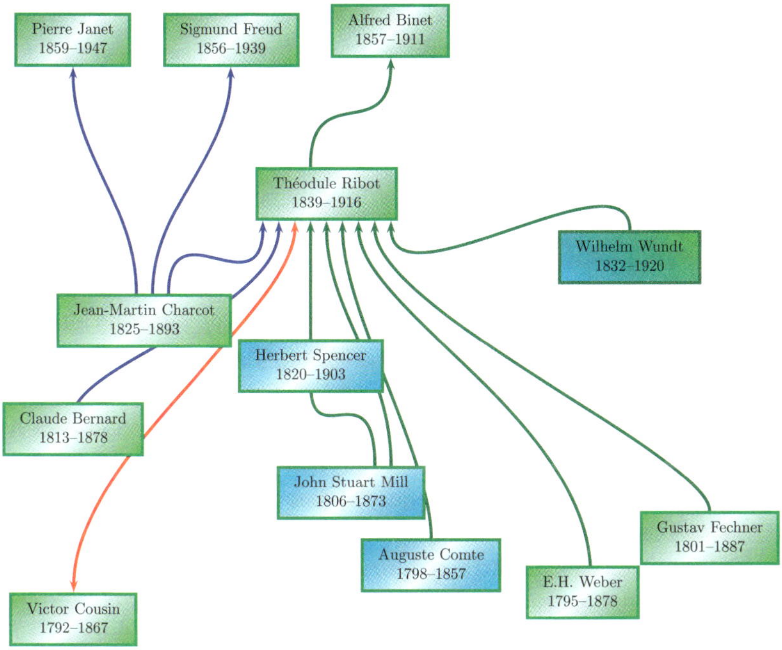

FIGURE 4.4. French psychology

meant a defense against the criticisms offered by Auguste Comte against that very position.[58] Ribot defended the view of *psycho-physiological parallelism*, by which "every psychical state is invariably associated with a nervous state," which not only allowed but obliged Ribot to study both external and physiological regularities and introspective reports.[59]

He founded the *Revue philosophique de France et de l'étranger* in 1876, which brought work from around the European continent to the attention of the French intelligentsia, and introduced his own work, and that of other French psychologists such as Jean-Martin Charcot, to a wider audience.

A year later, when he was 37, he published an article in one of the early issues of *Mind*, a British journal of philosophy devoted to exploring the scientific status of psychology. This article provides a perfect statement of how a generation feels when it looks back at the previous generation with disdain. Ribot's generation was the one with the new understanding of what makes something *science*, and the preceding generation was Cousin's, one that did not understand science, and did not understand that psychology did not want to stand on a foundation of principles offered by speculative philosophy. Psychology wanted to make claims about observations and let itself be challenged by the results of empirical tests:

> Undisputed master of a legion of disciples, Cousin watched over and strictly maintained a philosophical orthodoxy. . . . It was a doctrine without originality, and standing absolutely aloof from the discoveries of science. Its fundamental principle was this: In philosophy, everything has been said; the age of systems is past; all we have to do is to question history, to take what is true out of each system, and from all these elements to form a *perennis philosophia*. Without letting himself be stopped by the fundamental objection that, in order to choose, a criterion must first be determined, Victor Cousin fixed on Spiritualism, which seemed to him more congenial than any other doctrine to the political opinions and religious beliefs of the period and to the French mind. He leaned, above all, on Descartes, that he might give a patriotic and national character to his philosophy. The foundation was to be psychology, disclosing everything to man by mere reflection—his nature, the laws of his mind, morals, aesthetics, the nature and attributes of God. . . . The psychology of Eclecticism was, however, very superficial; it was only a literary expansion of the truths of common sense; the few facts to be met with in it were borrowed from the Scotch . . . the result [of his eclecticism] might be shortly described as "Christianity without miracles."[60]

If that was not bad enough, there was worse to come.

> Eclecticism had always a single criterion—common sense, a single aim—to maintain itself in power by a succession of skilful manoeuvers, especially in regard to the clergy. . . . Obedient to one impulse, and participating in the force which, thanks to centralisation the State possesses in France, the professional body was a real power, and formed a kind of lay clergy. Outside, there were but two classes dissenting from it: the Catholics . . . and the socialist, communist, and humanitarian schools, who were never weary of denouncing the bizarre invention of a State-philosophy.

It was a moment that called for the familiar metaphor of sweeping the stables clean: as two historians of the field have written recently regarding Ribot's task at this moment, "Throughout academia, there was a need to sweep away metaphysical speculations and verbose explanations."[61]

In chapter 2, we discussed the importance of the Ecole Pratique des Hautes Etude (EPHE) in the context of Saussure's appointment there in 1881. A chair was created there at the very same time in the history of psychological theory, and while Ribot was interested in the position, it was not offered to him. He did have the opportunity to teach some classes there beginning in 1885, and three years later, Ribot was offered a chair at the prestigious Collège de France. Ribot himself was not really an experimentalist at heart; in this respect he was much like William James. But there was an expectation that he would develop a laboratory, and so Ribot became engaged in an effort to establish a laboratory at the EPHE, and the government agreed to this in 1889.

Ribot's career was deeply rooted in the clinical treatment of patients, and in the early development of what came to be known as *psychopathology*.[62] He noted that in cases of amnesia, there was a strong tendency for recent memories to be lost first, and earlier memories to be lost only later, if at all. He called this the Law of Regression, but the world has come to know it as Ribot's Law. We will see an echo of this in chapter 9, when Roman Jakobson came to study aphasia some 60 years later; Jakobson's observation is better known in linguistics than Ribot's original statement is. Ribot's professional linkage between medicine and psychology was typical of the situation for psychologists in France in the late nineteenth century; it was in medical schools that research laboratories were developed in France for psychological research, and of course the need for clinical aid for patients led to an orientation towards questions of psychopathology.

Alfred Binet

Alfred Binet is best remembered today for developing the first practically useful examination for measuring intelligence, and his name is still found in what we now call the Stanford-Binet examination. He worked in a range of areas of psychology during the period of psychology's efflorescence, but the impact of his work would have been greater if psychology had developed in France within a university structure of the sort Wundt and his colleagues enjoyed in Germany at the same time.[63]

Binet did not have traditional academic training, either in philosophy or in psychology. In the late 1870s, he studied psychology and philosophy in books and journals, and was very much taken with John Stuart Mill's point of view. Théodule Ribot encouraged him to write papers, and he published them in his *Revue philosophique*. He worked for seven years with the world-renowned experimentalist and pathologist Jean-Martin Charcot at Salpêtrière, the major hospital in Paris, but this period ended very badly for Binet. One of the projects that Binet worked on involved hypnosis (Charcot's specialty), hysteria, and magnetism (the sort created by magnets). Over a period of years, Binet, Charcot, and another colleague named Charles Féré claimed that they could produce extraordinary effects on hypnotized subjects by using magnets. Other laboratories in France were unable to reproduce the results, but it was a researcher from Liège, in Belgium, who with a deft pen proceeded to argue that the results from the Charcot lab were almost certainly the result of suggestions made by the experimenters in the presence of the hypnotized subjects, or else were the result of the experimenter being trained by the subject![64] After several years of published barbs in the journals, Binet was obliged to admit that their results were largely the result of suggestion, and he left Charcot's laboratory with a renewed sense of how a person could be led to believe things by virtue of wanting to believe them.

We noted above that Ribot succeeded in establishing a psychology laboratory in 1889 as part of the EPHE, with Henri Beaunis as the lab's first director. Beaunis asked Binet to come on board with the project, making him associate director in 1892, and when Beaunis retired in 1894, Binet became director. The laboratory was never a great success in the way that Wundt's was in Leipzig. A number of explanations for this have been suggested, ranging from a domineering side of Binet's personality—but when was that a flaw in a laboratory director?—to the fact that students earned no official degree spending time in the laboratory. There was nothing to show for their work that they could take home and use if they were not French, and if they were French, there were hardly any teaching positions in psychology to be had.

Binet remained director of this laboratory for the rest of his career, and he was not able to obtain a chair as a university professor. He was a close friend of Paul Passy, an important linguist who was awarded a new chair in phonetics at the EPHE in 1894. Binet wrote to Passy in 1901, when Ribot had resigned his position at the Collège de France:

> You know perhaps that Ribot has just resigned, and that I am presenting myself against [Pierre Janet] to replace him. It will be a rough campaign, in which I am happily supported in the most vigorous manner, and if I lose, it will not

> be my fault. . . . It is over twenty years that I have been active in psychology, as you know; I educated myself all alone, without any teacher [*maître*]; and I have arrived at my present scientific situation by the sole force of my fists; *no one*, you understand well, no one has ever helped me. I have done experimental psychology—the title of Ribot's chair—and I am really the only one in France who has done so.[65]

But Janet was given the chair. When Binet applied for Janet's previous position, it was given to George Dumas. Both Janet and Dumas had been professionally close to Charcot in a way that Binet had not been; Janet had been Charcot's "substitute professor" for a number of years at the Collège de France.

Binet remained committed to the research we remember him for, the measure of intelligence. Early in his work, he tried to use the ideas of Paul Broca on phrenology, which offered the hope (if that is the right word) that a person's *mental* traits could be inferred from measurements of various geometrical aspects of the person's skull. Eventually he convinced himself that this approach had no future, and he went on to develop different sorts of tests. He had two daughters, and he learned a great deal by observing them in a naturalistic context, as Jean Piaget would do years later, and as Gestalt psychologists would do as well.

Working together with Théodore Simon, a colleague from his days at Charcot's lab, Binet created tests for children that would allow a measurement of intelligence in an experimental context.[66] As responsibility for education passed from the Catholic church to the French government, the usefulness of these tests became clear. Binet studied the nature of the learning deficits of the mentally handicapped, and in 1905 at the World Congress of Psychology, he and Simon presented a set of psychometric tests and a scale to mark the mental stage of a child as indicated by these tests. This was perhaps the high-water mark of the development of psychometric psychology, and Binet would work on this Binet-Simon scale for the rest of his life. Binet and Simon were clear on the point that they were not measuring the intellectual *age*, but rather the intellectual *level*, but this nicety was often lost. Just a few years later, the notion of intellectual quotient (better known as IQ, and sometimes called "intelligence quotient") was proposed (by Wilhelm Stern), and we all know how widely that measure has been adopted.

Some years back, Stephen Jay Gould dramatically brought out the uses to which intelligence measurements have been put, and he recognized

Binet's realistic qualms. He wrote, "Binet also had a social motive for his reticence. He greatly feared that his practical device, if reified as an entity, could be perverted and used as an indelible label, rather than as a guide for identifying children who needed help."[67] The importation of Binet's tests to the United States would also open up a veritable industry of testing, one which was of great use to educators, to employers, and to the military in a time of vast conscriptions during the world wars. As Gould documents, Binet's tests were useful to what he called "pioneers of hereditarianism," "a home-grown American product" that was not a part of the culture that lay behind Binet's life work.[68]

Paul Broca

We have seen certain parallels between the development of physical anthropology in the nineteenth century and the classification of languages and their morphological systems, which in turn led to a hierarchical view of linguistic systems. This is not the only influence that physical anthropology would have on the history of linguistics. In France, one of the important proponents of physical anthropology was Paul Broca, whom we have already encountered in connection with the cephalic index. Broca was a physiologist, an anthropologist, and a medical doctor, and he was naturally influenced by the positivist, empiricist, and scientistic atmosphere of the period—he was what was often called at the time a "free thinker," and he opposed conservative Catholic tendencies. He founded the Society of Freethinkers in 1848, and was denounced as a dangerous revolutionary agitator! Broca began his career as a physiologist working on hybridization in the animal kingdom—thus close in some ways to Darwin, and in other ways opposed to him. Banished from the conservative Biology Society, he founded the Société d'anthropologie de Paris in 1859, bringing together progressives and atheists opposed to the Ethnological Society of Paris, which was itself under the control of conservative Catholics, and which in 1863 saw a subgroup split off to become an organization we have already encountered: the Société linguistique de Paris. The Société linguistique de Paris progressively moved away from its founding conservatism and eventually became more generally the professional home for all French linguists.

As an anthropologist and doctor, Broca was not interested in typologies or hierarchical arrangements of languages of the Schleicherian sort. As a craniologist, he was interested in language as a species-specific characteristic of the human mind. He was associated as a doctor at the Bicêtre Hospital in Paris, where there was a patient, a certain Monsieur Leborgne, who had been there for 20 years. Leborgne's right leg was paralyzed, and while he understood what was said to him, he was able to utter only the syllable "tan"—which earned for him the nickname Tan-tan. He died on April 17, 1861, and when Broca removed his brain (which is still preserved in Paris), he discovered a cerebral lesion on the third left frontal convolution.

Broca drew the conclusion that the damage to this zone, which is now known as Broca's area, was responsible for Tan-tan's aphasia, and six months later, he presented a paper to the Société d'anthropologie de Paris on his discovery. He presented the material again before the Société d'anatomie de Paris, and this unleashed an enormous polemical discussion. The notion that cognitive functions were localized in the brain now had strong scientific support.

In chapter 3, we discussed Gall's theory of phrenology, in connection with Auguste Comte's general positive theory of mind and society. Gall's perspective represented one extreme end of a controversy that continues to the present time—with the goal of analyzing the powers of the mind into separate and autonomous faculties, a view that fit well with the spirit of nineteenth-century analysis and which would appeal to Paul Broca in his effort to understand the effects of lesions on the brain.

The polemics between the two views was fierce, and reached its apogee at the International Congress of Medicine in London (1881), where the German anatomist Friedrich Goltz presented surgical experiments he had performed on animals in support of his holistic and integrated view of the functioning of the brain, and where the Scotsman David Ferrier responded by presenting other surgical experiments supporting a strong localist position. Ferrier's position carried the day, and localism became the standard view for a good while after that.

But counterbalancing the importance of "Tan-tan" for Broca's localism was the case of Phineas Gage, a foreman who worked on the construction of railroads. In September 1848, Gage was at a worksite in Vermont when an iron bar went entirely through his left frontal lobe. It caused no significant damage to his cognitive or motor abilities, though he lost an eye, but his personality underwent significant changes. His case was an important one for the antilocalists, both for those espousing a totally holistic view and those advancing an associationist view.

In 1874, Carl Wernicke presented the case of a patient with damage to another area in the brain, one which would come to be known as

"Wernicke's area." Damage to this part of the brain is associated with aphasias of comprehension and reception, and the discovery appeared to support the localist position, since a specific area of the brain had been identified with a particular effect. But the defenders of the holistic position showed the importance of the arcuate fasciculus, a bundle of axons that connects Broca's area and Wernicke's area, playing an important role in linguistic functioning. Wernicke himself did not defend the localist position, arguing rather for an associationism linking a number of areas involved in the functioning of language, both in production and in reception.

The debate over localism, associationism, and holism would continue over the years to come, and we will see it next in the controversies surrounding Karl Lashley's holistic view during the heyday of behaviorism.

The Unity of Mankind—and the Differentiation of Types of Humans

Here is a question that nineteenth-century people asked themselves: What makes us different?

There is a lot packed into that question, to be sure: who does the word "us" refer to? And different from *whom*, anyway?

We noted earlier that the turn towards India in nineteenth-century linguistic thinking, along with its consequent fascination with things Aryan, was an impulse that permitted an alternative conception of how human beings across the globe were related to one another, a historical alternative that was most notably different from the conception offered by the Bible. The nineteenth century was a century in which the differentiation of mankind was explored and accounted for in a range of ways, often rejecting both traditional Western religion and the Enlightenment tradition that all men are one. Once again, the interest of this development for us is not antiquarian, but rather flows from the fact that we are still engaged today in trying to come to grips with conflicting systems of beliefs in this area. Whether it is in the context of a casual conversation, or a political speech, or an academic debate, people ask, who are we? And what makes us different?

In linguistics, one major school of thought today, the one associated with Noam Chomsky, sees the broadest horizon of explanation as deriving from universal grammar, a faculty of language shared by all humans, and this is an image that would feel quite comfortable for an eighteenth-century scholar, someone who was part of the Enlightenment, if he were to wake up today. Whether or not one adopts that particular ahistoricism—and many today, especially in the social sciences that are distant from linguistics, do not—it is impossible for a science of humankind to avoid the question: what makes us different from one another? It is impossible to avoid that question because people will not stop asking the question even if a prevailing theoretical framework has little to say about it; certainly different theoretical frameworks are equipped to different degrees to answer the basic questions of how we are all alike, and what makes us different.

The great comparative linguistics of the nineteenth century that we reviewed in chapter 2 naturally led to a sense that two peoples who have languages that were once one and the same must have a good deal in common, culturally and (if we accept the notion) spiritually. But that can leave the nagging feeling that peoples whose languages have no recognizable roots in common with ours are different.

In linguistics today, the question of who we are is largely understood with the interpretation that *we* means the entire human race, and the other, the one who does not have language, is constituted by the now extinct lines of descent from the original genus *Homo* who emerged on the scene two million years ago, whose descendants are no longer alive, and who probably died off because they did not have language as we know it.

Over time, the tacit understanding of just who counts as *us*, when we ask about what makes us different, changes too. And when we ask what makes us different, there is always a tacit phrase in there: who is it that we are different from?

Social forces of all sorts were responsible for confrontations that led people to think about *us* and *them* in the nineteenth century. Slavery was rapidly abolished—in 1830 in England, in 1840 in France, and in the mid-1860s in the United States. But colonial empires continued to thrive and expand, and the growth of nation-states only made more stark the confrontations of differences and inequalities that were found in cities. Everywhere the question was asked: is this man my equal, this man whose skin color is different, whose language, culture, religion, and everyday mode of existence are so dissimilar from my own?

The fact is that over the nineteenth century we observe a confrontation between the universalism of the Enlightenment and its opposite, a sort of *differentialism*, which ranged from an interest in how individual characteristics varied within social groups, to an interest in how they varied across groups, and the answers that arose to these questions could become potent instruments in the hands of those promoting a political or social view that disenfranchised those whose skin color was different, or whose great-grandparents were Jewish. As we will see in the next chapters, linguists have long weighed in on these issues, and usually—not always, but usually—in ways that seem judicious to us today.

Anyone who spends a lifetime studying cultures and societies that are not their own is passionately interested in the differences that they encounter, but the differentialism that this natural enthusiasm engenders can cross the spectrum from value-free typologies and rich ethnographies to the kind of heavy-handed judgmentalism that surfaces as speculation regarding the ways in which a society's language prevents its speakers from thinking abstract thoughts. Linguistics moved away from that latter way of conceiving of differences among cultures in the 1870s, with the changes associated with the third generation of linguists, in the 1870s, as we noted in chapter 2. But we have been struck by the ways in which valid and responsible studies of societies and cultures can be recruited in the service of more sordid ends, as has been described by such scientists as Stephen Jay Gould in the United States and Jean-Paul Demoule more recently in France.[69]

The strong inclination to measure and to classify during the nineteenth century also left its mark on the development of sociology and anthropology, two fields that are very close to the mind sciences we have focused on in this book. Anthropology arose as a science that took as its aim the development of a classification and an understanding of the physical and cultural characteristics of human groups, and their development over time, with attention to physical, cultural, and social characteristics. James George Frazer published in 1890 a monumental book, *The Golden Bough: A Study in Comparative Religion*, which would stimulate further interest in the study of myths, rituals, and religions. A number of anthropologists would have tremendous influence on linguists who followed them, such as J. R. Firth in England, who was greatly influenced by the anthropologist Bronislaw Malinowski, and Franz Boas in the United States, who was greatly influenced by Rudolf Virchow (Boas will figure prominently in chapter 6).

Anthropologists and biologists have not infrequently felt called upon to stand in judgment on the number and nature of the races of mankind. Darwin expressed his dissatisfaction with the question, never mind the answer offered to it:

> The question whether mankind consists of one or several species has of late years been much agitated by anthropologists, who are divided into two schools of monogenists and polygenists. Those who do not admit the principle of evolution, must look at species either as separate creations or as in some manner distinct entities; and they must decide what forms to rank as species by the analogy of other organic beings which are commonly thus received. But it is a hopeless endeavour to decide this point on sound grounds, until some definition of the term "species" is generally accepted; and the definition must not include an element which cannot possibly be ascertained, such as an act of creation. We might as well attempt without any definition to decide whether a certain number of houses should be called a village, or town, or city. . . . Those naturalists, on the other hand, who admit the principle of evolution . . . will feel no doubt that all the races of man are descended from a single primitive stock; whether or not they think fit to designate them as distinct species, for the sake of expressing their amount of difference.[70]

Darwin was skeptical whether the speciation that he observed, and applied in his account of the gradual unfolding of the tree of life over the course of countless generations, could be meaningfully applied to understanding the humans who are alive today:

> Man has been studied more carefully than any other animal, and yet there is the greatest possible diversity amongst capable judges whether he should be classed as a single species or race, or as two (Virey), as three (Jacquinot), as four (Kant), five (Blumenbach), six (Buffon), seven (Hunter), eight (Agassiz), eleven (Pickering), fifteen (Bory de St-Vincent), sixteen (Desmoulins), twenty-two (Morton), sixty (Crawfurd), or as sixty-three, according to Burke. This diversity of judgment does not prove that the races ought not to be ranked as species, but it shews that they graduate into each other, and that it is hardly possible to discover clear distinctive characters between them.[71]

The Era of Machines

A historian such as Eric Hobsbawm reminds us that at the heart of the nineteenth century are found revolution and great growth of capital, both of which played out on a stage that included the rapid development of cities and those who moved there looking for work. The metaphorical engine that drove this industrial revolution was, quite literally, the physical engine.

Civilization had used machines for 2,000 years and had effectively employed wheels, pulleys, and screws for all that time, making the efforts of humans and animals more effective. Leonardo da Vinci dazzles us even today with the mechanical devices his imagination created. But now in the nineteenth century, new sources of energy were being tapped for making bigger, faster, and more powerful devices. There were steam engines, and then there were internal combustion engines, and while they changed the face of the world, they are not the ones that we wish to focus on here. They were very important, but the ones we want to focus on are the machines that began to display intelligence, just a little bit of intelligence.

It was 200 years earlier, and more, that rapid changes had begun in how the world was seen by thinking people. The scientific revolution moved into high gear when at the end of the sixteenth century Galileo challenged the way that space, time, and motion were conceived. Gradually the view took hold that things in the world interacted mechanically, by bumping into each other in various ways. Look closely, these early scientists said, and you will see that things move until they are stopped by something else; objects interact and then move on. The gears that make up a clock provide a clean example of this: watches work because all the gears interact locally and immediately with each other, mostly two at a time. Strings and chains can be attached to gears and then pull other things: we knew that water mills worked that way, and maybe human muscles did also. These objects interacted with each other in *mechanical* ways, which means based on local interactions of contact between the two objects. And objects that can be understood and explained in such terms and nothing more were mechanical objects, or "machines."[72]

It is not difficult for the reader today to understand what René Descartes was getting at when he said that the human body was very much like a machine, nor when De la Mettrie famously declared that a human being *was* a machine. But what is significantly more difficult for the reader today is to recognize that we are still engaged in figuring out what we mean by machines and what aspects of the universe (including ourselves) can reasonably be described as "machines." As our understanding of the physical world has expanded and matured, we are able to account for more aspects of the universe by using the intellectual tools that evolved out of Descartes's and Newton's mathematics, and we are all inclined to accept the view that a process is mechanical if it can be thoroughly accounted for by such mathematical methods. But the science, the mathematics, and the logic that underlie what we mean when we declare a process to be mechanical have radically shifted over the past few centuries, and so has the engineering capacities of modern science to build tools to our own design. What counts, then, as a machine and as a mechanical process has evolved and changed, and the changes in how we view mind and language that we study in this book are part and parcel of those changes.[73]

For Descartes, an animal was a self-moving machine—and an observer from a distant planet might be excused for not seeing, at first blush, that a human being was nothing but a self-moving machine too. The most advanced machines of the age were the exemplars of machines which thinkers used to express their opinions about the human condition. If there could be an intelligent machine, it would have to be something like a clock, a watch, or a chronometer, built from springs, gears, or chains and possessing some internal source of energy. Unlike water clocks and sundials, clocks with gears kept their internal structure hidden from all but the initiated; this kind of mechanism, a mechanical structure built by humans, lent itself naturally to a perspective in which our universe was the handiwork of a divine watchmaker who built clocks whose face was visible to us but whose inner workings were at least for now hidden.

The middle of the seventeenth century was the moment when new ideas in this area exploded; it was the time of Descartes, of Newton, and of Blaise Pascal, all three philosophers and imaginative mathematicians. Pascal was particularly interested in building a machine to carry out arithmetic operations in a way that would be useful for people who needed to carry out many such operations quickly and accurately. The machine was later called Pascaline, in his honor, and its existence spurred others on. The philosopher and mathematician Gottfried Leibniz saw Pascal's machine and decided to go one better, and build a machine that would calculate multiplications as well. Over the course of the next century, watchmakers and inventors built successively more complex and impressive machines that imitated human intelligence and behavior. In retrospect, we can identify an important moment in this development, when the pattern of successive acts that the machine needed to perform could be built into a simple device whose sole function was to repeat the pattern, just like a music box that plays a familiar tune once we have transferred the notes to a rotating cylinder within it. The same technology was applied to the task of weaving inside of a loom, and Joseph-Marie Jacquard, later in the eighteenth century, went so far as to create a system in which punched cards were used to control the pattern of weaving, with each card used to define the pattern of one single row. By the middle of the twentieth century and well into the 1960s, this idea was used as a standard method to enter data into a digital computer. We used to call them "punch cards."

It was the Englishman Charles Babbage who deepened the theoretical reflections on how machines could in real and practical terms revolutionize the computation of mathematical tables and formulas, eliminating humans from the workflow, and it was Ada Lovelace, a talented mathematician and child of the poet Lord Byron, who was able to clearly see the mathematical character of what Babbage's machine could eventually accomplish.

Someone today wishing to get a sense of how Babbage and his contemporaries viewed this machine has no choice but to go back to an article published in 1842, in Switzerland and in French. It was written by an Italian engineer from Turin, Count Federico Luigi Menabrea, who had attended a lecture that Babbage had given two years earlier. Menabrea's paper was in turn translated by Ada Lovelace into English, and she and Babbage added copious notes to Menabrea's original. How much of the addition was due to Lovelace and how much to Babbage remains a point of contention.[74]

Let us look at the beginning of this paper, because it gives such a clear sketch of the ideas in the background:

> Those labours which belong to the various branches of the mathematical sciences, although on first consideration they seem to be the exclusive province of intellect, may, nevertheless, be divided into two distinct sections; one of which may be called the mechanical, because it is subjected to precise and invariable laws, that are capable of being expressed by means of the operations of matter [*physiquement*]; while the other, demanding the intervention of reasoning, belongs more specially to the domain of the understanding [*pensée*]. This admitted, we may propose to execute, by means of machinery, the mechanical branch of these labours, reserving for pure intellect that which depends on the reasoning faculties [singular, not plural, in the original]. Thus the rigid exactness [rigor, in the original] of those laws which regulate numerical calculations must frequently have suggested the employment of material instruments, either for executing the whole of such calculations or for abridging them; and thence have arisen several inventions having this object in view, but which have in general but partially attained it.

This clear division of mathematical thought into a part that could be called "mechanical" and a part that could not would become increasingly important, and it will be the focus of our attention in chapter 8.

There is much that is strikingly modern in the discussion of the way in which the "analytic engine" is designed in order to be as general as possible. Lovelace explained the way in which values of numbers could be held in a number of distinct places in the machine, but when they were to be used in the computation of a new value, they would be copied to the central calculating unit, which Babbage called the *mill*. The cards on which the sequence of intended operations have been punched are called "operation cards," and these "merely determine the succession of operations in a general manner. They in fact throw all that portion of the mechanism included in the mill into a series of different states, which we may call the *adding state*, or the *multiplying state*, etc. respectively." This is precisely what we say today, when talking about a finite-state device, or a Turing machine: it enters into a state, which consists precisely of the readiness to perform an operation (addition, multiplication, etc., according to which state it is on) on variables that exist in the mill—or as we say today, in a register. "In each of these states the mechanism is ready to act in the way peculiar to that state, on any pair of numbers which may be permitted to come within its sphere of action."

Lovelace was just as clear in explaining the importance of conceptualizing the mathematical operation as a set of recurring operations forming a cycle (we would say a "loop" today), a relation that can be extended to cycles of cycles, and so on:

> Wherever a general term exists, there will be a recurring group of operations, as in the above example. Both for brevity and for distinctness, a recurring group is called a cycle. A cycle of operations, then, must be understood to signify any set of operations which is repeated more than once. It is equally a cycle, whether it be repeated twice only, or an indefinite number of times; for it is the fact of a repetition occurring at all that constitutes it such. In many cases of analysis there is a recurring group of one or more cycles; that is, a cycle of a cycle, or a cycle of cycles.

At the same time she emphasized the fact that the design of a set of operations for the engine made clear the abstract character of each operation as a higher order entity:

> In studying the action of the Analytical Engine, we find that the peculiar and independent nature of the considerations which in all mathematical analysis belong to operations, as distinguished from the objects operated upon and

> from the results of the operations performed upon those objects, is very strikingly defined and separated.

Just a bit later, she added:

> But the science of operations, as derived from mathematics more especially, is a science of itself, and has its own abstract truth and value; just as logic has its own peculiar truth and value, independently of the subjects to which we may apply its reasonings and processes. Those who are accustomed to some of the more modern views of the above subject, will know that a few fundamental relations being true, certain other combinations of relations must of necessity follow; combinations unlimited in variety and extent if the deductions from the primary relations be carried on far enough. They will also be aware that one main reason why the separate nature of the science of operations has been little felt, and in general little dwelt on, is the shifting meaning of many of the symbols used in mathematical notation.

Lovelace left no doubt that Babbage's engine was a machine that reasons—that embodies reason—and that this machine, once it was made real out of disks and gears of metal, would change the way we understand thought and reasoning:

> In enabling mechanism to combine together general symbols in successions of unlimited variety and extent, a uniting link is established between the operations of matter and the abstract mental processes of the most abstract branch of mathematical science. A new, a vast, and a powerful language is developed for the future use of analysis, in which to wield its truths so that these may become of more speedy and accurate practical application for the purposes of mankind than the means hitherto in our possession have rendered possible. Thus not only the mental and the material, but the theoretical and the practical in the mathematical world, are brought into more intimate and effective connexion with each other. We are not aware of its being on record that anything partaking in the nature of what is so well designated the Analytical Engine has been hitherto proposed, or even thought of, as a practical possibility, any more than the idea of a thinking or of a reasoning machine.

In the end, Babbage was not able to build his analytical engine during his lifetime. We would have to wait for the skills of the engineer to grow and for the purse strings of governments to open, and it was not Babbage's machine that was first built, but those of others working in the 1930s and 1940s. But Babbage's and Lovelace's ideas fired the imaginations of generations to come, and today's computers are a direct result of their ideas in the first half of the nineteenth century.

We will return to this story in chapter 8, when we discuss the origins of the Turing machine (and its close relatives) in the 1930s.

Moving On

The last three chapters have prepared us for exploring and better understanding the development of the mind sciences in the twentieth century. We have looked just a bit at the political and social changes going on in Europe and the United States during the nineteenth century, but we have seen in much greater detail the ways in which thinkers of all sorts began to create new sciences: sciences of humankind, sciences of the mind, sciences of language. No one knew for sure the right way to create a new science: no one ever does. Some thought that doing science meant finding new questions to ask, and that decision often went along with dropping some useless old questions. Others thought that embracing science meant the development of new methods, and those methods frequently embraced measurements and other sorts of quantitative innovations. Some thought, perhaps darkly, that science was at its best when it could be pursued for its own sake, without any practical end in sight, while others were caught up in intellectual movements that loudly proclaimed their relevance for both the present and for the future.

In the next five chapters, we will do our best to pull apart the major trends in psychology, linguistics, philosophy, and logic, with a focus on those themes that were shared among these disciplines, themes that passed back and forth across, or beneath, the disciplinary fences. The anxiety over the nature of science, we will see, remained a major preoccupation throughout this period.

CHAPTER FIVE

Psychology, 1900–1940

Human consciousness? Up until the turn of the century in 1900, there would simply have been no controversy raised—nor even an eyebrow—if one were to say that psychology was the study of human consciousness, pure and simple. That consensus would soon be lost, and the focus on consciousness would become controversial. Eventually consciousness would be entirely lost to psychological study as it was conducted by just about the entire discipline.

In this chapter, we will take a bird's-eye view of psychology in the period from 1900 to 1940, following up on our briefer tour in the preceding chapter of the period at the end of the nineteenth century. We will look first at American psychology, but it is not possible to look at American psychology during this period without at the same time looking at European trends as well, primarily German, because as in virtually all academic fields, American scholars still viewed the Old World as more distinguished than their own. And we will see the start of a wave of European scholars who would come to the United States not just to visit or give a lecture, but to come, to stay, to work.

We will look first at the continuation of the trends we have already seen in the nineteenth century: structuralism and functionalism. Both continued to evolve over the first 40 years. But in 1913, a new voice in psychology was heard, one that was young, brash, impertinent, and impatient; it was the voice of *behaviorism*, and John B. Watson was the one who said that if we want to do this new kind of psychology, we are going to have to get rid of a lot of what we used to do, especially introspecting, and talking about perceptions, conceptions, and mind.

Watson's version of behaviorism was no doubt the most extreme. Behaviorism softened its rhetoric in its second formulation, the neobehaviorist moment, best exemplified by the work of Clark Hull and of Edward Tolman. And there were many psychologists who were turned off by behaviorism, even among those who were committed to studying learning in animals, such as Karl Lashley.

Some of the German scholars who came to the United States during this period brought with them *Gestalt psychology*, which was *very* different in outlook from most of what people were doing in American psychology. Gestalt psychology came to play a major role in the development of psychological thought in the United States, even if both American and European scholars implicitly grasped a stipulation that to be a real Gestalt psychologist, you must have come from Germany. Americans who admired Gestalt psychology, like Edward Tolman, embraced the ideas, but hesitated and in the end did not declare they were Gestaltists.

Structuralism and Functionalism

We saw in the previous chapter that Wilhelm Wundt was the central figure in German psychology at the turn of the century, and two of the major figures in American psychology in the twentieth century, Edward Titchener and G. Stanley Hall, both studied with Wundt in Germany, and brought back what they understood Wundt's teachings to be.

In the first decade of the twentieth century, most likely if you asked an American psychologist how his field was laid out, he would have said that it could be placed along a spectrum stretching from structural psychology, inspired by Wundt, on the one end, to functional psychology, a more American perspective, on the other. We have met James Rowland Angell already—a psychologist at the University of Chicago. He was president of the American Psychological Association in 1906, and he took the opportunity of his presidential address to explain to his colleagues how he viewed the disagreements in the field between these two camps. There was, he said, "a small but nutritious core of agreement in the structure-function apple of discord."[1]

Angell began his presentation with an expression of modesty; it is difficult to know in retrospect just how he meant it to be taken. "Functional psychology," he said, "is at the present moment little more than a point of view, a program, an ambition." Modesty, yes, but also an excuse, if one should ever be needed, for a lack of positive and lasting results. He was particularly aware that his discipline was rent by methodological disputes, and he explained that the functional psychology that he knew was one that "gains its vitality primarily perhaps as a protest against the exclusive excellence of another starting point for the study of mind": presumably he is referring to introspection.[2] To be a functional psychologist, then, is to be a protester carrying a banner down the Main Street of academic psychology, not a defender of any orthodoxy. Indeed, functional psychology "enjoys for the time being at least the peculiar vigor which commonly attaches to Protestantism of any sort in its early stages before it has become respectable and orthodox." We know what he meant. It is the feeling of liberation that comes about with a new perspective, before the bourgeois comforts have set in. (Angell was 37 years old at this point, and had been a professor at Chicago for nearly 15 years.)

In any event, Angell knew that the best way to explain his functional psychology would be to explain how it differed from structural psychology, which he undoubtedly saw as the dominant worldview in psychology. The functionalist seeks to understand *the operations of consciousness*, while the structuralist seeks to analyze and describe "its elementary and complex *contents*." The structuralist introspects, and seeks to analyze the experience that he discovers into component factors. The functionalist thinks that the structuralist is so focused on analyzing the introspective moment that he loses sight of how artificial that whole act is; the functionalist wants to analyze mental processes, which typically are efforts—that is, dynamic attempts to do something.

Angell underscored the fact that for the functionalists, it was natural to undertake experimental designs in which the subject was *never* asked to introspect: he was simply asked to do something, and the amount of work he successfully accomplished could be taken as the subject of a valid scientific protocol. The modern reader might be excused if he fails to note that what Angell is describing there could serve as a description of most of the work in modern cognitive psychology: an effort to learn how higher order thought works, not by asking the subject how it is done, but rather by giving him a task and seeing what measurable quantities are available that provide evidence for the internal processes that are operative.

John B. Watson and Behaviorism

The behaviorist movement in psychology was founded and named by John Broadus Watson, a psychologist trained at the University of Chicago—or so goes the story, which, as always, leaves out some not unimportant people. There were others, like Edward Thorndike and Max Meyer, who were also inventing behaviorism at the same time. The famous founding act occurred in 1913, when Watson published "Psychology as the Behaviorist Views It." Watson himself came from a dirt-poor family in South Carolina, of a father whose lot had gone steadily downhill in life, and who abandoned his family when it suited him. Watson managed to overcome early adversity; as a young man, he studied philosophy at the University of Chicago. Watson did not find Dewey's classes at Chicago appealing in the slightest, and he decided to work instead with James Angell and Henry Donaldson. He was also influenced by Jacques Loeb, a German professor of physiology at Chicago who was one of the founding faculty members of the University. But it was Angell who he was closest to; Watson knew he was *Angell's* student, Watson was indebted to *him* in every way, and Watson was not comfortable with thinking about moving outside of Angell's intellectual comfort zone while he was at Chicago.[3]

Watson stayed on for several years at the University of Chicago as an instructor after getting his PhD, with support—both professional and moral—from Angell. Watson then took a job, in 1908, as professor of experimental psychology at Johns Hopkins University.

In his letter of recommendation for Watson, Angell wrote that he would rather have Watson "twice over any man of his generation. He is better balanced, better trained and more effective as a university man than any other fellow of his generation."[4] And Watson recognized his own indebtedness, both to his Chicago mentors, but also to Edward Titchener at Cornell.

Titchener, we saw just above, was the dean of American structuralism, which Watson hoped to see jettisoned from the academic firmament. But on both a personal and a professional plane, Watson was close to Titchener for a good deal of his lifetime. Watson wrote to Titchener shortly after arriving at Johns Hopkins,

> I think I wrote you once about my regard for you. Angell and Donaldson have been like parents to me and I am sure that they will live in my memory as long as I live. My first debt is to them. It is an intellectual, social and moral debt. After these two men I have always placed your work and what I know of you personally. I am not so sure that I do not owe you as much as I owe them, I think if I had to say where the stimulus for hard persistent research came from I should have to point to you.[5]

The tone is honest, and the feeling was genuine. The letter is touching.

Watson thus saw himself, in the period just before his call for a new behaviorism, as deeply connected, both intellectually and personally, to the leaders of the two most prominent schools of psychology at the time, Titchener and Angell—a relationship that certainly played a distinct role in his obtaining a professorial appointment at Johns Hopkins University. We emphasize this point because it is not what one might have expected, considering the position he was about to take in the field—a position which seems like that of the outsider trying to break in, when he was anything but.

Watson's career over the next ten years went great guns. In 1913, he published the broadside and manifesto, "Psychology as the Behaviorist Views It," which catapulted behaviorism to the status of a major movement in the field of psychology; in 1919, he published *Psychology from the Standpoint of a Behaviorist*, which was to be even more influential, and in 1925, the book *Behaviorism*.

Watson married Mary Ickes Watson in 1904, and they had a daughter, also named Mary (but called "Polly"), in 1905, and then a son, John. This may not seem important, but Watson's views on psychology can hardly be separated from his views on child rearing, and his hugely successful book, *The Psychological Care of Infant and Child*, published in 1928, had an enormous impact on how children were brought up in the United States for a whole generation.

Polly and John were reared according to Watson's regime of strict discipline and no display of affection. They both led troubled lives. Polly had a daughter who is well known to many Americans who watch TV, Mariette Hartley; she is remembered by many for a series of delightful television commercials she did in the 1970s for Polaroid, in which she seemed for all the world like James Garner's wife (though she wasn't). In 1990 she published a memoir of her family life, *Breaking the Silence*, and wrote this stinging critique of John Broadus Watson, her grandfather, and his ideas:

> Grandfather's theories infected my mother's life, my life, and the lives of millions. How do you break a legacy? How do you keep from passing a debilitating inheritance down, generation to generation, like a genetic flaw?[6]

Watson's academic career came screeching to a halt in 1920, after it was discovered that he had begun an affair with one of his graduate students, Rosalie Rayner. His wife demanded a divorce, Johns Hopkins booted him out the door, and Watson left academia. Watson married Miss Rayner, he joined the advertising firm of J. Walter Thompson, and continued to write about behaviorism over the decade that came to be known as the Roaring Twenties.

The attraction of behaviorism

What is it about behaviorism that deserves our attention? There is the fact that it served as a focal point and a gathering place for young psychologists who were ready to find and adopt a new thing, one which would with great confidence announce that it was not just warmed up psychology of the 1890s. Casting behaviorism in this way made it appealing. It consisted of both a description of the Old, and an offer of the New. In some respects, the description of the Old was caricatural, and what was offered as New was not entirely new. It offered a clear and deliberate rejection of what had preceded, and an intellectual argument for it.

The call for behaviorism began with a familiar imperative: the old psychology is moribund, and the new psychology is ready to take over. This was an Augean challenge, announcing the need to wipe the world of psychological concepts clean of any notions that show the traces of past metaphysics; this is the same message that drove the appeal of positivism in the nineteenth century. The behaviorist meant by this that the new psychology would eliminate the notion of mind and everything that was tightly linked to it: perceptions, conceptions, and images, and that was just the beginning. Watson was clear: he wanted to completely replace the perspectives of William James, Edward Titchener, or James Angell, and change the very definition of psychology. His goal was to change it from the study of consciousness to the study of behavior.

And today, if behaviorism is a marginalized perspective, with few psychologists willing to say that they are behaviorists, we should note this: consciousness as such is no longer widely viewed as the principal object of study of psychology. Mind may well be center stage again in psychology, a fact that most behaviorists would object to (certainly Watson would have), but this return to a state of legitimacy was possible because a new meaning of "mind"—a hard mentalism, as we have called it, one in which mind is separated from consciousness—became possible and then became widespread. This new notion of mind was the product, first, of the cybernetics era, and then of the cognitive era, but the possibility of seeing pure consciousness as the center of psychological study, as Wundt, Brentano, and Titchener proposed, is barely an option in today's psychology. This much we may say is a lasting effect of behaviorism.

If behaviorism's first call was a radically new one—to utterly eliminate all reference to the mental in psychological description—its second was hardly new at all. Behaviorism claimed, just like almost every new wave and theory, to be the first approach to psychology that was properly scientific. Just what it meant to *be* scientific was a question that not everyone would answer the same way, and those differences would lead to different brands, kinds, and styles of behaviorism.

In the last chapter we saw that the role of introspection was hotly debated during the nineteenth century—Comte argued that it could play no role in a modern, positive science, and Brentano and Wundt worked out limited ranges within which self-awareness could play a role in scientific psychology. What joined together the nineteenth-century controversies with the behaviorists' concerns was the anxiety that came from the tension between acknowledging self-awareness and being a science. The behaviorists were ready—in fact, they were already committed—to jettisoning the awareness that the nineteenth-century scholars had worked so hard to establish.

One of the most important aspects of behaviorism was its deep commitment to the study of learning in non-humans. We have already remarked that American functional psychology, of the sort championed by Dewey and Angell, showed the imprint of Darwinian thinking, most notably through its emphasis on finding the connections between ways of acting intelligently, in the broadest sense, and successfully responding to the challenges of one's environment. That is, what we call intelligence and the ability to learn, when we focus our attention on an animal in an experimental situation, is ultimately little different from the animal's wherewithal to do what it takes to survive in nature—an adaptability that the Darwinian struggle is bound to reward.

We have not yet listened to John Watson explain what it was that he meant by behaviorism, and why he held it to be so different from psychology as he had learned it at Chicago. Let's turn our attention to Watson's reasons.

Watson's first declaration

We will take a look first at Watson's earliest behaviorist statement, the article from 1913 that started the movement. It began with an observation of how the establishment viewed psychology—the view that Watson would thoroughly reject:

> It has been maintained by its followers generally that psychology is a study of the science of the phenomena of consciousness. It has taken as its problem, on the one hand, the analysis of complex mental states (or processes) into simple elementary constituents, and on the other the construction of complex states when the elementary constituents are given. The world of physical objects (stimuli, including here anything which may excite activity in a receptor), which forms the total phenomena of the natural scientist, is looked upon merely as means to an end. That end is the production of mental states that may be "inspected" or "observed." The psychological object of observation in the case of an emotion, for example, is the mental state itself. The problem in emotion is the determination of the number and kind of elementary constituents present, their loci, intensity, order of appearance, etc.[7]

Yes indeed, this sounds like good Titchenerian structural psychology, or even much of what was being done at Chicago. And as far as psychologists today are concerned, how can we learn anything about these elementary mental constituents? Watson asked. Everyone seems to agree on the answer:

> It is agreed that introspection is the method par excellence by means of which mental states may be manipulated for purposes of psychology. On this assumption, behavior data (including under this term everything which goes under the name of comparative psychology) have no value per se. They possess significance only in so far as they may throw light upon conscious states. Such data must have at least an analogical or indirect reference to belong to the realm of psychology.

If psychology *is* the scientific study of consciousness by the very definition of the word *psychology*, then this does not seem so unreasonable. But this whole attitude creates a problem for psychologists like him, Watson said—which is to say, for any psychologist who studies animal learning, for what is the bearing of animal work on human psychology?

> I used to have to study over this question. Indeed it always embarrassed me somewhat. I was interested in my own work and felt that it was important, and yet I could not trace any close connection between it and psychology as my questioner understood psychology.

And there you have it: the results of Watson's psychology had indeed contributed little to human psychology, as he noted. And this left him on the horns of a dilemma: either it was wrong to expect psychology to focus on consciousness and nothing else, or else the study of behavior (which is what Watson studied, and all he wanted to study) was not psychology, and "behavior must stand alone as a wholly separate and independent science." Some psychologists might try to persuade us that even lowly paramecia have some sort of consciousness, and that with the right sort of experiments, we could shed light on consciousness in paramecia or in rabbits by running experiments on them. But where does that belief come from, anyway? We can make any assumptions we care to about where consciousness comes into the picture on the evolutionary scale of animals, and those assumptions will have absolutely no impact on the way we study these animals and their behavior. So why even ask the question as to when consciousness arises? We should restrict our questions to those that have a bearing on the work we actually do in a laboratory. That's where the science is, in Watson's view: study what animals (including humans) can do; decisions as to where consciousness arises in the great ladder of life are of no scientific interest.

Now, we have just read James Angell's presidential address, presented seven years earlier: Angell—John Watson's thesis advisor, his mentor—took pride in emphasizing that his view of functionalist psychology placed the study of animal behavior in an important position in psychology, not least because of our understanding of Darwin and biological evolution.

To study learning, Watson wrote, is to study learning in organisms—let's not get carried away with always focusing on humans. After all, think of evolution, and remember how students of evolution were side-tracked by their concerns about whether mankind did or did not descend from the apes.

> The moment zoology undertook the experimental study of evolution and descent, the situation immediately changed. Man ceased to be the center of reference. I doubt if any experimental biologist today, unless actually engaged in the problem of race differentiation in man, tries to interpret his findings in terms of human evolution, or ever refers to it in his thinking. He gathers his data from the study of many species of plants and animals and tries to work out the laws of inheritance in the particular type upon which he is conducting experiments. Naturally, he follows the progress of the work upon race differentiation in man and in the descent of man, but he looks upon these as special

> topics, equal in importance with his own yet ones in which his interests will never be vitally engaged.[8]

And why can't psychology just get with the program? That's the question that Watson was asking his colleagues to answer. Why can't psychologists realize that this obsession with humans is part and parcel of the prescientific frame of mind? Why should psychologists believe that "unless our observed facts are indicative of consciousness, we have no use for them, and unless our apparatus and method are designed to throw such facts into relief, they are thought of in just as disparaging a way"?

And this brings us right to the point that Watson wanted us to see—and we do not take the very next sentence very seriously; it is just a figure of speech (in particular, apophasis):

> I do not wish unduly to criticize psychology. It has failed signally, I believe, during the fifty-odd years of its existence as an experimental discipline to make its place in the world as an undisputed natural science.

No, actually Watson *did* very much wish to criticize psychology. Its infatuation with introspection, in Watson's view, was utterly unscientific—not because of the nature of the data, but because of the investment each introspectionist psychologist makes in his own, personal introspections:

> Psychology, as it is generally thought of, has something esoteric in its methods. If you fail to reproduce my findings, it is not due to some fault in your apparatus or in the control of your stimulus, but it is due to the fact that your introspection is untrained. The attack is made upon the observer and not upon the experimental setting.

And so, Watson concluded, it was time to draw a line in the sand and say, enough! We have had it!

> The time seems to have come when psychology must discard all reference to consciousness; when it need no longer delude itself into thinking that it is making mental states the object of observation.

If we were keeping score, we would note that Watson had now rejected two things: the *definition* of psychology (which immediately changed the nature of what a psychologist should be trying to accomplish), and the nature of the primary data that the psychologist should be trying to gather.

Watson then trained his sights on the dominant trends in psychology at this time: Titchener's structuralist view first, and then that of the Chicago functionalists with whom Watson had studied. As we have seen, Watson was close personally both to Angell, and to Titchener, and the friendship with Titchener lasted for many years to come. Watson was fed up with the disputes among Titchenerians who could not decide how many attributes a sensation has. One person says a sensation has extension and duration; another says there's also intensity. Someone else says that there's also clearness. Yet another introspector will say that he has found order among his sensations. Where will this all lead? When will it stop?

> Titchener, who has fought the most valiant fight in this country for a psychology based upon introspection, feels that these differences of opinion as to the number of sensations and their attributes [and a host of similar questions] . . . are perfectly natural in the present undeveloped state of psychology. While it is admitted that every growing science is full of unanswered questions, surely only those who are wedded to the system as we now have it, who have fought and suffered for it, can confidently believe that there will ever be any greater uniformity than there is now in the answers we have to such questions. I firmly believe that two hundred years from now, unless the introspective method is discarded, psychology will still be divided on the question as to whether auditory sensations have the quality of "extension," whether intensity is an attribute which can be applied to color, whether there is a difference in "texture" between image and sensation and upon many hundreds of others of like character.[9]

Titchener's introspection is a dead end, Watson said; don't waste the next 200 years on it. But there's nothing better in the field, Watson continued. Do you think functionalism is any better?

> The last fifteen years have seen the growth of what is called functional psychology. This type of psychology decries the use of elements in the static sense of the structuralists. It throws emphasis upon the biological significance of conscious processes instead of upon the analysis of conscious states into introspectively isolable elements. I have done my best to understand the

> difference between functional psychology and structural psychology. Instead of clarity, confusion grows upon me.

The functionalists replace the confusion in the mind's content with confusion in the mind's functions, as far as Watson can tell.

And now, nearly halfway through the article, Watson began to describe his own view:

> This leads me to the point where I should like to make the argument constructive. I believe we can write a psychology . . . and never go back upon our definition: never use the terms consciousness, mental states, mind, content, introspectively verifiable, imagery, and the like. . . . It can be done in terms of stimulus and response, in terms of habit formation, habit integrations and the like. Furthermore, I believe that it is really worth while to make this attempt now.
>
> The psychology which I should attempt to build up would take as a starting point, first, the observable fact that organisms, man and animal alike, do adjust themselves to their environment by means of hereditary and habit equipments.[10]

Watson's views on the role of heredity in explaining behavior changed over the years. Here, in his first paper on behaviorism, he leaves the door wide open to the study of hereditable behaviors, but that would change during the 1920s, after he had left academia, when he argued that the behaviorist should explain behavior based purely on what is happening in the environment surrounding the organism:

> Secondly, that certain stimuli lead the organisms to make the responses. In a system of psychology completely worked out, given the response the stimuli can be predicted; given the stimuli the response can be predicted. Such a set of statements is crass and raw in the extreme, as all such generalizations must be. Yet they are hardly more raw and less realizable than the ones which appear in the psychology texts of the day.

In this early paper, Watson was open to the possibility of a careful account of the inherited component of species behavior, as we see here:

> Some time ago I was called upon to make a study of certain species of birds. . . . When I reached there I found the animals doing certain things: some of the

> acts seemed to work peculiarly well in such an environment, while others seemed to be unsuited to their type of life. I first studied the responses of the group as a whole and later those of individuals. In order to understand more thoroughly the relation between what was habit and what was hereditary in these responses, I took the young birds and reared them. In this way I was able to study the order of appearance of hereditary adjustments and their complexity, and later the beginnings of habit formation. . . . Their food and water, sex and other social relations, light and temperature conditions were all beyond control in a field study.

Control the environment, and see how well you can control and predict the animal's behavior: this is what the sciences can teach us about how to study and analyze. At this point, Watson's scientific program took on some troubling aspects. What would all of this mean for the study of human beings?

"Had I been called upon to examine the natives of some of the Australian tribes, I should have gone about my task in the same way." We can be sure that the tribes that Watson was referring to did not include the anglophone residents of Canberra or Sydney; he had in mind darker-skinned people living in the outback. But still, he wrote,

> I should have found the problem more difficult: the types of responses called forth by physical stimuli would have been more varied, and the number of effective stimuli larger. I should have had to determine the social setting of their lives in a far more careful way. These savages would be more influenced by the responses of each other than was the case with the birds. Furthermore, habits would have been more complex and the influences of past habits upon the present responses would have appeared more clearly.

We see that not only has consciousness been removed from the conversation, but so has everything studied to this point by sociologists and anthropologists: talk of culture will be about as welcome in Watson's brave new world as talk of thoughts and consciousness. But Watson was confident that he could deal with Australian aborigines. Their system of behavior could not be very challenging.

> Finally, if I had been called upon to work out the psychology of the educated European, my problem would have required several lifetimes. But in the one I have at my disposal I should have followed the same general line of attack.

> In the main, my desire in all such work is to gain an accurate knowledge of adjustments and the stimuli calling them forth.

Studying Australian aborigines would require the same methods as studying animals, no doubt done more carefully. And studying Europeans would be no different, except that it would take longer.

Watson then explained that the goal of his new psychology was to *control* behavior, not to understand or predict it.

> If psychology would follow the plan I suggest, the educator, the physician, the jurist and the business man could utilize our data in a practical way, as soon as we are able, experimentally, to obtain them. Those who have occasion to apply psychological principles practically would find no need to complain as they do at the present time. Ask any physician or jurist today whether scientific psychology plays a practical part in his daily routine and you will hear him deny that the psychology of the laboratories finds a place in his scheme of work. I think the criticism is extremely just. One of the earliest conditions which made me dissatisfied with psychology was the feeling that **there was no realm of application for the principles** which were being worked out in content terms.
>
> What gives me hope that the behaviorist's position is a defensible one is the fact that those branches of psychology which have already partially withdrawn from the parent, experimental psychology, and which are consequently less dependent upon introspection are today in a most flourishing condition. Experimental pedagogy, the psychology of drugs, the psychology of advertising, legal psychology, the psychology of tests, and psychopathology are all vigorous growths.

And so Watson's call for a new behaviorism was sounded. In 1914, just a few months after Watson's article came out, Titchener wrote a bit dismissively about behaviorism, misjudging the climate of opinion in his field:

> There is now a flurry in favour of behaviourism; but that is largely because the thing is so far all positive, and no criticism worth mentioning has appeared. No doubt the point of view will permanently appeal to certain temperaments (as it has appealed in the past; it is no more new than pragmatism was!). But the present hullabaloo will quiet down after a few critical papers have made their appearance; and then we shall get our perspective again. I do not belittle

> behaviourism by hoping that it may soon be set in its right place! but I get a trifle tired of unhistorical enthusiasms.[11]

He would hear a lot more in years to come.

Among the chief merits that Watson claimed for behaviorism was that it was a better fit with what scientists were doing: behaviorism had a stronger claim to be a science than did the psychologists who had taught him and who dominated the academic scene, both in the United States and elsewhere. But there was also a powerful picture of how society worked, and how it could be made better, that behaviorism promised to make clear. By the time that Watson's most famous book, *Behaviorism*, came out in 1925, it was a movement with a strong social and political message. From today's point of view, and with our knowledge that the world would be turned upside down in the 1930s and 1940s by fascism and Nazism, it is not hard to be shocked by some of his conclusions. But we must also come to understand what it was that made Watson's behaviorism so attractive to so many people, and over such a long period. It certainly had an echo of Comte's anti-religious positivism: Watson made no bones over the alliance, as he saw it, between old-fashioned psychology and the "subtle religious philosophy" that, like Comte's first stage of human thought, was long past the time it should have been taken off the shelf, along with any use of its favorite term, *soul*. Religious leaders in days gone by were charlatans who had learned that they could frighten the gullible into doing work for them with tales of the supernatural. The next stage, if Comte were to be believed, was the development of dualistic philosophies, with both mind and soul, and though Watson gave no sign of ever having read Comte, he sketched a perfectly Comtean middle period for psychology in which the word *soul*, which was firmly rooted in its religious period, was exchanged for the new word *consciousness*, which was just as bad, from Watson's point of view. Because, Watson said, psychologists who wanted to study consciousness said that what we needed was to *look in on what goes on inside of us*. Now, this is certainly not what Wundt and Brentano had in mind, as we have seen; they expended considerable effort to emphasize that when we introspect, the experience that we analyze is not a *different* experience from that which we encounter when we are living life in the usual way. There aren't two different sorts of experiences of a cup of coffee, one that we have most mornings while we read the newspaper, and another one when we try to reflect on the subjective side of the experience. No, there is one experience, but it can be lived through without much reflection, or we can stop and reflect on it.[12]

Watson's behaviorism was in many respects a good deal more extreme than that adopted by other psychologists. If we are to make sense of the evolution of the field's thoughts, we need to recognize the differences among the behaviorists, and at the same time acknowledge the principles they held in common. In the discussion that follows immediately, we will be referring essentially to Watson's own views, which were considerably more extreme than those held by others such as Edward Tolman, a very influential psychologist of the next generation who had proudly called himself a behaviorist, even though his was a perspective that was much more open to aspects of psychology that Watson's original manifesto sought to eliminate.

Is it possible to give a *general* characterization of what behaviorism is or was, or what precisely it stood for? Probably not, if we insist on a definition that would satisfy all of those who wanted to view themselves as behaviorists. Perhaps as good an answer as any is the one given by Roger Schnaitter:[13] behaviorism is the study of animal behavior interacting with environment, without reference to events taking place at some other level, or some other place, or in some transcendent conceptual realm. That leaves a lot of room for different behaviorists to include some things, and others to include other things. This account also properly alludes to an intellectual connection to Comte's positivism, which was the original philosophical stance insisting that all there is is what you see, and you must stop hoping for the day to come when something else will turn up that will be the true meaning and explanation of what was always right there in front of your eyes. But this definition fails to give adequate attention to the Darwinian character of behaviorism, the strong belief that humans are an integral part of the biological world, endowed perhaps with more intelligence than most of our fellow species, but a product of the same pressures of selection that are responsible for all living things. That is how it is with movements that have more than one member and last for more than one day: it is not really possible to give a characterization that everyone will agree on.

We should also note that Schnaitter's characterization would not have satisfied Watson very much, because it is too intellectual. As Watson put it, the behaviorist "wants to control man's reactions as physical scientists want to control and manipulate other natural phenomena. It is the business of behavioristic psychology to be able to predict and to control human activity."[14]

Purpose is like phlogiston

In his most famous book, Watson gave a popular and direct account of what behaviorism was. "The Behaviorist," he wrote, "began his own formulation of the problem of psychology by sweeping aside all medieval conceptions."[15] This metaphor of *sweeping a stable clean* (typically of dung, but sometimes of dust) is recurring. What are the medieval conceptions, in any event? "He dropped from his scientific vocabulary all subjective terms such as sensation, perception, image, desire, purpose, and even thinking and emotion as they were originally defined." What will then replace these terms? Stimuli, and the responses that they occasion. "The Behaviorist finds no scientific evidence of any vitalistic principle," such as *purpose*.[16] "Purpose" was like *phlogiston*: stuff that educated people used to believe in, but do not anymore. Mainly, though, noted Watson, "We need nothing to explain behavior but the ordinary laws of physics and chemistry." Anyone who does more, and introduces consciousness into the discussion, "does so because of spiritualistic and vitalistic leanings." Watson also explained the behaviorist's interpretation of *thinking*: "thinking is behavior, is motor organization, just like tennis playing or golf or any other form of muscular activity. But what kind of muscular activity? The muscular activity that he uses in talking. Thinking is merely talking, but talking with concealed musculature."[17] Watson presented these words at a public debate with William McDougall, whose role in the affair was to criticize Watson's behaviorism. McDougall aptly began by pointing out one reason why Watson's views were "attractive to many persons, and especially to many young persons."[18] The reason was that "these views simplify so greatly the problems that lie before the student of psychology: they abolish at one stroke many tough problems with which the greatest intellects have struggled with only very partial success for more than two thousand years; and they do this by the bold and simple expedient of inviting the student to shut his eyes to them, to turn resolutely away from them, and to forget that they exist." McDougall thus called Watson and the students that heeded his call a group of Noah's Ark survivors (recall our discussion in chapter 1). The students no longer need to learn what had been said before: "This naturally inspires in the breast of many young people, especially perhaps those who still have examinations to pass, a feeling of profound gratitude to Dr. Watson. He appears to them as the great liberator, the man who sets free the slave of the lamp, who emancipates vast numbers of his unfortunate fellow creatures from the task of struggling with problems which they do not comprehend and which they cannot hope to solve. In short, Dr. Watson's views are attractive to those who are born tired, no less than to those who are born Bolshevists."

With these words, McDougall brought together two of the themes we discussed in chapter 1: the heady sense of intellectual liberation that can come from a new theoretical perspective, and the desire to encourage students not to read disciplinary literature before that moment of liberation. He saw the willingness of students to ignore earlier publications as a license to *not read*, which appealed to the laziness of the students. These kinds of perceptions of the younger generation we find repeated over the decades.

Edwin Boring was a student of Titchener, as we have seen. He became in time the elder statesman of the department of psychology at Harvard, and he is remembered today principally for his writings about the history of psychology. When he talked about the present, it was often with the perspective of a historian, though he knew full well that he was also a participant in the events he described. Boring was sensitive to the non-cumulative nature of psychology, and sensitive to the question of whether it was a science or not. In 1929, speaking of behaviorism, he wrote:

> With respect to scientific movements there seems to exist something like Newton's third law of motion: action equals reaction. You cannot move—in the sense of starting a movement—unless you have something to push against. The explanation of this law, I think, lies in the relation of movements to public attention. . . . A movement must move with respect to something, and progress must move away from something, if the movement is to command observational attention. It is therefore the business of the founders of new schools, the promoters and propagandists, to call **persistent attention to what they are not**, just as one political party is forever emphasizing the short-comings of the other.[19]

Boring, we can see, did not like behaviorism, and his interest in it waxed and waned over the course of his career, as did his attraction to Gestalt psychology. But the controversies that grew from the disagreements between those who found behaviorism appealing and those who found it repellant are of interest to *us*—for several reasons. In the first place, behaviorism undoubtedly marked an enormous change in the way that academic psychologists studied anything that might be called "mind," and it did so in what seemed like a great rupture from the structuralist and functionalist moments that immediately preceded. In addition, behaviorism had the allegiance of at least some of the important figures in linguistics over the following decades. Third, a decisive break with behaviorist psychology was part of the spirit of the rise of the cognitive movement in the late 1950s, in psychology first of all, but in all of the mind sciences more generally. Fourth, and perhaps most interesting of all, if behaviorism does not appeal to our sensibilities today, an effort on our part to understand *why* it was so successful in gaining adherents will tell us something about what makes people define their scientific premises as they do, and why they believe what they do. Boring's comment expresses his belief that behaviorism was more interested in taking cheap potshots at the structural and functional psychology that Watson had been trained in than it was in making scientific progress. But in our own day and age, it is astonishing to see how contemporary writers cannot even *describe* the basic principles that motivated behaviorists without making them sound like their thought processes were warped, distorted, and a card or two short of a full deck. This inability of cognitive scientists today to understand an earlier generation is a striking phenomenon which deserves our careful consideration. Something is not right.[20]

To better understand what work fell under the broad banner of behaviorism and what did not, it will be helpful to have some guideposts to identify the different groupings within the larger movement. We can roughly divide the observations of human activity in the broadest sense into three large domains: those involving *internal consciousness*, those characterized by *actions*, and those characterized by *muscle contraction and hormonal secretion*. These categories are in the first place different styles of describing, to be sure; many things that happen around us could be described in any of these three ways. Structuralist psychology took the first domain—that of consciousness—to be the primary responsibility of psychology, and functionalist psychology took the second—the study of human action—to be the only way to get a realistic handle on internal consciousness. Behaviorists were generally in agreement that the first area needed to be eliminated from professional psychology, and perhaps even from all of human discussion.[21] They were not in agreement on the relationship of the second and the third, that is, the relationship of human action to muscle contraction. The most extreme behaviorists took it that all meaningful discussion of human action could (and in the long run, should) be translated into descriptions of muscle contraction, while the neo-behaviorists (notably Edward Tolman) argued that human action was the right way to understand what it was that humans did with their muscles.

As we saw in the last chapter, one of Carl Stumpf's students, Max Meyer, came to the United States and developed a view of psychology that was focused on human *behavior.* His best-known work was a book entitled *The Fundamental Laws of Human Behavior,* which came out in 1911, when Meyer was 38 years old—and two years ahead of Watson's more famous publication. Unlike Watson's sort of behaviorism, Meyer's psychology focused on the neurophysiological: for Meyer, the study of behavior went hand in hand with studying the internal workings of the nervous system.

Max Meyer, as we noted earlier, had a single graduate student in his career, named Albert Paul Weiss. Weiss was a committed behaviorist, and a professor at the Ohio State University. His thinking would have an enormous impact on Leonard Bloomfield in the 1920s, as we will see in the next chapter.[22] Weiss had been born in Germany, and though he came to the United States as a young child, he spoke German at home, as did Bloomfield.[23] William Esper, one of Weiss's students, summed up the relationship between Weiss and Meyer:

> There were strong bonds between Meyer and Weiss: Weiss had been born in Germany and . . . spoke German in the home of his parents; his personality was most engaging: honorable, unassuming . . . eager in interest in all matters of scientific and humane import, humorous; well trained in physics, chemistry, biology, mathematics, and philosophy—subjects in which Meyer found most of his American students deficient; ingenious in devising and constructing apparatus. In his early publications Weiss followed Meyer in research on tonal intensity and "vocality," and in applying Meyer's hydraulic theories of the ear and of the nervous system to sensory discrimination and learning. In his later publications he enlarged upon Meyer's two main philosophical—or rather, methodological—doctrines: that psychology should deal only with objective data and only with behavior having social import. Meyer has said, "I have had very little—almost no—influence on American psychology directly, but perhaps a good deal through mediation by students of Weiss." Meyer produced one doctor of philosophy (Weiss), but Weiss produced twenty-five.[24]

The Second Generation of Behaviorists

Clark Hull, Edward Tolman, and Karl Lashley were three of the leading psychologists of their generation, those receiving their PhDs around the time of World War I. Hull and Tolman have both been called behaviorists, and with good reason; they were inspired by Watson's behaviorism, certainly, but they succeeded in shifting the nature of the questions posed by psychologists working under the banner of behaviorism. Karl Lashley, though, is harder to characterize. He has been called a behaviorist, and he was a student of John Watson, with whom he collaborated and published. But a behaviorist Lashley was not, and for several reasons. He began as a behaviorist, but by the mid-1920s, he was one no more. He was far too interested in what was going on inside humans and animals, especially in their brains, and that was not something that behaviorism countenanced.

If Tolman were to come back today, he might well decide that what psychologists are doing in their laboratories is very exciting, and a great advance over work in his lifetime. If Hull came back, in all likelihood he would shake his head, and feel that his efforts to make psychology scientific had been quite thoroughly ignored.[25] Hull and Tolman are frequently referred to as bringing in an era of *neo-behaviorism*, and the prefix *neo-* invites us not only to appreciate the continuity with the stem that follows it, but also to appreciate the differences.

Still, reasonable though it may be to put Hull and Tolman in the same category, we should not fail to note how different were their ideas of good scientific, psychological work. You can almost hear Hull muttering, "With friends like these, who needs enemies?" Clark Hull wanted more than anything else to do good science, and to take explicit lessons from the successes of the hard (and inorganic) sciences, which meant, as he saw it, the development of algebraic expressions for models of animal behavior. Tolman, on the other hand, was free-spirited and always ready to welcome a new idea. He left no doubt—in his own mind, at least—that he was a behaviorist, whose focus was on describing the ways in which animals (including humans) learn from a changing environment. But Tolman's scientific world emphasized both the *purposes* of the animals he was describing, and their *cognitive* processes. Purposive and cognitive: that was Tolman's world. He felt no attraction, as far as we can tell, to the perspective that was so important to Hull, the idea that quantitative expressions were the hallmark of science.

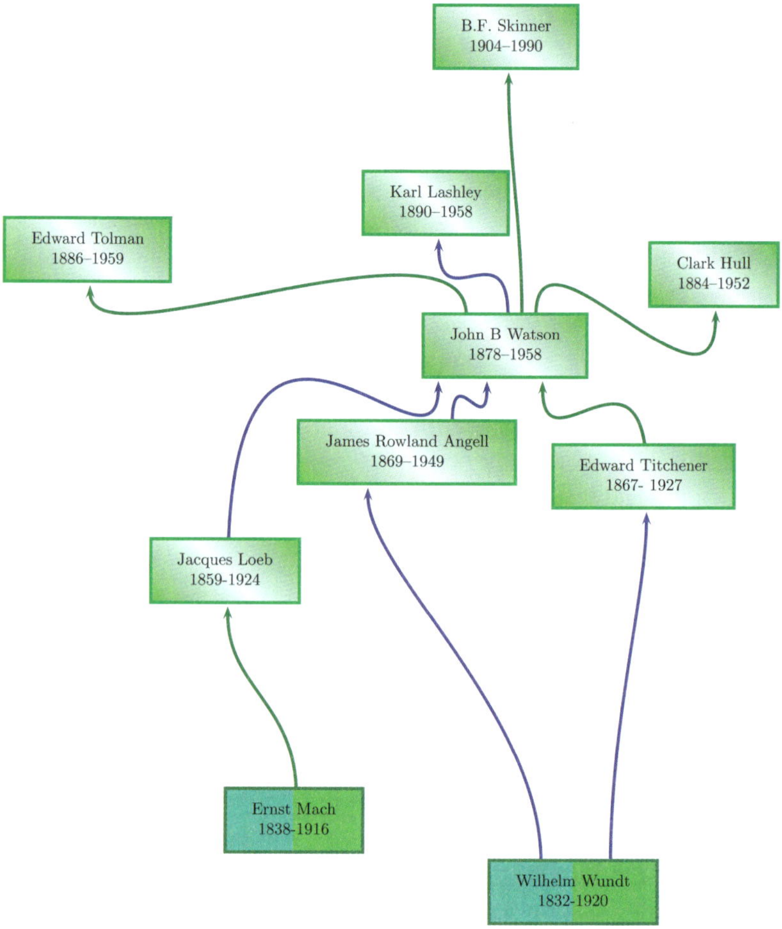

FIGURE 5.1. Behaviorism

Clark Hull

Clark Hull was born in 1884, received his PhD from the University of Wisconsin in 1918, and he remained there for a decade. In 1925, he taught a seminar on behaviorism, and found it extremely congenial, though he thought Watson's version of it was overly simple. In a note to himself, he wrote, "The Watsonian tradition would deny the existence of any such things [consciousness and will] and thereby dismiss the problems as non-existent. This is as vicious as to be content with a false solution—both inhibit further investigation."[26] Hull wanted to find general laws of learning, which he thought must exist, laws that would be of the same mathematical character as Newton's laws in physics. In his professional diary, his "Idea Books," he wrote while he was working on his dissertation that "it seems that the greatest need in the science at present is to create an *experimental* and a *scientific* knowledge of higher mental powers."[27] And what does that mean for the *person* who undertakes that project? Hull's goal was to be the "supreme authority" in the fields of psychology that he would work on: concept formation, abstraction, perhaps reasoning. He wrote that he would "both know the literature and create the literature on the subject." He stayed at Wisconsin until 1929, when a much larger wave came along and swept him east.

Looking back, Clark Hull noted that around this time,

> I came to the definite conclusion . . . that **psychology is a true natural science**; that its primary laws are expressible **quantitatively** by means of a moderate number of ordinary equations; that all of the complex behavior of single individuals will ultimately be derivable as secondary laws from (1) these primary laws together with (2) the conditions under which the behavior occurs; and that all the behavior of groups as a whole, i.e., strictly social behavior as such, may similarly be derived as quantitative laws from the same primary equations. With these and similar views as a background, the task of psychologists obviously is that of laying bare these laws as quickly and accurately as possible, particularly the primary laws.[28]

Indeed, in 1926 he had written in his journal something that would become a linchpin of cybernetics and modern cognitivism, but long before it would be any sort of commonplace:

> It has struck me many times of late that the human organism is one of the most extraordinary machines—and yet a machine. And it has struck me more than once that so far as the thinking processes go, a machine could be built which would do every essential thing that the body does (except growth) as far as concerns thinking, etc. And since to think through the essentials of such a mechanism would probably be the best way of analyzing out the essential requirements of thinking, responding to abstract relations among things, and so on, I may as well play with the idea, making assumptions as frugally as possible and of such a nature as are known probably to be true. In cases where I have to make a fundamentally new assumption this will be some presumption

> that some such new action of the nervous system is to be sought for, i.e., some organic mechanism must exist of that general nature. This will be a good start for all kinds of researches to check up the various hypotheses. In fact the whole thing can probably be reduced to a mathematical formula and it is not inconceivable that an automaton might be constructed on the analogy of the nervous system which could learn and through experience acquire a considerable degree of intelligence by first coming into contact with an environment.[29]

During the mid-1920s, Hull, like many American psychologists, heard about Gestalt psychology, and he found it intriguing. He had hoped to go to Germany to study with Kurt Koffka, but succeeded instead in bringing Koffka to the University of Wisconsin for the academic year 1926–27. Hull found his lectures interesting, but though Koffka was doing his best to criticize Watson and his behaviorism, Hull came away with the feeling that there was indeed a new kind of behaviorism, one that would incorporate laws of behavior and "deductive systematization," that ought to be developed.

Joseph Gengerelli, one of Hull's graduate students in the mid-1920s, recalled that Hull "wanted to create a theoretical structure that made no use of mentalistic faculties and entities to explain behavior. He was, in his aspirations, a 200% behaviorist, but he thought the then-current behaviorism did not meet the problems that had to be solved—the problems of motivation and purpose."[30] How can we understand the way in which humans and animals *persist* even in the face of setback and defeat? How can we understand the changes that take place when a human, or animal, stops engaging in an act which has not been successful in dealing with the current problem? How can either of these wide ranges of behavior be understood if all the scientist will be allowed to talk about is stimulus and response, that is to say, what there *is* and what *can be seen and measured*, rather than what is *not yet*? These were the questions that Hull wanted to deal with. Gengerelli recalled that Hull would say, "Watson is too naive. His behaviorism is too simple and crude."[31] Hull, like most of the leading behaviorists, felt that he was on a mission to clear out the confusions that haunted the thinking of too many psychologists, both among his contemporaries and among his predecessors. Those confusions were a sort of *mysticism*—just as Auguste Comte would have said. Hull, in a paper with a colleague of his, Robert Krueger, wrote that "the construction and study of models of the types described [here] will aid in freeing the science of complex adaptive mammalian behavior from the mysticism which ever haunts it."[32] The mysticism that they had in mind (they wrote) was a widespread belief that some kind of adaptation is possible only if something non-physical is added to the mix, something to which we have given a name but whose effects we have no way of clearly defining. This something is variously call *nous*, *entelechy*, soul, spirit, ego, mind, consciousness, or *Einsicht* (i.e., insight). Hull insisted on two points: first, that psychologists were responsible for confusing themselves by appealing to things that they do not understand, and second, the day was coming (or perhaps had already arrived) when we *can* give mechanical explanations for some of the phenomena that had seemed creative in the past.

Looking back at his life, Clark Hull wrote: "I recall the semi-fanatical ardor with which, at that time, some young people, including a few relatively ignorant undergraduates, would espouse the Watsonian cause with grandiose statements such as, 'Behaviorism has made a greater contribution to science than has been produced by psychology in its entire previous history.'"[33]

Now, the 1920s was a period with a great deal of energy in the development of the social sciences, with guidance and funding coming from the Rockefeller Foundation; this would have an impact on Hull's career, as we will see in a moment. In 1921, James Angell left the University of Chicago (he had been John Watson's advisor, you will recall, and had moved later into academic administration, serving as dean and as vice president at the University of Chicago), and after spending a year as the head of the Carnegie Foundation, one of the world's largest private foundations at the time, he became president of Yale University. The distinguished psychologist Robert Yerkes, then working at the National Resource Council, approached Angell at that point with an idea: why not create an *Institute of Psychology* at Yale? Angell liked the idea.

Now a *third* psychologist (and the second from the University of Chicago) came into the picture. Beardsley Ruml had just left Chicago in 1922 to become director of a new charitable foundation which had been founded in 1918 by John D. Rockefeller, a foundation named in memory of John D.'s wife: the Laura Spelman Rockefeller Memorial Foundation. Ruml liked Yerkes's idea for Yale, too, and he directed his foundation to provide $40,000 a year for five years to set up such an institute at Yale. Yerkes was hired the next year, 1924, as professor of psychology at Yale. Four years later, in 1928, a much larger project that would include the Institute of Psychology was funded by the Rockefeller Foundation; this new operation, the Institute for Human Relations, was the brainchild of the young dean of Yale's law school, Robert Maynard Hutchins, and the dean of the medical school, Milton Winternitz. In a stroke of irony, Robert Maynard Hutchins would move to become the president of the University of Chicago a year later, just in time to say goodbye to the linguist Edward Sapir, who would leave Chicago in 1931 to join Yale's Institute of Human Relations.[34]

This is where our digression comes back to Clark Hull: the Institute for Human Relations at Yale University made Hull an offer in 1929 which he accepted. He would be moving east to Yale, just as Edward Sapir would do after him.

Hull proposed four essential properties of a healthy scientific theory. It needs definitions and postulates, which must be clear and unambiguous, consistent with one another, and allow for a rigorous deductive system. All deductions must be made deliberately and explicitly. Not only that, the deductions (or theorems) of the system "must take the form of specific statements of the outcome of concrete experiments or observations."[35] And finally, theorems involving phenomena that have not yet been observed should be "submitted to carefully controlled experiments."[36]

Hull made two further points. Some psychological approaches merely *classify*, and do nothing more than add labels; such systems cannot be accepted as theories. A system must do more than just label phenomena as being of one sort or another. No mere taxonomies for Hull. The second point was that it was pointless to worry and argue ahead of time what the basic nature of the postulates ought to be with which we develop our theories. Should the postulates be limited to concepts that can be directly observed? Or should we insist that they include some entities that cannot be directly observed? Hull argued that these metaphysical arguments were beside the point. All that matters, as far as a set of postulates is concerned, is what theorems can be derived from them. "The history of scientific practice so far shows that, in the main, the credentials of scientific postulates have consisted in what the postulates can *do*, rather than in some metaphysical quibble about where they came from. If a set of postulates is really bad it will sooner or later get its user into trouble with experimental results. On the other hand, no matter how bad it looks at first, if a set of postulates consistently yields valid deductions of laboratory results, it *must* be good."[37] Choose your theory; Hull will not impede your choice. The laboratory and your ability to draw deductions will decide whether you have done good science.

Hull's image of a science is one with a number of variables that take on real values—that is, numbers along a continuous scale—and in which laws relate variables in fairly simple mathematical form, typically with nothing more than multiplication and addition. Consider a simple example, from a 1947 paper that Hull wrote with several coauthors.[38] The paper begins, "The central factor of a science of behavior naturally is behavior itself. It inevitably follows that the central problem of a systematic natural-science approach to behavior is that of determining from given antecedent events and conditions what behavior will follow."[39] This is the core positivist view: what the scientist looks at are events located in space and time, and the natural way to develop a science is to study cause-and-effect patterns that look from the past to the future.

"Such a determination," they continued, "amounts to a prediction, i.e., a statement of the behavioral potentiality (${}_SE_R$) which lies in the relevant antecedent events and states." It is not clear how far knowledge of the value that the variable ${}_SE_R$ takes on is from a real laboratory prediction, but what is important is that the positivist assumption we just made tells us what to do next: "Now, the critical immediately preceding event in the case of reaction evocation appears to be stimulation (S), and the critical immediately preceding states involved are (1) the strength of the habit (${}_SH_R$) and (2) the primary motivation or drive (D), e.g., hunger." These are two real numbers, according to the model; there is something in the animal that can be mapped to these two real numbers. How do these two numbers correspond to "behavioral potentiality"?

"There is some reason," Hull's group wrote, "to believe that the potentiality of the evocation of a reaction in a simple, i.e., a non-competitional, situation will turn out one day to be multiplication in nature." That intellectual modesty is disarming. They have certainly not inflated the grounds for their tentative belief! *Some reason to believe?* "More specifically," they continued, "it is believed that when these three functions are suitably quantified it will be found that ${}_SE_R = {}_SH_R \times D$." Hull and his coauthors note then two major difficulties they face in moving this belief forward towards legitimate scientific status. One is that the "objective manifestations" of ${}_SE_R$ "take so many different forms," and they will be measured in different units, or if in the same units, across different circumstances which make comparison meaningless. If that quantity is to be meaningful, they must find a form that abstracts away from conventional choices of units and the like. The second major problem is that the two factors on the right, which are habit strength and drive, are "never available as such to direct observation of the scientist."[40] But scientists have been able to determine the mass of individual atoms even if they are not directly observable, so this should not be taken as an insurmountable problem. That, in any event, is the Hullian perspective.

And what is *D* in the equation? *D* could be the amount of hunger that the rat has, if we are trying to predict what a rat's behavior would be. It would be related, in some fashion, to how long it had been since the rat had been fed. And perhaps to how many calories the rat had expended during that time. What would *D* be in the case of some higher cognitive function, like learning vocabulary items? Not so clear. But the strongest intuition that the Hullian theoretician has (other than the belief that quantities of this sort are meaningful in the realm of psychology) is that once they have been properly formulated, they enter into relatively simple quantitative relationships (as expressed by multiplication and addition).

Fritz Heider was a Gestalt psychologist who we will meet below. He wrote that

> Hull's name is now almost forgotten except for a small circle of older psychologists. In the 1930's and 1940's he was one of the most influential teachers. Every respectable psychologist who wanted to be taken seriously had to speak in terms of his concepts of drive reduction, goal responses, and so on. It is very hard to believe now, but at that time there were many who thought that Hull would be the American Newton of psychology. . . . Today one has difficulty remembering the degree to which the laboratory rat once dominated American psychology, when even such a humane and thoughtful man as Edward Tolman dedicated his book on the behavior of animals and men to the white rat.[41]

And so, on to Edward Tolman, the thinking man's behaviorist.

Edward Tolman

Edward Tolman was born just outside of Boston in 1886 to a well to do family, and went to MIT as an undergraduate. He entered the department of philosophy and psychology at Harvard University for a PhD in 1911 (which is to say, two years before behaviorism had learned its name), and completed his degree just four years later. He found Watson's work a methodological relief, because he did not want to rely on introspection.[42]

Tolman began doing research under Münsterberg, with direct supervision by Herbert Langfeld (who was only seven years older than Tolman). In the summer after his first year at Harvard, in 1912, he spent a month in Germany, at Langfeld's suggestion: going to Germany and learning German were what a young American *did*, after all, in Langfeld's day. Tolman got to know Kurt Koffka then. Koffka was the same age as Tolman, but he had gotten his doctoral degree—at 23—with Stumpf, in 1909, and Langfeld himself had been a student along with Koffka. One year later, at the age of 23, Koffka would become a significant leader in the creation of the new Gestalt psychology, as we will see shortly.

After getting his PhD, Tolman spent three years—1915 to 1918—at Northwestern, just north of Chicago, and then went to Berkeley. This was the point at which he began to view himself as a behaviorist.[43] Tolman would spend most of his professional life after that at Berkeley. He went back to Giessen, Germany, in 1923 for a few months to study more Gestalt psychology, again with Koffka, and he spent an important semester at Harvard interacting with social scientists there, after World War II.

Tolman's major book, *Purposive Behavior in Animals and Men*, brought together a clear statement of Tolman's understanding of psychology, detailing what it ought to be, and what it had succeeded in becoming by 1932. The first two fundamental elements are not strikingly different from those we would find in Hull's work, to be sure; Tolman's view centered around three points. First, the method of the science of psychology is controlled experiment, which by definition includes outcomes as measurable quantities (times, amount of food eaten, etc.), not discursive descriptions. Second, the central question for psychology is how animals *learn*, and whatever the right answer is to that question, it will be an answer that holds not just for humans, but for a wide range of species related to humans to varying degrees: hence we must develop a vocabulary, a conceptual toolset which allows us to formulate hypotheses which generalize across animal species.

The third characteristic of Tolman's work, and of his writing, was a bit more personal, a bit more a matter of his style. He was comfortable showing to the reader the real human being who was writing the book, and he was comfortable with a heartily fallibilist view of scientific activity. Like a player on a baseball team, he was ready to give his all, but at the same time he was prepared to be wrong and to learn from others, even if they came from other schools of psychology.

Towards the end of *Purposive Behavior*, Tolman remarked that his reader will have noticed that he has been trying to "offer a new 'system' of psychology."[44] And he felt—justifiably or not—that this might put off his reader: "But system-making is very properly open to suspicion." Now that is a telling remark, which sheds light both on the time in which he was writing, and on what separated him from someone like Clark Hull. Why the suspicion, though? System-making, he wrote, "is the resort of arm-chair hiders from reality." He offered no reason for this statement, and we can only infer from this that a suspicion of *systems* was partly his perception of his era, and partly a nagging voice he heard within. He went on: "Once set up, a system probably does as much harm as good. It serves as a sort of sacred grating behind which each novice is commanded to kneel in order that he may never see the real world, save through its interstices."

This is indeed the proper concern of scientists, who recognize that there is always a tension between the quite reasonable effort to sustain and extend a theory and the ever-constant responsibility of modifying or even discarding a theory when it becomes clear that it has departed too far from reality. And not just that, either: the responsible scientist has to keep alive the modesty that fallibilism demands: "And each system is so obviously bound to be wrong. It is twisted out of plumb by the special cultural lack of building materials inherent in the time and place of its origin, as well as by the lack of skill of its individual architect or architects."

Despite all of this perfectly sane reconstruction of the risks of system building, Tolman was about to embark on that venture. "An apology, therefore, is in order. We can, in short, merely hope that the propositions summarized in the succeeding pages, when set up in front of you as a pattern of mullions through which to observe the psychological landscape will serve (but only temporarily) to limn into prominence for you new areas for the gathering of data."

And then the greatest note of modesty:

> But may neither you nor we ever seek to hold up these propositions, save in a somewhat amused, a somewhat skeptical, and a wholly adventure-seeking and pragmatic behavior-attitude.

Let's step back and look at Tolman's view of how to do psychology. He called his theory *purposive behaviorism*, and by that very naming he brought into focus both his commitment to behaviorism, and to a kind of analysis that gave *purpose* in behavior a central role. That is, psychology was not so narrow as to demand that the psychologist study only the muscle movements of the rat or the human; quite the contrary. The object of study is what the rat (or human) as a *whole* was doing in order to achieve some end. Tolman asked us to distinguish *molecular* ways from *molar* ways of looking at what humans (or rats) do: the molecular view describes the actions of their muscles ("muscle twitches" was the phrase everyone liked to use when they wanted to take a certain rhetorical distance from John B. Watson), while the molar view describes actions in the terms provided by natural language. It is easy to provide examples of molar description: a cat getting out of a puzzle box, a man driving home for dinner, or a woman doing her washing or gossiping over the telephone. (These are not only good examples of molar descriptions, they bring vividly to life what the typical examples of human behaviors were in the imagination of a professor in the early 1930s when he thought about men and women. They are, of course, Tolman's examples, not ours.) "And it must be noted that in mentioning no one of them have we referred to, or, we blush to confess it, for the most part even known, what were the exact muscles and glands, sensory nerves, and motor nerves involved."[45] In short, for Tolman, doing behaviorist psychology really meant studying molar descriptions of behavior.[46]

But Tolman was at pains to emphasize that Purposive Behaviorism (which he would sometimes capitalize) had a third conceptual linkage, beyond behaviorism and what he called "purposivism." That third affinity is Gestalt psychology. The connection for Tolman was in the way his behaviorism identified molar behavior: for him, "behavior, as we have seen, is purposive, cognitive, and molar, i.e., 'Gestalted.'"[47]

In the academic and intellectual territory that constitutes a discipline, we recognize that different researchers do not make all of the same assumptions, but the fact is that there are general lines of agreement which make it both convenient and reasonable to refer to movements or factions such as behaviorism and Gestaltism. As a faction grows larger, it becomes useful, and at times imperative, to make further subdivisions—just as William McDougall, for example, proposed dividing behaviorists into strict behaviorists, near behaviorists, and purposive behaviorists.[48] But sociologists of knowledge will be quick to point out that these groupings are not simple tracings of how two researchers' assumptions happen to line up: the groupings are, right from the beginning, reflections of social relations between researchers, and decisions made not just about disciplinary assumptions but about what grouping, what team a researcher wants to be associated with, and what team wants to be associated with him.

Tolman was at pains to acknowledge all of this. By virtue of his intellectual pedigree, he knew that no one could deny him the right to call himself a behaviorist. But he asked himself whether he had the right to call himself also a Gestaltist. Here was his answer: "Undoubtedly, the final answer to this . . . question must be stated by the Gestalt Psychologists themselves. What is to be admitted as a brand of Gestalt Psychology, the Gestalists themselves, in the last analysis, alone can say. We, however, it should be noted, would be proud to be admitted to their fold. There are certain features of our system, however, which in all honesty and fairness should be brought to the Gestaltists' attention as possible blemishes, which may unite us, in their eyes, for being enclosed in their exclusive corral."[49]

Now, do not respond too quickly that Tolman was speaking with his tongue in his cheek. He was not. Yes, he chose his words with style: he talked about being "admitted to the fold," as if the Gestaltists were a flock of sheep (literally or by religious conviction); he spoke of theoretical positions as possibly being taken as "blemishes." But the simple fact was that short of turning himself into a sociologist, he had no other vocabulary to use to address the question of which team he wanted to be on. But that question is important for scientists, as we discussed at some length in chapter 1.

And the Gestaltists—what did they think? When Kurt Koffka published his major work just three years later (in 1935), he felt the field was fractionated in this way, too: "Psychology has split up into so many branches and schools, either ignoring or fighting each other, that even an outsider may have the impression—surely strengthened by the publications 'Psychologies of 1925'and 'Psychologies of 1930'—that the plural 'psychologies' should be substituted for the singular."[50]

And the blemishes that Tolman had referred to, that he feared would separate him from the Gestaltists? One was that he would have nothing to do with introspection, and he was not sure if that would be acceptable to Gestaltists. As we will see below in connection with Koffka—who was himself extending an olive branch towards the behaviorists, much as Tolman was to the Gestaltists—that decision was fine. (The other two blemishes were rather minor, and undoubtedly not problems for most Gestaltists.)

Tolman's behaviorism was one that did allow the scientist to speak of an animal's purpose, on the grounds that purpose was something that resided not only in the unreachable subjectivity of consciousness, but even in the total observed behavior of the animal. When we describe a dog as trying to reach a bone on a shelf, we are not guessing what is going on in his head, Tolman thought: we are giving a straightforward account of the behavior we see before our eyes. In Tolman's terms, the behaviorist was "trying to rewrite a common-sense mentalistic psychology in operational behaviorist terms."[51] E. G. Boring put it this way: "Some—Holt and Tolman first—were clear that behaviorism does not exorcise consciousness but absorbs it, reducing it to the behavioral observations by which it is observed."[52]

Towards the end of his career, he summed up what he thought the greatest influences had been on his thought: his students and his colleagues at Berkeley, of course, and also Kurt Lewin and the other Gestalt psychologists, and Egon Brunswik.[53] Next on the list he placed his colleagues at the Office of Strategic Services (OSS) Assessment Program during World War II, and the Social Relations Department at Harvard in 1949–50. These last two periods involve events that we have not gotten to yet; we will explore the OSS and Social Relations connections in volume 2; it is a fascinating story. That list of Tolman's is hardly the recollection of a good behaviorist: Tolman was quite the eclecticist, and his methodological philosophy was one which led him to reach out to research being done from a wide range of perspectives. His efforts were never trained on shrinking the world of the psychologist, and certainly not to reducing the size of the world of the mind. For Tolman, behaviorism was a philosophy of scientific behavior that introduced good common sense.

Categorizing Edward Tolman is not easy—and categorizing him as a behaviorist is especially not easy. Young psychologists working in the late forties and early fifties felt that behaviorism was a dominant, and in some ways oppressive, philosophy of psychology, and young people who wanted to publish had to mind their ways of speaking so that they would not sound like they were departing from behavioristic terminology. But despite the fact that he was as eminent a psychologist as anyone else of his generation, Tolman consistently defied this characterization of his age. Jerome Bruner was one of the first psychologists in the 1950s who would bring psychology into its cognitive period, and he certainly felt that behaviorism was a dominant presence during this time. And what was one to do if he wanted to bring in mentalistic talk of purposes and the like? Bruner wrote, years later, "The clearest way to espouse purposive mentalism in an age of behaviorism was to conceal it in learning theory, the chosen instrument of antimentalistic psychology. That was what Edward Tolman did. He did it so well that he forced even that most redoubtable antimentalist, Clark Hull at Yale, into such Byzantine absurdities (like 'pure stimulus acts' to stand for perception) that he finally come down of his own cumbersome weight."[54]

How did Hull and Tolman view each other? Hull's view was that he was developing psychology as a real theory, the way scientists since Newton had been developing scientific theories, and he was not at all sure that Tolman understood that point. We have indeed seen that Tolman did wrap his system building with a good deal of suspicion about theory. Hull made his point well, in a letter he wrote in 1934, describing a meeting that included Tolman, Hull, and Neal Miller. Hull wrote, "This was a real honest-to-goodness showdown. As we had expected, Tolman turned out to be very frank and good-natured about everything. Miller and I tried our best to be so also, though I fear we may have been somewhat more aggressive than Tolman was. The upshot of the bull session was that it became fairly obvious to everyone, I think, that the various things which we have been suspecting about Tolman's anthropomorphism and his lack of logical rigor were well substantiated."[55]

After some general discussion, Hull summarized what he took away from Tolman's remarks. "The upshot of the whole thing was that Tolman practically admitted that he had never seriously attempted to make logical deductions for his system, claiming for it little more than that it suggested to him a large number of interesting experiments. He seemed to be distinctly on the defensive, and said he thought he had a right to go on thinking in that way if he found it satisfying and if it suggested lots of interesting experiments to perform."

Hull was not impressed. Yes, Tolman's experiments were ingenious. But Hull thought there was an important difference between the "merit of fertility"—having a point of view that suggests interesting experiments to run—and "the merit of truth which must be possessed by a theory if it shall be entitled to any status in science as a theory."

Karl Lashley

Karl Lashley was John Watson's student at Johns Hopkins, beginning in 1911—just two years before Watson's declaration that he viewed psychology as the behaviorist views it. The two of them were close over the next 10 years, which is to say, during Lashley's twenties, but they moved away from each other as Lashley became his own man professionally, and Watson developed his thinking from outside of academia. Years later, when both were widowed, both were getting old and disabused of life, they picked up their friendship again, briefly, in their retirement years, when they could complain about society going to the dogs, and support each other's cranky opinions about groups in society that did not resemble them.[56]

But back when they were young, the two of them worked together closely. They published a paper in 1913 which explored the ways in which complex mammalian behavior has a strong innate component, a subject that would intrigue Lashley throughout his life. They concluded that there was no evidence "that the infant monkey ever gained a new activity by imitation. Walking, climbing, eating and even the different vocal sounds appeared as instinctive acts which were merely perfected by practice."[57]

Lashley's life work was an exploration of the neural basis of behavior, largely in rats and humans—which is to say, mostly in rats. A significant part of this question involved the degree to which genetically-based, inherited properties are responsible for that behavior, and the degree to which the environment is responsible for it. Lashley always came down strongly on the innatist side of the question, which was one significant reason why he criticized behaviorism over most of his career.[58]

In 1931, Lashley published a response to a critic, W. S. Hunter, that summarized well his reasons for not being a behaviorist. At issue was the behaviorist position that behavior can be explained through the reflex arc: that is the heart of the behaviorist theory. The alternative was to say that there was some kind of cerebral representation, though the word "representation" would not have been used; it might have been called an *engram*. Hunter argued that the rat's travel through the maze *cannot* be explained by specific cues in the rat's environment, nor from "kinaesthetic" sensations from within the rat. "The explanation," Hunter wrote, "seems to require the **assumption of some symbolic process or some neural engram**."[59] Hunter noted that there were two options open then: the neural engram could be (part of) a "central nervous control," or it could be something more peripheral. Neither Hunter nor Lashley could abide the thought of postulating some kind of symbolic representation in the rat, though. Hunter said it is an "intriguing but empty concept," and Lashley wrote that it is "very, very empty" as a hypothesis.[60] But then Lashley got personal—as personal, at least, as he would ever get in print. Listen:

> But the pot should not call the kettle black! I began life as an ardent advocate of muscle-twitch psychology. I became glib in formulating all problems of psychology in terms of stimulus-response and in explaining all things as conditioned reflexes. I had some small part in the formulation of the theory of implicit speech as a basis of thought which Professor Hunter has used in his theory of verbal symbolism. I embarked enthusiastically on a program of experiments to prove the adequacy of the motor-chain theory of integration. And the result is as though I had maliciously planned an attack on the whole system.[61]

No one could be clearer than Lashley with these words: he was *not* a behaviorist, even if he once had been one of those muscle-twitch psychologists. He had been in the in-group, but his science showed him eventually that there was nothing to it.

Lashley had a reputation for being opposed to the creation of general theories in psychology.[62] Hull wrote a letter saying exactly that about Lashley:

> I saw Lashley two or three times and told him that I would like to talk with him about his long article on visual mechanism. . . . At one of these brief meetings I told him I thought I could now explain practically everything in that article. He laughed in his characteristically hysterical manner and said that after I had done that he would go into the laboratory and dig up a whole lot of new things which I probably couldn't explain. . . . I am inclined to believe . . . he has a thoroughgoing distaste for any theory whatever, and would like very much to discredit every kind of theory regardless.

Lashley was sharply critical of Hull's theorizing:

> The spurious character of its quantitative and mathematical treatment of learning is illustrated by the definition of its units of measurement, the hab and wat, in terms of percentage of the practice required by a "standard" organism to reach the physiological limit of learning. Such limits are completely devoid of meaning. The physiological limit is in no case determinable and is a concept of questionable value. . . . [One might] as well define the standard rat for discriminative learning by length of tail as by performance in the maze.

It got worse. Lashley challenged not whether Hull was right or wrong, but whether he was doing anything at all by abandoning any responsibility for explaining how his abstract quantities could be grounded in what is known about the nervous system.

> No general laws descriptive of the processes by which a recognition of similarity is reached have ever been formulated. . . . Both associationist and holistic systems have fallen back upon the conception of a gradient of relations between perceptions as the basis of similarity and generalization, but this gradient is in a purely hypothetical medium having no substantial relation to the nervous system or to the transmission of nervous excitation. It amounts to no more than a confession that the basis of similarity and the mechanism of generalization are problems whose solution depends upon the discovery of principles of nervous integration which are as yet completely unknown.

Lashley had found behaviorism's Achilles' heel: behaviorism eliminated all talk of thoughts and intentions by eliminating any concern with what goes on *inside*—inside the body as well as inside the mind. Whether Hull saw that or not is unclear; he writes as if he did not, because in his defense, Hull wrote: "It is difficult to believe that any behavior scientist would actively oppose the placing of the behavior sciences on a secure quantitative, mathematical basis," though that was not Lashley's concern.

And then Lashley responded: "I certainly do not oppose the placing of behavior sciences on a secure quantitative mathematical basis. But the making of consistent mathematical constructs and determination of their fit to observable phenomena are different matters. *The implication of my* (I fear) thoroughly nasty *footnote is that the values or constants necessary for your equations are indeterminable.*"

In his published reply, Hull wrote, "we have in the italicized sentence quoted just above a confident and unequivocal, though characteristically unsupported, statement that the scientific goal in question is a human impossibility. Lashley thus deliberately puts himself on record."

Later, in a personal letter, Hull wrote to Lashley:

> Upon further thought I am inclined to believe that there is a difference between our approaches beyond that of the physiological versus molar. Perhaps it can be called, "loaf versus half loaf." You must solve the problem of attention or else you will not put forward your system. I, on the other hand, feeling much as you do in the matter of attention, get along without it as well as possible until a solution is found, meanwhile developing the rest of the system. I content myself, from this point of view, with an incomplete loaf and you

> insist upon a whole loaf or nothing. Each approach has its virtues. . . . Doubtless it is best for scientific progress for some to choose one extreme, some to choose the other, and some to compromise in different ways.[63]

We noted earlier that Watson was influenced by Jacques Loeb, a professor of physiology at Chicago during his time there. Loeb's writing described many aspects of behaviorism *avant la lettre*. He had been greatly influenced by Ernst Mach,[64] and strongly rejected the existence of objects whose existence was based on theoretical considerations. He became quite famous in the popular media; he was the basis of an important character, Max Gottlieb, in Sinclair Lewis's Pulitzer-winning *Arrowsmith*. He championed a mechanistic understanding of human behavior, and much of his work emphasized the notion that an animal's behavior was a direct response to stimuli in his environment. But mechanistic views were certainly possible that did not focus to that degree on external stimuli, and others—such as, notably, Loeb's colleague at Chicago, Charles Whitman, who was a zoologist. Whitman was among the first to declare that "instincts and organs are to be studied from the common viewpoint of phyletic descent."[65] Whitman was thus one of the first to give voice to the movement that would be known in later decades as *ethology*, the study of species-particular patterns of behavior. We will see later developments by researchers such as Konrad Lorenz and Niko Tinbergen. What behaviorism and ethology shared was an attraction to mechanical explanations of behavior, but the behaviorist pursued models in which the cause of the behavior was in the environment, while the ethologist pursued models in which the organism's behavior was largely determined from within—from innate properties—and where quite specific (and often surprisingly counter-intuitive) aspects of the animal's environment triggered the behavior.[66]

Edwin Boring was the president of the American Psychological Association in 1928, and delivered a presidential address which was sure to (and surely intended to) put everyone off just a little bit. We have already heard some of his remarks from this address. Where do things stand now? he asked. "In America, we used to have functionalism," he said.[67] We can imagine he is talking about the first 15 or 20 years of the twentieth century, before behaviorism had become such a strong current in the field.

"It was a revolt of the colonial psychologists against Germany, their mother-country," he said, speaking of behaviorism. The Americans needed something that was their own, and something to set themselves in opposition to. No doubt this was related to Americans' aversion to overt philosophical movements, quite the opposite of what might be found in Germany. Functionalism found its first toehold in Chicago, "but I think it was but symptomatic of what was quietly going on all over America, except in some protected places like Ithaca," he wrote. Ithaca was where Titchener defended what he took to be Wundt's structuralist point of view—and Boring himself had been Titchener's student there. In Ithaca, Boring added, "Penelope still remained faithful to her marriage vow." (Bear in mind that Penelope—which is to say, Titchener—had passed away just the year before Boring delivered this paper.)

> In those days the opposite of functionalism was structuralism, but nobody—except perhaps some graduate students—ever called himself a "structuralist." Titchener adopted the phrase "structural psychology" and abandoned it long before it went out of use. No, the functionalists had to have something definite to push against, and it was they only who talked about "structuralists."[68]

> We have this same phenomenon in behaviorism. For years the American tendency has been to have two behaviorists growing where one grew before. Any number of psychologists have been willing to call themselves behaviorists and to be proud of it, but they missed badly a definite opposition to set them off. Words have been coined for the opponent school, words like "introspectionism," or "introspectionalism," but I have never heard anyone apply such a term to himself.[69]

In these battle fields, there must be two armies if they are to engage in combat; who is the opponent of behaviorism in the mind fields?

> Behaviorism has been seeking an enemy so that it could disprove the charge that it is fighting windmills, for it must fight something; it is a movement.[70]

Now, Boring is a gentleman: he knows vituperation when he sees it, and he can talk about it without descending to it himself. But this is very dismissive talk about behaviorism, coming from the president of the American Psychological Association, in his presidential address. To put things in perspective, Boring is almost exactly the same age as Tolman and Hull, and all three are still in the upswings of their career. One can say at the very least that one does not need to be a behaviorist to be welcomed at the annual meeting of the American Psychological Association in 1928.

Boring continued: "I know it is not fair to leave behaviorism so casually, but I must do so. Behaviorism is not new; this has been shown more than once. Yet Watson is right in thinking that he founded it. He could not have founded it if it had been new; it would not yet have been ready to found. . . . Behaviorism is already past its prime as a movement, because movements exist upon protest and it no longer needs to protest."

Gestalt Psychology

> Mr. Köhler, the work done by the Gestalt psychologists is surely most interesting. But sometimes I cannot help feeling that you have religion up your sleeves. —Karl Lashley, in Köhler 1969: 48

The school of Gestalt psychology presents all of the complexities that we could possibly hope to find, as we proceed in our study of the conceptual continuities and the personal dynamics of the mind sciences. Gestalt psychology began in German-speaking Europe, arising in more than one place at more or less the same time, continuing ideas that were surfacing in philosophy and psychology, but simultaneously generating a sense of conflict, of newness and originality, and of divide across academic generations. In the 1920s and 1930s, it was uprooted from Europe because of the rise of Nazism, and part of it—what was called the *Berlin school*—resettled in the United States, but with a new agenda, one which was necessary because of a change of landscape and context. In the New World, Gestalt psychology had only one opponent it could attack, and that was behaviorism.

One of the simplest ideas at the center of Gestalt psychology is that recognizing a structure or a pattern in what is observed auditorily or visually is something different from perceiving a *set* or *bundle* of stimuli or sensations. Think of a musical melody. A melody has a unity and a quality that is not to be found if we take it apart into separate musical notes. And the melody remains perceptually *unchanged* if we raise or lower the key, which is to say, all of the notes by the same amount, despite the fact that all the individual notes will have changed, and it will be *different* if we change the order of the notes. This is not something special about music: the more we think about it, the more we realize that virtually everything we do as we interact with the world involves a structuration of our visual and acoustic world from the very beginning.

This perspective grew out of the passionate engagement in trying to figure out what individual experience in the world *was*, as we saw in the last chapter. There were people whose interests were equally rooted in philosophy and in psychology who turned to these questions, most notably Ernst Mach, Franz Brentano, and many of Brentano's students, including Carl Stumpf, Alexius Meinong, Christian von Ehrenfels, and Edmund Husserl.

Christian von Ehrenfels, as we saw above, was a student of Brentano and also of Meinong. We noted there that at the age of 31, he published one of his very most influential papers, called "On Gestalt Qualities," in which he revisited Ernst Mach's effort to understand what there is about knowing a *melody* that is more than acquaintance with a set, a simple "putting together" of a set of notes.

Ehrenfels was certainly not the first person to recognize the importance of understanding what it means to *hear* or *see* things as patterns, as *Gestalts*—Plato had thought deeply about the problem more than two thousand years earlier—but his paper brought the idea back to center stage, and his name was associated with the challenge of understanding the centrality of the perception of pattern. Now, Ehrenfels's teacher Meinong, you will recall, spent most of his career in Graz, in south-eastern Austria, and it was in Graz that serious psychological experiments were first done to analyze Gestalts. Meinong's notion of Gestalt was one that was produced in a two-step fashion, in a way that the Berlin Gestalt psychologists, the more famous ones, would later reject: for Meinong, external factors, the *stimuli*, were responsible for the first step of a two-step process, and the active contribution of the perceiver, which includes the perception of the Gestalts, is limited to the second step.[71] This would be the crucial assumption that the Berlin Gestalt psychologists rejected: for them, the active participation of perception cannot be removed from sensation.

Meinong's colleague in psychology in Graz was Vittorio Benussi, and looking back many years later, Fritz Heider saw that it was Benussi who first engaged in laboratory experiments in Gestalt psychology, publishing his work starting in 1902—a decade before the others whose celebrity would eventually eclipse his.[72]

What the world came to recognize as Gestalt psychology ten years later came from a group of young psychologists who had studied in Berlin, largely students of Carl Stumpf, as well as Christian von Ehrenfels and Oswald Külpe. This group of Gestalt psychologists included Max Wertheimer, who began the work, Wolfgang Köhler, and Kurt Koffka; in later years, another of Stumpf's students, Kurt Lewin, would often be considered a member of the same Gestalt group. They shared a view and a set of assumptions, and they would cooperate a great deal throughout their careers. All four would be among the European academics who emigrated to the United States and have an important impact on the American intellectual scene. We have already heard from Tolman about his debt to Gestalt thinking, which began well before the Germans had moved to the United States.[73]

And we will not forget that there were other styles and approaches to Gestalt psychology being pursued in Germany at the same time beyond this Berlin school–type of Gestalt psychology; the forgotten Gestalt group led by Karl Bühler will be important for us as we turn to Vienna in the 1930s, which was Trubetzkoy's home.[74]

Gestalt psychology of the Berlin school was presented to the German academic audience in the 1910s and 1920s in three ways: first, as a new way to think about old psychological questions; secondly, as an attack on the older, established schools that had failed to address those questions and had restricted themselves to problems that were simply less interesting for real human beings; and third, as a proposal to use the newest mathematical tools from the physical sciences in order to create a scientific model that was far more sophisticated than the ones that had been developed to date. This third aspect of Gestalt psychology has received little attention, largely because it was not translated in English, and because the Gestalt psychologists did not emphasize it once they had come to the United States. But Köhler was well trained in physics, and understood the ways in which thermodynamics explains how a system settles into a position of minimum energy, and how stable states of a dynamical system can be identified with eigenvectors of a dynamical system—two notions that reemerged in discussions of psychological systems in the 1980s, but without an awareness that they were pursuing lines originally outlined by the Gestalt psychologists.

The Berlin Gestalt psychologists defended the position that "experiences are usually organized wholes whose parts are co-ordinated in a hierarchical system around a central point. Such structures [*Gestalt* in the original] are in no way less immediate than their parts; indeed one often apprehends a whole before anything regarding its parts is apprehended."[75] We cannot view perception, the Gestaltists said, as an activity that begins with unstructured sensation: we do not first hear notes, and then from that derive a melody; we do not first see small spots of light and color, and then from them derive a visual shape; we do not see an object first here, and then there, and then there again, and from those differences derive an awareness of motion. Sensation does not *precede* perception, and the sensed-and-perceived world is always dynamically organized, and it is always in a condition of dynamic equilibrium. As Wertheimer put it, the inner structural laws of the whole may determine what happens in the parts, and not the other way around. Elsewhere he put it even more simply: *pieces* almost always appear *as parts* in whole processes.[76] The global structure of perception is not secondary; it is essential.

The Gestalt psychologists knew that they were challenging the mechanistic views of their age. In the previous chapter and in this one, we highlighted some of the concerns associated with a rejection of mechanical modes of thought. Few thinkers actually described *themselves* as defending a mechanistic view of the human condition, so we have to work out more subtly what people had in mind when they talked about the mechanistic view that they were rejecting, and we do this in part by listening to which conceptual metaphors they were rejecting.[77] The two principal metaphors that critics placed at the center of the mechanical view of the universe was the image of a complex piece of clockwork, of an intricate complex of gears turning and engaging with each other as they moved, and the view of billiard balls moving on a frictionless table at a constant velocity until they collide with another billiard, at which point an instantaneous transfer of momentum and energy leads to a new direction and speed for each of the balls.[78]

And so by rejecting mechanistic models, the Gestalt psychologists were taking the position that models built on a billiard-ball-impact metaphor, or a gears-meshing-with-gears metaphor, were inadequate. As Wertheimer put it, they rejected the view that the part-activities that surround us are "independent, piecemeal, fortuitous and blind."[79] Yes, we can create inventions that work that way, but that is *not* the way that natural phenomena develop. Mechanists see nature "as something essentially blind in its laws, where whatever takes place in the whole is purely a sum of individual occurrences. This view was the natural result of the struggle which physics has always had to purge itself of teleology."[80] The Gestaltists argued that the teleology of psychological dynamics emerges from the overall laws of Gestalt organization, and they appealed both to notions of fields of force and of entropy.[81]

Teleology was not the specification of a goal that was somehow *outside* of present reality, but there was rather a natural direction in which a system will tend, based on its overall structure, a fact which, they noted, is true in physics as much as it is in perceptual psychology. If it was expressed in the way that the Gestalt psychologists were trying to make explicit, it was perfectly legitimate to think of the mental world, just as the physical world, as having goals. So the teleology of the Gestaltists was *immanent*: it was in the world, it was part of the world, and it had as much scientific right to be there as any other concept did.

Wertheimer identified two central views of mainstream psychology with which he and his colleagues most strongly disagreed: the first was that what was complex in the psychological sphere could be viewed as a mosaic or bundle of more elementary items, and the second was that the primary bond between the elementary items in a psychological complex was fashioned from continuity in space, and closeness in time. These were the two ideas at the core of Hume's view of ideas, and they have fascinated everyone who has been an empiricist, or flirted with empiricism in psychology. They were the target of the Gestalt psychologists' assault.[82]

What could the Gestalt psychologists offer as an alterative? They offered the conviction that all perception is an activity that spontaneously provides organization and grouping of sensation, without which perception cannot occur.[83] This conviction was tested and elaborated with a very large number of experimental paradigms that showed in the visual domain that points, lines, and figures were subject to what could only be described as perceptual forces, and that the functional equivalent of a force field could be identified in the visual field, one which reflected general Gestalt principles, on the one hand, and the content of the field, on the other. The general principles were based on notions of figure and ground, of perceptual closure, of forces of attraction among similar forms, and several others.

Koffka's highest principle of organization, though, was this: "Every Gestalt is as good as possible: that is, under the given conditions the greatest possible simplicity (*Einfachheit*); and further, what is together 'belongs together.'"[84] This statement is intimately linked to what Wertheimer called *Prägnenz*: the preference of the system for overall simplicity. Different Gestalt psychologists seemed to place different degrees of emphasis on whether the Gestalt pattern—the *whole*—had indeed a logical priority over the parts, or whether (as Köhler would emphasize) the right way to view things was as a dynamic interaction between the forces of Gestalt formation and those derived from the parts. We are always subject to a panoply of influences from without and from within, but the resultant of these influences is not mere chaos, and the reason for these influences are constantly subject to forces seeking *agreement* and resolution, always tugging individual influences in the direction of a resolved and structured organization. From our viewpoint today, these questions seem strikingly modern, and they also seem to be matters that could be adjudicated only once explicit mathematical models have been spelled out to make explicit what the models would consist in.

The Gestalt psychologists devised many visual examples that illustrate the way in which we perceive Gestalts, and how that perception wins out over other perceptual observations even when those alternatives are more familiar than the winning Gestalt. It is impossible to give a sense of what Gestalt psychology is like without considering at least a few of the many compelling Gestalt examples, but there is a danger that must be avoided of thinking that clever examples are the heart of Gestalt psychology.

Here is an example that illustrates why Gestalt psychologists soundly rejected the notion that the patterns that we see are those that are familiar, those that we have experienced often. They argued that the principles of preferred patterns very easily override any advantage that familiar patterns have. Consider the figure H, which is very familiar. Here we see two capital H's:

But if we devise a more complex but well-organized image like this next one, we cannot see that pair of H's, even though it is perfectly well still present (we expect that the reader can find the H's if she really looks hard):

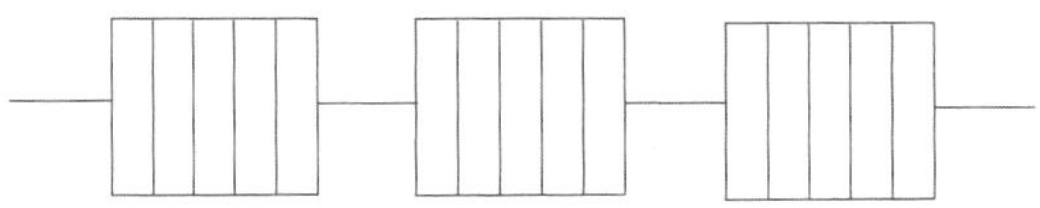

We can ask fundamental questions about the characteristics of preferred patterns. For example, if we see eight vertical sticks, do we organize the whole thing as eight vertical sticks, or as four groups of two sticks? If they are equally spaced, we see them as eight separate, but similar, objects:

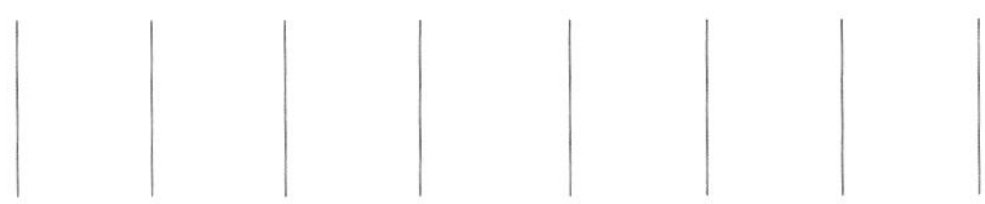

But if we make the spaces between sticks alternate between larger and smaller, we organize them as four pairs of sticks:

If we shade in color, then we can force one grouping or another. In the first example below, the group extends from odd to even numbered sticks, while in the second, it is the reverse, and in both cases, we try our very best, so to speak, to view the vertical lines as the edges of a rectangle.

The colors strongly prefer to define a region which is a "figure" perceived against a neutral, white background.

Because we have the additional stimuli of color, the grammar of visual patterns insists on representing the colored areas as a particular object.

We have gone into this in a bit of detail so as to illustrate what Koffka was getting at when he said that the highest principle of organization is one that seeks the greatest possible simplicity, given the visual stimulation, and with a preference for internal structure made of regular geometric shapes.

The focus on finding a simple principle to account for the visual data is an important thread in the development of the ideas that we are tracing throughout the nineteenth and twentieth centuries. In the nineteenth century, this idea's strongest defender was Ernst Mach, but in the twentieth century, it became central in several contexts: here, in psychology, but also for the Vienna philosophers who wanted to understand how the simplicity of a statement could be related to probability, and later in the context of generative grammar, the notion of simplicity of grammatical description would become the centerpiece of Chomsky's *Logical Structure of Linguistic Theory.*

Wertheimer, Koffka, and Köhler form a team

Max Wertheimer, Franz Koffka, and Wolfgang Köhler were the three best known of the Gestalt psychologists, and the trajectories that their lives took were both complex and unexpected. Wertheimer was the oldest—just two years younger than the behaviorist John B. Watson, in fact, and both of them published calls to arms in the psychological literature at exactly the same time, to the psychologists of their own countries. Wertheimer was born in Prague in 1880, a member of the Jewish community there, and took courses at the University of Prague (today's Charles University) from the philosophers Anton Marty and Christian von Ehrenfels, who, we have seen, had first referred to "Gestalt qualities," in a widely read publication in 1890.[85]

In 1902, Wertheimer went to Berlin to study psychology with Carl Stumpf, who also viewed the whole as more than the simple sum of its parts, and Wertheimer's work fell under the influence of the great interest in musical structure that was dominant in Stumpf's lab; Friedrich Schumann was Stumpf's assistant at the time, and he will become important to our discussion in just a moment.[86] Wertheimer, like Ehrenfels and Stumpf, took his music seriously, and Wertheimer's interest in musical and numerical systems in aboriginal cultures also grew during this period in Berlin.

Two years later, Wertheimer went to Würzburg, and he received his degree with Külpe in 1904. He spent the next six years thinking about a range of topics, including important early work on free association, on *Völkerpsychologie*, and we know that he was seriously reflecting on his readings of Husserl, Mach, and Ehrenfels. He was preparing for a *Habilitation*, which would permit him to progress in an academic career, and he went to Frankfurt in 1910 to this end. There he could work with the professor of psychology, Friedrich Schumann, who had just taken the position but who was someone that Wertheimer already knew personally. Schumann was, in fact, interested in tachistoscopic experiments leading to the perception of apparent motion, which Wertheimer wanted to turn to now. Three years earlier, for example, Schumann had noted his own work on tachistoscopically induced apparent motion: if "one first exposes the vertical beam of a cross for a moment and immediately thereupon the horizontal one," he had written, "the impression of a turning of the vertical beam will present itself. It has to be particularly taken into account that it is neither the afterimage nor the primary memory picture of the vertical beam that really execute the turning in consciousness. This picture rather keeps its vertical position until its disappearance, and nevertheless there is the impression of the turning."[87]

This was the subject that Wertheimer turned to: the impression of movement induced by tachistoscopic visual presentations. This work would be the basis of his habilitation dissertation, successfully defended two years later. Wertheimer discovered something quite unexpected. He found that if he carefully controlled the length of time between the tachistoscopically presented lines, he could induce a scenario in which the observer paradoxically saw movement without seeing a particular object that had moved, and Wertheimer called this the *phi* phenomenon. He saw it as evidence that the observation of motion was fundamental and basic, rather than an inference drawn from the perception of several distinct still lifes. Wertheimer believed that these experiments showed that movement was an aspect of sensation itself, and *not* something that was inferred from a prior sensation of position and time. Movement was a part of the sensation, and not an illusory inference from the sensations. He started to work out a new picture of how sensation and perception work.

Once he had published his results, Wertheimer encountered some problems in obtaining a professorship. Those problems forced him to deal with the credit-attribution problem. How much credit should be his for his work on the allusion of vision, what he called the *phi* phenomenon, and how much should go to Schumann, to Wertheimer's teachers, or to others? "I have never denied the contributions (of the Brentano school), especially in relation to logical exactitude and clarity, and I hope myself to be essentially indebted also to my Prague schooling in logical exactitude." But the work he did on the illusion of movement was his: "it is in no sense the case that the basic thinking of my motion paper came from Schumann . . . the theory I developed has nothing to do with Schumann's ideas."[88] This is a difficult judgment call to make on one's own work; one rarely if ever gets it right.

This early and formative period for the movement was the beginning of the close cooperation of Wertheimer with Kurt Koffka and Wolfgang Köhler, who, along with Wertheimer, would remain close intellectual partners until the ends of their lives.

Kurt Koffka and Wolfgang Köhler were two research assistants in psychology that Wertheimer met when he arrived in Frankfurt. Both were a few years younger than Wertheimer. Kurt Koffka was a native of Berlin, and he entered the University of Berlin in 1903. Three years later, he discovered Stumpf's psychology, and was thoroughly taken by it. This happened during a period when Stumpf was discussing Ehrenfels's notions of *Gestalten* in his seminar, and Koffka worked on the notion of rhythm in both sound and vision. In 1909, when Koffka finished his studies in Berlin, he was able to get an assistantship at Würzburg with Külpe's group (which included Karl Bühler, who will return to our story below). But as we have seen, German academics continued to move from one university to another, and in the middle of the year, Külpe (and Bühler) moved to the University of Bonn. Koffka had, it seems, little choice at that point, and moved to Frankfurt, where he could work with Külpe's former assistant, Friedrich Schumann. That turned out to be a fateful encounter, and Koffka was present in Frankfurt just in time to start working with Wertheimer.[89]

Wolfgang Köhler was the third Gestalt psychologist. He was the same age as Koffka, and studied philosophy and psychology as seriously as he studied physics—first in Tübingen, then in Bonn, and then in 1907 in Berlin, where he went to work with Stumpf. As soon as Wertheimer arrived, Koffka and Köhler began serving as subjects in his experiments, and as he revealed the thinking that lay behind them, the two younger men became more and more enthusiastic about the larger project. Köhler wrote, "I was aware of what Wertheimer was trying to do and found it not only objectively interesting but also most refreshing as a human endeavor. He observed important phenomena regardless of the fashions of the day and tried to discover what they meant. I had a feeling that his work might transform psychology, which was hardly a fascinating affair at the time, into a most lively study of basic human issues."[90]

We have noted that relief from the shackles of the past is part of the tremendous feeling of liberation. Köhler wrote:

> We were excited by what we found and even more by the prospect of finding further revealing facts. Moreover, it was not only the stimulating newness of our enterprise which inspired us. There was also a great wave of relief—as though we were escaping from a prison. The prison was psychology as taught at the universities when we still were students. At the time, we had been shocked by the thesis that all psychological facts (not only those in perception) consist of unrelated inert atoms and that almost the only factors which combine these atoms and thus introduce action are associations formed under the influence of mere contiguity. What had disturbed us was the utter senselessness of this picture, and the implication that human life, apparently so colorful

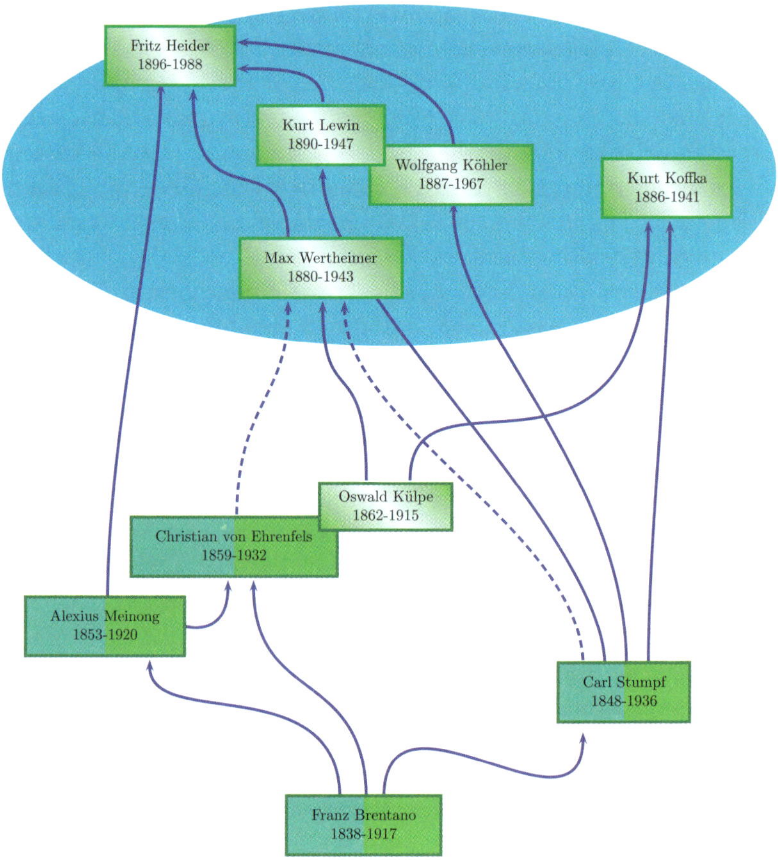

FIGURE 5.2. Berlin Gestalt psychology

> and so intensely dynamic, is actually a frightful bore. This was not true of our new picture, and we felt that further discoveries were bound to destroy what was left of the old picture.[91]

Kurt Koffka shared his recollection as well:

> I remember the actual moment perfectly well, when I learned of this new view. It was in Wertheimer's room in Frankfurt, when he told me, who had been his perfectly submitting subject for several months, of the result of his work and of his conclusion. I can still feel the thrill of the experience when it dawned on me what all this really meant. Of course, at that time I had the merest inkling

> of it, none of us saw as yet very far, but I saw that much, that now at least form had become a subject that could be handled; it [had] made its final entry into the system of psychology.[92]

As the three of them were formulating and presenting their ideas, the world outside academia shuddered and shook; academic priorities were tossed overboard when war overtook everyone's professional hopes and plans in 1914. Even before World War I broke out, Köhler had gone to direct a research station on Tenerife in the Canary Islands, and the outbreak of war prevented him from returning home until 1920. He was able to study psychological processes in apes that led to an important book that he published in 1917.

When Köhler came back to Germany in 1920 he was made acting director of the Psychological Institute in Berlin, and soon came to replace Stumpf as the professor in the department, and permanent director of the Institute. The 15 years that followed were the golden age of Gestalt psychology. Fritz Heider, a student of the first generation of Gestaltists who went to the University of Berlin in 1921, recalled some of the magic of the time:

> I went every day to the psychological institute in the palace which the Kaiser had had to leave at the end of the war just three years before. The institute . . . was an intricate warren of rooms of all sizes. Every so often Koffka came for a visit from Giessen and then Wertheimer and Köhler raced with him from one room to another and he was shown all the new apparatus and phenomena. Everything was fresh and growing and of great importance. One had the feeling that something remarkable was happening there, something that would influence the history of psychology for a long time to come. The big courses by Wertheimer and Köhler were popular and fashionable with the young intelligentsia of Berlin.[93]

MAX WERTHEIMER

Wertheimer spent the years from 1918 to 1929 at the Berlin Psychological Institute as a valued colleague of Köhler, and his participation was a major component of the magic of the era. Fritz Heider remembered Max Wertheimer's lectures well: "Wertheimer was a short, intense man. He gave his lectures in one of the larger rooms of the university and was very popular with the young intellectuals of Berlin. . . . He had a unique style of talking as of writing. He operated by fits and starts in a way that produced the impression that his ideas were fresh and pungent. One felt that this little man with the walrus mustache really believed what he as saying and that it must be something new since it made him so excited."[94]

Heider went on to recall the two groups of Gestalt psychologists in the 1910s: those in Berlin, and those in Graz, centered around Meinong and Benussi, and they were on good terms with each other. Heider started out in Graz, and only later moved to Berlin.

> The Berliners, [though,] always seemed to be engaged in a sort of holy war against nonbelievers and were much more belligerent than the Graz group. They were fervid partisans of all the ideas that had to do with configurations, with every kind of super-unit or totality. For them the very thought of attempting to derive these noble whole-quantities from despicable elements or pieces was unspeakably sinful, a cleansing of deplorable and corrupt thinking. Certainly, Gestalt theory as I knew it at that time in Europe was mainly shaped by its opposition to elementarism, but later in America, the opposition to behaviorism was more important.[95]

One of the Berlin students in these years had a recollection of the heady feeling that came from understanding the new ideas that Wertheimer and his colleagues were floating: "I . . . had the impression when I was his student that most of us had no idea what he was talking about. . . . When we did catch on, we were delighted! Our whole lives changed, our whole outlook on life changed. All of a sudden, everything became colorful and lively and had meaning."[96] In 1929, Wertheimer went back to Frankfurt, but this time as the professor of psychology and Schumann's successor, and when Hitler came to power four years later, he had to leave Germany with his family. He became one of the first of the European expatriates hired at the New School for Social Research in New York.

Wertheimer's last book was entitled *Productive Thinking* and was published in 1945. It is a lively discussion of how people think creatively about new problems, and it scintillates with insights. "Thinking does not, as many believe, necessarily proceed merely by passing successively from one item to another, by formulating successive propositions," Wertheimer wrote. "It does sometimes, but in the very act of thinking, in genuine processes, often it does not."[97] Later, he picked up that train of thought again. "The successive habit—and so the widespread theory that thinking is by nature so—is due to its adequacy in summative situations in which the performance of one operation is merely additively connected with others. It is due further to the fact that we cannot say two sentences at the same time, that we cannot write down two propositions simultaneously, that in reports we have to proceed with one thing after the other."

One of his studies in *Productive Thinking* involved asking people, both children and adults, to make certain structures out of blocks, and in most of the situations, he would build a structure first, and the subject would be expected to do the same thing—and sometimes understanding what counted *as the same thing* was precisely what he was interested in seeing. In some of the most interesting cases, he was working with deaf children, so his interaction with them was not linguistic (the reader gets the impression that the children do not know sign language, and in any event the experimenter does not).

What is so interesting about the study in the first place is that it sets up an ideal situation for gaining some insight into what is going on in the subject's head; an example or two can make that clear. In many cases, there is something that we want to say the subject "gets" or "doesn't get"; we want to see he gets a new idea, or understands something. And the cases are important, if only because this kind of description is what the behaviorist wanted to both deny and eliminate. The behaviorist will countenance a description of the behavior, and will allow including speech among the behaviors, though speech is specifically excluded in the context of Wertheimer's experiments with deaf children; the behaviorist will not allow any inference about what the subject understands, or sees, or gets, on the grounds that speaking in that way is obscure or animistic. But that was exactly what Wertheimer was interested in. And his empathy with the deaf children that he was observing gives us insight into how a mind achieves a new level of understanding.

Max Wertheimer passed away in 1943, in his tenth year in the United States. His book, nearly ready, was published posthumously.

WOLFGANG KÖHLER

Wolfgang Köhler remained at the University of Berlin until the rise to power of Adolf Hitler, and he left Germany in 1935. He was not forced to leave because of his ethnicity, and others have said that he probably could have remained in Germany. But he publicly criticized the Nazis for taking away academic liberties, and it became clear that he had no future in a Nazi Germany.[98]

Köhler came to the United States and took a position in the Psychology Department at Swarthmore College, a college in a suburb of Philadelphia, and many influential psychologists visited or studied with him there until he retired in 1955.[99]

KURT KOFFKA

Kurt Koffka moved to the United States in 1924, still relatively young; he was the first of the Gestaltists to do so. Just as fluency in German was an essential skill for earlier American linguists and psychologists, the time had come when it was becoming important for academics who sensed the growing importance of the United States to gain fluency in English. Koffka's early education was cosmopolitan, and he had learned English as a child from an English governess and spent a year in Edinburgh while still a student, so English was not a challenge for him. He published a long introduction to perception as it was seen by Gestalt psychology, which appeared in October 1922 in the *Psychological Bulletin. Gestalt-Theorie*, as he called it, "is more than a theory of perception: it is even more than a mere psychological theory." He saw it as a theory that "has not yet won its way in Germany," and the theory "has met with serious difficulties, and numerous misunderstandings [which] have occasioned a great deal of the disapprobation which the theory has met."[100]

Early on in the United States, Koffka had two visiting positions, one at Cornell and one at the University of Wisconsin, but in 1927 he was fortunate to receive a quite spectacular offer from Smith College that allowed him five years of research before he needed to begin teaching. The fruit of those years was a book entitled *Principles of Gestalt Psychology*, which was published in 1935. In this book, as with almost everything the Gestaltists wrote while in the United States, there was a heightened awareness of the nearness and the dominance of behaviorist psychology. The modern reader cannot help but be struck by the fact that Koffka chose to define psychology as *the science of behavior*, rather than that of mind or consciousness—a sign of where he took psychology to be at that point, in the United States—but at the same time, he cited William McDougall, not John B. Watson, as the first psychologist to declare behavior to be the correct starting point for a scientific psychology. Throughout he wrote about behaviorists in general, but mentioned Watson's name only once in 700 pages; by contrast, Edward Tolman is cited 20 times. And Tolman is cited in such a way that the reader cannot be entirely sure that Koffka took him to actually *be* a behaviorist, because Koffka takes behaviorists to be committed to studying "molecular" levels of description, and as we have seen, Tolman could not be more clear about the fact that his version of behaviorism is one that is situated at a molar level of description and analysis. Koffka himself was comfortable using Tolman's terms *molar* and *molecular* in his explanation of Gestalt psychology, but he used them in order to emphasize that to the Gestalt psychologist, it was critical to analyze not only behavior but also physiological action at the molar level (we just saw that Tolman wanted to be accepted as a Gestalt psychologist). The target of Koffka's criticism was the notion that the appropriate way to understand the physiology of the nervous system as the sum of a large number of small parts (neurons) with a large number of scattered moments of activation. The whole of the nervous system had a structure to it which was not visible when we look too closely at the lowest level.

Koffka emphasized that, along with Köhler and Wertheimer, he saw the Gestaltist insight as one that was by no means limited to psychology—today we might call it a "systems approach." He pointed out that in chemistry, it was well understood that the nature and behavior of a molecule could be conceptually linked to the behavior of its component atoms only in a very limited way. While the *mass* of a molecule is essentially the sum of the mass of its component atoms, the chemical properties of H, H_2, and H_2O are wildly different, for various deep reasons: the stability of a molecule is determined by how well the atoms find a way of coming together to achieve alignments among them. The stability of a molecule is closely related to the outer shell of its electrons, which has a natural, preferred count of electrons; if we combine atoms that have too few electrons with those that have too many, the result can be a molecule with an overall lower energy level, which amounts to being in a preferred and stable state. Koffka reflected a bit on the origins of the Gestaltists and their distinctive mode of thinking. The central idea they shared had two parts: first, that physiology should be studied from a molar point of view, not a molecular one, and second, that there was an isomorphism between the molar level of physiological analysis and the molar level of description of behavior. Precursors of this can be found, Koffka noted, in early work of Ernst Mach, as early as 1865, and Koffka found Mach's formulation much the same as Wertheimer's, which had come 50 years later. But Mach's work "played no role in the development of our science,"[101] Koffka noted—though we would say that he should be careful, and not take absence of citation of Mach's work for lack of influence. That Mach's work was little known, Koffka wrote, was apparent from the fact that "Köhler fails to mention Mach" in connection with isomorphism, and he himself (Koffka wrote) found Mach's earlier work "by mere accident." "We need not look far to find the reason for this apparent injustice of history," wrote Koffka: Mach perceived fundamental problems in psychology that others "a whole generation later . . . failed even to understand." Koffka's use of *even* suggests he thought that under normal circumstances, if a scientist perceives a problem at one point in time, it should be easier for later scientists to see it; but this is far from obvious, in our opinion. Koffka also wrote that Mach "had a philosophy which made it impossible to give fruitful solutions" to these problems. Koffka was at pains to emphasize that natural science is able to advance only by means of empirical hypotheses that cannot be verified by direct inspection until a good deal of work has been done employing the hypothesis: for a good long while, the empirical support for the hypothesis will remain indirect, and at times elusive. "Physics and chemistry," Koffka noted, "would have been condemned to a permanent embryonic state" if hypotheses were not warmly received even before they could be directly tested.[102]

GESTALT PSYCHOLOGY IN THE UNITED STATES

Americans were aware of the emergence of a Gestalt psychology, even if little was available in the published literature before 1922. E. G. Boring recalled his first trip overseas in 1923, to an international conference in Oxford, England. "The most exciting persons were . . . Köhler and Koffka, for Gestalt psychology was just getting to be known in America . . . and we Americans—like William James in the 1870's and J. McK. Cattell in the 1880's—were all eager to know just what magic German psychologists had contrived."[103] Most American psychologists had first become aware of the Gestaltist movement with the publication in 1922 of the article by Koffka in *Psychological Bulletin* that we mentioned earlier. Two years later, Harry Helson's dissertation, which he wrote with Titchener's advice and submitted to Harvard University, appeared in the *American Journal of Psychology*. What Helson saw, when he looked at Gestalt psychology in Germany in the early 1920s, was a combative academic movement. "In its beginnings," he wrote, this new psychology "has inevitably been destructive; for it has arisen as a protest against the assumptions underlying the methods and hypotheses of older schools."[104] We should have expected no less; that is how new schools arise, as rejections of their teachers' assumptions, with the excitement that that brings with it. "It has brought together from different quarters," he went on, "the criticisms leveled against the traditional psychologies, and it utilizes them in a frontal attack on association, attention, sensation, conditioned reflex, trial and error, and a whole host of concepts which inform current psychological systems. And it has also resurrected many historical concepts which have either fallen into the discard or play but a minor rôle in current psychological theory."

Helson was not entirely happy with the attitude that the early Gestalt psychologists took towards the field of psychology as they found it. Rather than building on what had preceded them—as Henson felt early psychologists had typically done—the Gestalt psychologists "insisted upon starting out afresh, demolishing old systems, and building anew from the foundation up."[105] That sounds familiar, all too familiar. Ralph Barton Perry, a philosopher at Harvard (and a former student of William James), made a similar remark in 1925: "The theory is as yet on what might be called a war-footing. It enjoys the *esprit* of an armed revolution."[106]

Clark Hull recounted an interaction with Köhler in 1941, about six months before the United States entered World War II. After a meeting, Tolman, Hull, Köhler, and Köhler's assistant went to a beer joint (as he called it) to talk shop (as he put it). Hull spoke with Köhler, and

> tried to persuade him that it would be better for the prestige of psychology if we did less fighting and cleared up, so far as possible at least, the pseudo-differences which stand between us. He countered with the suggestion that the behaviorists had been the ones who had been doing the attacking. I was naturally somewhat astonished at this remark. . . . Köhler then went on and made a remark something like this: "Also, I have heard it said that a professor in one of the prominent eastern universities is accustomed, whenever he refers to the Gestalt psychologists, to call them, 'those goddamned Gestalters.'" I must confess that my face was pretty red. The whole crowd gave me a good horse laugh, and of course I had it coming to me. There again my uptake was so slow that it didn't occur to me to tell them that I always smile when I use that expression, which, it seems to me, does make a difference.[107]

Perhaps. Does it depends on the kind of smile one uses? Probably not. Now, all this is well and good, and one can interpret such stories as one likes (and Hull's letter is considerably more detailed than what we have provided in this excerpt). But more interesting is the final comment that Hull made. Köhler "came out with this remark: he said that he was willing to discuss most things in a logical and scientific manner, but when people try to make man out to be a kind of slot machine, then he would fight. And when he said the word 'fight,' he brought his fist down on the table with a resounding smack, and he did not smile when he said it, either."

Hull recalled replying to Köhler that whether one fights or not over such a question, the fighting will not settle the scientific question. "At this point he began telling me about the trouble he had had with the Nazis in Germany and commented on how stupid the English had been not to prepare for war, and so on."

Köhler had begun his side of the conversation with the behaviorists long before he came to the United States. In 1929, he published a book in English and very much targeting American psychologists, entitled *Gestalt Psychology*. His book offered a personal view of how psychology was developing at that moment, and the direction in which it should go in the future. The reader today cannot help but be struck by the fact that Köhler maintains a constant conversation with an off-stage behaviorist, whose reactions and responses Köhler occasionally repeats for us to hear. Regardless of whether behaviorism was dominant in the entire discipline of psychology, there can be no doubt that it set the parameters of the ongoing discourse as far as Köhler and his Gestalt psychology colleagues were concerned.

Let's listen to some of what Köhler wrote.[108] "We know a few things," he wrote, "about the effects of stimulation upon the sense organs of our subjects." And we can watch their responses, too, but that does not mean that we have any understanding at all about the vast terra *incognita* that stands between stimulus and response. The goal of psychology must be "to invent hypotheses about the events which here take place . . . [the] response cannot possibly be understood in terms of peripheral stimulation alone." Just a few pages earlier, Köhler had been discussing the lessons for psychology that could be drawn from the history of physics, and he remarked at this point, "those who know the history of physics will be inclined to believe that this task of finding fruitful assumptions about the hidden antecedents of behavior is about the most important of all." Phys ics has engaged in an enterprise that has lasted for centuries with the goal of determining the existence of unseen objects and causes. "The whole future of psychology may depend on" engaging in just such a project. "On this point all the creative force of Behaviorism ought to be concentrated in a fine emulation of physics."

But Köhler saw nothing but negativism among the behaviorists. He wrote, "I feel a trifle disappointed by the work of Behaviorism," but the sentiment was quite obviously a good deal more charged than that. Behaviorism challenges the legitimacy of direct experience, and then eliminates from the reach of psychology any theory that employs concepts beyond reflexes and conditioned reflexes, as if this were the second commandment handed down by Jehovah. But anyone with even a "modest knowledge of organic events which is now at our disposal" would understand that this constraint is incompatible with what we know about organisms. Behaviorism is *narrow*: "while the school claims to be revolutionary, it actually is dogmatically conservative."[109]

A finer example of ideological critique in the mind fields we will not find. Köhler had many criticisms to level at the work of the behaviorists, but mixed in with that is his dismissal of how the behaviorists view themselves: it is, Köhler believed, quite simply inaccurate. The behaviorists were not revolutionary: they were conservative. "Physics permits itself at least one new idea about the nucleus of the atom per year," Köhler wrote, but the behaviorists seem all too content with the functional ideas that they already have. "Apparently, when Behaviorism was born, somebody put final truth about the functional possibilities of the organism into the cradle."

The behaviorist does little but count the number of occurrences of various types of behaviors that are classified into bins that make it nearly impossible for the researcher to see the rich range of possibilities that humans (and animals) are capable of exploring. "Narrowness in observation," Köhler wrote, "protects narrowness in theory." Why not be grateful for ideas regardless of where they come from, if they will be useful in developing testable hypotheses about what lies between stimulus and response? "Virtually any assumption will be more useful than mere waiting. The hypotheses of empirical science are often based on somewhat meager evidence. Since such assumptions will be tested and continually corrected, they can surely do no harm. If they prove to be wholly or partly right, no one will have scruples about the legitimacy of their origin. If they are shown to be wrong or sterile, they can always be discarded and replaced by better ideas." Did Köhler imagine that he was lecturing to behaviorists, or to young people thinking about becoming psychologists? One thing is certain: whoever he was talking to, he felt that they needed a remedial course on the philosophy of the natural sciences.

FORCES, EQUILIBRIA, AND CAUSATION

There is a striking richness and depth to the Gestaltists' overall conception of how psychology should be done. We have already noted that in their publications in Germany, in German, they rely on the reader's background knowledge of physics. We need to review some of the concepts that they were able to take for granted which derive from basic Newtonian mechanics. This will allow us to reflect a bit more on how the notion of causation and explanation evolved and differed in different scientific contexts—an important point for us, because these notions are still changing, not the least in the mind sciences.

Newtonian mechanics refers to the view of dynamics that emerged in the late sixteenth century and the seventeenth century with the work of Galileo, Descartes, Newton, and others, based on the principle that the language of nature is mathematics. Their work led to new standards for successful theories, and to new conceptions of what counted as an explanation.

The creators of the modern mechanics were all schooled in Aristotelian philosophy, which developed four kinds of explanation, each of which could be viewed as depending on a kind of causality. The four types of causality were material, formal, efficient, and final, and these corresponded to different ways of providing an answer to why something exists or happens. The material cause of something is the *stuff* that already existed, as stuff, and which became the thing we are asking about: why is there a puddle of water on the floor? It is because I dropped an ice cube there and it melted, and the water you see there is what was provided by the ice cube.

The formal cause is the form, or structure, of the object. Elementary explanations of Aristotle's views may choose to offer a *mold* as an example of a formal cause of a statue which has been cast in that mold, and while that is a fine example of what Aristotle had in mind, it suggests too little. To pursue our example of a puddle of water on the floor, to explain that puddle is also to explain what properties this thing that we call water has. The material cause gives an account that involves continuity over time (though Aristotle would not have put it that way); the formal cause is a way that stuff can be, and this is not in any essential way related to time.

Let's begin with the paragon of the modern scientific success, Newton's work on forces and dynamics. In the third quarter of the seventeeth century, Isaac Newton created a new kind of mathematics, differential calculus, in order to express his new conception of motion. He looked at the trajectory of an object in the physical world as a function which specifies the object's position at each point in time. His calculus allowed him to define the first derivative of this position, which we call today the object's *velocity*, and it allowed him to define the second derivative of the position, which today we call the object's *acceleration*. While position can be defined at just a split-second, a moment in time, both velocity and acceleration are meaningful only over an interval of time (though that interval may be as small as anyone might want it to be). These three values (position, velocity, acceleration) are the key elements of Newtonian dynamics: velocity at a given moment will tell us where the object will be just a moment later (to use a poor metaphor), and the acceleration at a given moment will tell us what the velocity will be just a moment later.[110] As far as simple motion is concerned, that is it; dynamics involves mathematics with those three values.

If we think of *cause* and *effect* as the labels that can be associated with two events, the cause strictly preceding the effect, then these two terms are not comfortably attached to describing an object's trajectory in space as we calculate the course of its position, velocity, and acceleration. All three values evolve, and as quantities they are profoundly attached to one another, but none of the them is the *cause* of either of the others.

These three quantities comprise the first conceptual element of Newtonian dynamics, and the second element is the notion that all of the complexity of motion that we observe in the world is due to the presence of *forces* present at every point in the universe. The force that Newton understood best was the force of gravity, for which he had worked out the quantitative statement, and Newton declared that the acceleration that each object possesses at a given time and location is linearly proportional to the force present at that spot (the coefficient of linearity is the mass of the object in question).

This is the revolutionary Newtonian idea of motion, but if you follow the mathematics (which is to say, the content) of the theory, it is not at all a model like a complex watch or set of meshing gears. The intellectual power and magic comes from the fact that the real interactions in the world are all to be found at the second derivative of the location, so to speak, and then in a totally continuous fashion, the second derivative affects the changes in the first derivative, and in a totally continuous fashion, the first derivative affects the changes in the location.[111]

There are other physical models that Newtonian mechanics explored in which the traditional notion of cause and effect still had a natural footing, such as how a set of billiard balls will bounce around on a billiard table, or a collection of molecules will bounce around in a gas. Such a dynamic obeys Newton's laws as regards the motion of each ball (or molecule) when it is in free flight (its acceleration is zero, and its velocity is constant). Newton's laws provided no natural account of what happens in collisions, but Christian Huygens developed an analysis involving what we call today conservation of energy (conservation of *momentum* is essentially Newton's third law of motion) at the same time that Newton was working on his laws, and the final years of the seventeenth century produced a common doctrine concerning the basics of mechanics that we still teach in high school and college today.

The point that is important for us here is that the analysis of a system as a sequence of one event followed by another, each caused by one or more events that preceded it, was one kind of scenario that physicists were able to analyze in the post-Newtonian world. *But* there were other scenarios that were *not* of that sort which were also pure Newtonian situations. One scenario involved collections like the solar system, a set of discrete objects in free motion in a gravitational field, and another was a set of objects (each with its own mass) attached to one another by springs.[112]

Systems that evolve not through instantaneous collisions but through forces which influence the objects' accelerations are simple examples of what we call today *dynamical systems*. This way of understanding the evolution of a system played a major role in the way Gestalt psychology understood cognition, and we think that it played a role (though a less clear and explicit role) in the thinking of the Prague structuralists that we will look at in chapter 9.

The Gestalt theories proposed that psychological systems are dynamical systems, composed of a very large number of small parts, each part interacting with a large number of other parts, with the system as a whole affected both by the environment in which the organism finds itself and the overall set of "forces" that constitute the organization of the system. The Gestaltists very consciously aligned themselves with a kind of dynamical analysis that had emerged in a number of quantitative sciences over the nineteenth century. In each of these cases, the steady state or equilibrium state of a system could be best understood as one which *maximizes* (or equivalently, *minimizes*) some explicitly defined quantity, but typically we must calculate this maximization subject to some further condition, generally the condition that mass, or energy, or both, must be conserved (that is, held constant so as not to vary). Modern students of physics often learn this way of thinking in connection with what is called Fermat's principle, which says roughly that light travels along a path that minimizes the total time it must travel, a principle that is helpful in calculating the path of light through a lens or paths including mirrors. (The insight, though not the modern formulation, can be found in Euclid, and was developed by other pre-modern scholars.)

This way of thinking of physical phenomena was enhanced by the development of the notion of a *field*, like a gravitational or an electrical field, and at the heart of the notion of a field is the fact that an object can be assigned a *potential*—which is, in simple cases, just a number—based on what its location in the field is, and the object senses a force in the direction in which the potential decreases most rapidly. Gravitational potential increases with height, and so an object (suspended in the air or not!) senses a force pushing it downward: any object will try to minimize its gravitational potential, which in normal life means getting closer to the center of the Earth. If it is suspended from the ceiling on a spring, it will have a potential which is the sum of the gravitational potential and the potential energy of the spring, and the combined system seeks an equilibrium in which the potential is minimized.

Much more powerful extensions of this mode of thinking have been developed. A gas, for instance, and the molecules that compose it will quickly find an equilibrium which maximizes the total entropy, subject to the conditions that mass and energy are conserved.

What these systems all have in common is that at any given moment they are in a dynamic equilibrium: the very small parts of the system keep moving and rearranging, but they do so in a way that keeps the total system roughly in stasis, as far as the measurable properties of the system as a whole are concerned. The molecules in a gas keep bumping up against one another, so that at any given moment some are moving very fast and others very slowly, but the total distribution over the speeds and directions of the molecules remains constant for a system in equilibrium.[113]

The Gestalt psychologists drew people's attention to an enormous range of phenomena which was amenable to study in a laboratory, and yet which showed the active role played by the perceiver—the human, in most cases. In case after case, they established clear experimental evidence of the ways in which people consistently saw structure: what we see is a structured image, not a mosaic of separate tiles. The central *fact* of perception, whose recognition must be the starting point of psychological research, is that the way in which relational structures are put together is more important, more basic, than the pieces that put them together.

The empiricist perspective was based on the idea that perception was based on a prior stage of sensation, and sensation was the direct response of the sensory organs to the way in which sound, light, and external objects affected the sense organs. Perception, in turn, was a classification and organization of the signals of the senses. This picture was integrated into the Bloomfieldian model of the phoneme, as we will see in the next chapter. The term *phones* was used to indicate the sensory content of a linguistic signal, and the phones were classified by the part of the perceptual system that was the phonology of the language in question. Gestalt psychologists rejected this view, and thus rejected the Bloomfieldian view of phonology.

Karl Bühler

Wertheimer, Köhler, and Koffka formed a tight team of Gestalt psychologists, and they developed students and followers both in Germany and the United States. But they were not the only psychologists in Germany pursuing the analysis of Gestalts, and the picture that we have sketched of psychology in this period would be incomplete if it were limited to the American behaviorists and the group centered around Wertheimer, Köhler, and Koffka. These three psychologists were described as the Berlin school of Gestalt psychology. But there is another part of the story of how Gestalt psychology was received by both the older and the younger psychologists in Germany at the point when Gestalt psychology was coming to be the dominant perspective, in the heady decade of the 1920s when Berlin was the home to the Gestalt psychologists we have discussed so far.[114] Karl Bühler was another leading psychologist—in Vienna, during the period we are discussing—and there was little love lost between Bühler and the Berlin Gestalt psychologists. Bühler would arguably have considerably more impact on the development of linguistics, as we will see in the chapters to come, due in large part to his cooperations and friendship with Nikolai Trubetzkoy and with Roman Jakobson.[115]

The Berlin Gestalt psychologists had to confront the generation of psychologists who preceded them, many of whom wanted to make it clear that they had already established an important part within their theory for what the young Gestalt psychologists were focusing on. The landscape of theories and positions in psychology during this period was complex, more complex than we have time and space to explore, and the reader who wishes to know more can find an outstanding account in Kusch (2005).[116] But there was also a more disturbing accusation (if that is not too strong a word, and it is not) that the early Gestalt psycholo-

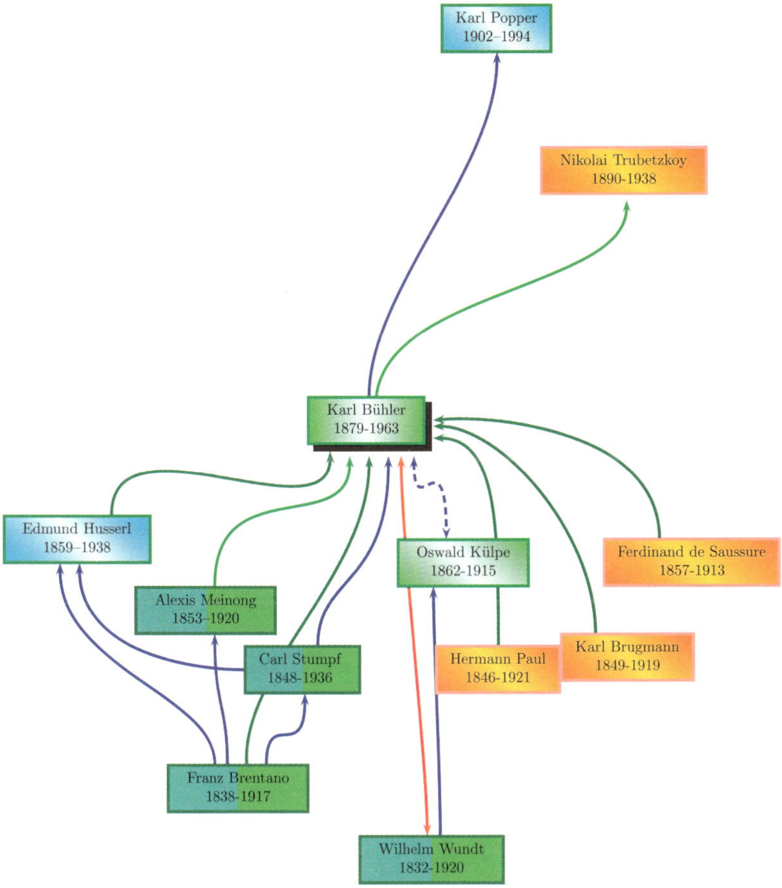

FIGURE 5.3. Karl Bühler

gists had adopted without attribution much of their most important ideas from other young psychologists of their generation, Karl Bühler and Otto Selz. The story is interesting, in part, because the causes that led to the charges were, if not inevitable, then at least unsurprising in a broader context, and the entire set of charges and countercharges is by no means unique in our story.

Bühler's story is also interesting because it is so poorly known today.[117] While the case is complex, Bühler's falling into the black hole of forgotten memories was caused in part by Hitler's rise to power. And it is most of all important for us because while the Berlin group of Gestalt psychologists did little work on language, Karl Bühler was a major figure in the reconfiguration of linguistics.

Karl Bühler was born in 1879, and after earning a medical degree, he became an assistant to the psychologist Oswald Külpe in Würzburg in 1906. Over a span of nine years and several universities, he made important contributions to the Würzburg school of psychological thought. Bühler's name first became known to the world of psychology around 1908, when he became embroiled in a controversy with Wilhelm Wundt, who was nearly 50 years his elder.

The controversy involved a question that is not at all resolved today, the degree to which the matching of utterance and meaning can be viewed in the abstract, and viewed as the result of a pairing that could be performed by a speaker regardless of what social situation she was in. Wundt's view, which is largely the view of linguists today, was that this pairing of form and meaning could be done in the abstract: the meaning of this present sentence, one might argue, is independent of just who wrote it, and even more independent of who might eventually read it. Bühler's position was not that; Bühler argued that language is always embedded in a social interaction that accounts for what must be said, and also for what does not need to be said. Indeed, the strongest arguments for Bühler's view very often rested on the nature of *ellipsis*, the pervasive nature of leaving unsaid what is evident to both speaker and hearer.[118] As we just noted, neither side is obviously wrong, and Wundt's perspective appears every time someone thinks of communication as an encoding of a message in the speaker's head which then passes to the hearer, who decodes the sounds into a message. Bühler's view appears every time we think about how a person decides to formulate a message, to make her point to the person listening to her, perhaps to avoid or encourage ambiguity, or for any of a thousand other reasons.

In 1922, Bühler moved to Vienna, where he became professor of philosophy and psychology, as well as the head of the Psychological Institute there. During the halcyon days in Vienna, he received substantial financial backing from the Rockefeller Foundation[119] and was able to give a number of lectures in the United States as well as in Europe. He would remain in Vienna until 1938, when the Nazi *Anschluss* forced him to emigrate to the United States. Let us be clearer: after the annexation of Austria, Bühler was arrested on March 23, 1938, and held in custody for more than six weeks by the Gestapo, saved only by dint of efforts of friends and family abroad, and Bühler's life was never the same afterwards.[120] Thomas Sebeok later wrote that his treatment by the Gestapo had "incurably shattered his personality: realization that his own people would mistreat him in this way wounded him for life. He became withdrawn and suffered from such an abiding depression that the rich wellsprings of his creativity were effectively consumed."[121]

Vienna in the 1920s and 1930s was the home of the scientific philosophers who formed the Vienna Circle, and Bühler and his students were engaged with the Vienna Circle.[122] Bühler was also involved with the linguists of the Prague Circle, such as Roman Jakobson and the Russian prince Nikolai Trubetzkoy—Trubetzkoy, like Bühler, was thoroughly engaged with the questions of the Prague Circle, but he lived and worked in Vienna, and was thus Bühler's immediate colleague. Bühler, like Trubetzkoy, published works on language in the *Papers of the Prague Linguistic Circle*.

Bühler published his most famous book, *The Theory of Language (Sprachtheorie)*, in 1934, the synthesis of two decades of thinking about language and linguistics. The most famous and influential theme of his book was the proposition that human language has three functions: one is the expression of the speaker, who may wish to inform or something else—like request an action, or perform an action. The second is an expression directed at the listener, an *appeal* to the listener. The third is the part that we have always focused on, the *representation* (*Darstellung*) or the content of the linguistic message. For Bühler, this meant that everything we know about language should be considered in three different lights. A linguist who views an utterance only in its ability to represent a state of affairs will see only part of the story, because often—not always—some part of the utterance must be understood as the expression of a specific speaker in a specific context.

That belief gave rise to Bühler's long-standing interest in deictic elements in grammar, words like *here*, *there*, *now*, and *then*, and also *you*, *me*, and *her*: words whose meaning and function is ineluctably tied to the context of speech, and thus to the speaker and the addressee. Roman Jakobson would later follow up on this and add the word *shifters* to the vocabulary of linguistic analysis for such words. Indeed, Roman Jakobson cited Bühler some 20 times in the 1971 *Selected Writings*.[123]

Bühler left Vienna in 1939 and spent the rest of his life in the United States. He had turned down an offer from Harvard in 1930, too committed at the time to leave Vienna, but he did not succeed in revitalizing his intellectual career when he did come to the United States.

Bühler was perhaps the most interdisciplinary of all of the scholars we have encountered in this study: his work continued a conversation with the Neogrammarians, with the psychologist Wilhelm Wundt, the philosopher Edmund Husserl, with Saussure, with the philosophers Ernst Cassirer and Anton Marty, and with the work of Franz Brentano. Bühler's work, notably in *Sprachtheorie*, developed many of the central ideas in what has come to be known as pragmatics.

The reader will recall a brief discussion of Otto Selz in chapter 1 in connection with the question as to whether he had gotten enough credit for his influence on Karl Popper's thought. Bühler and Selz were outraged in 1925 when they read a survey article that Kurt Koffka published called "Psychology." The three of them had all worked with Oswald Külpe. Bühler thought that Koffka had imported wholesale ideas from his own book on psychical development. He himself had published work on spatial relations from a Gestalt perspective in 1911. There were more ways than just those of the Berlin psychologists to carry out the promise of Gestalt psychology, Bühler argued; the term should not be inappropriately linked to a particular take on the problem.[124] Irritated at Koffka's publication, he made a comment on scientific continuity: "A philosopher might strive for a new philosophy, but it would never occur to a physicist to present a new physics. In psychology it should not be otherwise. Even 'revolutions' do not eliminate the continuity of an advanced science."

A philosopher who is familiar with late nineteenth-century philosophy will see in Bühler's work a continuation of the theory of part-whole relations that Franz Brentano developed in front of his students, as we saw in chapter 3. That theory was developed in at least three ways that are relevant: by Marty, by Ehrenfels, and by Husserl. One philosopher writes, "It seems to me to be most likely that Bühler became familiar with this [part-whole] theory by studying its applications to linguistic and psychological phenomena in the writings of Brentano and pupils of the latter such as Stumpf, Marty, Ehrenfels, Husserl, Twardowski and Meinong—'the older Austrian school of psychologists,' as Bühler calls them."[125]

Bühler developed ideas about the psychology and phonology of linguistic sounds that are difficult to separate from the ideas of Trubetzkoy that we will explore in more detail in chapter 9. He viewed a linguistic word as a Gestalt, a whole, which is composed of phonemes which are ordered, but the word is more than just its phonemes, in precisely the same way that a melody is more than just its notes. At the same time, he emphasized that what is essential about phonemes is their distinctiveness from one another—we noted in chapter 2 that he used the term *diacrisis* to refer to phonological contrast, in his comparison of Trubetzkoy's structures of phonemes to Mendeleev's periodic table; he did not use the term "distinctive" in this context, though Brentano and Marty had used the term "distinctional" (in the phrase *distinktionelle Teile*, "distinctive parts").[126] As early as 1922, Bühler noted that sounds were decomposed into what today we call features: "our plosives, for example, are acoustically determined by three or four independently variable moments that in their particular accord (*Zusammenklang*) constitute the *p*, *k*, or *d* character of a sound . . . which is a Gestalt."[127]

The Period Comes to a Close

Jerome Bruner was a graduate student at Harvard studying psychology at the end of the 1930s, and in 1983 he looked back on what was happening in psychology as he first saw it, and as he saw it many years later. Bruner would be one of the leading figures in the development of cognitive psychology in the 1950s and beyond. When he was a student, there was no doubt that the dominant theme in psychology was empiricism and physicalism. But he himself reflected a world that was changing, because despite the leanings which he perceived in the world around him, he found his heroes among the Gestalt psychologists, Sigmund Freud, and the cultural anthropologists. He liked the questions those folks were asking, and he hoped to bring the methods of physicalism to bear on the more interesting questions. He was, as he put it, juggling two sets of maps.[128]

He entered the PhD program at Harvard in 1937, and he found that "nearly every one of my fellow graduate students at Harvard was in opposition" to the objectivist orthodoxy.[129] That should have clued him in to the fact that the times were changing, he said, but he did not see it yet.

Bruner felt that at that point there were two directions a young psychologist could chose: it would be perception or it would be learning. If he chose "perception," then he would work in a world that was "mentalistic, phenomenological, essentially European." If he chose "learning," then the world would be behaviorist and deny any role to anything subjective. He chose the first of the two: Bruner's advisor was Gordon Allport.[130]

Bruner had very clear recollections of Edwin Boring, and how he treated the students. "He took us seriously," he wrote. "He was accessible for talk, and when the talk was done, you would often receive the very next day one of those long, single-spaced letters, restating his points and reflecting on yours. . . . He did not suffer fools at all. And lazy scholarship angered him into gruff silence. He was one of the 'older men' in my life for whom I felt genuine love. And a certain respectful fear."[131]

Bruner remembered how he and his fellow graduate students would argue with Boring about the priority of sensations compared with the world of ordinary experience. Boring thought that the sensations were the foundation on which the world of experience was built; Bruner thought it was the other way around, that sensations were somehow an abstraction. The Gestalt psychologists formed a "protest movement," it seemed, and they "gave each other courage in standing up to Boring." And then Bruner stopped his pen for a moment, and wrote, "Standing up to Boring—what an odd way to put it. For all the twenty-five years I knew him, I never saw him *act* the bully or the browbeater. Some people are just formidable."[132]

The four decades of psychology that we have looked at in this chapter constitute a relentless search for a way to establish psychology as a science. No matter what suggestion would be made to rethink psychology as a science, critics and skeptics would join forces to dispute the new proposals. And if the challenge to psychology was to be a science, no one disagreed that the way to be a science was to apply the scientific method. Method was what distinguished science from non-science.

Edna Heidbreder's classic study of the history of the themes in American psychology, published in 1933, made this point with great clarity. She wrote,

> Psychology, especially in the United States, has risked everything on being science; and science on principle refrains from speculation that is not permeated and stabilized by fact. Yet there is not enough fact in the whole science of psychology to make a single solid system.

Harsh words! But it got worse:

> No one knows this better than the psychologists themselves. They see with the eyes of familiar association not only the undeniable poverty of their science, but the flimsiness and shoddiness of much of the material they are asked to accept as genuine fact.

Heidbreder listed the main factions of the warring psychologists: there were the younger psychologists focused on animal and comparative psychology, "most of whom pride themselves on being hard-headed and realistic and on having discarded the airy nothings of a psychology that deals with minds." In short, the behaviorists, who think that they know the best method for becoming a science: discard the mind. "They wish above all else to be severely scientific, and some of them seem convinced that they can best realize this ambition by resembling as closely as possible their near neighbors, the physiologists."[133]

The second group of psychologists are those who see themselves as the heirs of Wilhelm Wundt and, more directly, Edward Titchener. These are the trained introspectionists, who believe they have a method for inspecting consciousness. They form "a less aggressive group, but one that is no less assured, no less conscious of the integrity of its science. . . . They represent the established aristocracy of a generation ago. Theirs is a psychology, they are willing to maintain, that has stood and will stand the test of time."[134]

The next group are the experimental psychologists who focus on quantitative methods and experimental methods that aim to measure intelligence and the realm of the mental; they are strongly allied with (and indeed, often indistinguishable from) industrial psychologists. Ten years later, these psychologists would become an essential part of the government's war effort, as the government in Washington, DC, ramped up to create the largest war machine ever created, for which it needed effective methods to evaluate the skills and weaknesses of millions of men and women hoping to join the armed forces. But in 1933, this was still unknown.

How can psychology be formed by three such disparate groups as Heidbreder described, the groups who disagree so fundamentally on such important questions? They share something very important:

> In the course of the struggle there has grown up a vast respect for science as such. In the presence of the older sciences, psychology feels something of the awe of the novice for the master, something of the abashed admiration of the nouveaux riches for an established aristocracy. It feels too the same anxious concern lest its mode of life fail to conform to the standards of the caste, and in its very zeal for maintaining those standards it is sometimes betrayed into a swaggering superiority to the practices it has so recently learned to scorn . . . beneath its bluster there is a desire for solid fact and sure technique.
>
> The arrogant intolerance it sometimes displays toward anything on which the seal of science has not been set, is in part at least a jealous concern for the integrity of its knowledge, not wholly unlike the fine scorn of Bacon for the pretensions of the Aristotelians of his time.
>
> For in the few years of its existence, psychology has acquired not only diligence but the skepticism that for science is the beginning of wisdom . . . it

> knows that its greatest virtue is its determination to follow the scientific method, and that at its best, it attempts to push that method into a region which hitherto the inquiries of science have not penetrated.[135]

In the end, Heidbreder took a remarkably modern perspective on the social and intellectual value of the warring factions, and their proclaimed beliefs: "systems of psychology are to be regarded not as statements of scientific knowledge, but as tools by which scientific knowledge is produced; not as accounts of scientific fact, but as means of acquiring scientific fact. . . . They offer a specific and sometimes glamorous program of action, but the program is not to be confused with accomplishments."

And yet, she could not hide her concern that there was a powerful anti-intellectual theme in contemporary psychology:

> Psychology, with its determined devotion to the scientific method, becomes at times almost bitterly anti-intellectualistic. . . . Perhaps because of the recency of the separation of psychology from philosophy, perhaps because of an acute sense of need for observed facts, many psychologists regard speculation with suspicion and distaste, almost with resentment and dread. For many of them are still somewhat afraid of being metaphysical, and by being metaphysical they mean, whether rightly or wrongly, spinning theories out of the head without verifying them by fact.[136]

And she urges a change in perspective, for

> frequently the victories of science are won through the use of conjectures not yet established by fact, conjectures that become the basis of active and ingenious research especially directed toward that particular body of evidence which will prove or disprove the point at issue. Guesses on the basis of inadequate evidence have proved to be powerful and, in actual practice, indispensable tools, which science regularly employs.[137]

Heidbreder's was a remarkable voice at that moment, and the fact that her book is still in print 80 years later bespeaks a resonance which her perspective found among readers. It is still a book worth reading.

We turn now to what was happening in American linguistics during this same period.

CHAPTER SIX

American Linguistics, 1900–1940

The development of linguistics as a profession accelerated during the first four decades of the twentieth century. More people were thinking of themselves as members of a discipline whose focus was the nature of language as it is found in human societies around the world. In the United States, the energy captured in this development flowed out of two traditions. The first was the particularly American tradition of anthropological linguistics, and the second was the continuation on American soil of European, but especially German, linguistics, with scholars of the old Indo-European languages, but there were also phoneticians, and philosophers of language.

It was during this period that American linguists came to the conclusion that they needed to create an organization of professionals who cared about language and linguistics, a movement that would give rise to the Linguistic Society of America in 1924.

We will focus on the founding father of American anthropology, Franz Boas, and two of the most influential linguists of this period, Edward Sapir and Leonard Bloomfield. Sapir, a few years older than Bloomfield, was as much an anthropologist as he was a linguist, or nearly so, and he was the student of Franz Boas. We are fortunate to have the accounts of several of the younger linguists who worked with both Sapir and Bloomfield, such as Zellig Harris, a linguist whose role in the development of linguistics would in time be equally influential.

We hope that you will see that Bloomfield and Sapir were linguists whose worldviews were worlds apart. Both were inspiring and impressive researchers who greatly influenced the younger linguists who followed them. But they were at opposite ends of more than one spectrum. The shy scholar and the guru: linguistics was something that they shared, but there was little else. They both left their imprint on the field as we know it today, but for very different sorts of reasons.

To appreciate the work of Bloomfield and Sapir, it is necessary to understand the larger context in which anthropological linguistics was established in the United States. We will review first the relevant historical background in the United States during the nineteenth century.

Early American Anthropology

Manifest destiny

The nineteenth century was the period during which the United States expanded from a small country hugging the East Coast to one of the largest countries in the world, stretching across a wide continent. The expansion was carried out in part by purchases (the reader will recall the Louisiana Purchase), in part by military seizure, and it was accomplished with an explicit policy to see English replace Native American languages. We can only allude here to depredations carried out by the American government during this period against the Native Americans, such as the Creek War, fought on a western front at the same time as the War of 1812 against the British, the Seminole War to rid Florida of its native population, and increasing levels of military conflict before the American Civil War as a result of growing numbers of white settlers, miners, and ranchers. The close of the Civil War brought with it a firm policy from Washington that Native Americans would either be assimilated into the society that was brought by Europeans, or they would be required to live on reservations defined by the government in Washington.

In March 1824, Congress approved the establishment of the Bureau of Indian Affairs, placed within the Department of War. Maintenance of Native American languages was discouraged in the strongest possible terms. The Commissioner of Indian Affairs in 1887, J. D. C. Atkins, wrote, "Barbarous Dialects Should Be Blotted Out . . .":

> I have to advise you that the rule applies to all schools on Indian reservations, whether they be Government or mission schools. The instruction of the Indians in the vernacular is not only of no use to them, but is detrimental to the cause of their education and civilization, and no school will be permitted on the reservation in which the English language is not exclusively taught.[1]

During this period, however, there were a number of intellectuals, as well as scholarly and academic organizations, who were concerned with the circumstances of Native Americans—their personal survival, and the maintenance of their languages, cultures, and civilization. The Smithsonian Institution played a crucial role in this. It was created by Congress in 1846, responding to a bequest of more than $500,000 left by a wealthy Englishman, James Smithson. From its very inception, the Smithsonian has supported scientific expeditions, geographical exploration, ethnological expeditions, and the collection of materials that have resulted from those activities. In 1879, Congress established the Bureau of American Ethnology, and placed it within the Smithsonian Institution. Its first director was John Wesley Powell, and under his direction, Franz Boas and other linguists were supported in their work on Native American languages. Boas and Sapir, along with their colleagues and students, would conduct fieldwork with native tribes and describe their languages, their rites, and their cultures. The annual reports of the Bureau to the Secretariat of the Smithsonian and to the Congress contain innumerable linguistic descriptions, texts, and grammars of Native American languages that were still spoken or, in some cases, were near extinction.[2]

Franz Boas

For more than half a century, from the beginning of the time that Franz Boas was active in the United States, American linguistics was indelibly marked by a descriptive and anthropological orientation. With few exceptions, American linguists worked on Native American languages, many carried out ethnological studies as part of their field work, and they analyzed linguistic material that they had gathered—or that someone they knew had gathered.

Franz Boas was born in 1858 in Westphalia, part of Prussia. He was Jewish by family background, nonpracticing by faith. His family felt the influence of the revolutionary movements in 1848 and the culture of the Enlightenment that was widespread in Germany at that time. As a young man Boas was very interested in Humboldt and Darwin, but also in Linnaeus's work on classification. As a student at Heidelberg University, he studied physics, mathematics, and psychophysics under Hermann von Helmholtz; he also engaged in the classical dueling matches among students of the day, a proclivity that left scars on his face and head. After defending a doctoral dissertation in physics in 1881 on color perception, his interests shifted towards psychophysics, and Helmholtz supported him in his *Habilitation* in Berlin in 1883 for ethnographic research involving Baffin Island in the north of Canada. Boas continued to work then with Friedrich Ratzel, the founder of anthropogeography.[3]

He studied in Heidelberg, Bonn, Berlin, and Kiel (where he got his PhD in physics), and then he came to the United States, at the age of 29. We have already encountered Boas's name in connection with G. Stanley Hall, who was the first president of Clark University, and who tried to build a research university from scratch. Boas was one of the first appointments that Hall made to the faculty there, in 1888. In that same year, Boas helped found the American Folk-Lore Society, and was the first president of the association.[4]

During these years, Boas was influenced by Kant's views on anthropology, and by Wilhelm von Humboldt's "Outline of a Comparative Anthropology." It was in Humboldt's work that Boas encountered the idea that each language defines a particular vision of the world, an idea that Boas would later teach, and which would be articulated by students of his, notably under the description of "the Sapir-Whorf hypothesis."

Boas was also influenced by the violent debates that shook German anthropology at the end of the nineteenth century involving race and anti-Semitism. As we have seen, the impact of Darwin's work at this time was

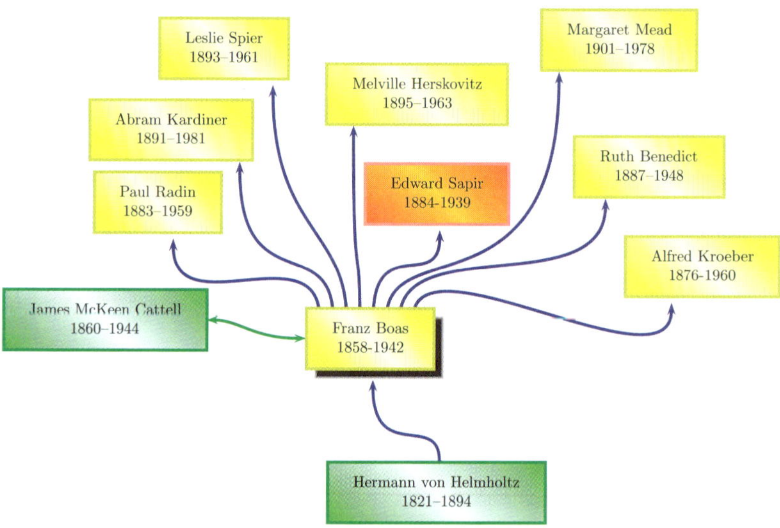

FIGURE 6.1. Franz Boas

felt across the spectrum of thought about society, from racist and eugenicist positions to anti-racist and universalist positions.

Declaring a certain skepticism with regard to the Darwinian model, Adolph Bastian and Rudolph Virchow founded the Berliner Gesellschaft für Anthopologie Ethnologie und Urgeschichte in 1869, and supported the development of a strong anti-racist position.[5] Boas would later acknowledge Bastian's influence on his thinking, and the struggle against racism would remain throughout his life a major concern for Boas. In 1883, Boas spent a year among the Inuits of Baffin Island, in the northern reaches of Canada, studying the relationship between the environment and the ways of life he found there. This experience led to his decision to devote his life to anthropology, and it had a considerable effect on how he viewed anthropology. He lived with the Inuits, learned their language, myths, and culture, and he described their customs and their rites. All of the premises of the American cultural anthropology that he would found 20 years later, and of his understanding of the relationship of fieldwork and the cultures of Native American groups, can be found in the work that he carried out during this formative year.[6]

Boas felt a connection to the United States, and from the beginning of his career, he published in English, even though his oral English was not very good at the time. In Berlin in 1885, he met Chief Tom Henry and nine Bella Coola entertainers who were on tour in Europe, and this led to Boas organizing several visits, and ethnological expeditions, to British Columbia and Oregon. He worked on Chinook, beginning an interest that would continue throughout his lifetime. Attracted by what he had found in the United States, and troubled by the racism and anti-Semitism he had found in the German academic world, he spent the summer of 1887 in New York, after completing his work in British Columbia, and reached the conclusion that he would remain in the United States. Soon after that, Boas was recruited to Clark University by G. Stanley Hall to continue his research and to teach one course a year as a docent.

We saw briefly in the last chapter that Hall's efforts to create a research university at Clark along the lines of Johns Hopkins University did not result in the institution that he had hoped for. The failure was the result of a number of causes, one of which was a disenchantment felt by many faculty members for Hall himself as a leader. Boas would certainly have counted himself among them, and while he was not one of the large number of faculty members who left Clark for the new University of Chicago in 1892, he did leave Clark that year, and took a position at the World's

Fair in Hyde Park, right next to the University of Chicago's campus. That position led to another in New York at the American Museum of Natural History, and then he was offered a position as lecturer at Columbia University. The psychologist James McKeen Cattell was instrumental in obtaining this position for him, and the two would remain friends and colleagues for many years. Boas was promoted to professor in 1899, and he spent the rest of his career at Columbia.

During his career at Columbia, he developed around him a virtual army of gifted and enthusiastic graduate students who pursued his ideas, and extended them in many directions. Among his students at Columbia were the outstanding and influential anthropologists of the next generation: there was Alfred Kroeber, Paul Radin, Abram Kardiner, Leslie Spier, Margaret Mead, Robert H. Lowie, Zora Neale Hurston, Ruth Benedict, Melville J. Herskovits, Edward Sapir, and Otto Kleinberg. This list of Boas's students is at the same time a list of the superstars among anthropologists in the generation that followed Boas. One scholar noted later on that "if we drew up a genealogical table, the *guruparamparā* (as the Hindus call it), the succession of teachers and pupils, would come clear. It is only necessary, however, to mention Sapir and Kroeber, Jacobs and Andrade as his own pupils, and Leonard Bloomfield as a devoted admirer of his, and to note that practically without exception the younger generations are pupils of one or more of these men and often of Boas himself as well, due to the habit of young linguistic students in this country . . . indulging in *Wanderjahre*."[7]

Boas did not retire until 1936, when he was in his late seventies. The linguist Roman Jakobson came to New York in 1941, and the friendship that arose between Boas and Jakobson was important for both of them, and the same was true for the friendship that Boas shared with Claude Lévi-Strauss, who was, like Jakobson, a refugee in New York during this same period. Boas died in 1942 of a heart attack while dining at the Faculty House with Lévi-Strauss, at an event organized in his honor by his former student and founder of French anthropology, Paul Rivet. The necrology that Jakobson published in 1944 of his friend Boas is a moving and touching tribute.[8]

Boas's method of studying the native languages of North America had a tremendous impact on the field that was aborning. He was not the first European to learn an indigenous language, nor the first to analyze one. But he established the notion that the serious study of a language included learning to speak the language, developing a grammar, and collecting and transcribing myths and other texts in the language.

Boas's conception of anthropology included a partition that is still very much with us today in the United States, based on the notion that the field was comprised of four subfields: archeology, linguistic anthropology, physical anthropology, and cultural anthropology. He had no place and no tolerance for the view, often espoused during the nineteenth century, that cultures would pass through a series of stages as they progressed.

Karl Teeter wrote about him, nearly 25 years after his death:

> Franz Boas approached language with the training of the nineteenth century natural scientist, a training emphasizing the respect for empirical fact which had been so instrumental in that century in removing harmful vestiges of aprioristic views of the natural world. When he began to study real languages in the New World, he was immediately impressed by how little use all of the linguistic theories then prevalent were. Much in them, as with then current theories of the natural world, was aprioristic and normative, based upon a general grammar which often unwittingly universalized what were specific features of familiar languages. . . . Both training and field experience showed Boas that these theories could not begin to account for what he found, that it was necessary to start anew from each language to discover the functioning units in each. . . . In all of this prodigious labor his working assumption was that linguistic data must be approached without presuppositions. The relevant reality is their own inner structure. We must adopt the prejudices of the native to understand it, and eschew our own.[9]

Teeter continued with the remark that taken too literally and too far, this enterprise—of beginning with no presuppositions—is impossible to carry out. In this, Teeter was speaking as a man of his times, and as such he was suspicious of anyone claiming to work without presuppositions. "Selectively applied, however, it is possible for the notion to be psychologically fruitful. Such was in large part the case with Boas' actual practice. . . . Though he often seems to be advocating a more extreme view, it was not in fact all presuppositions about language he avoided, but only those which his experience had shown were mistaken ones."[10]

Those who knew him during his lifetime had no doubt about the power of Boas's efforts to approach a new language without presuppositions. Murray Emeneau wrote, "He brought to his work an almost complete freedom from preconceptions, at a time when linguistic scholarship had not yet quite freed itself from the last preconceptions of pre-scientific linguistic study." He bought himself a *freedom of thought*: "He was able, because of this freedom of approach, to admit that in linguistics anything is possible and, consequently, to analyse exotic material without forcing it into the straitjacket of the familiar. **This was perhaps the most valuable single lesson in analysis that he taught his pupils and they theirs.**"[11]

Robert Lowie was one of his students, one who became a leader in American anthropology as a professor at Berkeley, and in his obituary of Boas, he made some remarks that bear on the Noah phenomenon that we raised in chapter 1. He mentioned something that another student of Boas had written in another obituary:

> Boas's service to anthropology were so great and manifold that occasionally enthusiastic disciples unfamiliar with history talked and wrote as though his predecessors and contemporaries were negligible. One obituary article declared: "He found anthropology a collection of wild guesses and a happy hunting ground for the romantic lover of primitive things; he left it a discipline in which theories could be tested and in which he had delimited possibilities from impossibilities."[12]

This other student was Ruth Benedict, speaking of her teacher.[13] Lowie did not go along with this.

> This is to parade Boas as a mythological culture-hero creating something out of nothing. The conception would have been intolerable to Boas, who fully esteemed what had been done by E. B. Tylor, Lewis H. Morgan, Eduard Hahn, Karl von den Steinem, and others. Indeed, he was especially appreciative of men who had achieved what he himself never attempted—an intimate, yet authentic, picture of aboriginal life. I have hardly ever heard him speak with such veritable enthusiasm as when lauding Bogoras's account of the Chukchi, Rasmussen's of the Eskimo, Turi's of the Lapps.[14]

Marvin Harris noted that for many of Boas's students, it was a sign of the health and maturity of the field as Boas lived in it and left it that there *was no Boas school*. Kroeber, for example, saw partisan behavior in the social sciences of the sort not seen in physics as a sign of the field's immaturity. He observed, in words that we will see repeated in the context of Bloomfield, of Chomsky, and of others:

> The image of Boas that had the greatest appeal to his students was that of a professional scientist who had raised anthropological research methods and standards of proof to a level with which even a physicist would feel comfortable.[15]

And Marvin Harris noted that Robert Lowie "identified Boas with the philosopher-physicist Ernst Mach and associated both of them with the development of 'higher standards of proof' and the perfection of the scientific method."[16] Lowie noted, "Boas was singularly unexacting in regard to a student's factual information. Probably there is not nowadays a single undergraduate major in any of our large anthropological departments who does not control a wider range of data than I did when Boas deemed me fit for the doctorate. It was enough that I had worked in the field, gained a theoretical conception there, and thrashed out the issue in a formal paper."[17]

No one who was following the academic scene in anthropology was unaware of Boas's influence or of his engagement in strategizing the course of the profession. Lowie described this in remarkably sympathetic terms, writing that "estrangements from one-time students were in part merely the familiar phenomenon of filial revolt, but in part they resulted from Boas's taking a rational point of view that clashed with the disciple's emotional urges. He was wont to survey the chessboard of anthropological jobs and figure out how science could be best served, then he would try to move anthropologists about like the pawns in a game. His judgment was usually right, but some men and women resented the impersonality of his strategy."[18]

Boas was one of the most prominent opponents of the then popular ideologies of scientific racism, the idea that race is a biological concept and that human behavior is best understood through the typology of biological characteristics. In a series of groundbreaking studies of skeletal anatomy, he showed that cranial shape and size was highly malleable and depended on such environmental factors as health and nutrition, arguing against the claims by racial anthropologists of the day that held head shape to be a stable racial trait. Boas also worked to demonstrate that differences in human behavior were primarily not determined by innate biological dispositions, but rather were largely the result of cultural differences acquired through social learning. In this way Boas introduced culture as the primary concept for describing differences in behavior between human groups, and as the central analytical concept of anthropology.[19]

Edward Sapir

Edward Sapir was born in 1884 in Lauenberg, not too far from Hamburg, and his first language, the language that was spoken in his home, was Yiddish. When he was four, his family moved to England, then two years later to Richmond, Virginia, and finally to New York City. Sapir came from a poor family, but his precocious intelligence was evident. With the help of a Pulitzer scholarship, he was able to attend Columbia University.

At Columbia, he met Franz Boas, who headed the anthropology department and who allowed Sapir to take an advanced course on American Indian languages. Sapir was hooked. He completed his undergraduate degree in three years, and a master's degree a year later, with a thesis on Herder's essay on the origin of language. The next year, Sapir began graduate study with Boas,[20] all the while continuing to study traditional linguistic subjects, such as Sanskrit. Sapir finished his coursework for his PhD in 1906, and over the next few years, he had several positions that enabled him to do linguistic description and analysis, and comparative linguistics in the context of the North American continent. Boas enabled

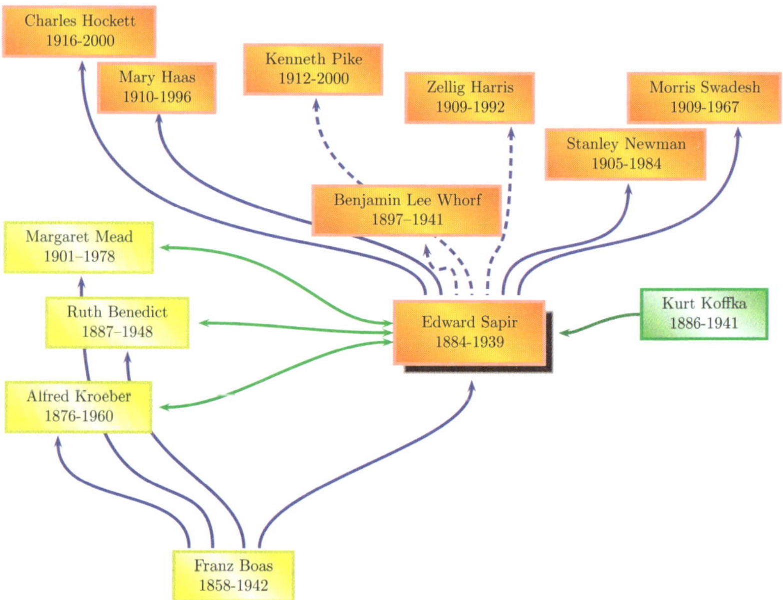

FIGURE 6.2. Edward Sapir

Sapir to do fieldwork with Takelma speakers in Oregon, and in 1909, Sapir defended and published his dissertation on Takelma. In 1910, he took a position in Ottawa as part of the Geological Survey of Canada, and he remained there for 15 years. He wanted to come back to the community of American academics, and in 1925, he accepted a position at the University of Chicago, where he began teaching anthropological linguistics in the Department of Sociology and Anthropology (which would divide into two departments four year later).[21]

During his Canadian years, Sapir had established a reputation for himself in his field, and the publication of his book *Language* in 1921 added significantly to his profile. Alfred Kroeber had urged him to write the book: "The decadence of linguistics is largely your own fault," his friend had written him. "You're an individualist and haven't built up a school. Do something general in character."[22] A large part of Sapir's professional work had gone into establishing the validity of Neogrammarian styles of sound change as valid for the establishment of North American language families. That was brilliant work, but Sapir was capable of much more. Sapir's book attracted both linguists and a more general readership. It was relatively short, but provided a brilliant overview of the nature of language as he saw it.

Sapir was a charismatic man, an inspiring teacher in every sense, and there seems to be something cosmically unfair about the fact that he had no students until he was well into his forties. But that changed once he began teaching at the University of Chicago, and many of the leaders of the next generation of linguists were trained by him during this period. Over the six years he was in Chicago, his interests and his ideas developed in three areas: linguistics, of course, but also cultural anthropology and the relationship between personality and culture. He had always been close to the whole family of anthropologists who had studied with Boas, and he was especially close to Margaret Mead and to Ruth Benedict. But during his years in Chicago, he developed strong relationships of friendship and intellectual closeness with Harry Stack Sullivan, a psychiatrist, and Harold Lasswell, a political scientist.

Sapir became deeply interested in the psychology of the individual, and how an understanding of individual psychology could lead to greater insight into the construction of culture. Margaret Mead, who he was close to during the years just before he moved to Chicago, introduced him to the work of Kurt Koffka, who he would later met in 1927, when both participated in a meeting in Chicago on the role of the unconscious in the various social sciences. In a letter to Ruth Benedict, Sapir wrote of Koffka's work:

> I've been reading Koffka's *Growth of the Mind* (Margaret's copy) and it's like some echo telling me what my intuition never quite had the courage to say out loud. It's the real book for background for a philosophy of culture, at least your/my philosophy, and I see the most fascinating and alarming possibilities of application of its principles, express and implied, mostly implied, to all behavior, art, music, culture, personality, and everything else. If somebody with an icy grin doesn't come around to temper my low fever, I'll soon be studying geometry over again in order to discover what really happens when a poem takes your breath away or you're at loggerheads with somebody. Nay more, unless a humanist like yourself stops me, I'll be drawing up plans for a generalized Geometry of Experience, in which each theorem will be casually illustrated from ordinary behavior, music, culture, and language. The idea, you perceive, is that all you really need to do to understand—anything, is to draw a figure in space (or time) and its relevance for any kind of interest that can be discovered just by noting how it is cut by the place (=context) of that interest.[23]

This passage from Sapir's letter is fascinating at several levels; it illustrates the way that effects across disciplinary boundaries happen on their own, and it also allows us to understand that this kind of effect will never vanish. Sapir's familiarity with Koffka seems to have begun with a suggestion by Margaret Mead, a close friend who was a whole generation younger than he was, who shared a book that had just come out that she found exciting. Sapir's interest in psychology can be traced to any number of circumstances, including the illness that his wife suffered from in the first half of the 1920s, as well as the widespread discussion in the media of psychological approaches from Watson's behaviorism, at one end of the dial, to Sigmund Freud and his psychoanalysis at the other.

Two decades after Sapir's death, Alfred Kroeber, one of Boas's first students and by then himself a distinguished anthropologist at Berkeley, wrote that "Edward Sapir, I should say, is the only man I have known at all well, in my life, whom I would unreservedly class as a genius." Boas was an extraordinary person, Kroeber acknowledged, fired by great passions both intellectual and ethical. But Boas did not measure up to the genius that was Sapir.[24]

In 1928, Yale established the Institute for Human Relations at Yale University, as we saw in the previous chapter—Clark Hull was brought to

Yale to participate in that new institute. As part of that same expansion in the social sciences, Yale University's president, James Angell (whose psychology we have already considered), was able to convince Edward Sapir to take up a position at Yale, and in 1931, Sapir became the Sterling Professor in the Institute of Human Relations, with the charge to create a new anthropology department, and to work with the three linguists who were already there (Franklin Edgerton, Eduard Prokosch, and Edgar Sturtevant).[25] A number of students went to Yale from Chicago along with him, including Morris Swadesh, Stanley Newman, and Mary Haas. The linguists working with Sapir later included Charles Hockett, George Trager, Benjamin Lee Whorf, Charles Voegelin, and Zellig Harris—an extraordinary group of creative linguists.

Clyde Kluckhohn was a graduate student in anthropology at Harvard who was interested in Navajo, and who participated in Sapir's Athabaskan work during the late 1930s.[26] He also was a participant in Sapir-led seminar on the relation of individual and society.[27]

Sapir's health began to decline rapidly in 1937 after suffering a heart attack while he was at the Linguistic Institute. He died in early 1939, at the height of his intellectual powers. It was a tragic loss to the field.

The Phoneme

We have observed that the phoneme is the single most important concept that has emerged in the field of linguistics. Its definition has changed over time, and there has never been complete agreement on just what the precise, technical definition should be. It is also fair to say that it is a development of the folk concept that led to our Western alphabetic manner of writing, and that the *word* "phoneme" has been at times a charged and loaded term. There are two closely related notions that lie at the heart of the concept of the phoneme. The first is that each language has a particular inventory of sounds—somewhere between a dozen and perhaps 60—out of which its words are built. The second is that when we count up these "sounds," we realize that each of these is *not* a particular sound, but a whole *class* of sounds that is treated as a single functional unit by a language—partly, but only partly, because the sounds are similar to each other from an acoustic or articulatory point of view. There is a large element of arbitrariness as well in how a language will decide (so to speak) to take a large class of sounds and treat them as functionally equivalent.[28]

Controversy—especially healthy controversy—is encouraged by careful attempts at definition, and in the American context, this began to happen in the 1920s. Edward Sapir published a very influential article that helped set the terms of the debate, "Sound Patterns in Language," the lead article in the first issue of *Language* in 1925; he published another, this time in French, called "La réalité psychologique du phonème," in 1933, in Pierre Janet's journal, *Le Journal de psychologie normale et pathologique*. Sapir's student Morris Swadesh published an article in *Language* in 1934 in which he attempted to synthesize what was agreed upon at that point, and we will turn to it now.

Swadesh's paper in *Language* was called "The Phonemic Principle," and it was one of the first comprehensive attempts to clarify just what American phonologists meant when they talked about phonemes. Swadesh was a PhD student at that point at Yale, working with Sapir, Stanley Newman, and George Herzog, and he was married to Mary Haas, herself an influential linguist of her generation.[29] Swadesh wrote in this paper that in addition to the conversations that he had at Yale, his understanding was heavily influenced by Bloomfield's book *Language*, two articles by Jones, two by Trubetzkoy, and one by Ułaszyn, all five in *Travaux du cercle linguistique de Prague*, and the two papers by Sapir: "Sound Patterns in Language," and "The Psychological Reality of the Phoneme." From Swadesh's perspective, the Americans who were engaged in an effort to define the phoneme were equal partners with their European colleagues in Prague and elsewhere. We will see in chapter 9 that the European structuralists, such as Trubetzkoy and the Prague linguists, viewed Sapir as an important influence on their thought, and an equal on the intellectual level.[30]

Swadesh's paper began,

> The phonemic principle is that there are in each language a limited number of elemental types of speech sounds, called phonemes, peculiar to that language; that all sounds produced in the employment of the given language are referable to its set of phonemes; that only its own phonemes are at all significant in the given language. The phonemes of a language are, in a sense, percepts to the native speakers of the given language . . . [but if] the phonemes are percepts to the native speakers of the language, they are not necessarily percepts that he experiences in isolation.[31]

In short: "The phoneme is the smallest potential unit of difference between similar words recognizable as different to the native."[32] Why did

Swadesh write his paper in the first place, and why was it published? The reason is simple: at that moment, linguists who were working out the definition of the phoneme realized that finding the right phonemic analysis for a language meant dealing with a fine balancing act. When looking at data from a particular language, there were *some* reasons that pushed the analyst to increase the inventory of phonemes in the analysis, and *others* that pushed the analyst to decrease the inventory. All other things being equal, the analysis with the fewest phonemes was the best, but this is only one factor to be considered when we compare different analyses. It was relatively easy to identify the reasons that supported having more phonemes or fewer in the analysis of a particular language, but it was very difficult—and ultimately, perhaps impossible—to give a precise account of how these reasons played off of each other, which is to say, to decide when an argument for increasing the number of phonemes would be allowed to supersede an argument for decreasing the number. For those who believed that a commitment to phonemic analysis meant assuming that there was one and only one correct analysis, nothing would be satisfactory until one could provide a method either for *finding* that analysis, or else for *testing* it once it had been found.

Swadesh also emphasized a very Sapirean point: that "each language has a characteristic word and syllable structure."[33] In one language, we might find that all words begin with a single consonant, or never end with a long vowel, even though these conditions fail to hold for syllables inside the word. And often a phoneme will have a distinctive realization that occurs only word-initially, or word-finally. In German, a word which phonemically begins with a vowel will be realized with a glottal stop before the vowel, and that element is not a phoneme, Swadesh said, but rather a "mechanical sign" of the edge of the word.

Swadesh offered an important guideline that he called "the criterion of complementary distribution": "If it is true of two similar types of sounds that only one of them normally occurs in certain phonetic surroundings and that only the other normally occurs in certain other phonetic surroundings, the two may be sub-types of the same phoneme."[34] Even two sounds which are quite different may be members of the same phoneme, as long as those sounds are found in different *contexts*. At the same time, Swadesh did hedge his bets: he did say that the two sounds in a single phoneme had to be "similar." But holding off on exactly what it might mean for two sounds to be similar and what not, this interpretation of complementary distribution is one that *widens* the range of analyses, by allowing analyses in which we decrease the total number of phonemes, and associating more than one sound to a particular phoneme.

Swadesh is clear in saying that it is "phonetic" surroundings that we look at when we want to determine if two different sounds occur in complementary environments. This is much more important than it might seem at first blush, and would be challenged later by Zellig Harris (1951).

What if you have a sound which is in complementary distribution with more than one sound? The *p* in *speech*, for example, is in complementary distribution with the *p* in *pea*, but also in complementary distribution with the *k* in *key*, and the *t* in *tea*. In that case, Swadesh said, the original sound should be "identified with"—that is, be made part of the same phoneme as—the sound to which it is most similar, which in this case would be the *p* of *pea*.

Let's summarize the major points so far. If we are trying to propose a phonemic analysis of a language, then we are looking for as small a set of phonemes as possible, subject to the condition that each sound in the utterances of a particular language has to be analyzed as representing one phoneme. We are, therefore, looking for ways to set up phonemes, each of which can be realized in more than one way (that is, each can correspond to different sounds). Sometimes a phoneme is said to represent a range of different sounds because that range of different sounds can appear in a particular position in a word without any change in the meaning of the word: this multiple possibility is called free variation of the phoneme.

Dynamic rules

One of the striking characteristics of much of Sapirean linguistic analysis is its *dynamic* character. Phonemes are not merely *there*; they line up and they affect each other, sometimes aggressively. There is no better way to explain this than to look at an example, such as Stanley Newman's description of Yokuts. Newman was Sapir's student, and his work hews close to the Sapirean model.[35]

Newman described the Yawelmani dialect of Yokuts, a language spoken in California. He published a very short and preliminary paper about Yokuts in a journal in 1932, and then, a decade later, two important works: a book which he finished in 1943, and an article in a important collection in 1946. These last two have been extremely influential in American phonology in the decades that have followed, and while there is certainly consistency across the two major analyses, there are some points that differ that are of interest to us.[36]

In the Yawelmani dialect, there are 10 different vowels, and there are many, many cases where the linguist looking at the data wants to say that one vowel has changed into another; there is a very regular pattern, but it lurks underneath the surface.

The ten vowels are {i, e, a, ɔ, u} plus the long versions of these vowels {i:, e:, a:, ɔ:, u:}. He referred to these as the "light" and "heavy" versions of each vowel. But, Newman continued, it would be wrong to line them up this way, which is what we would do if we were focusing on the physical character of the sounds:

light	i	e	a	ɔ	u
heavy	i:	e:	a:	ɔ:	u:

Rather, he noted, the long i: and u: and the short e are really not part of the same game as the others. The right way to describe the inventory of vowels is like this, with the ɔ: occurring twice in the diagram:

light	i	a	ɔ	u
heavy	e:	a:	ɔ:	ɔ:

And he gave names to each of the four columns:

	i-series	*a*-series	*o*-series	*u*-series
light	i	a	ɔ	u
heavy	e:	a:	ɔ:	ɔ:

Newman gave several reasons for this analysis, though it would take us a bit afield to reproduce the quite technical arguments here. The three vowels that he left out of the second diagram (i:, u:, e) only occur in stems, not suffixes, and they only appear as the result of a process that he called *ablaut*. So the reader must note that there are two odd, or remarkable, things going on: first, some vowels appear only as the result of a process but do not appear as what Sapir would call an *organic* part of the stem or suffix, and second, there are two different phonemes /ɔ:/. One is part of the *o*-series, and one is part of the *u*-series.

In Newman 1946a, he did not actually say that these are two different *phonemes*; there are ten phonemes, he said, as in the first chart above, and the long ɔ: is just *one* of them. He wrote, "it is a striking feature of Yawelmani that the . . . two phonological entities (ɔ:, of the o series, and ɔ:, of the u series)" are "actualized in a single vowel phoneme." There are ten phonemes, but only eight " fundamental vowels of roots," and of the ten phonemes, one of them, ɔ:, corresponds to two different "fundamental vowels."

In Newman 1946b, however, after introducing the ten vowel phonemes, he wrote that there is a "pattern of eight vowels [which] underlies the vocalic processes. . . . [they are] morpho-phonemes . . . the utilization of seven vowel phonemes for eight basic morpho-phonemes results from the splitting of the phoneme into two morpho-phonemic entities." This new way of analyzing sounds would eventually become the standard view.

There are thus two valid answers to the question of what the vowels are in Yawelmani: there is what the ear hears (which are the ten phonemes), and there is what the brain of the speaker knows (which is that there are eight vowels, from which the ten spring).

There are a wide range of stems with two vowels which display what Newman called *vowel harmony*, by which he meant that both vowels in the stem are chosen from the same series. In such a case, an ɔ: can appear in two different kinds of stems: "the vowel series membership of an ɔ: vowel is defined by the base in which it occurs. Thus, the second vowel of *dɔ:sɔ marks this base as being composed of ɔ series vowels; in the same way, *ʔɔ:ṭˀu is identified as a base containing vowels of the u series. There is no base which is ambiguous in its vowel series membership."[37] And then there is another kind of vowel harmony, Newman explained. There are suffixes with a single vowel in them, and each of them has two forms. The passive aorist is realized as *-it* or *-ut*: it is *-ut* if the stem vowels are of the *u*-series, but *-it* otherwise. Another suffix, marking the dubitative form, is realized as *-al* or *-ɔl*: *-ɔl* if the stem vowels are from the *ɔ*-series, and *-al* otherwise. And we can find stems that have a long ɔ: from both vowel series: the speaker has to know *which vowel series* the stem vowel is from in order to predict (so to speak) which form of a suffix will appear. Newman speaks of "rules" as well: at one point, he wrote of "the rule that shortens long vowels in closed syllables" (1946a: 227).[38]

The beauty of the linguistic worldview that Newman presented is in one way the beauty of a fine watch made of flawless gear mechanisms, though in another way it is quite the opposite of mechanical. There is something machinelike about the picture, one in which a complex set of stem forms can be accounted for with a small set of generalizations, but the generalizations require knowledge of the underlying system of the language. The generalizations apply with clocklike precision, but the conditions that set them in motion are not so much what is *heard* by the hearer, but rather *known* by the speaker.[39]

We cannot find a clearer articulation of the dynamic vision of linguistic structure than this: "Stems are formed from roots by processes of vowel change. These processes are to be regarded as operating on two planes: on the one hand, dynamic vowel processes effect ablaut changes that are to be defined in terms of morphological conditions; on the other hand, a few phonetic processes introduce additional vowel changes of a mechanical nature." And not only that—one process of this sort can feed into another. Newman put it this way: "In the formation of stems these two planes interact; a stem which has undergone dynamic vowel changes may, in turn, be subjected to secondary phonetic changes."[40]

There is something special about the set of consonants {m,n,w,y,l}, which today we would refer to as the sonorants of the language. Each of them has a *glottalized* version, written {m̓,n̓,w̓,y̓,lˀ}. (A glottalized consonant is produced by closing and opening the vocal cords during the pronunciation of the sonorant.) The difference between a consonant that is glottalized and one that is not is a phonemic difference in the language, which means that the language uses the difference between the two as a basic phonological building block. But there are conditions under which a morpheme which we would normally identify as a plain sonorant is realized as a glottalized sonorant; this happens when a suffix is added which is of the set of suffixes that always glottalize a preceding sonorant. The agentive suffix /ˀa/ has this effect (here we have surrounded the symbol by slashes, which came to be the standard way of indicating that a phoneme had been posited). For example, the root /huls̩/ means 'sit down,' and the agentive form is /hulˀs̩a:/; likewise, the root /diy/, 'lead,' is realized with a glottalized /y/ when the agentive suffix is added: /diy̓a:/. Newman wrote, "When occurring as second consonants in stems, the [sonorants] absorb the floating glottal stop of suffixes."[41] In another context, Newman described the way in which each root has alternate forms, and he added that a suffix can "demand" that the root to which it is attached be of a particular form: "thus, to add the agentive /-uc̓/, a stem is normally demanded whose first and second vowels are respectively weak and strong."[42] Perhaps the best way to describe this is as a purely mechanical but far from obvious analysis: it is fine to call this "mechanical" as long as we don't mean by this that it is either plodding or obvious. It is mechanical like the Neogrammarians' analyses which related the phonemes of one stage of the language to those of the next. Only now, this kind of analysis was integrated into the synchronic treatment of the morphology of the language. This was a step forward of great moment.

Linguistics as a science

Sapir devoted an entire article in 1929 to the question of "The Status of Linguistics as a Science." He began by noting that linguistics is arguably the most developed of the *Geisteswissenschaften*, the mind sciences, due to its finely honed methodology:

> Linguistics may be said to have begun its scientific career with the comparative study and reconstruction of the Indo-European languages. In the course of their detailed researches Indo-European linguists have gradually developed a technique which is probably more nearly perfect than that of any other science dealing with man's institutions. Many of the formulations of comparative Indo-European linguistics have neatness and a regularity which recall the formulae, or the so-called laws, of natural science.[43]

He then notes that there are those who have raised questions about the psychological plausibility of the "regular phonetic developments" that linguists have uncovered.

> Why such regularities should be found and why it is necessary to assume regularity of sound change are questions that the average linguist is perhaps unable to answer satisfactorily. But it does not follow that he can expect to improve his methods by discarding well tested hypotheses and throwing the field open to all manner of psychological and sociological explanations that do not immediately tie up with what we actually know about the historical behavior of language. A psychological and a sociological interpretation of the kind of regularity in linguistic change with which students of language have long been familiar are indeed desirable and even necessary. But neither psychology nor sociology is in a position to tell linguistics what kinds of historical formulations the linguist is to make.[44]

It is hard to read that and not feel the irritation in these words 75 years later. But he went on, not without irony:

> It is very encouraging that the psychologist has been concerning himself more and more with linguistic data. So far it is doubtful if he has been able to contribute very much to the understanding of language behavior beyond what the linguist has himself been able to formulate on the basis of his data.[45]

But there may be some reason to expect changes:

> But the feeling is growing rapidly, and justly, that the psychological explanations of the linguists themselves need to be restated in more general terms, so that purely linguistic facts may be seen as specialized forms of symbolic behavior. The psychologists have perhaps too narrowly concerned themselves with the simply psychophysical bases of speech and have not penetrated very deeply into the study of its symbolic nature. This is probably due to the fact that psychologists in general are as yet too little aware of the fundamental importance of symbolism in behavior. It is not unlikely that it is precisely in the field of symbolism that linguistic forms and processes will contribute most to the enrichment of psychology.[46]

And so Sapir inevitably comes back to how linguistics can help psychology better understand the mind, rather than looking forward to assistance from psychology for linguistics.[47]

We have barely touched on the intellectual magic that Sapir seemed to bring to whatever he was working on. We have looked briefly at the development of phonemes, and the style of insight-rich analysis that he and his students brought to the analysis of complex Native American languages. But he was also a poet, and he was deeply interested in the ways in which depth psychology was shining light on the dark passageways of the human mind that remain so often beyond the ken of our superficial individual awareness. As an anthropologist, he knew that much that defines us as humans goes well beyond the individual as well. Like many of his age and like others since then, he looked forward to a deeper understanding of both the individual and society that would emerge out of a marriage of depth psychology and the study of culture. We will turn our attention now to the other great American linguist of this age.

Leonard Bloomfield

Family and career

Leonard Bloomfield was born in 1887 in Chicago, and grew up there and in Elkhart Lake, Wisconsin. His father, Sigmund Bloomfield, had come to the United States as a child in 1868, from Bielitz. Bielitz is not a large city; it is now part of Poland, and called Bielsko, but it had gone back and forth between Poland and Austria-Hungary over the course of modern history. The majority of the people of Bielitz spoke German or Yiddish: in 1910, more than 84 percent of the population spoke one of these languages, and the Jewish community at that point was estimated at 16 percent. The family name had been "Blumenfeld," and it was part of the Jewish community, but when they came to the United States and settled in Milwaukee, Wisconsin, they changed the name to "Bloomfield." Not long after, Sigmund's father moved his family to Chicago, where he opened a dry-goods store. As with many German-speaking immigrants, the language of the home continued to be German, a language which would play an important role in Leonard's life, as well as in that of several other members of his family.[48]

The Bloomfield family belonged to a not insignificant community of Jewish immigrants from German-speaking Europe whose attachment was more to Jewish culture than to its religion, who identified with the Enlightenment, and who maintained a firmly patriotic sense towards Germany, at least before the United States entered into World War I in 1917.

Growing up, Leonard had an aunt and an uncle on his father's side who became quite famous. The uncle was Maurice Bloomfield, who we have already met—he had studied with the Neogrammarians in Leipzig, and was a distinguished professor of Sanskrit at Johns Hopkins University. His life and career would have a significant impact on Leonard's.

Leonard's aunt Fannie had a distinguished career as an internationally recognized concert pianist. She had begun playing the piano at 6, and at the age of 11, her mother had taken her to Austria for five years to study with the great piano teacher Theodor Leschetizky. Her career was brilliant; she played in concert with Paderewsky, and toured both in the United States and in Europe. She married Sigmund Zeisler on October 10, 1885, a successful lawyer who, like her parents, had came from Bielitz, and who was involved in several important cases, including the defense of the militant anarchists in the Haymarket riot affair of 1887.

Leonard's uncle and aunt were thus Maurice and Fannie; his father, as we have said, was Sigmund Bloomfield, who had neither an intellectual nor an artistic career, and who married Carola Buber (who was, it is true, from the same family as the philosopher Martin Buber). They had three children, two boys and a girl. Leonard was the one in the middle. The oldest was Grover, who became a chemist, and the youngest was Marie.[49]

In 1895, when Leonard was eight years old, his parents bought the Elkhart Lake Resort, and the neighboring Hotel Schwartz, where the Bloomfield family would remain for three decades. Elkhart was a resort area where people from Chicago and Milwaukee would come to stay during the summer, in northern Sheboygan County. It also lay in the heart of the land that as little as 60 years earlier had been the home of a number of Algonkian tribes, including the Menominee and the Potawatomi.

Leon Despres was a Chicago lawyer who knew Bloomfield for most of his life, and looking back on Leonard Bloomfield the man, he wrote, "My recollection of Leonard is that he was a very quiet person. He talked quietly. He smiled quietly. In a conventional social group, he tended not to talk very much at all."[50] Like others who knew them, Despres remarked on how strong the bond was between Bloomfield and his wife Alice.

> I used to think that the ordinary social engagements that Alice compelled him to endure were very painful to him. I am sure he begrudged every minute that was taken away from his work. Nevertheless, he would go out to dinner and sit through an evening, as if this were a duty he owed to Alice and to the world. When he came home, no matter how late, he would work one hour before going to sleep. . . . I was impressed by the people from out of town who would come to visit Leonard. Alice would entertain them for dinner or for an evening, and I noticed that they all viewed Leonard with considerable respect. They were usually academicians, many of whom had been colleagues of Leonard at Ohio State University or University of Illinois.[51]

Like his uncle Maurice Bloomfield, Leonard went to Harvard University for his undergraduate studies, and he received his bachelor's degree in 1906 after three years of study. In some ways, uncle and nephew were poles apart: one was open, brilliant, and very sociable, while the other, the younger, was shy, reserved, and closed to most people. But Maurice had a strong intellectual relationship with Leonard, and his influence is not hard to discern, especially early on in Leonard's career. Leon Despres wrote, "Of course, at the beginning, people thought that uncle Maurice

Bloomfield was a more distinguished linguist than Leonard. I remember his coming to see Leonard once and being treated with great respect and surrounded with silence so that he might confer with Leonard."[52]

One of Bloomfield's students later recalled, "Bloomfield in 1940 told me of a time his uncle, Maurice Bloomfield, had overheard him worrying about how a certain sound could have changed into a certain other sound, and had said, 'Don't worry about that, any sound can turn into any other sound.'"[53]

After graduating from Harvard, Leonard Bloomfield went back home, and enrolled as a graduate student at the University of Wisconsin, which is not too far from Elkhart. In Madison, he met Eduard Prokosch, a young professor who made a great impression upon him. Prokosch had just received his degree at Heidelberg studying under Eduard Sievers (though he had already spent some time in the American Midwest, and he had studied linguistics at the University of Chicago before going back to Germany). When Bloomfield arrived, Prokosch taught in the Germanic department at Wisconsin, though he would later take a position at Yale University. The young Bloomfield always considered himself indebted to Prokosch, as we can see in Bloomfield's own recollections of his youth. "In the summer of 1906 I came, fresh out of college, to Madison, to be looked over for an assistantship. Desiring to earn an academic living, I had developed no understanding or inclination for any branch of science. The kindly professor Hohfeld delegated Prokosch, one of his young instructors, to entertain me for the day."[54] Prokosch was only nine years older Bloomfield. "On a small table in Prokosch's dining room there stood a dozen technical books (I seem to remember that Leskien's Old Bulgarian grammar was among them) and in the interval before lunch Prokosch explained to me their use and content. By the time we sat down to the meal, a matter of perhaps fifteen minutes, I had decided that I should always work in linguistics. At the end of the two years of study that followed, I knew no greater intellectual pleasure than to listen to Prokosch."[55]

Bloomfield thus became a Germanist, and remained engaged in the teaching of Germanic languages for most of his career. William Moulton, years later a distinguished scholar himself, wrote, "It is not usual to speak of Bloomfield as a Germanist. We think of him rather as one of the founders of linguistics as a discipline, as a scholar in the field of American Indian languages, and—above all—as the author of an immensely influential book, his 1933 *Language*. . . . Yet Bloomfield was also a Germanist: primarily so during the early years of his scholarly life, and at least partly so even in his later years. . . . Throughout his life, Bloomfield taught many Germanists."[56]

Bloomfield stayed at Wisconsin for two years, and then moved to Chicago, where he earned his PhD from the University of Chicago in 1909. For his doctoral work, he chose two advisors, Francis A. Wood and Carl Darling Buck. Wood had done his doctorate at Chicago in 1895 on Verner's Law in Gothic, while Buck had been a distinguished student of William Dwight Whitney at Yale, where he had received his PhD in 1875, and by the time Bloomfield was his student, he was arguably the most influential voice for the German tradition of Indo-European studies in the United States. Buck had been one of the original faculty members when the University of Chicago was founded, and he was officially a professor of Sanskrit and Indo-European comparative philology. Buck, too, had studied in Germany after his PhD: he had gone to Berlin and to Leipzig to study under Brugmann and Leskien, and to meet the outstanding young linguists of the day. And not just to *meet* them; he was engaged in their most serious intellectual concerns. George S. Lane would write of Buck later,

> [Buck] brings into bright focus the seriousness with which Indo-European studies were being pursued in Germany at that time, just when so many old cruces were giving way before the renewed onslaught by the Jung-grammatiker [the Neogrammarians], now in their maturity, under whom Buck was studying. For example, there is young Buck's vigorous defense of his methodology, which calls for RECONSTRUCTION not mere COMPARISON as the final goal of comparative grammar. . . . One cannot but tremble for and at the same time admire this young American who states his position so clearly in matters which were the subject of heated controversy at that time, especially between Berlin and Leipzig.[57]

In 1909, then, Bloomfield was a young man with little standing in the field of his own, but with the credentials that came from being the student of the most outstanding professor of Indo-European in the country, and the intellectual capital that accrued from the connections he had made through his uncle, and his cultural and linguistic command of German. He taught German for a year in Cincinnati, and then became assistant professor of comparative philology and German at the University of Illinois, about 100 miles south of Chicago.

This was still the time when an American with a fresh PhD would go to Europe to study with the men who had actually *written* the texts that he had read. To advance in his career, Bloomfield needed to do just that, and so he spent the academic year 1913–14 in Leipzig and Göttingen, almost literally following in the footsteps of his uncle. Bloomfield studied with August Leskien, Karl Brugmann, and Hermann Oldenberg, and he also attended Wundt's lectures in Leipzig, lectures which had an enormous impact on him as well. In Leipzig, he rubbed shoulders with other young linguists who would become famous in their time. Prince Nikolai Trubetzkoy, who we will meet in chapter 9, was there, and the French student Lucien Tesnière, himself a student of Antoine Meillet, was also; all three of them were part of the linguistic community at the University of Leipzig in this most remarkable year.[58] One of Bloomfield's students, Robert Hall, later observed,

> At Leipzig, Bloomfield attended courses given by August Leskien and Karl Brugmann in Indo-European and by Hermann Oldenberg in Vedic and Sanskrit. He can thus be considered as one of the last pupils of Leskien and Brugmann, and therefore as a direct link between the "Junggrammatiker" of the 1870s and American linguistics of the mid-twentieth century. Undoubtedly this contact would have strengthened Bloomfield's adherence to the "Neogrammarian hypothesis" (as he was wont to term it) of regularity in sound-change, which had already been upheld by his uncle Maurice.[59]

In Göttingen, he studied under the noted Sanskritist Jacob Wackernagel, whose name is remembered today in connection with Wackernagel's Law. Bloomfield would later call him his teacher: "My models are Pāṇini and the kind of work done in I.E. by my teacher Professor Wackernagel of Basle. No preconceptions, find out which sound variations are distinctive (as to meaning) and then analyze morphology and syntax by putting together everything that is alike."[60] Wackernagel thus introduced him to the grammar of the ancient Sanskrit grammarian Pāṇini, and this would have an enormous influence on his linguistic thought. He told one of his students that Paṇini's grammar was one of his "bedside books."[61] Murray Emeneau, Bloomfield's student, later commented that "anyone who has been exposed to Pāṇini will recognize the Pāṇinian-like character of Bloomfield's style in his Algonquian descriptions and, in fact, in *Language*."[62] Bloomfield would have agreed; he wrote this, himself:

> The descriptive grammar of Sanskrit, which Pāṇini brought to its highest perfection, is one of the greatest monuments of human intelligence and (what

> concerns us more) an indispensable model for the description of languages. The only achievement in our field which can take rank with it is the historical linguistics of the nineteenth century, and this, indeed, owed its origin largely to Europe's acquaintance with the Indian grammar. . . . The comparative grammar of the Indo-European languages got its start only when the Pāṇinian analysis of an Indo-European language became known in Europe.[63]

Bloomfield wrote his first book, *An Introduction to the Study of Language*, just before leaving for his year in Germany—but the book was published by Teubner in Germany, so he could correct the proofs there on the spot during that year. This book has everything that we might expect of the most brilliant young master of his generation, who would extend the Neogrammarian view from Indo-European to language as a whole. Robert Hall wrote,

> The 1914 *Introduction* is a remarkable achievement for a man in his middle twenties. It is basically a summa of what was known about language at the time. It contains a fantastic amount of information drawn from over eighty languages. . . . In some respects, it was more in line than was his 1933 book with widely accepted ideas concerning language, especially with regard to psychology. In this respect, Bloomfield accepted then the dominant approach and used the terminology of the *Völkerpsychologie* of Wilhelm Wundt, with its emphasis on the "rôle" of mental factors in human use of language. He also assigned to the individual speaker of a language a more important role than he tended to do at a later stage in his thinking.[64]

Thus when Bloomfield's book appeared in 1914, it would be read as the work of a talented student of the Neogrammarians, and as an account of the relevance of Wilhelm Wundt's thoughts about language for linguistics.

Due to the tremendous success of the German academic system in the nineteenth century, science *was* German at this beginning of the twentieth century. German was the language of scientific culture, and our survey of the mind sciences in the nineteenth century could leave little doubt that Germany, and the countries in its cultural penumbra, were home to the dominant intellectual and scientific activities during this time, and quite naturally the most brilliant of young scholars from the rest of Europe and the New World did everything they could to become a part of the social fabric of German scholarship. By the same token, the most important literature in the linguistics world had been published in German, and the German-language journals continued to dominate the field. If we were to fast-forward a hundred years, we would find much the same thing today, though now the language on top is English, and it is to the United States that students go, and it is English that they do their best to publish in.[65] Bloomfield was perfectly positioned to be a leader in an American setting—he was, to be sure, as American as one might be—in a profession dominated by German thought and culture, to which he had access as a bilingual speaker of German.

Bloomfield's career begins

In 1921, Bloomfield moved to the Ohio State University, where he would stay for six years. He was greatly influenced during his years there by the behaviorist philosophy of Albert Paul Weiss, a psychologist at Ohio State University some nine years his senior.[66] Weiss had been born in Germany, and though he came to the United States as a young child, he spoke German at home, as Bloomfield did.[67] A close friendship grew between Weiss and Bloomfield. Even before Bloomfield arrived at Ohio State, Weiss was suffering from serious heart ailments, and this obliged him to stay at home much of the time. Bloomfield and his wife Alice were frequently guests of the Weiss family, and Weiss's views of behaviorism and how to do science had an enormous impact on Bloomfield.

Weiss was a firm and convinced behaviorist; he had been a student of Max Meyer, the early behaviorist (and student of Stumpf), who we met in the previous chapter. Weiss's behaviorism greatly appealed to Bloomfield at this point in his life—though it was very different from the Wundtian perspective that Bloomfield had known to that point.[68] Here is Weiss's general position on what psychologists should do, when faced with the fact that people speak:

> The behaviorist raises the question as to whether a subject who is introspecting is actually describing mental states. Instead of maintaining that introspection reveals the character of some mental process, it is simpler to say that it reveals only the fact that the experimental stimulus, in addition to producing the experimental response (pressing a key for instance) also produces an oral response (the introspective report). All that is actually observed is the fact that the energy of the response is not a simple function of the energy and

> character of the stimulus. The behaviorist regards introspection as the behavior of a very special and limited class of individuals. Human laws, institutions, social customs are developed by nonintrospecting individuals, and it is the behavior of this type of individual that engages the primary interest of the behaviorist. While he may of course investigate the introspective reaction, he regards it as merely one way in which a psychologist may react to a special situation. The method of the behaviorist thus reduces itself to a statistical, genetic and mechanical analysis of those movements that form the basis of human interaction.[69]

We have already looked at behaviorism. Listening to Weiss, we hear a firmly convinced behaviorist: it is as if when the behaviorist looks at someone who is pointing at the Moon, he looks no further than the tip of the fellow's finger. Is he really pointing at the Moon?, asks the behaviorist. *I* have no need of such hypotheses, of such conjectures; I see his finger, but that's all. Surely you don't think that branches of *a tree* can point to the Moon, do you? I don't see any reason why a full scientific description of a man's posture and the location of his arm, hand, and fingers needs to be supplemented by the purely superfluous remark that the hand is pointing towards the Moon. His hand is what it is, and it is located where it is, and the Moon is so far away that it is perfectly unreasonable to bring it into the discussion.[70]

Bloomfield considered Weiss to be a great scientist with a considerable influence, and he published an obituary for Weiss in *Language* in 1931. "There had always been students who refused to believe in the spectres of our tribal animism (mind, consciousness, will, and the like)," he wrote, "but these students had never given a clear-cut and satisfactory explanation for the super-biological actions of man—the actions which transcend the possibilities of the animal world. In our time these students are the behaviorists—an ugly name, said Weiss, but accepted it for want of better."[71]

Leon Despres, the friend of Bloomfield's family, recalled that "once [Leonard Bloomfield] told me that he thought the greatest book written was Albert Paul Weiss's *A Theoretical Basis of Human Behavior.*"[72]

In 1922, when Bloomfield was 35 years old, he put together a list with George Bolling, his colleague at Ohio State University. On this list were 50 linguists who they felt would join them in the formation of a new professional organization of linguists. The tragic and unexpected suicide of Leonard's sister Marie in February of 1923 took the wind out of his sails.[73] Along with Edgar Sturtevant, who had finished his PhD at Chicago just before Bloomfield, Bolling and Bloomfield formed the core of an organizing committee that developed a group of linguists over the months of 1924 who would be the charter members of the Linguistic Society of America.[74] The three of them sent out a letter to some 200 linguists, and so the LSA was born. It had its first meeting in New York on December 28, 1924, and it launched the journal *Language*.

To Chicago, and then to Yale

In 1927, Bloomfield moved to the University of Chicago to take a chair in Germanic philology. Edward Sapir had already been at Chicago for two years, and there they would be colleagues for four years, until Sapir moved to Yale in 1931. But they never got to know each other, never really worked together; they were too different in temperament, in personal style, in their view of scientific method and where linguistics was headed. They moved past each other with respect, but between them, nothing clicked. Bloomfield was now a confirmed behaviorist, and Sapir was more and more interested in depth psychology, and the relationship between personality and culture: how could they open up to each other? And what would they say if they did?

As we have seen, Edward Sapir passed away in 1939, just a year after Edward Prokosch's death—the same Edward Prokosch whose teaching had meant so much to Bloomfield at Wisconsin, and who had taught at Yale since 1929. Yale University reached out to Bloomfield and made him an offer in the summer of 1939 which he eventually accepted. We have access to the correspondence that Bloomfield had with his immediate superior at the University of Chicago, who was the distinguished philosopher Richard McKeon, then dean of the humanities. The written record leaves little doubt that Bloomfield's ultimate decision to leave Chicago was based on his sense that Chicago did not support linguistics strongly enough, and that Bloomfield himself felt that too much of his time was spent on administrative tasks that did not feel important to him. But while Bloomfield was dissatisfied with the status quo, McKeon's correspondence suggests that Bloomfield was unable to specify just what it was that that the University could *do* that would lead him to turn down Yale's offer. The two did not communicate well.

The upshot was that Bloomfield moved to Yale in 1940. While his personal library was eventually left to the Yale Library at the end of his life, it seems that he left some of his books behind in Chicago, including his copy of Wundt's *Grundriss*, filled with his copious marginal notes. That book, marked "Property of Leonard and Alice Bloomfield," sits on the stacks at the university library, a book that any student can take out and read; it is hard to avoid the conclusion that by the time he left Chicago, Bloomfield saw no reason that he would ever have for looking at Wundt's book again in his office.

Leonard and Alice Bloomfield moved to New Haven, Connecticut, but it was not the step up that he must have been expecting. Alice's health went into a serious decline, and the world was turned topsy-turvy by the war that had already begun in Europe at the point when they moved to Yale. During the war, Bloomfield was involved with the effort down in New York City in which pedagogical grammars were being developed by linguists at 165 Broadway, a subject we will return to in volume 2. He suffered a stroke in 1946, one from which he never recovered, and he passed away in 1949.

Bloomfield the man

Those who revered him—Bernard Bloch, for example, and Robert Hall—saw Bloomfield as a man with little charisma, an unassuming man shy in the presence of others, but friendly in his own way—self-effacing and taciturn. William Moulton recalled Bloomfield the advisor as "warm, friendly, immensely patient, and perhaps even more shy than I."[75] Bloomfield's influence was not through his personal engagement with students and other linguists, but through his writing. He was aloof, and detached, apparently uncomfortable when allowed to learn of others' personal issues, to the point of seeming (though perhaps only to those whose acquaintance with him was superficial) callous and unfeeling. Kenneth Pike was one of those who revered him, but even he recalled in print that Carl Voegelin had told him

> something to the effect that he had had instructions from Bloomfield (with whom I think he was staying at the time) to avoid making arrangement for him to meet people. And on one occasion, when I tried to ask him about orthographies, he turned away, saying "You know more about that than I do"—although I was less than a beginner in that area.

> But Sapir was quickly open to meet people. . . . Sapir, in meetings of the LSA, would seem to me to be almost always seated in the front row, and the first to comment on a paper—whereas I seldom heard Bloomfield comment in public.[76]

Bloomfield once wrote,

> A linguist shouldn't marry. He should spend long summers in the field, and the rest of the year working up his data. That way, one could perhaps produce adequate descriptions of three languages in a lifetime.[77]

It is not hard to imagine whose idea of a perfect life Bloomfield was describing.

Charles Fries wrote about Leonard Bloomfield:

> In spite of his modesty and avoidance of public controversy, he exerted a tremendous influence upon American linguists and American linguistics. The channel of that influence was not his lectures to students as their professor . . . he had very few students in linguistics. His powerful influence worked through his reviews, his articles, and his books,—especially his *Language* (1933), which was used as a text and studied widely throughout the universities in the United States. It was thus the younger generation of those just beginning their academic careers, rather than the older established linguists and language scholars, that constituted the group most thoroughly influenced by Bloomfield.[78]

Charles Hockett, in an interview in 1995, said of Bloomfield's behaviorism:

> Of course, after 1921 or so Bloomfield openly espoused physicalism. But at the same time (and this must not be forgotten) he recommended a linguistics free from dependence on any particular brand of psychology. . . . In principle his espousal of behaviorist psychology and his rejection of reliance on any one psychological school for linguistics were two different stances, and the latter, which is of primary concern to us, was neither behaviorist nor antibehaviorist. True enough, he didn't manage to keep these two stances quite as separate as one might expect—in particular, his general discussion of language and of linguistic methods, as over against his way with a language, tended always to bend toward behaviorism.

But, Hockett went on to say, later in his life, Bloomfield's treatment was really quite different, and he pointed to several passages in "Menomini

Morphophonemics" which did not sound at all like a behaviorist. And surely something new and different was brewing in that paper.

The first chapter of Bloomfield's work

We have spent quite a bit of time discussing Leonard Bloomfield's character and context, and all of it allows us to better understand someone who was perhaps the most important and influential linguist in the first 100 years of American linguistics. Bloomfield tried very hard to make everything simple, clean, and neat, but there is an irony in that, because we will see that he passed through three quite different periods in his life, and each one was important. Do we think that he viewed his life that way? Actually, we do, though we cannot prove it, as he is not here to answer our questions. But we will do our best to make the case that there were three Leonard Bloomfields, and each has something to say to us. The first Bloomfield was a German linguist trained in classical Neogrammarian linguistics but very much taken with Wundt's development of his ideas about psychology. The second Bloomfield was the one who rebelled against that intellectual heritage, and who found a new home in the positivist and behaviorist world of the 1920s in the United States when he took up his professorship at Ohio State University in 1921. The second Bloomfield published *Language* in 1933, and spent more than a decade as a professor of Germanic philology at the University of Chicago. The third Bloomfield came to the surface when the man moved to Yale, but this third phase had been developing throughout the mid- to late 1930s. And this third Bloomfield would see the light of day in the remarkable paper, "Menomini Morphophonemics."

The best image we have of the first Bloomfield comes from his first book, *An Introduction to Language*, which was written during the Illinois period. It was a great success—and while much of it sounds modern to our ear today, a great deal of it is grounded in a nineteenth-century conception of psychology, and *that* part of it sounds very dated.

When we read *An Introduction to Language*, we meet a young man: he is eager to make it clear that linguistics is a modern science—but that does not distinguish him from many other linguists. This young Bloomfield saw language as a system of expression linking an inner mind along the lines proposed by Wilhelm Wundt, and he prized clarity of thought and expression. He saw Steinthal and Whitney as having cleared the way for a psychological interpretation of language that speaks to linguists—and he underscored his debt to Whitney, who

> applied to the historic phase a remarkable clearness and truth of comprehension, to be appreciated in a field from which mystic vagueness and haphazard theory have been slow to recede.[79]

The psychology of the late nineteenth century and the first decade of the twentieth century as well as the growth of social anthropology (what he called ethologic studies) "have been, of course, of the highest benefit to this phase of the science of language."[80] For the young Bloomfield, there was no way to do linguistics which avoids psychological interpretation, and if there is no way to avoid that, then better the linguist should have as strong a professional understanding of what current psychological wisdom is than to fall prey to what Bloomfield termed "popular psychology."[81] Of all the mental sciences, linguistics "is most in need of guidance at every step by the best psychologic insight available."[82]

This young Bloomfield looked to psychological explanations nowhere more than in his treatment of sentences, and how their structures yield insight into the ways in which humans analyze the total experience of their living world into the bite-sized chunks we call words, and how they decide on the order in which words will be uttered, in their language: there is a "transition of the attention from the total experience, which throughout remains in consciousness, to the successive elements, which are one after another focused by it."[83] Bloomfield used the term *attention*, but he explained that he has in mind here the technical term *apperception*, which was Wundt's term, as we have seen.

Bloomfield described the inner act of speech in terms that are hard to take too seriously today, but they were the terms in which all of the influential writers of the time wrote—and when we realize that less than 10 years later, he would repudiate all of this way of speaking, the wonder is all the greater that he was ever satisfied with saying things like:

> The relation of the elements of a sentence to each other has a distinctive psychological tone. It is called the *logical* or *discursive* relation. It consists of a transition of the attention from the total experience, which throughout remains in consciousness, to the successive elements, which are one and another focused by it.[84]

This is exactly the kind of talk that the second Bloomfield, the famous and influential Bloomfield, would try to banish from the speech of linguists. Does it say anything meaningful? We would be hard-pressed to answer that question. This first Bloomfield reasoned as follows: Our attention can manage to focus on but one thing at a time; hence something being considered by consciousness is divided at any moment into that which is attended to and that which is not. In the case of a sentence (though Bloomfield does not use the word yet), the initial division is *called* division into *subject* and *predicate*. The classical terms are recycled and given new theoretical meaning, derived from an act of division proper to consciousness.

The same division can happen again; if the subject is the focus of our attention, then if it contains two words (Bloomfield's example is *lean horses*), the two words are divided up into the attended to and the background, and that too corresponds to the subject *horses* and the attribute *lean*.

The human mind is fundamentally binary in its analytic procedure, because it separates everything in its field into that which it attends to, and that which is left over, that is, *not* attended to.[85] This is the ground of the *subject-predicate* distinction, and the relation is called *predication*. We saw that this classical view had already been challenged in logic by Gottlob Frege, and Frege's new view had been championed by Bertrand Russell some years before Bloomfield's first book. But this news had not yet reached the field of linguistics.

At this point in Bloomfield's life, there was no talk of the phoneme: he talked about *sounds*, even where it would be extremely clear to the him 20 years later that he should be using term *phoneme*. The first Bloomfield talked about subphonemic differences as *automatic variations*,[86] and he gave the example of English vowels being longer when they are word-final or before a voiced consonant compared with their pronunciation before a voiceless consonant, a difference that "is never significant": "It depends solely on the following sound and can never be determined by the meaning of the word: it is an *automatic sound-variation*," as we see in the pair of words *bid*, *bit*. We are not aware we are making the vowel longer in the first word, and it is purely automatic. Once we have dealt with the varieties of this sort associated with each sound, then we can say that each language "has a limited sound-system, which, if only significant distinctions are counted and non-significant variations, whether automatic or merely casual, are ignored, is never very great."[87] If there are no cases in which the shorter and the longer vowel are in contrast, then the difference is one that can simply be attributed to this contextual variation, which would later be called an instance of *allophony*—but that term had not been coined yet.

Later on, Bloomfield mentioned another automatic sound variation, the voicing in early (or pre-) Germanic fricatives after unaccented vowels, and he gave the example *wása* 'I was' and *wēzumún* 'we were.' When, at a later stage of the language, the stress shifted to the first syllable, the variation between *s* and *z* was "of course no longer automatic."[88] But he does not raise the question as to *why* the language, after the stress shift, still shows the *z* in the plural form: why did the next generation of speakers not simply revert to using *s*? We have already met George Bolling: he would shortly become Bloomfield's colleague at Ohio State, and the two worked together to found the LSA. But before they were colleagues, Bolling published a review of Bloomfield's first book.[89] He was quite unhappy that Bloomfield talked about "automatic sound variation," even in the case of the purely allophonic differences seen in pairs of words like *bid*, *bit*. Bolling thought that this terminology suggests that "such changes have no sufficient causes but just happen of themselves—automatically. But even worse than this is another suggestion, that each speaker continually makes these variations (according to the conditions involved) in each production of the sound; that we, for instance, start always with the short vowels and automatically leave them unchanged when we say *bit*, *beat*, but automatically lengthen them when we say *bid*, *bee*, or *bead*. Such a position hardly requires refutation."[90]

Over the next two decades, as the Sapir-Bloomfield view of the phoneme matured, the position that Bolling found off-putting, and which in his view hardly deserved refutation, would become the mainstream phonemic view. The variations in pronunciations that fell strictly within a single phoneme was, in some fashion, part of the synchronic grammar that was shared by speakers at a moment in time. What Bloomfield would do in the years to come was to emphasize that this range of variation did not require, and for the most part did not deserve, a dynamic style of description in which one sound is changed into another.

The second Bloomfield

The second Bloomfield was the confirmed behaviorist and physicalist, who thought it would be possible to eliminate all talk of thoughts and ideas from the science of language. The world of three-dimensional space and everything that lies within it is all that there is. There is no hidden or inaccessible world that explains what we see and hear; there is just what we see and hear.

Is it possible to study the analysis of a language into phonemes even as one denies meanings and intents? Bloomfield thought it was. The phoneme was *not* a mentalistic notion, Bloomfield and his students insisted. George Trager was viewed by some as the methodological dauphin of Bloomfieldianism, and after Bloomfield's death he wrote, with some asperity,

> There still exists among many linguists, and especially among persons who, while not linguists, are peripherally concerned with linguistics, the idea that phonemics is some kind of mentalistic or psychological exercise, and has little if anything to do with science. This idea, of course, is completely mistaken. . . . A phoneme is a selectional class of complementary distributed, phonetically similar, and congruently patterned sound-types; it contrasts and is mutually exclusive with every similar class in the language, with some or all of which it enters into juxtapositional classes.[91]

Archibald Hill looked back on the development of the phoneme:

> The first revolution, then, was that which came to maturity in the mid-thirties. It began earlier, with the founding of the LSA, and it found its roots in writings by men like Sapir, who gave us the phoneme. His disciple Swadesh, too, was important for his article in 1925,[92] which gave us concepts like the morphophoneme and morphophonemics. Above all, however, the first revolution goes back to Bloomfield, and his resolute purging of mentalist concepts. He is the scholar who firmly pointed out the danger of circularity in saying that meanings were identified by form, and that forms were identified by meaning. The result, of course, has been the concept which seems to me most fundamental in linguistics, the concept of contrast. As Bloomfield put it, language consists of "sames" and "differents." In the area of sounds, for instance, before Sapir and Bloomfield, we operated with gross sound types. We lumped sounds together as dental spirants, dental stops, labiodental spirants, and so on. What happened was that as we began to observe sounds closely, we found that we could not group them into physical types with any certainty. Sounds existed in a continuum of difference. For instance, the tongue can close off the oral cavity at any point along the roof of the mouth which the tongue can reach. That is, the point of closure is along a line, and we all remember enough geometry to know there are an infinite number of points along a line. Since sounds are in fact in a continuum we need the concepts of contrast and noncontrast which give us the phoneme. We needed then and still need the phoneme.[93]

TELEOLOGY

Bloomfield was one of the few people we encounter in this book who explicitly espoused mechanism in his account of humans and society. In a generous review of a book on syntax and style, he dismissed the author's effort to explain a phrase *on the basis of* the beauty of the expression. "If the reader is a mentalist," Bloomfield wrote, "he can probably accept this as an explanation." But Bloomfield was not one of *them*. If the reader "is a mechanist, he will object that Havers has not answered the question 'Why did the speaker use this and not a different speech-form?'" Bloomfield wrote that "a mentalistic pseudo-solution can only discourage the search for real answers."[94]

Bloomfield gave further insight into his view as he remarks how poor a choice he thought Havers had made in selecting an epigraph:

> In the evolution of language "causal" relationships cannot be found; rather, it is a teleological view that is offered.

No, said Bloomfield: "teleology does not stand in contrast with 'causality,' but represents merely a more primitive form of the same age-old popular notion. A teleological 'explanation' can be given without difficulty for any and every happening. . . . Teleology cuts off investigation by providing a ready-made answer to any question we may ask."[95] This was Auguste Comte speaking, we might almost think.

There was no concept that Bloomfield hoped more to purge from linguistics than *teleology*. In "Linguistics as a Science," he compared linguistics with the hard sciences, where science, he wrote, "has been successful."[96] The hard sciences would not be satisfied with the ways of speaking that are used to discuss human affairs. "The physicist and the biologist do not content themselves with teleologic formulae. Teleology is a form of wording which says that things happen because there is a tendency for them to happen. Water seeks its own level; nature abhors a vacuum; trees strive towards the light." These are the examples Bloomfield offers as typical examples of teleological explanation, and for us, trying to understand what he thought was objectionable about these notions, it is hard to avoid the conclusion that it is the vagueness of the proposals rather than their attempt to provide an explanation based on a goal-state. We would have learned more if Bloomfield had taken on an interesting case, like Fermat's principle, which says that the path that light takes between two points is the path that takes the least time, a striking principle that allows for an elegant way to predict the path that light will take through a series of mediums in which light passes at different speeds, a principle that was formulated in the eighteenth century. Light, we might say, strives to find the path that allows it to reach its goal in the shortest time. Is that a teleological statement? It seems to be much on a par with the statements that Bloomfield dismisses, but for three things: Fermat's principle is mathematically precise, it can be used to make precise predictions, and its predictions are correct.

"Physicists and biologists have long ago ceased to accept such teleological pseudo-explanations," Bloomfield continued, "having recognized them as mere roundabout statements of the event. It is only when we deal with man that we are satisfied with teleological formulae: men do things because they 'want' or 'choose' or 'have a tendency' to do them." Linguists have dispensed with them, too, Bloomfield suggested, at least since the days of the Neogrammarians: "for more than half a century linguists have been studying a fundamental phase of human activity without recourse to teleologic or animistic formulae. To be sure, linguists, like other people, are finalists and mentalists in their explicit opinions about human affairs. It is only in their working methods that they have abandoned these forms of discourse. You may strip the teleologic and animistic verbiage from any linguistic treatise, and the effect is only an improvement in style, with all the technical procedures and all the results unchanged."

We will see in chapter 9 an enormous difference between these views and those defended by Trubetzkoy and Jakobson in Europe.

Bloomfield's syntax is not easy for us to read today. Perhaps that should not be taken as a criticism of Bloomfield's approach; perhaps any approach to syntax is difficult to get accustomed to. But in the 1940s, Bloomfield's students converged on the conclusion that his account of syntax was not as clear as his treatment of phonology and morphology, and a number of papers were published, mostly in the journal *Language*, with the hope that a clearer formulation of syntax would emerge.

Bloomfield's system was based on two principal ideas: first, that a sentence can be hierarchically subdivided into immediate constituents, and second, that the way two immediate constituents were put together to form a constitute could be described with a small number of descriptors. The first part is familiar to us today, because it is similar to the model of phrase-structure developed by Chomsky in the 1950s, which most linguists still use today; the second part is not at all so obvious. Let's take a closer look.

Given an utterance, it will usually be clear how it can be (or should be) divided up into two consecutive subparts. Bloomfield wrote, "Any English-speaking person who concerns himself with this matter, is sure to tell us that the immediate constituents of *Poor John ran away* are the two forms *poor John* and *ran away*; that each of these is, in turn, a complex form; that the immediate constituents of *ran away* are *ran*, a morpheme, and *away*, a complex form, whose constituents are the morphemes *a-* and *way*; and that the constituents of *poor John* are the morphemes *poor* and *John*."[97] It is striking that Bloomfield is perfectly happy to leave the evidential basis of this in the hands of the educated and involved native speaker! This is surprising to us today, because we think that appealing to native speaker intuitions would have struck the behaviorist that Bloomfield was as illicit. But it did not.

We will come back to the question of how immediate constituents should be determined, and how we should respond when we see conflicting indications that leave the analyst betwixt and between. We will just assume we know what the constituents are. But a syntactic analysis for Bloomfield was more than a hierarchy of constituents; each pair of constituents that together formed a constituent was an example of a *construction*. And the grammar of the language consists of all of the ways that the language allowed constituents to be put together to form a larger constituent. Like what? The most obvious was *order*; an adjective and a nominal *big clouds* could form a nominal sort of constituent, but *clouds big* could not, because that was not the order that English required.

Bloomfield saw four ways that a language could impose conditions on the constituents that it put together: one was order, the second was intonational (he called it *modulation*), the third was phonetic modification, which could emerge when there were special phonological effects that occurred when two morphemes, words, or phrases (such as *liaison* in French), and the fourth was what he called *selection*. Selection was quite an open concept. A constituent like *cold rain* selected two words, from two form-classes, which we might call *adjectives* and *nouns*, and something parallel could be said about a nominative substantive followed by a finite verb, in the case of a sentence such as *John ran*. But in addition, there will sometimes be agreement conditions that must hold: we say

John is, but *they are*; singular subjects go with singular verbs, at least in the present tense.

Bloomfield then defined a *taxeme* as anything that a language might declare as part of the definition of a grammatical constituent—which form-classes are allowed, and in what order, for example. And a *tagmeme* was the combination of these taxemes that a language used in a meaningful grammatical way.

Ten years later, Kenneth Pike would come back to Bloomfield's account, and in the most polite and uncontentious of ways suggest that a good deal of what Bloomfield had said about tagmemes did not make too much sense.

Four publications

We will look at four more of Bloomfield's publications to get a better sense of what his contribution was. The first is a terse statement of how Bloomfield thought linguistics could be improved by laying out a clear set of postulates: "A Set of Postulates for the Science of Language," and the second is his most extreme statement of his behaviorist and physicalist view of linguistics, a paper entitled "Language or Ideas?" In light of the fact that this paper was published in *Language*, we can draw the conclusion that Bloomfield intended this paper to speak to linguists, though no one took up the banner of behaviorism within mainstream linguistics with anything near Bloomfield's fervor. Both papers present the austere, dust-free environment that Bloomfield hoped linguists would come round to accepting. The point of the second paper was explicitly to draw linguists' attention to the rise of the Vienna Circle, a group of philosophers we will look at in detail in the next chapter. Bloomfield saw the Vienna Circle as a group that had arisen totally independently of the American behaviorists, among which he included himself, and he thought that this parallel evolution of thought was a good omen of the validity of their train of thought.[98]

The third piece that we will look at is his most important and influential work, his book simply entitled *Language*, and the fourth is Bloomfield's last major publication, a remarkable paper called "Menomini Morphophonemics" which developed an analysis of an Algonkian language, an analysis that was heavily influenced by the dynamic methods of Edward Sapir's view of phonology. This last paper is the most substantive part of the third chapter of Bloomfield's career.

"POSTULATES"

Bloomfield's "A set of postulates for the science of language" was published in the second volume of the journal *Language*, and is just barely 12 pages long. It was an unusual paper, and one would be hard-pressed to find another paper in a mainstream linguistics journal that could be said to be a similar sort of paper. It began with an explanation of why it came to be, which was a belief that an effort to make linguistics explicit, through a set of axioms, would ultimately serve everyone's best interest by making explicit what we are assuming, and by helping us avoid errors. The method "saves discussion," too, "because it limits our statements to a defined terminology; in particular, it cuts us off from psychological dispute."[99] There were linguists who were saying things, in print, that sounded substantive and even interesting, but Bloomfield was convinced that if you really push them on the correct definition of their terms, they would not be able to provide a satisfactory account. For Bloomfield, that meant that they were not saying anything at all, from a scientific point of view. Bloomfield gave an example: there were linguists who were saying that "some forms have less meaning than others and are therefore more subject to phonetic change," and he cited one linguist who was making such claims. But Bloomfield thought this was mere verbiage: "I, for one, can discover no workable definition of the terms 'meaning' and 'phonetic change' under which this notion can be upheld. The whole dispute . . . is at bottom a question of terminology." For Bloomfield, now, the road to making a *better* science of linguistics was going to be the careful meta-analysis of the terms and axioms of the field, and a patient insistence on introducing a term only when the proper safety precautions have been taken.

The paper began with three short, punchy paragraphs, each one sentence long.

> The method of postulates (that is, assumptions or axioms) and definitions is fully adequate to mathematics; as for other sciences, the more complex their subject-matter, the less amenable are they to this method, since, under it, every descriptive or historical fact becomes the subject of a new postulate.
>
> Nevertheless, the postulational method can further the study of language, because it forces us to state explicitly whatever we assume, to define our terms, and to decide what things may exist independently and what things are interdependent.

> Certain errors can be avoided or corrected by examining and formulating our (at present tacit) assumptions and defining our (often undefined) terms.

It is almost as if Bloomfield had Saussure in mind, the Saussure who wrote to Meillet about the "immensity of the work that would be necessary to show the linguist what it is that he is doing."[100] The linguist must understand what he says, why he says it, and what he means by saying it. We can also hear a faint echo of Francis Bacon's famous observation that truth arises more readily from error than from confusion.

In this paper, Bloomfield began by noting that use of language in the world can be conceptually divided into three stages. First, some stimuli causes a person to speak; second, something is said by the person; and third, some response made by someone else to what was just said. The first and the third are proper to the domain of psychology, not linguistics; linguistics cares only about that which was said. The linguist studies communities of speakers who share a way of constructing things that *can be said* in various contexts and cause various responses, but the linguist only cares about those speakers' possible utterances. And the linguist is "obliged to predict," Bloomfield noted immediately; the utterances of a language will always go beyond what was observed so far by the linguist; indeed, the need for prediction "constitutes the greatest difficulty of descriptive linguistics."[101]

Bloomfield made it clear at one point that it was not his intention to provide (or to encourage others to provide) a set of postulates of this sort that would work once and for all, and for all languages; rather, the act of making these postulates was an activity that would have to be done frequently, and there was no reason to exclude the possibility that the set of postulates we were led to would differ from language to language.

There is one more thing to notice about this paper of Bloomfield's. Like most, it has text and it has footnotes: it has seven footnotes, of which six occur on the first two pages. The *text* of the paper is written in an abstract and timeless way, with essentially no appearances of real human beings; it is largely abstract elements of a linguistic analysis which are the subjects and objects of all of the verbs. But it is in the footnotes that the human character of the linguistic analysis resides. He refers the reader to Albert Paul Weiss, and reminds the reader about the "difficulties and obscurities"[102] of Humboldt and Steinthal, and the disputes of Paul, Wundt, and Delbrück. And he acknowledges the important steps taken in the right direction by Ferdinand de Saussure and his soon-to-be colleague, Edward Sapir.

This paper of Bloomfield's fits perfectly into the dominant view of science of his time, the view that science only stands to gain by the development for each science of a formalized account of each of the terms of art that it employs, how each depends for its meaning and use on terms already defined, and how some of them depend on observations that lie outside of the formalism of the theory. This view was being developed in mathematics, as Bloomfield noted, and Bloomfield's friend Weiss had done something very similar for psychology. As we will see in the next chapter, this perspective on science was being developed by the logical positivists, who talked about *constructive systems* in their effort to accomplish what Bloomfield had in mind as well. The most significant attempt to achieve this goal would be undertaken by Noam Chomsky in the mid-1950s.

LANGUAGE

Bloomfield's book *Language* was published in 1933. It is not a book that feels revolutionary to a reader today. It feels *dry*, and its prose gives no hint that there was a human being who had written it. P. H. Matthews wrote not too long ago:

> It will be obvious from this analysis that Bloomfield's *Language* was designedly not a revolutionary manifesto.[103]

But that's not what was said at the time. Perhaps Bloomfield would have been comfortable with the remark that his book was not a revolutionary manifesto, but the people learning linguistics from his book for the first time found it new and utterly refreshing. Bernard Bloch, in his capacity as a linguist, as a professor of linguistics at Yale, and as editor of *Language*, had as much right as anyone to be called the dean of American linguistics after Bloomfield's death, and in his necrology, Bloch wrote that *Language* was a "work without equal as an exposition and synthesis of linguistic science."[104] Noting that while Bloomfield was shy, his impact on others through his writings was spectacular, and most of all with his book *Language*:

> It was a shocking book: so far in advance of current theory and practice that many readers, even among the well-disposed, were outraged by what they thought a needless flouting of tradition; yet so obviously superior to all other

treatments of the subject that its unfamiliar plan could not be dismissed as mere eccentricity.[105]

P. H. Matthews observed more recently that Bloomfield's *Language* had different effects on two different sorts of readers: we must look at

> the way that *Language* was read by [Bloomfield's] successors. We have seen . . . how it was taken as a fresh start in linguistics. It was therefore one thing to have read it as a colleague or reviewer, with a knowledge of the issues that were debated and the conclusions that had been reached before its appearance, and quite another to work through it as a beginning student, coming to it from scratch. As historians, we must try to read it in both modes. For things which in one reading seemed essential to the argument may well have seemed far less important in the other.[106]

Matthews comes back to this point in trying to read Bloomfield's *Language* as a young student might have in 1933: "It is through this reading that we may perhaps uncover the theoretical authority for distributionalism, if it is to be found at all in Bloomfield's writings. But it is the reading of someone who was young in the 1930s, who did not fully understand the background against which Bloomfield was writing, and who responded instinctively to what was most radical."[107] Matthews is surely right, but the fact is that if we are to understand the dynamics of a discipline, we have to understand that this kind of reading is nonetheless of the utmost importance.

Charles Hockett described a moment of just the sort that Matthews was referring to, in remarks made in 1979, 30 years after Bloomfield's death, in a most revealing way:

> Remember that I cut my professional eye teeth on Bloomfield's book back in 1933. Bloomfield himself assumed no "eclipsing stance": the very opposite, for his respect for his predecessors was profound and he tried to inculcate the same attitude in his students. But I found Bloomfield's synthesis so satisfying (except in some minor technical details) that for a long time I simply couldn't bring myself to read much of the work of those predecessors. That was the price I paid for my largely superb induction into our discipline. Then, just a few months ago, I finally had reason to undertake a serious study of William Dwight Whitney's general writings. I knew that Bloomfield had overtly acknowledged his debt to Whitney; nevertheless, I was overwhelmed to discover

> the extent of that debt (and thus of our own), and amazed at the variety of topics on which Whitney's remarks, allowing for a difference of terminology and style, are as valid and profound now as a century ago.[108]

He wrote something similar in a letter to one of us:

> I began my training in the spring of 1933, the third quarter of my freshman year at Ohio State, under the Homeric scholar George Melville Bolling, using Leonard Bloomfield's new book, hot off the press. I was very young (17) and the experience was very powerful: for a long time I simply assumed (I guess) that Bloomfield had successfully absorbed and integrated all earlier findings, so that it would just be a waste of time to attempt any independent reading of his predecessors.
>
> A weird thing about that is that Bloomfield himself preached the very opposite. He revered his predecessors for their accomplishments, even when critical of them for what he saw as their mistakes (or, more often, the mistakes of their times), and insisted that science must be cumulative. This was obvious both in his writings and in his comportment as teacher and colleague. Once, in 1939 or 1940, I said to him that it seemed to me we had managed to learn an amazing lot about language in the last decade or so. He replied that he didn't think we knew anything of importance that hadn't been known to his masters thirty years earlier. He meant Wackernagel, Leskien, Prokosch, and so on. His comment was partly a put-down aimed at my brashness, but it also accurately reflected his attitude.[109]

But Bloomfield's own view of his intellectual predecessors is not simple. *Language* is not heavily burdened with footnotes and citations, but Bloomfield did include a bibliography that ran to 20 pages. We were quite surprised as we read through it, more for what it did not include than for what it did. The important work of William Dwight Whitney is included. Of Sapir's work, only his popular book *Language* is mentioned. Of the nearly 300 scholars who are cited, there are only a handful of Americans; the vast majority are Europeans, almost all of them German. What is most astonishing is that no articles at all are cited from the journal *Language*. A *very* small number of scholars working in America are cited (these include C. C. Fries, and also G. O. Russell, who was his colleague at Ohio State University). The scholars who are cited more than three times are mostly German linguists closely linked to the Neogrammarians (Brugmann, Leskien, Schuchardt, Meyer-Lübke, Kluge,

Meillet, Hermann Hirt) and International Phonetic Alphabet linguists (Jones, Jespersen, Sweet, Passy). Judging only from his bibliography, we would not imagine that this was the work of an American developing a new and American approach to linguistics, and certainly not that he had been Sapir's colleague for four years, or that he was aware of the Prague school of linguistics. Reading the bibliography of Bloomfield's language as a stand-alone text requires hermeneutic skills, and it does not easily give up its message.

"LANGUAGE OR IDEAS?"

To a reader *today,* Bloomfield's short paper in *Language* entitled "Language or Ideas?" is an odd publication, perhaps the oddest of all of Bloomfield's writings, though we did remark that Bloomfield's "Postulates" was also a bit odd. This one was in all likelihood an odd paper to many of its readers even at the time it was published. "Language or Ideas?" set out to show to linguists that a group of philosophers, the Vienna Circle, was just becoming known for an important new view of science and its relation to knowledge, a view that Bloomfield took to be very close to the kind of physicalism that he had adopted under Weiss's influence, and what he thought was the spirit of the times. Bloomfield insisted that the parallel development of these two views, so similar in their core, was a significant indicator that what they shared was both important and true. The Vienna Circle, and the origin of their ideas, will be a large part of our concern in the next chapter, and without the background that the next chapter provides, it is difficult for a contemporary reader to figure out where Bloomfield is coming from.

All forms of positivism urge some sort or other of *cleansing.* There is a difference between *cleaning* and *cleansing*: in *cleansing*, the cleaner has to first figure out what counts as dirtiness, and then undertake a process that is specially designed to remove that foreign matter. *Cleaning* takes it for granted that the dirt that has to be removed is just the familiar sort. Most forms of positivism have been concerned with what kinds of *explanations* should be permitted in our conversation about the causes of things that exist or that happen in the world. This kind of positivism is less concerned with descriptions of what exists, or what has happened, and more concerned with accounts that may be offered about why these things happened. But when positivism is brought to psychology or linguistics, it may take as its target not just what counts as an explanation, but even what stands in need of an explanation (scholars who can point to a classical education in their youth often use the term *explanandum* to refer to what needs to be explained, and to *explanans* as the sort of explanation that can be offered). Positivist *cleansing* is thus a kind of decision to remove from the discussion a sort of style of expression that human beings find appealing but which is, in the positivists' view, inappropriate.

Bloomfield's version of positivism shared with the Vienna School's version a slippery tendency related to how the term "science" would be used, though we will focus for now on Bloomfield's remarks. Bloomfield's principal effort in this paper is to show that his linguist reader should accept the idea that terms such as "mind," "perception," and "idea" are terms that will have to be reanalyzed and replaced *as far as science is concerned*; they will need to be replaced by terms that come from physiology, or by terms that come from a distributionalist sort of linguistics—or else simply be replaced by nothing at all, which is to say, to be eliminated. What is slippery about all this is that Bloomfield and his Vienna counterparts only make this suggestion for *science*, knowing that the battle would be lost before it had begun if they were to propose that people in general should stop talking about ideas and minds. All of this puts the linguist in a difficult situation, because he has one of three options in front of him: he can, first of all, accept what the positivists say, stop talking about minds, and establish linguistics as a science with the positivists' blessing; or he can object to the positivists' restrictions on what counts as a science, continue to talk about minds and ideas, and insist that linguistics is a science despite what positivists say; or, finally, he can shrug at what the positivists say, continue to talk about minds and ideas, and say that it does not matter to him whether linguistics is a science or not. Obviously positivists were hoping that linguists would not take the third option.[110]

The kinds of talk that Bloomfield wanted to eliminate from linguistic usage were the sorts of things that fell under what he called "mentalism" and "animism," and he hoped that we would discard them in the way that we have discarded Ptolemaic astronomy: we still say that the sun rises in the morning and sets at night, and we certainly find it convenient to think in those terms, even if we know perfectly well that if we have any scientific interests at stake, we would say that the Earth goes around the Sun. But Bloomfield, like many positivists, sailed right at the edge: he wrote, "statements that are not made in [physicalist] terms are either scientifically meaningless or else make sense only if they are translated into statements about language."[111] But the phrase "scientifically meaningless" passed into "entirely meaningless" in the next sentence, and then back to "scientifically meaningless" a few words later. And then a bit later, he wrote, "If we are right, then the term 'idea' is simply a traditional obscure synonym for 'speech-form.'" There is another side to the story that Bloomfield adumbrates, a side which he just barely pointed to at the end of the paper; the allusion is so quick that it is not at all clear how far he might have wanted to push it, were he given the chance. As we will see in the next chapter, the Vienna positivists were committed to classifying all true (and hence meaningful) sentences into two categories: empirical statements, which are those that are true by virtue of the world as it happens to be, and other truths which are the consequences of how we establish our language. In "Language or Ideas?," Bloomfield makes the point that the Vienna positivists have an unsophisticated view of language: they have not thought at all deeply about what it means to say that "redness" is a noun, for example. To assign a word to a lexical category requires considerable linguistic sophistication, and must be done on a language-by-language basis.[112]

Bloomfield ends his paper with the sentence, "If this is true, then linguistics in the future will deal with far wider problems than today." The word "this" here appears to effectively refer to his entire thesis, and his point was that to determine whether sentences were true by virtue of their linguistic form requires real linguistics, and not arm-chair pontification. The reader today could imagine that Bloomfield might say that there is real linguistic work that needs to be done in order to know how the truth conditions (or any sort of felicity conditions) differ, for example, between sentences like *Mary is not here* and *Mary is not here yet*; more generally, Bloomfield leaves the door wide open to the position that the linguistic analysis of what has been called meaning is an extremely important part of linguistics: it will no longer be called meaning, but the work remains to be done nonetheless. Unfortunately this is all left unstated and lurking just under the surface, and it would be reading too much into what Bloomfield wrote to say that he had this in mind when he wrote this paper.

The third Bloomfield

The third Bloomfield was the Bloomfield who took what he had done during his mature career, and went back to the ideas of his youth (what we called *the first Bloomfield*) to see how it all fit together. One of the connections was the style of grammatical analysis that he had found in

Pāṇini, taught to him by Wackernagel.[113] Bloomfield had never forgotten Pāṇini—far from it. He published a paper in 1927, "On some rules of Pāṇini," in which he discusses how the two major commentators on Pāṇini had interpreted a passage in Pāṇini's grammar. But certainly he took some distance from Pāṇini's style of analysis in his own mature work during the Chicago period, the period of *Language*, at least as far as of the issue of ordered rules, or of abstract basic forms, were concerned. The Bloomfield of the third period was ready to rethink that distance which he had taken from Pāṇini, and to subject his own work to revision.

"MENOMINI MORPHOPHONEMICS"

"Menomini Morphophonemics" was one of the last publications of Bloomfield's career. It appeared in 1939, just as World War II was starting, and during the war, most American linguists, including Bloomfield, were engaged in the war effort. But "Menomini Morphophonemics" was a paper that Bloomfield published in a volume to honor the memory of Nikolai Trubetzkoy, and the volume in which it appeared had a star-studded cast, including contributions from Emile Benveniste, Marcel Cohen, Louis Hjelmslev, Lucien Tesnière, Tomás Navarro Tomás, Morris Swadesh, George L. Trager, and André Martinet. Publication in this volume was a major event in a linguist's life.

Bloomfield's main point in this short paper was that in the analysis of phrases, compound words, and simple words, Menomini presents little analytic difficulty—it is easy to see where to make the cuts to identify those pieces. But simple words, or the members of a compound, can be resolved (as he put it) into identifiable subpieces which vary considerably in their surface form, depending on what combinations they appear in. These complexities are what Bloomfield called the *internal sandhi* or *morphophonemics* of the language, and they would be the central point of this paper. These morphophonemes were precisely the sort of units that Sapir's student Newman employed in the account we looked at of Yawelmani Yokuts. This is where the real intellectual action would be found, now that the phoneme had been tamed and defined.

Bloomfield's proposal was that a satisfactory description of Menomini can be achieved if we posit what he called a *base form* for each morpheme, and a sequence of statements that describe "deviations." If these statements are applied in the order that Bloomfield presented them, then "we will arrive finally at the forms of words as they are actually spoken."

Most of the ideas that Bloomfield developed in this paper are found as well in his own *Language*.[114] But the *whole* of the paper was different in style and in tone. Bloomfield had no way of knowing it, but the type of analysis that he developed in this paper would come to play a central role in phonological theory when it was adopted by Chomsky and Halle in generative phonology. Still, there are textual indications that Bloomfield was quite aware of the change in the style, the direction of his phonological analysis here. One indication of this: someone who has read a good deal of Bloomfield's written prose cannot fail to be struck by the way he uses first-person pronouns in this paper. He actually *uses* them, as we saw in the quotation just above, in this paragraph! Bloomfield's style had always been one in which the author's voice was suppressed and disembodied, and we have already seen this in his earlier papers. His paper on "Postulates" that we looked at just above is typical. When Bloomfield used the first-person plural pronouns, it is not to include him (or the reader) in an engaged way with what we may want to do: it is typically in order to *remove* our options. He wrote, for example, in such a disembodied tone, "the postulation method can further the study of language, because it forces us to state explicitly whatever we assume." But the Bloomfield in *MM* really is different, because he wrote sentences like "if one starts with the basic forms and applies **our** statements," we arrive at the current phonemic forms. "**Our basic forms** are not ancient forms . . . and **our statements of internal sandhi** are not historical but descriptive." At one point, he admitted that he did not have a relevant form in his fieldnotes, and wrote, "as I am not a native speaker, I cannot guarantee words which I form, but I should not hesitate to say *ε̄hkw-ōhtah*."[115]

All in all, Bloomfield proposed about 30 ordered rules. Here is an example of a typical rule: "if an element ending in vowel plus *w* precedes an element with intial *w*, the [connective] -e- is used: *kaw-e-wε̄p-*" (illustrating the morphemes *kaw* and *wε̄p*). But in some cases (and Bloomfield did not specify which ones), this insertion does not apply, in which case the next rule applies, to a cluster of consonants abutting at morpheme boundary: "in such forms, if the first consonant is other than n, it is replaced by h," and Bloomfield gave the example *sε̄nak-at-k*, which surfaces in */sε̄nakāhken/*.

Thus in this final chapter of his life and work, Bloomfield joined Sapir and his students in developing complex modes of analysis and explanation in the treatment of Native American languages.

Sapir and Bloomfield

Sapir and Bloomfield: these were two inspiring figures in their day, and while they were very different in all sorts of ways, they did lead the field in certain directions over the course of the 1920s and 1930s.[116] Zellig Harris was a leading linguist of the next generation, who reflected on his memories of Sapir and Bloomfield. Long after the death of both, he wrote:

> I have to refer to one other misconception. I am sometimes asked if there was any antagonism between Bloomfield and Sapir. These were two agemates with very different backgrounds, who did not know each other until each had appeared on the scene as major theoreticians at the birth of a science. The question is posed from the competitive values of this society. The answer is no. Publicly and in print they always spoke with great respect about each other, and praised and used each other's work. Privately, they knew each other very little and had no particular warmth for each other; their styles and personalities were indeed exceptionally different—but each one, and his students, spoke with respect for the other and above all with appreciation for the other's linguistics. I was close to each of them in their last years and never heard a derogatory comment.[117]

They knew themselves and they knew each other, and recognized the distance that separated them. Bloomfield gave an address to the Modern Language Association of America at the end of 1929 on the subject of linguistics as a science, and in it he expounded (for neither the first nor the last time) upon his confidence that the new linguistics was one that avoided all mention of goals and intents. He felt a bit isolated among linguists as he enunciated these views, he admitted: "If in any sense I here represent linguists in general, I must add that very few of them share the beliefs to which I shall give voice." What linguists agreed on was that linguistics is scientific, but many would go no further in the direction that he was sketching, and he mentioned just one by name: "Edward Sapir, an excellent scholar who would probably agree with very little of what I am saying tonight."[118]

Let's return to what Zellig Harris wrote about Sapir and Bloomfield:

> Neither competed, or saw his scientific achievement as a matter of personal aggrandisement. And this was not for lack of a sense of history about their

> work. Both men knew that they were creating—or rather participating centrally in the creation of a science. There was an excitement around them, in their ideas, among their students and colleagues. Each of them pushed for his ideas—Bloomfield by incisive argument, Sapir by brilliant exposition—though without seeking to pre-empt the field. Each was, to the good fortune of those who knew them and I hope of themselves, an extremely decent person of high integrity; each had utter and explicit contempt for the posturings and status in this society as well as for its vast injustice and inequality. They were people not with ambition, least of all with ambition in the terms of this society, but rather with satisfaction in what they were producing. Those who remember Bloomfield and Sapir know this about them.[119]

Kenneth Pike and Yakov Malkiel were both major figures in the generation that followed Bloomfield and Sapir, just like Zellig Harris. Pike wrote,

> Leonard Bloomfield contributed to me greatly as a *person*. He, along with Edward Sapir and Charles C. Fries, became a role model in terms of personal relations to the academic community. He always showed to me quiet kindness, with a willingness to listen to a total beginner. His personal consideration for others was combined with an intensity for scientific truth which transcended his own presumed (by me) human potential, dreams of reputation or of having been proved to be "right." . . .
>
> Yet how different Bloomfield was from Sapir! Bloomfield seemed to me to be retiring in disposition, and almost aiming to avoid contact with people in general.[120]

And Yakov Malkiel wrote,

> Bloomfield cut a very impressive figure in those years; to me and to others, he appeared as the perfect gentleman, doubly thoughtful and polite in dealing with underdogs, whom another person placed in a position of comparable influence might have been tempted to brush off. Not all his prognoses proved right; however, I vividly remember this one excellent bit of advice he gave to me (it has stood me in good stead ever since): not to attempt to adjust myself to changing winds, but to listen to the voice of my conscience. Such young American linguists as I used to meet in those days referred to the unique blend of scientism and mysticism in Bloomfield's personality; perhaps they should have substituted hard moral core for mysticism. Sapir, in contrast, was remembered as a magician.[121]

H. A. Gleason looked back at those times, and wrote,

> It was usual to think of these two men, Sapir and Bloomfield (with Franz Boas standing behind them) as the founders of North American linguistics. There is some justice in this. Each was, himself, a broad-gauged highly competent scholar. Each was deservedly recognized for work in more than one of the converging traditions. Differing as they did in attitudes, ways of working, and objectives, they complemented each other . . . it is far too simple a view to ascribe the building of North American linguistics to Bloomfield and Sapir, either individually or jointly. Neither of them dominated the field.[122]

When he wrote about Sapir and Bloomfield, Zellig Harris went on to consider the relation of their work. At the top of the list, of course, was the phoneme:

> Sapir and Bloomfield were the final developers of the concept of the phoneme, which had grown out of the work of the English phoneticians and of the investigators of sound change, and whose recognition had been initiated in the work of de Saussure, Baudouin de Courtenay, Boas, and Trubetzkoy. Sapir was among the creators of the concept, in his *Language* (1921), and in his eophonemic *Sound patterns in language* and in some of the ideas which appeared in Swadesh's *The phonemic principle*. Bloomfield, in turn, presented the first comprehensive view of phonemics in *A set of postulates for the science of language* (1926) and in his epoch-making book *Language* (1933).[123]

This is certainly true, and Morris Swadesh's paper on the phoneme, as we have seen, begins with an explicit accounting of the intellectual roots of the notion of the phoneme that are as generous as this statement of Harris's. But Bloomfield's discussion of the phoneme in *Language* gave no hint that he was part of a larger community of scholars working out this concept. Chapter 5 is devoted to the phoneme, and though Bloomfield discusses Bell, Jespersen, and Sweet in connection with their proposals for systems of phonetic transcription, Bloomfield includes no mention at all of Sapir, Trubetzkoy, or anyone else interested in the development of the phoneme. We observed Charles Hockett recalling that he and his fellow students thought it all began with Bloomfield's *Language*—and it is not difficult to see how Hockett and his friends got that impression. Trubetzkoy is mentioned *nowhere* in the book, and the only item in the bibliography by Sapir is Sapir's own book called *Language*. For now, we will leave this observation without an explanation, but it is worth bearing in mind: Bloomfield in *Language* made explicit his indebtedness to Whitney and to the Neogrammarians in a way that he did not regarding his contemporaries. Think about that.

Zellig Harris continued,

> Phonemics was the crucial advance; without what it yielded one could not move on to a science of language, toward which Bloomfield was consciously working. . . . In the service of these more general interests, Bloomfield presented impressive arguments against the use of teleological explanations for language change, such as the view that semantic convenience fosters change, or blocks change in particular words. . . . He also argued against the use of semantic explanations for syntactic phenomena . . . : he discusses (p. 407) "the prescientific and indeed barbarous character" of "the notion that, beforehand and independently of any particular act of observation, one could formulate, by virtue of some sort of philosophic acumen, a realistic outline of the universe which would serve as a frame of reference for statements of the meaning of linguistic forms."
>
> . . . Just as the phoneme was defined with the aid of the relative occurrence (complementary, free-variant, contrastive) of phones, so Bloomfield was here defining syntactic elements by their relative occurrence, which linguists somewhat confusingly called "distribution." Although Sapir, and indeed the Boas students generally and all makers of satisfactory grammars of "native" languages, worked explicitly on such grounds, it was a new and rather courageous claim on Bloomfield's part to say that distribution and not meaning was the criterion for syntactic analysis.

That remark should give us pause today. Virtually every allusion made to American distributionalism today portrays it as a sort of petulant methodological conservatism. This is not historically correct; distributionalists were rather saying that meaning-based criteria were being used sloppily, and there had to be a better way.

> Even Bloomfield stopped short of using, or seeing, the distributional relation in cases where the phonemic difference between complementary or free-variant morphs was too great. Thus he accepted, in addition to the distributional criteria, also such "processes" as suppletion (although suppletion is nothing more than a complementary alternation between morphs which have no appreciable phonemic similarity). In a somewhat similar way, the early

phonemicists balked at putting into one phoneme two phones which had no phonetic similarity to each other. These were, however, only the hesitations of novelty. Bloomfield himself noted that there was no essential syntactic difference between morphemes and intonation or contrastive stress; that things which were expressed by intonation in one language might be expressed by a morpheme in another. The generality of his views may be gauged from the fact that when I once asked Bloomfield whether he would agree that the distribution of an element was or should be the sole criterion for the syntactic characterization of that element, he answered yes. Perhaps it is relevant that on another occasion Bloomfield said to me that these ideas—his ideas—were hard for him to understand fully, and that he expected they would be easy only for the next generation.

The Creation of Linguistics as a Profession

When a new field develops, as linguistics did in the United States in the early twentieth century, it begins with relatively little structure, and the rules of entry are tacit and uncertain. What counts as outstanding academic excellence is not yet established, and the very lines that separate the field from its nearby neighbors are soft, porous, and hazy. New members of the discipline very quickly find reasons to establish some order and organization where there was none, to strengthen the principles used to evaluate work, to define professional boundaries, and to ensure that professional recognition is limited to those who bring some proof of their professional qualities—all the more so if they are young, well-educated, and possessed of significant symbolic capital.

Bloomfield explained the need for the LSA in an article that appeared in the first issue of the journal:

> Students of language do not need to ask Why a linguistic society? but many laymen have asked this question. . . . The layman-natural scientist, philologian, or man in the street—does not know that there is a science of language. . . . Such a science, however, exists; its aims are so well defined, its methods so well developed, and its past results so copious, that students of language feel as much need for a professional society as do adherents of any other science. . . . Not only the furtherance of our science, but also the needs of society, make it the duty of students of language to work together systematically and with

> that sense of craftsmanship and of obligation which is called professional consciousness. For this they need a Linguistic Society.[124]

Years later, Martin Joos would share his recollections of the times in a publication that that LSA published in 1986.

> The Signers of the Call were not rebels. They were continuity men. Their research, teaching, and publication continued unbroken the patterns of linguistic thinking defined in the neogrammarians . . . movement long before Leonard's birth on April 1, 1887; but now after 1918 that was (at least in the thought of schoolmen generally) viewed as the least promising of many competing treatments of languages in publication. . . . The neogrammarian tradition itself was being carried on by relatively few persons, for instance by far too few young students of those 29 fully mature scholars whose average age was above 50: Leonard Bloomfield at 37 was the youngest of them. . . . The 1914–1918 Great War had suspended the normal participation of Americans in European culture.[125]

In short order the major steps were taken to establish a functioning profession. The first step was the establishment of a regular annual meeting at Christmastime, which would guarantee that ideas of the members could be regularly presented to the entire membership. The second was the creation of a journal with an established editor who would see to it that the quality of material published was up to the association's expectations. Bolling himself undertook this responsibility for the first 14 years of the journal's life.

The degree to which the linguistics of the LSA was a development of the German tradition can be illustrated by a genealogical tree that sets out the presidents of the LSA who were either educated in Germany (their names are placed in green boxes), or were academic descendants of others who had. We have placed William Dwight Whitney in the figure, though of course he died long before there was an LSA—but his connection to the group of scholars presented here is transparent. To be sure, this tree does not include all the presidents; we can see that some years are not represented in the tree.[126]

The LSA, as the society came to be known, had its first regular meeting in New York, and 69 people attended. As we have noted, its first purpose was to make it possible for people interested in the same questions to

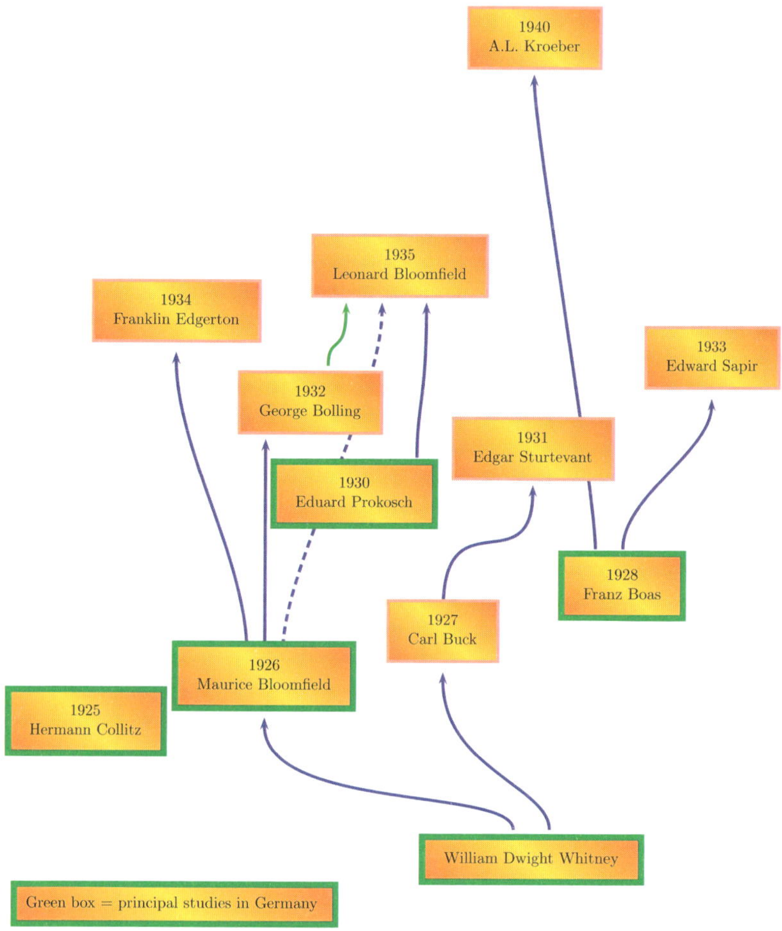

FIGURE 6.3. Whitney, the German tradition, and the early presidents of the LSA

have a chance to meet and communicate with other people—something that simply would not happen on a regular basis under other conditions. There were not a lot more than 69 linguists in the country at the time, and how would someone interested in linguistics in New York or New Haven get a chance to meet and talk face to face with someone from Chicago? That is what a professional organization made possible.

It quickly became clear that one annual meeting was not enough. The country was too large to get people together for meetings during the academic year: it took days to cross from one coast to the other, in those days before affordable air flights. In 1928, the LSA began sponsoring institutes every summer at which linguists, both young and established, would get together to give or take courses, and to talk shop 24 hours a day. Edgar Sturtevant was in charge of this, from its inception up until 1940.[127] From the beginning, an invitation to be a faculty member at an LSA Linguistic Institute was a badge of honor whose significance was lost on no one.

There was nothing that came close to the these summer institutes when it came to developing a single consciousness of American linguistics. Carl and Florence Voegelin remembered, many years later, what it was like to be at the Linguistic Institute of 1937. And they give us a sense, too, of how Leonard Bloomfield was viewed at the time.

> We do not know when the close knit membership of the LSA, inhospitable to European theory, began to realize that Bloomfield had given them a wholly American and wholly explicit linguistic theory. We do, however, know that they could talk about nothing else at the half-dozen Linguistic Institutes preceding Word War II; and, more importantly, they could talk to Bloomfield, who was present at every one of these LI's. Having virtually no Ph.D. students himself, Bloomfield taught the postdoctoral representatives of the LSA who were teachers and visitors at the LI's. They admired Bloomfield above every living linguist.[128]

That summer of 1937, in Ann Arbor, Michigan, was a transformative moment for a whole generation of linguists. We can listen to other linguists talk about what that Institute meant to them. Raven McDavid, for example, noted that in 1937,

> I went to the Linguistic Institute. Wayne Tyler lured me into Bloch's seminar; and Bloch discovered in me a talent for phonetics. . . . Sapir spent an hour with me on the "M" bench, sounding me out and suggesting that I study Catawba—a suggestion shortly reinforced by Swadesh. . . . The Institutes of 1938, 1940, and 1941 strengthened this foundation, especially in phonetics, dialectology, and lexicography. I met most of the great people in the profession—Bloomfield, Sturtevant, Kent, Malone, Haugen, Trager, Kepke, Hockett, Harris, Voegelin, Whorf, Adelaide Hahn—as was easy when a whole Institute could fit into a large classroom. From them I learned two lessons, not in any formal curriculum. With the leaders accepting a beginner as one of themselves—before the day when linguistics was torn by ideological *Schrecklichkeit*—I could do no less for those who later came to work with me; learning

> first hand about their own professional tribulations, I was prepared for what I encountered later.[129]

John B. Carroll has given us an even clearer picture of the impact of the 1937 Institute on him, at a time when he was just 21. Later, in his professional life, he became a psychologist, and a study that he undertook on the field of linguistics in the early 1950s would become important. Carroll had a close connection to linguistics his whole life. He grew up in Hartford, Connecticut, and as a boy he met Benjamin Lee Whorf, who gave a talk about the Aztec and Maya Indians that made a big impression. That meeting in 1929 was the beginning of a friendship that lasted till Whorf's untimely death in 1941. Carroll attended the 1937 Linguistic Institute at the University of Michigan, and later wrote about it:

> I regarded my attendance at the Linguistic Institute, in the summer of 1937 at the University of Michigan, as a kind of "last fling" with linguistics, but I wanted to take it very seriously because I felt that this represented a unique opportunity, as indeed it was (more so than I realized at the time). Al Marckwardt, on hand to aid new students, helped me decide what courses I should take, though he told me I was over-ambitious in wanting to take four—two from Sapir, his Introduction to Linguistics and his Field Methods in Linguistics, and two from Franklin Edgerton, his courses in Sanskrit and in Pali. Sapir's lectures, my notes on which I still retain, were brilliant, and Edgerton was a master at leading one through the complexities of Indo-European comparative linguistics by way of studying Sanskrit and Pali.[130]

Imagine taking an introduction to linguistics from Edward Sapir! And Sapir taught a "field methods" course as well that summer, which is normally a course offered in which the structure of a less-documented language is studied by the students, together with the teacher, based on judgments given by a native speaker:

> Sapir used himself as the informant in Navaho in his field methods course, and gave me an A- (I think it was) on a paper I turned in on the psychology of Navaho grammar. What was most memorable about my experience at the Institute was the intellectual climate and camaraderie that were much in evidence there. Here was a group of people—I was not always clear about who were faculty and who were students—all with intense personal interest in a wide variety of issues about language and languages, all ready at the least provocation

> to discuss almost anything within their competence. Most were already well acquainted with each other; the group included Bernard Bloch, Martin Joos, Kenneth Pike, Raven McDavid, Norman McQuown, Henry L. ("Haxie") Smith, Jr., Morris Swadesh, and others. [*Such as Fred Householder and Zellig Harris. See (Murray 1991)*: J.G./B.L.] I felt myself to be rather an outsider and a neophyte, despite my long association with Whorf, who was known by many of these people. The evening lectures were memorable (one by Zellig Harris on the rebirth of Hebrew in Israel made much impression on me), but of most interest were the colloquia that took place every other day or so at the lunch hour at the Michigan Union. Here Sapir's competence and intellectual authority showed up most strongly. Typically, someone would make a formal presentation on some topic, and there would be a discussion afterward. Sapir was usually very quiet during the presentation and for the first two-thirds of the discussion period. Toward the end, however, he would start talking, in his modest but engaging way, wrapping up the issues, settling all the disputes, and giving some wonderfully apt illustrations of his points from his rich store of knowledge—no matter what the topic might have been. This was always a time when the previously vocal, argumentative discussants would fall silent to hear the master's word.

In this chapter we have looked at the linguistics of the American Great Depression, the decade before the Second World War. It was a small community, but an active and enthusiastic one. In volume 2, we will look at its next phase, as such linguists as Zellig Harris and Charles Hockett became the movers and shakers in this academic and intellectual world.

CHAPTER SEVEN

Philosophy, 1900–1940

Chaque siècle a la grammaire de sa philosophie. —Antoine Meillet (1926)

We turn now to a chapter in the development of philosophy which is dominated by three concerns: the nature of language, of logic, and of science. The principal focus in this chapter is the period of high logical positivism—the period between the two world wars, intellectually dominated by Central Europe, notably Vienna and Berlin.

There are four threads running through the ideas that we will want to pay close attention to. The first is the elimination of what we have called *soft logic*, and its replacement by *hard logic*. Soft logic is based on the premise that rationality can be reduced to some kind of judgment, like that which is aided by Descartes's *lumen naturalis*, an ineffable power to obtain truth by seeing it with the mind's eye. With the elimination of soft logic came the rise of a more powerful family of logics which allowed a formal account of a far broader range of propositions, what we have called hard logic. Closely tied to the development of hard logic was a strong interest in developing a formal account of *syntax*. In short, a transformation from soft logic to hard logic, with a renewed interest in the formal syntax of language. This topic will continue into the next chapter, where we will look at formal logic and mathematics, and the *soft logic* versus *hard logic* will reemerge there in the controversies regarding intuitionism in the interpretation of mathematics. There will be some inevitable overlap between the material we look at in this chapter and the next, but this chapter will focus on the philosophers, and the next on the logicians and mathematicians.

The second theme that we will look at involves a distinction that Carnap and Reichenbach emphasized between the logic of justifying a scientific hypothesis and an explanation of how the idea arose, in terms both personal and historical. Reichenbach coined the terms "context of justification" and "context of discovery" for these two. This distinction fit naturally—at least, it seemed to—into the world view of American structuralists, but when Chomsky, Halle, and Postal began developing arguments for generative grammar in the early 1960s, they would argue that the American structuralists had really missed the point. And the issues raised by Carnap and Reichenbach would morph into the questions about whether it made sense to look for discovery procedures in American linguistics.

The third theme is the rapid rise in the importance of syntax, for philosophy and logic. The idea that the central components of language are syntax, semantics, and pragmatics was first clearly stated by Charles Morris in 1938, and this was itself just a slight restatement of what Rudolf Carnap had written a few years earlier. This was radically out of step with what linguists, either in Europe or in the United States, were saying—and yet today, as a statement it would hardly raise an eyebrow. What philosophers said during this period changed the nature of the game for linguists in the years that followed.

The fourth theme is the most important of all. The main point of Rudolf Carnap's first major book, *The Logical Structure of the World*, was that philosophy had to be completely rethought and reformulated, and in that new form, it would be no more and no less than this: the explicit formulation of the logical structure of each and every specific science. Philosophy, as the study of knowledge, would focus on empirical knowledge, and the sciences are the domains in which empirical knowledge is responsibly gathered; the goal and purpose of philosophy is to shed as much light as possible on the fundamental steps taken in each science, to help ensure that logical error is not built in. Empirical error is guaranteed to exist, but it is easier to uncover, and *philosophy* cannot be charged with uncovering empirical error. That has to be left to the special science itself.

This project had its roots in work that was being done in logic and mathematics early in the century, and even at the end of the nineteenth century, work that was associated with David Hilbert more than anyone else.[1] In the late 1890s, Hilbert went back to Euclid's axioms and Euclid's project to ground all of geometry in a small set of axioms and a set of principles of permissible inference. All this had to be rethought in the light of the reality of non-Euclidean geometries. Once he had accomplished this,

Hilbert went on to see if a similar project could be undertaken for the other sciences. We saw Leonard Bloomfield's attempt to do just this, under the direct influence of David Hilbert's work. And this vision played a major role in motivating the Vienna Circle. We will hear more from David Hilbert, primarily in the next chapter.

We begin our story a bit earlier than the rise of logical positivism; we will start with a brief look at some aspects of the philosophy of Edmund Husserl. Although his work is rarely cited in the linguistic literature, or in Anglo-American philosophy or psychology, his thinking on the nature of language and logic had an enormous impact both on logical positivism and on European structuralism. And we must also look, if only briefly, at the impact of the work of Bertrand Russell and the early Ludwig Wittgenstein, crucial work in the rise of logical positivism, and also on the Polish logicians in the next chapter.

It's an unfortunate fact that Husserl is not discussed much in modern Anglo-American philosophy. Michael Dummett is an influential philosopher who has objected to the easy equation of analytic philosophy with *Anglo-American* philosophy, because such

> terminology utterly distorts the historical context in which analytical philosophy came to birth, in the light of which it would better be called "Anglo-Austrian" than "Anglo-American." In central Europe, that is to say, in the great cultural region defined by the use of the German language for the purpose of publication, there were throughout the nineteenth century a great many diverse currents in philosophy, which did not, however, flow along isolated channels, but collided with each other because of the communication between representatives of the different trends in the universities. More than one of these currents contributed, in the twentieth century, to the formation of analytical philosophy, which, before Hitler came to power, was to be viewed as more a central European than a British phenomenon.[2]

We will turn then to logical positivism, a movement that began in Vienna and Berlin in the period around World War I and blossomed during the 1920s. It continued after a fashion when several of the most prominent members moved to the United States, philosophers such as Rudolf Carnap and Hans Reichenbach. Noam Chomsky's first great linguistic effort, his massive *Logical Structure of Linguistic Theory*, was clearly drawn up as a working out of a Carnapian project, the development of a metalanguage which would provide a logical presentation of linguistic theory.

The events that we look at in this chapter also continue one of the important waves of European influence on American thought. This influence began with young American philosophers, Quine and Goodman among others, traveling to Europe to study, and then continued a few years later when a large group of the philosophers came to the United States, forced to flee Nazi-dominated Europe. This ebb and flow of the tides of academic researchers replayed itself in psychology, in linguistics, and in philosophy.

But before we discuss logical positivism, we will turn to the work of Edmund Husserl, an enormously influential philosopher himself, and two other philosophers who inspired the logical positivists, Bertrand Russell and Ludwig Wittgenstein.[3]

Edmund Husserl

Edmund Husserl is remembered today as one of the founders of the phenomenological movement in Europe in the early twentieth century, a movement that included German philosophers like Martin Heidegger, and French phenomenologists such as Maurice Merleau-Ponty and Jean-Paul Sartre.[4] Phenomenology is a philosophical effort to understand the most basic structure and organization of subjective human experience.[5] In the split between analytic philosophy and phenomenology, Husserl is generally viewed today as firmly grounded in the phenomenological side of the discipline, but that is really only part of the story, and Husserl's influences on philosophy as a whole in his day are becoming better known. From his own perspective, a large part of what he did during his lifetime was *psychology*.[6]

Edmund Husserl was born in 1859 in what is today the Czech Republic. At 17, he went to study in Leipzig, where he attended Wilhelm Wundt's lectures (we have, of course, discussed Wundt) and studied with Tomáš Masaryk (who we will get to know better). He spent the next three years—1878 to 1881—in Berlin, which was at that point the center of the mathematical world, and there he studied mathematics with two of the great mathematicians there, Karl Weierstrass and Leopold Kronecker, whose names are well-known to mathematicians today.[7] He completed a PhD in mathematics, on the calculus of variations, in Vienna in 1883, and then went back to Berlin. Just a short while later, on Masaryk's suggestion, he returned to Vienna to study with Brentano for two years, and Husserl submitted his habilitation thesis with Carl Stumpf in 1887, because Brentano was not permitted to supervise dissertations.[8] Georg Cantor was also in Halle at this time; Cantor was a controversial and highly original mathematician—he is still well known today as the man who tamed infinity and is also the hero of a recent book by David Foster Wallace. Cantor and Husserl became close friends in Halle. Carl Stumpf was close to Husserl during this period: Stumpf wrote that Husserl "was my first student, later an instructor, and became intimately associated with me scientifically and as a friend."[9] Husserl published his book *Logical Investigations* in 1900–1901 and dedicated it to Stumpf.

Husserl remained in Halle for nearly 15 years.[10] He gradually moved away from mathematics. Towards the end of his life, he was getting ready to meet L. E. J. Brouwer, an important mathematician we will meet in chapter 8, and he wrote that he was certain he would disappoint Brouwer by no longer being able to discuss the foundations of mathematics. (We

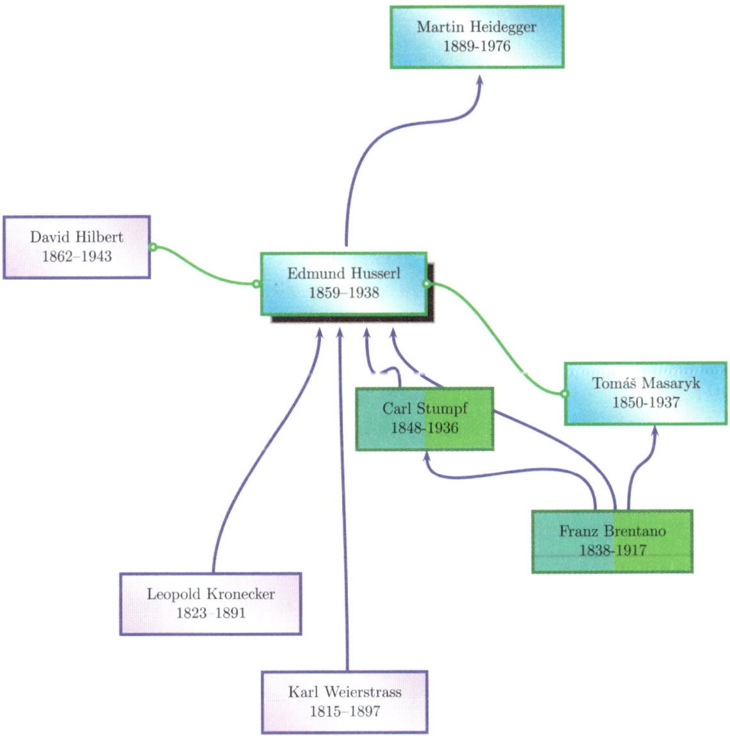

FIGURE 7.1. Edmund Husserl

do know that Husserl wrote back to a former student that he was having long conversations with Brouwer, who he found "wholly original, radically sincere, genuine," and "entirely modern."[11])

In 1901, he took a position as an *ausserordentlicher Professor* at Göttingen, offered due to an interest in his work by David Hilbert, and Husserl would remain there until 1916. Hilbert was one of the most prominent and influential mathematicians in all of Europe, and someone who thought deeply about foundational issues in mathematics. Husserl was influenced by Hilbert—notably by a lecture that Hilbert gave on November 5, 1901, which must have been just after his arrival.[12] Husserl's notes after that contain his thoughts on how axiomatic systems could be used to define objects, after the fashion that Hilbert was using to define integers. Husserl's ideas about mathematics influenced Hilbert's very best student, Hermann Weyl, and influenced him in the direction of intuitionism, which was a direction that Hilbert did not appreciate. The young American prodigy, Norbert Wiener, who would years later become the proud father of cybernetics, went to study mathematics with Hilbert and phenomenology with Husserl in Göttingen in 1914, following a recommendation made to him by Bertrand Russell. (Wiener did find it a particularly stimulating environment, he remarked in a letter to Russell at the time.)

How did Husserl himself see the connection between his work and that of his contemporaries? Lindenfeld makes the case that Husserl saw himself "as a lonely prophet, struggling with the most profound issues of his day without recognition,"[13] and saw his effort to bring philosophy to the Promised Land (phenomenology, in Husserl's view) as parallel to Moses's unfulfilled desire to get to the land of Canaan. From our perspective today, when a research library has shelves that groan and bend under the load of books written by and about Husserl, it may be difficult to imagine Husserl's fear that he would remain unrecognized, but as Lindenfeld points out, Husserl did not have the speedy career path that Alexius Meinong, his contemporary, did. Husserl remained an unsalaried *Privatdozent* for 14 years (1887–1901), and his work was not well received by his colleagues in Göttingen in 1905. We cannot emphasize enough how important it is to bear in mind that most of the creative and influential work we have looked at has been accomplished by young people who had little evidence in front of them that their work would eventually become recognized and influential.[14]

From Brentano, Husserl adopted the notion of *intentionality*, the property of psychological phenomena of *being about something*. A thought is a thought *of* something; it not only can be a thought about something else (an apple, one's sister, etc.), it *must* be a thought about something; there is no such thing as a thought that is not actually about something. Objects in the world are not like that: things can be, quite simply, things: a rock, for example, or perhaps a shoe. A photo is, indeed, a photo of something, but most things are not representations of something else. Early in his intellectual life, Husserl's view changed about the relationship of psychology and mathematics. His early work heavily leaned towards a psychologistic view that exploring psychologically how the human mind deals with a subject is the way to understand the very nature of that subject. He came to reject that view, in part because of a strongly negative review that Gottlob Frege published of Husserl's first book, and the varying perspectives on *psychologism* (most often without citing the term, and often without realizing what is actually at stake) is an important leitmotiv throughout this book. (As we have seen, psychologism is the view that the study of specific aspects of mind—linguistics, say, or logic—is a subdiscipline of empirical psychology.)

While Husserl's place in the history of philosophy is well known, the importance of his work for the rise of some aspects of modern syntactic and semantic theory is much less familiar.[15]

Over the next three chapters, we will have the opportunity to see the influence of Husserl's ideas on Carnap and on the Polish logicians in the 1930s as well as on the Prague linguists at much the same time. We will look at Carnap in this chapter, and the Polish logicians in the next chapter. This work would have a clear impact on Bar-Hillel's work, and thus on Chomsky's; the development of categorial grammar, by Joachim Lambek and others, is a very direct development of these ideas as well.[16]

Equally important is the impact of Husserl on the young Roman Jakobson. Recently, Zahavi and Stjernfelt have written about Husserl's influence:

> . . . via Roman Jakobson who very early on read Gustav Spet's Russian translation of the *Logical Investigations*. Even though Jakobson did not himself conceal this inspiration, for a long time it passed largely unnoticed that the very founder of linguistic structuralism imported crucial parts of his conceptual apparatus from the *Logical Investigations*, especially from the First, Third, and Fourth Investigations. Jakobson used, among other things, the idea of foundation in his phonology (the "distinctive feature" as a moment of the phoneme). From Jakobson, this inspiration passed—often unnoticed

> or via inscrutable connections—into mainstream linguistics and semiotics; thus it is striking to see Husserl's three basic dependency relations between parts surface in exactly identical form in Louis Hjelmslev's *Prolegomena* of 1943. The Third and Fourth Investigations also inspired and influenced the Polish logicians—in particular Leśniewski's idea of a formal mereology, which was initially formed in order to dismantle Russell's paradox, and Ajdukiewicz's idea of categorial grammar, defining word classes by dependency relations.[17]

Jakobson's work will be a major concern of ours in chapter 9, and there we will make further connections with the work we look at here.

The main work of Husserl's that will interest us is his *Logical Investigations*, published first in 1900, and revised in 1913. This is his first mature work, and it had an enormous impact on any number of scholars. It was in part a continuation of a conversation that had been ongoing among logicians from the mid-nineteenth century with regard to the notion of *meaning*, and the difference between *intentional* and *extensional* understandings of meaning. Views and opinions have varied so greatly that it would be foolhardy to try to give a simple account of how these two differ from each other—but we will try to do so anyway. An extensional account of the meaning of a phrase makes its case by presenting those objects that are denoted by the phrase. The extentional meaning of the phrase *the prime numbers* is a set which begins with the numbers 2, 3, 5, 7, and 11. An intensional account of the meaning of a phrase is a list (or something of the sort) of those properties shared by all the objects in the phrase's extensional meaning. The intensional meaning of *the prime numbers* is a positive integer which cannot be evenly divided by any integers other than 1 and itself.

There are four important points that link Husserl's work to the development of linguistics and the mind sciences at this point.

(I) Priority of terms and sentences. The first point is that only a relatively small fraction of the subparts of things we say have a meaning, because most words, and most combinations of words, are able to contribute to the meaning *only* by virtue of the particular context they are found in. There are several kinds of examples we could use to illustrate this point; Husserl is unfortunately not very good at giving examples, even when an example would make it much clearer what he was getting at. He seems to mean that a phrase such as *my friend Scott* is meaningful, as is *the door slammed shut*, but the phrases *slammed shut* or *remarkably tough* are not; they will contribute to the meaning, once placed in an appropriate context.[18]

While any actual use of language is an act carried out in some real context, the analytical urge to understand how language works encourages us to take smaller pieces and say something about them in a fashion that is less context-dependent. Although it is artificial, we can conjure up a situation in which we open a bottle that lands on a beach with a paper inside bearing a word or a phrase: "Macadamia nuts," or "$F=ma$," or "the Moon revolves around the Earth." In each of those cases, we would likely feel that we were able to get all of the meaning out of the expressions that the writers had put into them: we know what macadamia nuts are, and though we don't know why the writer wanted to draw our attention to them, we know what he was writing about. As for Newton's equation, or the fact about the Moon, we know too what the writer meant, and we probably agree that these statements are basically true, too. But the same would not be true if the paper had said "good" or "the" or "extremely." We could recognize that they are English words, but we just don't know what function those words were going to play in some message. Hopefully you share that sense; that is what Husserl was counting on.

We may see Husserl's work here as the first modern study of language in which the problem is to clearly specify the boundary between the strings of words that are *in* the language—that is, grammatically well-formed—and those that are *out* of the language, now called "nonsensical."[19] This was the project that was handed down in the genealogical tradition we mentioned just above, ending in generative grammar.

(II) Types of grammatical well-formedness. This focus goes with Husserl's more general concern about how things are put together: not just what they *are*, but how they get fit together. And this leads us to Husserl's second main point, that there are at least three classes of word strings. There are perfectly normal strings of words, like *the door slammed shut*; there are strings of words that by their very meaning cannot be true, like *I have invented a round square*; and then there are those that are so bad that they cannot even enter into the testing ground for whether they are meaningful or not: *the under might bluish*.

Husserl drew a sharp distinction between what he called, in German, *Unsinn* and *Widersinn*. These terms have sometimes been translated into English as "nonsense" and "absurdity." The idea of a distinction of this sort has gone on to be of primordial importance, so it is important for us to be as clear as we can about the difference that Husserl had in mind here. We will continue to use the term "nonsense" for Husserl's *Unsinn*, but a more literal translation of *Widersinn* as "counter-sense" might serve us better than "absurdity."[20]

The term "nonsense" was intended to cover strings of words that cannot be integrated into a sentence: *the the to of when*, for example, while "counter-sense" included any talk about round squares, or wooden iron, or a statement that a particular square has five distinct corners. These two categories arise out of two entirely distinct considerations: nonsense violates grammar, while counter-sense violates logic. You can try to argue against counter-sense, but when faced with someone speaking "nonsense," you look to see if you can help them to figure out how to get it out in English. Or some other language.

This was a major step, primarily because with this, Husserl offered us a way to distinguish between *grammar* and *logic*. Are we willing to use the tools that we have in order to sharpen the distinction between these two, or should we do our best to show that one derives in some fashion from the other? And if the two exist and are distinct, how different are grammar and logic? Do they belong to different metaphysical realms? Is perhaps one of them specific to human beings, and if so, which one? Scholars who have looked at Husserl have not all agreed on just what Husserl's strategy was here. One strategy could be that we can simplify the problem of how we ought to analyze meaningless sentences by setting up two different kinds of conditions—one semantic, say, and one syntactic—and a sentence that fails either of these conditions is meaningless. As a variant on that, we could imagine an analysis in which it is only sentences that are well-formed (=not meaningless) in one sense that are even eligible for being considered sensible in the other: for example, if one of the senses could be called "syntactic," we could well imagine an account in which some sentences are semantically well-formed (that is, sensible), but only sentences that are syntactically well-formed can even be tested to see if they are (semantically) sensible.

Husserl used the division of sentences into three sets (nonsensical sentences, absurd sentences, and all the rest, which are the fine sentences that we use in daily life) in order to define what he called "semantic categories," which are sets of words (or expressions) that can be substituted for each other without a change in the status of the resulting sentence. If any occurrence of "Paul" can be replaced by "Richard" (and vice versa) without changing the grammatical status of the sentence, then "Paul" and "Richard" are in the same semantic category. We see here the very basis of a distributionalist approach to linguistic analysis.

One reason to separate grammar from logic is that we might find—indeed, this seems extremely plausible—that we can much more easily set up the rules of logic if we decide that they apply only to strings of words that are grammatically well-formed. From such a perspective, the goal is to find the right division of labor between the grammarian and the logician. The grammarian may have her work simplified for her if some problems can be left to the logician, and the logician (Husserl is pretty sure about this) will have his work simplified for him if many strings of words can be passed over in silence by the logician on the grounds that they are accounted for by the grammarian, who can establish that they are *nonsensical*.

In retrospect, we know now that quite a few modern scholars think that the distinctions Husserl has drawn are not *enough*, though to be fair to Husserl, his main point was that an important distinction needs to be introduced, not that there were no others to be drawn. Consider four ways to classify English sentences:

1. There are normal sentences, which are grammatical and meaningful, and much of the time, are used to make statements about our world. Most of the sentences in this book are good examples of these normal sentences.
2. There are ungrammatical sentences, or ungrammatical sequences of words that do not together form a grammatical sentence. These can be divided into two sorts: *failures* and *artifacts*. Failures are the sentences that any of us produce, typically in an effort to do or say something, that fail some grammatical constraint, while artifacts are either strings of words we produce as examples of ungrammatical sequences of words, or are produced by some breakdown in technology. For example, *John should have schedule the meeting* is ungrammatical (the verb should be *scheduled*), while *Of Philo is or of Philos is not* is an example constructed (from Plato, given in the translation of Gardies [1986]) to be ungrammatical. Here we have to be careful with the differences of opinion: speakers and experts can disagree about certain cases, and tempers can flare.
3. There are grammatical sentences which contain some words or morphemes that most speakers of the language may not be familiar with (perhaps because they were used for the very first time in that sentence!). The two classic examples of this type of sentence are Lewis Carroll's *'Twas brillig and the slithy toves Did gyre and gimble in the wabe*, and Rudolf Carnap's *pirots karulize elatically* (which was originally penned in German: *Piroten karulliern elatisch*—how remarkable that such word salad can be translated from one language to another).[21]

4. There are grammatical sentences which we know cannot be true by virtue of some kind of contradiction in the sentence's meaning: sentences like *This square is round*.

Over the years that followed, right down to this very day, this set of distinctions has played an important role in many people's thinking about the nature of language. For now, we just notice this: the idea that language could be studied by providing a formal characterization of the boundary between the syntactically well-formed strings and the other, ill-formed strings is an idea that Husserl first formulated clearly. It would become a central fixture in generative syntax 50 years later.

(III) Return to *grammaires raisonnées*. Husserl's third point in his *Investigations* was that it was time to bring the seventeenth-century *grammaires raisonnées* back and to explore the nature of universal grammar. Early in the Fourth Investigation, Husserl wrote, "Modern grammar thinks it should build exclusively on psychology and other empirical sciences."[22] We can only imagine that Husserl had Wilhelm Wundt in mind. But Husserl himself did not agree with that view. "As against this, we see that the old idea of a universal, or even an *a priori grammar*, has unquestionably acquired a foundation and a definite sphere of validity,"[23] because there are laws which are themselves *a priori* and which determine the range of possible meanings. Husserl then declared himself ready to take up arms to defend "the old doctrine of a *grammaire générale et raisonnée*, a philosophical grammar, for its obscure, undeveloped intention aiming at the 'rational' in speech, in the true sense of the word, and in particular at the 'logic' of speech, in the true sense of the word, and in particular at the 'logic' of speech or its semantic *a priori*."[24] We propose to offer just a brief observation. Along lines that he shared with Kant and the neo-Kantian tradition, Husserl believed that there were elementary structures of thought that were not grounded in anything empirical in the universe in which we live, and which would be true in all universes. Some aspects of human language reflect those elementary structures, and any effort to explain those generalizations about a language (English, say, or Latin) which failed to tie them to this logic would be missing the important point—and that failure was almost certain to befall anyone whose commitment was to a highly empiricist form of psychology. Less clear in Husserl, but still there, we believe, is the suggestion that the ways in which these logical principles form aspects of language may differ from language to language, or at least give that impression. Husserl was not given to providing examples, but we will offer one. Suppose that there is a logical foundation (rather than empirical foundation) for *break* as a predicate to demand two arguments. One language may require those arguments to precede the verb (as in Japanese), while another may put one before the verb and the other after (as in English). For Husserl, the explanation of English and Japanese requires three steps: one is part of universal grammar, and the second and third are parts of the accounts of English and of Japanese.

Husserl was making two points that are worth noting: first, he was developing a new logic (*mereology*, the study of parts, wholes, and their logic, was a part of it), which meant that it was important to realize that the scope of logic had *not* been fixed by Aristotle once and for all; logic was going to be advancing its boundaries as Husserl and other logicians expanded the domain that logic covered. The second was that even if language can be fully understood as a system which links meaning and outer form, it will give rise to the appearance of grammatical impossibility if we put words together that logic insists cannot go together. In today's terms, what appears to be ungrammaticality may be the result of semantic ill-formedness.

(IV) Recursion and unbounded sets of grammatical sentences. Husserl's fourth point involved the importance of both recursion and composition. Husserl emphasized the notion that it was essential to develop a project in two stages: a first stage in which primitives were fixed, and a second in which patterns of compounding and modification—in short, of composition—were defined; he wrote of obtaining "a systematic survey of the boundless multitude of further forms which are derivable by way of repeated compounding or modification."[25] Husserl went so far as to provide examples of what he meant, describing conjunction with *and*, and connection of propositions with *if . . . then*.

Husserl's Investigation IV, section 13, begins this way:

> The task of a fully carried out science of meaning would be to investigate the laws of meaning-composition (and the closely related meaning-modification) and to derive them from a minimal number of independent axioms. And it would obviously also be necessary to separate the essential categories of meaning which appear in these laws as unknowns (or, in the mathematical sense, as variables). What formal laws of combination may achieve can be made fairly plain by arithmetic. There are definite forms of synthesis, through which,

> quite in general or in certain definite conditions, two numbers give rise to new numbers. The "direct operations" $a+b$, ab, a^b yield resultant numbers unrestrictedly, the "inverse operations" $a-b$, a/b, b^{th} root of a, only under certain conditions. . . . Even in the light of the small amount that we have considered so far, it is clear that similar laws regarding the existence (or non-existence) of meanings exist and that in these laws the meanings are not free variables, but rather are rather limited to the scope of this or that category which is grounded in the nature of the semantic domain.[26]

Husserl suggested that just as these mathematical operations make new numbers from old (to put it very crudely!), so can similar operations produce new meanings from old: "we must note," he wrote, "that complete propositions can become members of other propositions," giving us what is translated in Husserl 2001b: 69 as "boundless multitude of further forms, all derivable by way of repeated compounding or modification." The "boundless multitude of further forms" is Husserl's "unbegrenzte Mannigfaltigkeit weiterer Formen," which in this context would better be translated as "infinite set," making this sentence sound much more familiar to the contemporary ear, who would understand that Husserl is talking about a set of rules that generate an infinite set of propositions or sentences.[27]

How do we establish that the set of sentences is unbounded, or infinite? Husserl did this by exhibiting several processes that amounted to a recursive definition of sentence. He wrote,

> Any two propositions yield, when combined in the form *M and N*, another proposition . . .

and the same argument can be made at the word level:

> Any two adjectives [yield] another adjective . . .

And the list is not hard to extend:

> To any two propositions, M, N, there belong, likewise, the primitive connective form *If M then N, M or N*, so that the result is again a proposition.[28]

And this can be repeated ad libitum, so that "an infinity of complex form is" regularly generated. And the same goes for putting an adjective in front of a noun, and so on. In short, grammars grounded in Husserl's conception of logic will provide explicit principles that generate infinite classes of meaningful sentences, based on rules that generate complex sentences from simpler sentences.

Years later, Roman Jakobson wrote,

> At the beginning of our century the thought of Husserl . . . developed in the second volume of his *Logische Untersuchungen* and particularly in the chapter where he treats "the difference between independent and dependent meaning and the idea of pure grammar," became a powerful factor for the first steps of structural linguistics by superimposing "the idea of a general and *a priori* grammar" on "the exclusively empiric" grammar which at that time was the only one accepted. Husserl advocated the idea of universal grammar "as it was conceived by the rationalism of the seventeenth and eighteen centuries." Anton Marty, the critical adept of Husserl, noted in this connection the valuable contribution to general grammar made by the Stoics, then by Scholastic science, later by the Cartesians, such as the authors of the Port-Royal grammar, and finally, by Locke's *Essay Concerning Human Understanding* and by Leibniz' *New Essays*.[29]

Bertrand Russell, Ludwig Wittgenstein

Bertrand Russell

Bertrand Russell's work as a mature thinker lasted many decades, from the end of the nineteenth century up until his death in 1970 at the age of 97. He was born into the British upper class, but actively defended social causes and anti-war movements throughout his life, as late as the American war in Vietnam in his last years. His thought had considerable impact in philosophy (it is in this context that we shall look briefly at his work in this chapter), and also in logic, where he tried to show that mathematics could be reduced (in some sense) to logic, and we will return to that aspect of his work in more detail in the next chapter.

Russell must be read with the understanding that many generations of thinkers have found his attitude towards knowledge and towards tradition to be thoroughly liberating. Here is an example of how he encouraged his reader to think about philosophy:

> The problems and the method of philosophy have, I believe, been misconceived by all schools, many of its traditional problems being insoluble with our means

> of knowledge, while other more neglected but not less important problems can, by a more patient and more adequate method, be solved with all the precision and certainty to which the most advanced sciences have attained.[30]

A very large broom or two and a crew of stalwarts is all that it would take to start the task of cleaning up philosophy—let's get going! Russell was quite comfortable with the sweeping-it-all-away metaphor:

> But [Galileo's] few facts sufficed to destroy the whole vast system of supposed knowledge handed down from Aristotle, as even the palest sun suffices to extinguish the stars. So in philosophy: though some have believed one system, and others another, almost all have been of the opinion that a great deal was known; but all this supposed knowledge in the traditional systems must be **swept away**, and a new beginning must be made, which we shall esteem fortunate indeed if it can attain results comparable to Galileo's law of falling bodies.[31]

This very famous book of Russell's ends with a clarion call:

> The one and only condition, I believe, which is necessary in order to secure for philosophy in the near future an achievement surpassing all that has hitherto been accomplished by philosophers, is the creation of a school of men with scientific training and philosophical interests, unhampered by the traditions of the past, and not misled by the literary methods of those who copy the ancients in all except their merits.

Sweep the Augean stables of philosophy clean, Russell urged; begone, traditional disputes in philosophical garb; make way for logic.

Rudolf Carnap as a young man responded to this passage and indeed to the whole book, later writing, "Some passages made an especially vivid impression on me because they formulated clearly and explicitly a view of the aim and method of philosophy which I had implicitly held for some time. I felt as if this appeal had been directed to me personally. To work in this spirit would be my task from now on! And indeed henceforth the application of the new logical instrument for the purposes of analyzing scientific concepts and clarifying philosophical problems has been the essential aim of my philosophical activity."[32] Rudolf Carnap resonated directly to Bertrand Russell's call.

Ludwig Wittgenstein

Ludwig Wittgenstein was born into one of the wealthiest families of Austria in 1889. But it was not a happy family. He was one of nine children: four girls, and five boys, of whom Ludwig was the youngest. Of the boys, three committed suicide, and the two who did not were to become famous: Ludwig as a philosopher, and Paul as a concert pianist (he lost his right arm during the fighting in World War I, and Maurice Ravel wrote a very special piano concerto for him, his "Piano Concerto for the Left Hand"). Until Ludwig was a teenager, he was tutored at home, and for him, home meant a spectacular mansion in which Johannes Brahms and Gustav Mahler would come and play for the family.

Wittgenstein was of the same age as the psychologists Wolfgang Köhler and Edward Tolman, and the linguists Leonard Bloomfield and Nikolai Trubetzkoy (and only six days younger than Adolf Hitler, another actor in this story—and hard as it is to believe, Ludwig and Adolf went to the same school for a year). Wittgenstein went to visit Gottlob Frege in the summer of 1911, after he had become passionately interested in the philosophy of mathematics, and Frege made a suggestion that would have great consequences. He said that Wittgenstein should go to Cambridge to study with Bertrand Russell.[33] Wittgenstein followed that advice, and in the years that followed, he was widely seen as Russell's most brilliant student, though it is not at all clear that Wittgenstein believed that Russell understood the larger intellectual context in which Wittgenstein's own work was framed.

Wittgenstein's philosophical career divided into two parts, and of the two, we will mention only the first here. The first period of the younger Wittgenstein was defined by a short book that he wrote called the *Tractatus Logico-Philosophicus*, published just after World War I. The *Tractatus* is a small book, broken up into a sequence of small paragraphs which bear numbers, like "3.1.2," giving the impression to the reader of book organized in a most extraordinary fashion. Some of it was written while Wittgenstein was fighting in the trenches on the Italian front. Each sentence seems to be intended to carry an enormous amount of weight, from the first: *The world is all that is the case;* right up to the final: *Whereof we cannot speak—thereof we must remain silent.* "We must remain silent": much of the *Tractatus* is about what we must be silent about. The phrase "remain silent" is a poor translation from the German *schweigen*; in German, *schweigen* is a verb, much like the French verb *se taire*, and easily suggests a more active participation of the subject than we see in the English phrase "we remain silent"; it suggests at least the possibility that we actively decide to say nothing. This is not unimportant, because the silence that Wittgenstein insists on is not the silence that comes from turning an cold shoulder to things, but rather the silence that we feel when we enter an imposing cathedral. As we will see shortly, not all of his readers read him that way; the logical positivists wanted to stop people from talking about things that, for philosophical reasons, they shouldn't be talking about, because they were kidding themselves and confusing themselves by doing so. The logical positivists took the *Tractatus* to be telling people not only to stop talking about those things, but to stop thinking that there was anything very worthwhile over there. This was not what Wittgenstein said, but there you are.[34]

David Bell draws an apt parallel between Husserl and Wittgenstein: Husserl's grounding in the philosophy of Brentano is as deep as Wittgenstein's is in Frege's, and we cannot hope to understand either Husserl or Wittgenstein without some exposure to Brentano or Frege. Why? Because "neither Wittgenstein nor Husserl enjoyed a widely-based, formal education or training in philosophy; on the contrary, their respective philosophical outlooks were initially formulated in the context of, and in response to, an extremely narrow set of philosophical concerns—the concerns, predominantly, of a single philosopher."[35] If we try to understand what Husserl was talking about, much of what he wrote would "remain at best arbitrary and unjustified, and at worst inaccessible."[36] This is actually a common situation, it seems to us: when a person does not know the larger context of a field, he cannot understand how much he depends, intellectually and contextually, on his teacher, and the result of that lack of awareness is less fine control over one's own intellectual position.

Logical Positivism, Logical Empiricism

Logical positivism

Logical positivism was a philosophical movement that brought together a number of important intellectual trends of the nineteenth and twentieth centuries, and both its principles and the impact of its efforts form an important part of the story that we trace in this book.[37] We will look at some of the events and people who played major roles in this movement and then return to consider some of the ideas in greater detail.[38]

Logical positivism was born of the Austrian nexus that we discussed in chapter 3 in connection with Franz Brentano and his students. The logical positivists all saw the development of physics and of modern logic as one of the greatest accomplishments of the world as they knew it—perhaps *the* greatest—and they hoped that the triumphs of physics and of logic could serve as models for other disciplines and fields of human endeavor. Philosophy was the domain that they put at the top of the list: that is, they argued that a new philosophy should be formed that could *learn* from the sciences how the world was.[39] In this important respect, they were following in the tradition of Kant and the positivists: they saw science as the philosopher's best friend. They also shared a complete and utter rejection of all discussion that they saw as containing, or even hinting at, obscurantism, mysticism, metaphysics, and claims of a hidden reality that could be inferred but never observed.[40]

The positive view of the logical empiricists was an explicit rejection of the view that philosophy should, would, or could serve as the *foundation* for empirical science; if either one were to be the foundation for the other, it should be philosophy that depended on empirical science rather than the other way around. In 1915, even before there were people identifying themselves as logical positivists, Schlick had written that "the only fruitful method of all theoretical philosophy consists in critical inquiry into the ultimate principles of the special sciences." He emphasized that it is the revolutionary new ideas coming from the sciences that lead to new perspectives in philosophy: "It is primarily, or even exclusively, the principles of the exact sciences that are of major philosophical importance."[41] His meaning was clear: a philosophy which has not learned and digested the latest scientific discoveries is dead—even if it is not aware of it yet. It may still be walking, but it is zombie philosophy.

In line with a philosophical strategy seen in G. E. Moore's writings, and in Wittgenstein's—and very much against the spirit of German idealism (but let's not forget that the Viennese were Austrian, not German)—the logical positivists hoped to banish much of the discussion that earlier philosophers (like Hegel and Schelling) took quite seriously and which the logical positivists themselves found nebulous and obscure. This was the second plank in the positivist platform that we mentioned above. They took both logic and mathematics to be reflections of a purely formal system, and hence statements of logic and mathematics were viewed as having no content that went beyond the axioms with which the systems began: they were, in this sense, analytic.

What this typology of knowledge quite intentionally left out was what Kant had found most interesting: to wit, the synthetic *a priori*, that is, things which could be known to be true but which were not based on experience. Factual, or empirical, propositions are known from experience, hence *a posteriori*; logic and mathematics consist of analytic statements. There is no room left, in this typology of propositions, for the metaphysics against which Auguste Comte, and other anti-metaphysicians of the nineteenth century, railed. And logic was moved to a more revered and austere position for the logical positivists: it was not empirical, not a subject that could be treated as contained within psychology, as Mill and other nineteenth-century positivists had argued.

In addition, logical positivists put a strong emphasis on the ultimate unity of all scientific theories, and the belief that in the long run the language of physics is sufficient for the statement of all scientific truths. Their method of integrating the sciences was inspired by David Hilbert, the German mathematician we met earlier, who from the very end of the nineteenth century developed an idea of how the various sciences should be thought of in their relation to mathematics. His idea was to have his cake and eat it too: he wanted room for scientific theories to be fully mathematical, and yet there had to be a way in which they related to the empirical world. He developed his ideas originally in the context of geometry, which was the poster child of the field that both was and was not about the real world. The rise of non-Euclidean geometry had persuaded all mathematicians that Euclidean geometry was only one of several possible geometries that they could study as mathematicians, but there was still much that Euclidean geometry could do to make very accurate, correct predictions about the real world. Hilbert proposed that the theory of geometry should be founded on a small number of axioms—just like Euclid had, two thousands of years earlier—but the everyday meaning of the words (in fact, all meaning of the words) should be abandoned as long as one was doing mathematics. Hilbert is often said to have explained what he meant by saying, "one must be able to say at all times—instead of points, straight lines, and planes—tables, chairs and beer mugs" (though he never wrote that himself). Terms like *point* and *line* will appear in our proofs, and if we choose to interpret them as we do in normal speech, then we can carry the mathematical conclusions over to the real world, where they become *claims* of an empirical theory: the theory that says that the axioms of the mathematical model are true for the world in which we live.

The anti-metaphysical leanings of the logical positivists eventually led to serious problems, both social and philosophical. Part of their inheritance from Hume was the strong need to split off all talk of morality and obligation from what they considered meaningful language. All talk of what should be done fell into the *metaphysical*, and in the early days of logical positivism, that was never a good thing. Later, when American philosophers, notably John Dewey, were confronted with logical positivism, they found that position off-putting, and even repugnant; they wanted a philosophy that could learn from science and have real moral lessons to give to society at large. Dewey was always clear on that score. Carnap would later write to Dewey, "But Schlick and the others of us do not mean to say that value expressions have no meaning at all, but only that they have no cognitive content. . . . Certainly we do not deny, but rather admit explicitly the great psychological and historical effect of metaphysical statements."[42] For most outsiders, this disclaimer on Carnap's part would leave them scratching their heads, wondering how that statement could be squared with the much clearer statements of the 1920s, when the charge that a statement was metaphysical was cause enough to have it taken out the rear door and shot. Part of the problem grew out of the fact that there was a wide spectrum of political views across the founders of the Vienna School.

Context

We need to be a bit careful when making generalizations about *all* logical empiricists; when we look at the their writings and teachings, we find a range of views, positions, and interests, since it was a movement spread over space, time, and quite a number of people with different backgrounds and aims. The movement began informally in the years before the First World War, and its first roots were in Vienna and the cities of Central Europe that maintained close contact with Vienna, notably Berlin and Prague. In Vienna, Philipp Frank, Hans Hahn, and Otto Neurath were regularly meeting on Thursday nights to discuss philosophy, and what began as an informal discussion group grew into an organization, eventually with international meetings and its own journal. They found inspiration in work done throughout Europe: there was the work of Ernst Mach first of all, and the scientific and mathematical advances made by Albert Einstein, Henri Poincaré, and Bertrand Russell.[43] During the halcyon days of this Vienna group, the acknowledged leader was Moritz Schlick. Schlick spent a year as a visitor at University of California, Berkeley in 1931, where he (apparently) influenced the psychologist Edward Tolman.[44] Schlick had earned a PhD in theoretical physics, studying with Max Planck in Berlin, but moved into philosophy soon after that. He wrote an influential book on Einstein's physics, *Space and Time in Contemporary Physics* (1917). In 1922, Schlick was offered the chair in "the philosophy of the inductive sciences," at the University of Vienna, the same chair that had been earlier held by Ernst Mach and Ludwig Boltzmann but which had not been held by anyone for a number of years. In 1924, Schlick established a more official group called the Vienna Circle, where both students and professors met to discuss issues of scientific philosophy. Hans Hahn, a member of the Circle and a mathematician, gave informal talks about the philosophical content of Whitehead and Russell's *Principia*; in 1925, they read Wittgenstein's *Tractatus* and an early version of what would become Carnap's *Logical Structure of the World*.

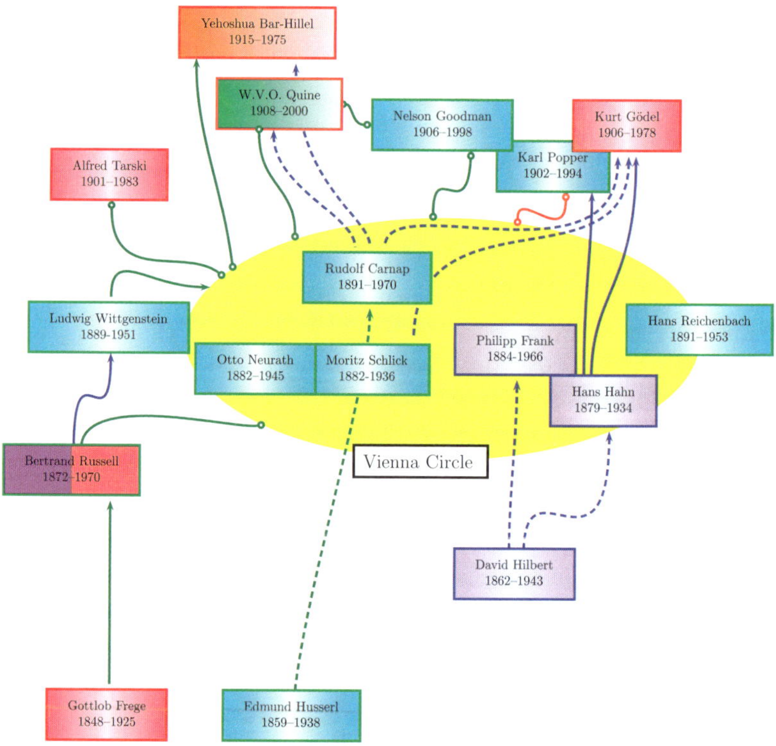

FIGURE 7.2. Logical positivists and friends

Over the course of the 1920s, this group grew to include Rudolf Carnap, Kurt Gödel, and a number of others, drawn from the ranks of philosophy, mathematics, and the physical sciences. In 1926, Carnap was offered the position of *Privatdozent* in Vienna (over Reichenbach). In the late 1920s, several students who would become famous were active participants in the Circle: Kurt Gödel, Gustav Bergmann, and Karl Menger. Schlick was killed by a deranged former student of his in 1936. In the end, the most influential and best remembered of the larger circle of logical positivists were Rudolf Carnap, a member of the Vienna group, and Hans Reichenbach, who would become the center of the Berlin group.

RUDOLF CARNAP

Rudolf Carnap was born in 1891, and he studied philosophy, mathematics, and physics early in his university studies. He studied Kant with Bruno Bauch and attended courses with Gottlob Frege (we saw his recollection of Frege's unhappiness above).

After fighting in World War I, he went to the University of Berlin and then the University of Jena, originally intending to write a dissertation in physics, but he found that the physics he was interested in was considered to be philosophy in the eyes of the academy, and he ended up defending a dissertation on space.[45]

There was a natural connection between what the early logical positivists were interested in and Kant's philosophy of space and time. Kant's treatment of space could be interpreted either subjectively as an account of how space is perceived by the human mind, or it could be interpreted as an account of any conceivable framework within which the spatiality of objects in the world are understood. The subjective interpretation was challenged by recent work in psychology, notably Gestalt psychology, and the second was challenged by Einstein's novel ideas of space, time, and ultimately of gravitation. If experimental science has some kind of epistemological priority over philosophy, then the 1920s was a period in which philosophy was in dire need of a change.

Space and time were thus enormously interesting topics at that moment for people interested simultaneously in physics, philosophy, and psychology. Albert Einstein had completely changed the scientific understanding of what space and time are: on the new relativistic view, space and time are intimately bound to one another, and from the point of view of the Einsteinian universe, the notion of distance-between-two-objects-at-a-given-moment no longer makes sense: one must talk about the distance between two *events*: an event *occurs* at a specific time and place, and of two events, we can measure the distance *and* the time between them, the *space-time interval* that separates them. But even then, there is no unique right answer for that pair of measurements, of space and time: there is a wide span of physically real answers to the question, how far apart are two specific events, in time and in space? That was the physics, but then there was the philosophical side: had not Kant persuaded us that classical space and time were part of what we presuppose as minds when we encounter the world? How could that whole picture have been overturned? What did we do wrong, when we persuaded ourselves that absolute space and time would remain, absolutely, forever? Was Kant simply wrong, because he mistook something superficial—our workaday assumptions about organizing perceptions—for something deep and abiding? Those were the questions that were *hot*.

We have already seen (chapter 5) that in Berlin at this time, there were young people who would become famous as psychologists—Max Wertheimer, Wolfgang Köhler, Kurt Lewin, Egon Brunswik—who were worrying seriously about these questions, and spending long hours in conversation with each other, developing Gestalt psychology. Carnap wrote, in his autobiography, "I recognized, under the influence of the Gestalt psychology of Wertheimer and Köhler that the customary method of analyzing material things into separate sense-data was inadequate—that an instantaneous visual field and perhaps even an instantaneous total experience is given as a unit, while the allegedly simple sense-data are the result of a process of abstraction."[46]

Many years later, he recounted the impact that he felt when he read Bertrand Russell's call for a new philosophy (we read his words above) built by men trained in science and interested in philosophical questions. He wrote, "I felt as if this appeal had been directed to me personally. To work in this spirit would be my task from now on!"[47]

Carnap published *The Logical Structure of the World* (the *Aufbau*) in 1928 and *The Logical Syntax of Language* in 1932. His goal was to show how scientific knowledge can and should be organized, emphasizing the importance of the logical analysis of each of the terms that play an important role in the discipline, with the intent of making it easier to define what today we might call the *interface* between pairs of disciplines, so that the ultimate unity of all scientific knowledge becomes a practical truth, and not just a slogan.

If he was to be a major philosophical figure in the twentieth century—and indeed, he was—he brought with him the tools and the techniques of the mathematician that he had trained to be. Herbert Feigl put it this way:

> I remember vividly Carnap's first lecture to the Vienna Circle [in 1925]. He presented his Space-Time topology . . . in the manner an engineer might explain the structure of a machine he had just invented. To the non-logicians Carnap indeed seemed to be no philosopher at all. Some of this sort of misunderstanding was Carnap's fate throughout most of his long life. But this did not seem to disturb him at all.[48]

Indeed, he often felt out of sync with other academic philosophers. At the University of Chicago in the 1940s, when there was a dissertation defense on the history of philosophy, Carnap would say, "I am not much interested in the history of philosophy."[49]

Carl Hempel went to a meeting in Prague in 1929 to meet Carnap, whose most recent books had struck him as "singularly powerful and illuminating."[50] There was something about the style that was unpretentious and yet bold, lucid and rigorous that Hempel found greatly appealing. He asked Carnap if he could come to Vienna to study with him, and Carnap agreed immediately to the request. Carnap eventually invited Hempel to come to the University of Chicago in 1937 as an assistant.

In 1931, Carnap moved to Prague, where he wrote *The Logical Syntax of Language*, a book that would have considerable influence. The German invasion cut short this chapter of his life, and with the help of Charles Morris and Willard Van Orman Quine, he came to the United States, where he became professor of philosophy at the University of Chicago. Later, in 1952, after Reichenbach's death, he moved to UCLA, where he would stay there till his death in 1970.

HANS REICHENBACH

Hans Reichenbach was a leading philosopher in the logical empiricist movement in Europe. Like Carnap, he was born in 1891, and he too had been a student of mathematics and physics, as well as of philosophy. He studied with Ernst Cassirer, as well as with David Hilbert and Max Planck. Reichenbach went then to Berlin, where he led the philosophical movement in Berlin until Hitler's rise to power. He left Germany in 1933, and spent five years in Turkey, where Mustapha Kemal Atatürk's modernization efforts had opened up opportunities for Turkish students to study Western philosophy. Reichenbach then took up a position at the University of California at Los Angeles. His influence in the United States was already considerable; his book *Experience and Prediction*[51] was published by the University of Chicago Press in 1938, and was widely read. His writing style is quite the opposite of Rudolf Carnap's: Reichenbach's prose is direct, clear, and in its way quite brilliant; the reader senses his gift for coming to the heart of the matter, regardless of what he is talking about, in a way that makes everything really very simple.

As we have seen, Carnap kept natural language at arm's length when he developed his artificial language; he was concerned about the vagueness, ambiguity, and opacity of natural language. Reichenbach's attitude was different, perhaps in part because of his exposure to the non-Indo-European differentness of Turkish. Whatever the reason, Reichenbach seemed to enjoy looking rather more closely at the ways in which natural language and the study of logical representation could shed light on each other.

In chapter 7 of *Elements of Symbolic Logic*[52] and after 250 pages of logical analysis, Reichenbach took the reader on a short tour of what he found in natural language—in English, in this case. He noted that there are three major word categories in grammar: nouns, adjectives, and verbs, and while he saw that nouns form a separate category, adjectives could be assimilated to the category that includes the present participials of verbs. Just as important is the number of variables or arguments, and he pointed out (as Frege had, some years earlier) that verbs, adjectives, and nouns can all have up to three arguments. One-place functions are nouns and adjectives or verbs (*house, red, sleep*); two-place predicates can be found (*see, father, taller*, and other examples); three-place functions are primarily verbs (*give*) but also nouns (*gift*) and possibly the preposition *between*.

> Language has developed ways [Reichenbach wrote, perhaps thinking of German] to mark the argument positions of elements, sometimes with linear order, but "he loves her" can be transformed into "her he loves" or even "her loves he"; although this order of words is unusual the meaning remains clearly the same. Similarly the sentence "he gave the book to John" can be transformed into "to John he gave the book" or "to John the book he gave."[53]

Reichenbach castigated traditional grammar for its obtuseness in the face of this interesting structure and, echoing Frege and Russell, he found the emphasis lain on the subject-predicate dichotomy in traditional grammars unmotivated and artificial. In "Peter is taller than Paul," he saw no difference between the logical status of the positions of the subject *Peter* and the object of *than*, which is *Paul*.[54]

Indeed, he made it clear that he recognized that there *is* a structure of natural language, and that this logic was superior to the account of language which was offered both by traditional grammar and by logic as it was then understood. He was quite explicit on this point, and he went so far as to actually use the word "superior." Traditional grammar stood on the shoulders of traditional logic, and even intelligent students in grammar school or college were smart enough to know that traditional grammar was, as he put it, filled with artificial classifications, gratuitous constructions, and obvious misunderstandings of the structure of language.

But he did not stop at chiding the grammar that was taught to students in school. He took on the professional grammarians as well: "It seems to us that the deficiencies of traditional grammar are equally visible in the science of language in its present condition. The high level of historical and psychological analysis in philology is not matched by a similar level in the understanding of the logical side of language."

Reichenbach's "logistic grammar" had three components: an account of argument structure, of functions, and of logical terms. Arguments included deictics and tenses, as well as traditional nominal arguments.

> The instrument of language as it has been developed in the course of human civilization is superior to the theory of the instrument constructed by logicians. Traditional grammar reflects the primitive stage in which logic remained up to the beginning of logistic. We should not be astonished when the instruction in syntax, in grammar schools and colleges, meets with antagonism, in particular on the part of intelligent students. Our present grammar as it is taught, with its artificial classification and gratuitous constructions, is based on obvious misunderstandings of the structure of language. We should like to hope that the results of symbolic logic will some day, in the form of a modernized grammar, find their way into elementary schools.[55]

This is one of those very clear shots across the bow, where a researcher in one of the mind fields calls out to his colleagues in another: Hey! Wake up!

> If philologists would try to make use of a modernized grammar for linguistic purposes they might discover new means of elucidating the nature of language.

And just in case the linguists were not imaginative enough to see what Reichenbach was hinting at, he added,

> In the hope that our appeal will be heard outside the camp of the logicians and will be taken up by the few linguists who are aware that a science of language cannot be constructed without a scientific logic, we present, in the following sections, the present status of logistic analysis of conversational language and indicate the outlines of a logistic grammar.[56]

This call would be heard by others—perhaps first and most clearly by Yehoshua Bar-Hillel, who we will encounter shortly, and would be echoed even more clearer in Rosenbloom's book in 1950, which we will mention briefly in the next chapter and discuss at greater length in volume 2.[57]

It is unusual to find such a clear and explicit cry from someone working in one field directed at workers in another. Despite the relatively porous barriers that separate philosophy and linguistics, it is rare for the philosopher to invite the linguist with open arms to come join him and do some better work in their study of language.

Carnap and Gestalt psychology

Carnap first met Reichenbach at a conference whose organization they had both participated in. It took place in 1923 at Erlangen, and Reichenbach introduced Carnap to Köhler at the same meeting. Reichenbach was working on getting a journal started that would focus on the ideas that he and his colleagues were engaged in; this journal would eventually be given the name *Erkenntnis*. Reichenbach was eager to have Köhler participate in the publication project. Köhler was a participant in Reichenbach's seminar "Gesellschaft für wissenschaftliche Philosophie," which began in 1927.[58]

Feest (2007) proposed two ways in which Carnap's work was able to profit from interaction with Gestalt psychology: on the one hand, Carnap could try to show that Gestalt psychology could be fit into the total picture of a unified science, and on the other, it could offer a way to think about human knowledge acquisition, as it begins from experience with the world.

The Gestalt psychologists emphasized that the elements of an experience may be the result of a process, or effort, of abstraction rather than necessarily being pieces out of which the whole is formed.

Feest wrote,

> It seems clear why Gestalt psychology might have been attractive for Carnap's project: This psychological school was offering a scientific conception of subjective experience, according to which phenomenal experience is structured in accordance with laws, which describe the functional relationship between types of stimulus configurations and types of Gestalt experiences. These laws can be determined experimentally and hold across different individuals. Alan Richardson has suggested that "Carnap . . . looks to . . . Gestalt psychology . . . for an account of the structure of human experience in human agents and uses this as the basis from which to start his constitutional system" (Richardson 1998, 9). . . . Carnap's account of how experience is structured was in fact contrary to that of Gestalt psychology. This incompatibility of naturalistic description and rational reconstruction speaks against an interpretation of the *Aufbau* as attempting to follow psychology in providing an account of the real cognitive processes of knowledge acquisition.[59]

Feest also remarks,

> The Berlin/Frankfurt school, itself, has to be historically situated before the background of this older epistemological context. This does not mean that Carnap's reference to Köhler and Wertheimer is not significant in its own right. For example, a remark Carnap makes in Paragraph 71 make it quite clear that his notion of "basic experience" is that of the Berlin/Frankfurt school (more specifically: their rejection of Ehrenfels's version of Gestalt quality). There, he writes that even though we think that we hear the c in the c-e-g chord, this apparent sensation of the c is a quasi-element, not a real element. Otherwise, one would come to the conclusion (which has indeed sometimes been maintained) that the chord c-e-g consists of the individual tones c, e, g, and, in addition to them, of something which comprises the actual character of the chord. (Carnap 1967 [1928]).[60]

Feest observed that Carnap wrote an unpublished manuscript entitled *From Chaos to Reality* in 1922 in which he pointed out that many epistemologists seem to pose the question as to how we arrive at an orderly world, given that we begin with an experiential chaos. But Carnap raised the question as to whether that assumption—that there is an original chaos—is justified.[61]

> It is important to note, though, that those philosophers who proposed a scientific analysis of the mind were by no means all in favor of experimental methods in psychology. Two philosophers who exemplified the different approaches (empirical and non-empirical) to the analysis of consciousness were Carl Stumpf and Edmund Husserl. . . . For example, the philosopher Hans Cornelius, to whose 1903 work, *Einleitung in die Philosophie*, Carnap refers in §67 of the *Aufbau*, had previously published a book, entitled, *Psychologie als Erfahrungswissenschaft*, which had the explicit aim of laying the epistemological foundation for "a purely empirical theory of mental facts, free of any metaphysical presuppositions." . . . But in this book he draws on assumptions about experience that do not appear to be the results of experimental analyses. In other words, the call for an empirical psychology that was guided by an anti-metaphysical sentiment did not necessarily imply the use of experimental methods. . . . I believe that if we want to do full justice to what Carnap was doing with his phenomenal basis, we need to take seriously his references to the older philosophical literature, i.e., a literature that viewed the (largely introspective, non-experimental) analysis of consciousness *as a weapon against metaphysics*.[62]

A time for manifestos

The Roaring Twenties was the era of the manifesto: everywhere that politics and intellectuals could be found, a manifesto was being written—recall our discussion of manifestos in chapter 1.[63] A manifesto was the intellectual's way of telling himself (and anyone else within earshot who might care) who he was by virtue of being a member of the group in question. It was the explicit construction of a group's story of who it is, where it came from, and why it existed. A list of the important manifestos of the period would include the one that the Italian Fascists published in 1919 and the one that the Dadaists wrote in 1920. The Realistic Manifesto of the Constructivists came out that year too. The Pan-African Conference published their manifesto in 1921; André Breton published the first Surrealist Manifesto in 1924.

In 1929, it was time for the Vienna positivists to draft their own manifesto. Neurath had already referred to a "Vienna circle" scientific conception of the world in a book review in 1928, just the year before, and the manifesto of 1929 was the place where this conception would be spelled out. Carnap had sent him a note, writing, "So you wish to gain the historical merit to name and proclaim the 'Vienna School' for the first time.

You are right, incidentally, that a slogan, a summary name, is important for the reception even if it does not say anything on its own accord."[64]

The Vienna Circle's manifesto was called "The Scientific Conception of the World"—*Wissenschaftliche Weltauffassung*, in German. It was written by Otto Neurath, with post-editing by Carnap and Hahn, and it was dedicated to Schlick. It hinted at a good deal of politics at the socialist end of the spectrum, and this aspect alienated some of the Circle, including Schlick himself. Some, like Neurath and later on Carnap, referred to a left-leaning wing of the Circle, and it appears that this label described both a philosophical and a political tendency.[65] The group of Carnap, Hahn, and Neurath "was sometimes called the left wing of the Vienna Circle, in contrast to the more conservative right wing, chiefly represented by Schlick and Waismann."[66]

It takes some effort—though it is effort that is well rewarded—to see the platform of logical positivism as a call to arms. Uebel notes a recollection of Reichenbach concerning a moment when a student said his work was "exclusively concerned with logic and epistemology, not with social reform." To which Reichenbach replied, "But no! That is not true. The whole movement of scientific philosophy is a crusade. Is it not clear that only by ending the dogmatism of irresponsible claims to know moral truth, that only by clarity and integrity in epistemology, people can attain tolerance and get along with one another?"[67] Carnap wrote to Reichenbach in 1930: "All of us here are of the opinion that philosophy at present is at a decisive turning point, that it is not a matter to continue the philosophy done up to now in a somewhat improved, more careful fashion."[68] Something breathtaking was happening. A. J. Ayer, a British philosopher who visited the Circle in the early 1930s wrote that, "The members of the Vienna Circle, with the notable exception of Otto Neurath, were not greatly interested in politics, but theirs was also a political movement. The war of ideas which they were waging against the Catholic church had its part in the perennial Viennese conflict between the socialists and the clerical reaction."[69] The beginning of the manifesto laid out the skein of influences that the members of the Circle chose to offer as important influences in the development of their thought. They pointed first to anti-metaphysical thinkers of various sorts, including William James, and the Berlin colleagues, such as Reichenbach. In the context of Vienna culture, they mention thinkers like Theodor Gomperz, who had translated John Stuart Mill, and Ernst Mach, who was "especially intent on cleansing empirical science, and in the first place, physics, of metaphysical notions. We recall his critique of absolute space which made him a forerunner of Einstein,"[70] as well as Ludwig Boltzmann, who was named to Mach's chair.

They also cite Franz Brentano: "As a Catholic priest Brentano understood scholasticism: he started directly from the scholastic logic and from Leibniz's endeavours to reform logic, while leaving aside Kant and the idealist system-builders. Brentano and his students time and again showed their understanding of people like Bernard Bolzano, who were working toward a rigorous new foundation of logic in the nineteenth century. In particular, Alois Hofler put this side of Brentano's philosophy in the foreground before a forum in which, through Mach's and Boltzmann's influence, the adherents of the scientific world conception were strongly represented." We should note parenthetically that Alois Hofler had been a student of Ludwig Boltzmann and of Alexius Meinong, who himself was one of Brentano's closest students. The manifesto continues, citing Alexius Meinong, whose theory of objects showed similarities to the Vienna Circle's views.

At the moral heart of this movement was the strong belief that the natural consequence of clearing away centuries of conceptual debris would be enormous strides *forward* in many ways. This image shares something in common with Comte's view, which formed the basis for nineteenth-century positivism: the world needed the help of conceptual midwives to get it out of its metaphysical period and attain its positivist period, and a sizable part of that rebirth would consist of sweeping away the need for a belief in metaphysical objects. At one and the same time, the manifesto could make the program first sound simple and austere, and then just a few paragraphs later, part of a major shift in world view. First it said, "No special 'philosophic' assertions are established, assertions are merely clarified . . . there is no such thing as philosophy as a basic or universal science."[71] Philosophy is not the basis of anything: *experience* is. This is transformed, then: "The representatives of the scientific world-conception resolutely stand on the ground of simple human experience. They confidently approach the task of removing the metaphysical and theological debris of millennia." This is a clear allusion to Comte, of course. "Or, as some have it: returning, after a metaphysical interlude, to a unified picture of this world which had, in a sense, been at the basis of magical beliefs, free from theology, in the earliest times."[72]

Arne Næss, a Norwegian philosopher, participated in the Vienna Circle for about a year shortly before its end—that is, shortly before the murder of Moritz Schlick in 1936. Sixty years later, he took the opportunity to think back to the style of Schlick's circle in 1934 and 1935. The first thing he noted was the feeling that the group of people in the seminar were engaged in a joint project, with what he called an atmosphere of eager cooperation. There was a shared sentiment that something was afoot, something new was being created, and everyone could contribute to it. Consensus was sought, and no one's opinion was forced. He likened the mode of interaction to a "Gandhian nonviolent approach," which meant looking for a *charitable* interpretation of one's opponents' statements:

> Looking back, I feel sorry that the combined analytical and social initiative of the logical empiricists petered out. It constituted in the 1930s a cultural force and a threat against fascist and authoritarian regimes in general. The authoritarian Austrian government did not underestimate this threat, and the newspapers—which on the whole were on the authoritarian side—expressed relief when Professor Moritz Schlick was killed on the doorstep of the university. Logical empiricism was proclaimed to be a blot on Austrian culture. I was asked to protest against this in Scandinavian newspapers, but I regret today that at that time I did not feel able to engage in public debate. When Quine and others took over the analytical leadership, the movement was largely robbed of its social and political aspects.[73]

He added:

> I entered the shabby seminar room of the Schlick seminar in 1934 with a peculiar philosophical background. When I was seventeen I had become deeply fascinated by both Spinoza's *Ethics* and Whitehead and Russell's *Principia Mathematica*. It was impossible for me not to talk and talk about those texts. The first chapters of the *Principia* were enough for me to become a friend of symbolic logic and a permanent user of simple propositional and functional calculus—even using it to make Spinoza's conceptual structures clear to me. The *Ethics* has remained for the rest of my life the supreme paradigm of philosophy. There has been no *Wende der Philosophie* since the *Ethics*!

Wittgenstein's *Tractatus* was read by members of the Circle with great interest—indeed, with reverence—and discussed in groups beginning in 1922. Wittgenstein himself appears not to have been interested at first in participating in any way, shape, or form. He finally agreed to meet with Schlick in 1927, but came away from the meeting with the impression that both of them must have thought the other crazy. Nonetheless, he began to meet with the Circle.[74]

Central tenets

What were the central tenets that the logical positivists held?

The first was that language can be used deceptively, in a way that makes it seem that we are saying something when in fact, we are saying nothing at all. All philosophical questions, therefore, demand that we look carefully at the language in which they are posed. Not simply in order to better understand what is being said, but in order to rule out of order a good deal of talk that the positivists thought was meaningless and that we would be simply better off without. In its heyday, logical positivism drew on its authority (such as it had) to eliminate certain kinds of speech. It did this by labeling it *metaphysical*, and it did its best to eliminate all such language. (At one point, Herbert Feigl aptly referred to the positivists' "persuasive prohibitionism."[75]) A sentence can work if it is meaningful, and to know the meaning of a sentence is to know the claims that it makes about what can be seen, measured, or observed.

Its commitment to looking at language had two edges to it. On the one hand, there was the hope that a careful analysis of language would eliminate bad philosophy, and eliminating bad philosophy was likely to make the world a better place: perhaps people could be dissuaded from giving their lives in the service of a fraudulent metaphysical political system if they could be persuaded to better analyze the logical form of the propaganda that they were subjected to. But in addition, a careful analysis of language might help us to do better science.

Richard von Mises wrote that language "is a tool whose application shows many shortcomings," and therefore "it seems imperative to us to test the tool critically over and over again in so far as it is used."[76] Philosophers in the past had not done this, and we all pay the price for that.

Their second tenet was that an enormous part of all of the philosophy of the past falls into the category of language that seems to mean something but in fact means nothing. It must be recognized as poetry: pretty, but meaningless, or else it must simply be discarded. Old philosophy is dead: long live the new philosophy!

Another was that there are only two kinds of true statements: those that are true by virtue of how the observable world happens to be, and those that are true by virtue of the meaning of its words and how the words are put together, which is to say, its grammatical construction.

A fourth tenet was that just as Russell and Whitehead's *Principia Mathematica* created a new foundation for mathematics, based on logic, so too must philosophers, working hand in hand with scientists, lay foundational platforms for all sciences.

> The alleged peculiarly philosophical point of view, from which the objects of science are supposed to be considered, is abolished, just as the alleged peculiarly philosophical stratum of objects was already previously eliminated. Aside from the questions of the individual special sciences, the only questions that remain as genuinely scientific questions are those of the logical analysis of science—its sentences, concepts, theories, etc. We will call this complex of questions *Wissenschaftslogik*. . . . Taking the place of the inextricable tangle of problems that is known as philosophy is *Wissenschaftslogik*. Whether, on the basis of this conception, the designation "philosophy" or "scientific philosophy" should be applied to this remainder, is a question of expedience, which is not to be decided here.[77]

And finally, the logical positivists held that there had been unfortunate confusion in the past regarding how scientific theories are justified. Just to set the record straight: we have to distinguish clearly between what leads a scientist to a new idea and what constitutes a substantive argument for a scientific idea. The former is a creative act, and while we want to encourage scientists to be creative, that is not something we can legislate. The second stage—justifying the scientific idea—is an activity of an entirely different order. Philosophy can play a major role in gaining clarity with regard to what it means for a scientific idea to be supported.

Of these, there are four positions that had a direct impact on modern linguistics, and we will focus on them in the next section (p. 412).

As we look back at the proposals of the logical positivists, we find that the distinctions that they emphasized have remained with us, and that the work that they left to us is important—indeed, that it stimulated, both directly and indirectly, many intellectual movements that have been fruitful. But when we enunciate the specific planks of their platform, in every case we find that we have to shake our heads ruefully and say that they went too far, and they placed their hopes in a view of the world that turned out to be much too simple.

To be sure, much the same can be said about the best of the movements that we explore over the course of this book.

Virtually everyone who was involved in logical positivism had advanced training in science or mathematics—and by no means did they all have a comparable training in philosophy. This was not a gap in their education that any of them worried about. They all knew that the sciences, and most notably physics, were undergoing revolutionary changes, and Albert Einstein's contributions were among the most revolutionary. The logical positivists took Einstein's theory of special relativity as grist for their mill. They saw the big message behind Einstein's early work on special relativity as proving the ultimate value of metaphysical skepticism. Einstein's theory was based on asking the simple question: what if these fixities of everyday life and of Kant's *a priori*, time, and space—what if they are not what we have assumed, and what if we have to look much more closely at our measuring rod and our stopwatch? In fact, there was no "what if?": Einstein's account established beyond any reasonable doubt that Newtonian physics had erred by making the terribly natural assumption that space and time existed independently of each other. It was a natural assumption, but it took a genius to see how to uncover the hidden assumptions and to replace them with alternatives that jibed with the observed facts.

The logical positivists took that as an example that all scientists needed to bear constantly in mind: make a clear distinction between the observations that you make of nature (for nature does not lie, if we listen to what it has to say) and the way you embed your account of your observations into a theoretical scaffold. It is not always easy, but who ever said science was supposed to be easy?

Philosophy is dead; long live philosophy!

Moritz Schlick, the founder and guiding light of the Vienna School, displayed a marvelous sense of irony in his contribution to the first issue (1930) of *Erkenntis*, the flagship journal of the movement. Even the title says it all: "The Turning Point in Philosophy" (*Die Wende der Philosophie*).

He began by making an observation slightly different from the one we have just been considering. He pointed out that in philosophical discussions, there is a tendency to see all progress as having occurred during the march of time up to the moment that some particular philosopher—Plato, Descartes, Kant—appeared on the scene and started to philosophize, and that since that time, progress has been spotty and dubious. And so philosophers end up talking more about historical influence than about substance and results.

Conclusion: philosophy, up until now at least, has not produced *results*. "It is just the ablest thinkers who most rarely have believed that the results of earlier philosophizing, including that of the classical models, remain unshakable. This is shown by the fact that basically every new system starts again from the beginning, that every thinker seeks his own foundation and does not wish to stand on the shoulders of his predecessors."[78] As we might put it, they would rather have a pact with God of the sort Noah worked out. Already we sense the irony, because we know that the gap between what a person *wants* to do and what he *does* is the starting point of irony, and often of tragedy, too. Descartes, Spinoza, Kant—they all thought they were starting over afresh, Schlick reminds us, and "further examples are superfluous, for practically all great thinkers have sought for a radical reform of philosophy and considered it essential. This *peculiar fate of philosophy* has been so often described and bemoaned that it is indeed pointless to discuss it at all. Silent skepticism and resignation seem to be the only appropriate attitudes. Two thousand years of experience seem to teach that efforts to put an end to the chaos of systems and to change the fate of philosophy can no longer be taken seriously."

Schlick knew full well that many had swept the Augean stables of philosophy clean before him and still the stables filled up with dung, and he wanted us to know that he was aware of this. But still he was ready to rush in where he *just reminded* us that angels fear to go:

> I am convinced that we **now** find ourselves at an altogether decisive turning point in philosophy, and that we are objectively justified in considering that an end has come to the fruitless conflict of systems. We are already at the present time, in my opinion, in possession of methods which make every such conflict in principle unnecessary. What is now required is their resolute application.

So *this time* we have really got it right, Schlick said. And it was because we finally had insight into the nature of logic itself.

The tools and methods needed to accomplish this had come from Bertrand Russell, Gottlob Frege, and Ludwig Wittgenstein. They developed new logics, but what was more important than the particular details of the new logics that they created, they developed new insights into what logic is. For Schlick, the consequence was that any question about *knowledge* could be characterized as being a question for an empirical domain such as psychology, or as one regarding the nature of meanings (something that is essentially a linguistic investigation in the broadest sense of "language"), or else the question is simply shown the door: it is booted out. "What we have up till now taken to be questions are not real questions, but meaningless strings of words which look like questions but which in fact are composed of empty sounds, because they violate the deep, inner rules of logical syntax, which this new analysis has discovered."[79] What remains for philosophy, as the guardian (Schlick uses the metaphor of "queen") of true knowledge, is to *explain*, or better, to *reveal*. Philosophy reveals the meanings of statements, and then the various sciences determine if the statements are true or not.

Schlick suggested that history was on his side. In times long ago, philosophy consisted of any sufficiently abstract question; but one by one, specific domains split off from the mother domain of philosophy—physics, for example. For Schlick, this could happen when the concepts had been clear enough that real work could begin. Schlick pushed this idea as hard as he could, to the point where he suggested that if any domain still views itself as part of philosophy (ethics or aesthetics, for example, or who knows, maybe even psychology), then that is a sign that the fundamental concepts are not yet clear enough. The most important task for the philosopher is to make the concepts that a particular science needs clear enough that it can get on with its real work.

The ultimate task for the philosopher, then, is to uncover the logical structure of each science.

The logical positivists attempted to understand science as a fabric of propositions, each of which was either a basic observation—what Carnap would call a protocol sentence—or a proposition which could be derived from other propositions within the larger fabric. Carnap proposed a simple solution for what has come to be, over the course of ensuing decades, a very difficult problem for understanding the nature of science: the statements of a science are expressed in language (a combination of natural language and mathematical language), and we can study the way statements of varying degrees of generality do or do not support each other—for example, we can determine whether the reported results of an experiment support an author's claim. But when we try to drill down to the least theoretical and most observational sentences, at some point we arrive at sentences that somehow simply report what someone observed. That, in any event, is the intuition behind Carnap's notion of protocol sentence, which is one that reports what someone observed.

From this perspective, understanding the nature of science can be divided into two parts, both of which involve language in essential ways. The first part crucially involves protocol sentences and the way in which language comes to grips with reality: where the rubber meets the road, so to speak. How does a sentence mean something, and is knowing how it means something enough to allow us to understand what it really says about the world? The second part involves the relationships between the sentences (or propositions that the sentences express) and is concerned with the relationship between the bigger ideas of a science and the more pedestrian observations.

There was a problem right from the start—or a disagreement that would not go away—which had to do with the nature and definition of these protocol sentences. The very term *protocol* suggests what gets jotted down in a laboratory notebook: the facts, nothing but the facts, but still, facts expressed in some language or other. There were three major candidates for what these protocol sentences might look like, and we can get a pretty clear idea of the difference of the three by these descriptions: there was, first, the "red here now" school of thought, which is certainly Bertrand Russell's, and largely Carnap's as well; next, there was Otto Neurath's view; for example: "Otto's protocol at 3:17 o'clock: [At 3:16 o'clock Otto said to himself: (at 3:15 o'clock there was a table in the room perceived by Otto)]."[80] And then finally, there was the laboratory-notebook view: "In the laboratory in room 257B at 1100 E 58th St., on January 1, 2013 at 10 A.M., a fluorescent bulb shattered," which was essentially the view defended not by logical positivists, but by operationalists, inspired by Percy Bridgman (1927).

Each of the three views gives different answers to the question as to what the outer frame is within which the protocol sentences are formulated. The first is in some ways solipsistic: it takes the limits of the protocol language to coincide with the limits of my subjective, my *me*-centered world, while the second tries to turn a private observation into something that others can treat as public.

If something like the logical positivist project is to be made to work, which of these three understandings of what a protocol sentence is must be involved? They are so different: could it be that a choice really did *not* need to be made? Could it be that we could accept more than one of these views, along with a reasonable translation guide between pairs of protocol languages?

And whether a translation scheme was acceptable or not, which of these three views we take as fundamental has an impact on how we want to understand introspection in psychology. The first view seems to leave the door wide open to introspection: all knowledge is stated in a language that is nothing but pure subjectivity. The second would also seem to leave a lot of space for introspection: the first embedded clause there expresses Otto's introspection. It is only the third, the operationalist view, that seems to be doing a good job of leaving no room for introspection.

After sweeping out quite a bit of detritus from the stable of human knowledge, what is left? To be sure, there is logic and mathematics, but what else? The Vienna Circle generally shared the hope that something that could be called a foundation could be found—an empirical foundation for knowledge. But while two candidates for this role seemed suitable for a while, nothing really worked. The two candidates were private sense-data, and very simple observation statements, something so basic that there was nothing added to it by the observer's set of beliefs: "the grandfather clock in the living room just struck two."

Of the two, Carnap's choice, in the *Logical Structure of the World*, was that protocol statements are reports of private sense-data. In this, he was following in the footsteps of Ernst Mach and Bertrand Russell.[81] The two positions are very different: the first is private, first person, and subjective; the second is public and inter-subjective. But each had its appeal if one were to hope to formulate a level of description that could be viewed as the foundation of what we know.

Let's step back and get a broader view of what Carnap's project was. His goal was to develop a logical construction of our knowledge, and this project had two halves. Simply put, we need to understand both things (or concepts) and propositions. In a way, it was a Cartesian project: we start afresh with no assumption that we know or understand anything, and then we attempt to build up our understanding of the world from a minimal set of assumptions—and this building up of our understanding includes both the construction of objects (along with their properties and relations among them) and the construction of what we know about these things. People had worried more about the second part of this: just what do we need to know in order to be justified in concluding some statement or other? That is, after all, the domain of logic. But Carnap's project was to include the logical construction of objects, as well, a problem that was more than a little off the beaten path.

Four areas that affect linguistics

There were four areas where the logical positivists particularly influenced the future development of linguistics.

CONTEXT OF DISCOVERY, CONTEXT OF JUSTIFICATION

Reichenbach proposed a distinction that would come to be of a great deal of importance for linguistics; this was the difference between what Reichenbach called the *context of discovery* and the *context of justification*. He called upon his reader to separate out questions of epistemology—questions of the justification of knowledge—from those of psychology. "Epistemology does not regard the process of thinking in their actual occurrence; this task is entirely left to psychology."[82] Epistemology constructs, or reconstructs, a picture of reasoning that leads directly from starting point to finishing point with no unnecessary steps, and a clear justification for each intervening step. Real thought never works that way, but this human fact is of no interest to the epistemologist. The epistemologist demands of himself a rational reconstruction of the process of arriving at a conclusion (the term *rationale Nachkonstruktion* was due to Carnap).[83] For Reichenbach, this task is downright beautiful:

> It is even, in a certain sense, a better way of thinking than actual thinking. In being set before the rational reconstruction, we have the feeling that only now do we understand what we think; and we admit that the rational reconstruction expresses what we mean, properly speaking.[84]

Reichenbach then drew his famous distinction:

> The way . . . in which a mathematician publishes a new demonstration, or a physicist his logical reasoning in the foundation of a new theory, would almost correspond to our concept of rational reconstruction; and the well-known difference between the thinker's way of finding his theory and his way of presenting it before a public may illustrate the difference in question. I shall introduce the terms context of discovery and context of justification to mark this distinction. Then we have to say that epistemology is only occupied in constructing the context of justification.[85]

Karl Popper had written something similar just shortly before Reichenbach as well, in *Logik der Forschung* (1935), which Reichenbach could be viewed as developing:

> The question of how it happens that a new idea occurs to a man . . . may be of great interest to empirical psychology; but it is irrelevant to the logical analysis

> of scientific knowledge. This latter is concerned not with *questions of fact* but only with question of *justification or validity*. . . . I shall distinguish sharply between the process of conceiving a new idea, and the methods and results of examining it logically. As to the task of the logic of knowledge—in contradistinction to the psychology of knowledge—I shall proceed on the assumption that it consists solely in investigating the methods employed in those systematic tests to which every new idea must be subjected if it is to be seriously entertained.[86]

Carnap had also expressed this idea in the preface to the first edition of *The Logical Structure of the World*:

> It must be possible to give a rational foundation for each scientific thesis, but this does not mean that such a thesis must always be discovered rationally, that is, through an exercise of the understanding alone. After all, the basic orientation and the direction of interests are not the result of deliberation, but are determined by emotions, drives, dispositions, and general living conditions. . . . The decisive factor is, however, that for the *justification* of a thesis the physicist does not cite irrational factors, but gives a purely empirical-rational justification. . . . The practical handling of philosophical problems and the discovery of their solutions does not have to be purely intellectual. . . . The *justification*, however, has to take place before the forum of the understanding.[87]

We shall see that Reichenbach's distinction came to be influential in linguistics in two distinct moments of interpretations. Its first application is in its Harrisian form: Zellig Harris's goal was the application of this distinction to linguistic work. Harris saw the goal of the methodologist (such as he himself was) as the development of an epistemology specific to language, and thus a coherent notion of a rational reconstruction of linguistic grammar construction. The actual steps a linguist passed through on the way to finding a grammar were like the context of discovery, but once the grammar was achieved, it was necessary for the linguistic epistemologist to re-present the grammar, only now in the context of justification. Harris's work was to make explicit what a context of justification would be for linguistics.[88]

The second great influence of Reichenbach's distinction was its application by Chomsky and Halle in the late 1950s, when they tried to make the case that any theoretical statements about the method by which a grammar is derived from data involves the context of discovery, and not the context of justification—and hence is of interest to philosophers or historians of science, perhaps, but not to linguists as such. This was a point of view that never made sense to the post-Bloomfieldians, as we shall see; let us consider just a single, but typical, example at this point.

In his 1962 review of Halle's *Sound Pattern of Russian* in *Language*, Charles Ferguson wrote:

> When Halle says that the identification of the method by which Newton discovered the principles of gravitation belongs to the philosophy of science and not to science itself, this seems quite reasonable. Similarly it seems quite reasonable to turn over to psychologists or philosophers of science the investigation of the methods by which a given innovating linguist has arrived at a concept like the phoneme, the language family, glottochronology, or transformational grammar. But this is quite different from trying to provide techniques for linguistic analysis. Once the innovator has arrived at a new concept it is presumably his duty or the duty of his followers to communicate it to others and then to help devise techniques for putting the concept to work in actual linguistic analysis. The more detailed, the more explicit these techniques are, the better, i.e., the more success investigators will have in applying them and in testing the usefulness and validity of the new concepts. Halle himself sometimes carries out this duty very well. For example, he gives three pages of careful instructions (34-6) on how to construct a branching diagram before announcing (37) that "The phonological system of a language will be presented by means of a branching diagram." The reviewer for one is grateful for this kind of helpful explanation of a new discovery procedure and cannot understand why Halle does not want such help from others.[89]

SYNTAX

Rudolf Carnap's *The Logical Syntax of Language* was published in German in 1934, translated into English in 1937, and reviewed in 1939 in the *Journal of Symbolic Logic* by Stephen Kleene. One cannot overemphasize the impact of this book on young scholars.

To read *The Logical Syntax of Language*, even today, is to experience a feeling of encountering brilliant insight into language for the first time. It is like hearing Mozart for the first time: one has the feeling, So this is where it came from? The book begins like this:

> By the logical syntax of a language, we mean the formal theory of the linguistic forms of that language—the systematic statement of the formal rules which

> govern it together with the development of the consequences which follow from these rules.[90]

In case the reader is unsure, Carnap then explains what he means by formal:

> A theory, a rule, a definition, or the like is to be called formal when no reference is made in it either to the meaning of the symbols (for example, the words) or to the sense of the expressions (e.g. the sentences), but simply and solely to the kinds and order of the symbols from which the expressions are constructed.

Carnap had certainly not read Bloomfield—*Language* had not come out while Carnap's book was being written, after all—but here were two minds going in the same direction, as Bloomfield was quite aware. Carnap lays out the major point of his book: that the principles that explicate valid inference—which is the domain traditionally called logic—should be analyzed in a purely formal way, every bit as much as the principles of grammar or syntax of a language should. "In this way," he notes, "logic will become a part of syntax, provided that the latter is conceived in a sufficiently wide sense and formulated with exactitude." And it is perfectly clear that Carnap thinks it should be. The next sentence makes us sit up with a start: "The difference between syntactical rules in the narrower sense and the logical rules of deduction is only the difference between formation rules and transformation rules, both of which are completely formulable in syntactical terms. Thus we are justified in designating as 'logical syntax' the system which comprises the rules of formation and transformation." The original German, of course, did not speak of transformation rules; the original used the word *Umformungsregeln*, while the formation rules were *Formregeln*.[91] But Carnap himself was not interested in applying his notions to what we today call natural language: "In consequence of the unsystematic and logically imperfect structure of the natural word-languages (such as German or Latin), the statement of their formal rules of formation and transformation would be so complicated that it would hardly be feasible in practice." We have already seen that Hans Reichenbach, whose perspectives were in many ways parallel to those of Carnap, had no such compunctions and was quite interested in what insight these methods inspired by logic could provide regarding natural language. And Zellig Harris would be disappointed by Carnap's discouragement, but that would come later.

Carnap develops a pair of related notions: *general syntax*, which deals with the syntax of all languages, and *special syntax*—special, or specific, to a given language. Together, these form the logical syntax of the language. Given how similar this is to Husserl's proposal in the *Investigations*, it is remarkable that Carnap does not indicate the relationship of the ideas. As Carnap put it,

> By the "logical syntax" (or also briefly "**syntax**") of a language we shall understand the system of the **formal (i.e. not referring to meaning) rules of that language**, as well as to the consequences of these rules. Therein we deal first with the **formative rules** (*Formregeln*) which decree how from the symbols (e.g. words) of the language propositions can be built up, secondly with the **transformation rules** (*Umformungsregeln*), which decree how from given propositions new ones can be derived. If the rules are set up strictly formally they furnish mechanical operations with the symbols of the language. The formation and transformation of propositions resembles chess: like chess figures words are here combined and manipulated according to definite rules. But thereby we do not say that language is nothing but a game of figures; it is not denied that the words and propositions have a meaning; one merely averts methodically from meaning. One may express it also thus: **language is treated as a calculus**.[92]

There is much to unpack in these words. We will perhaps never know if Carnap realized that his reference to chess was an allusion to Saussure, who famously described the similarities between the rules of chess and the rules of language. The final note is crucial, though, for understanding the development of linguistics, because Carnap was describing a perspective that Zellig Harris, and the early Chomsky, thoroughly subscribed to: the importance of meaning is not denied, but it is bracketed, put on hold, and tentatively ignored, so that formal analysis can proceed.

Carnap then calls outside of his discipline, to linguists:

> That the formal, calculus-like representation of the formative rules is possible is evident. What linguists call rules of syntax are indeed such formal (or at least formally expressible) rules for the formation of propositions. We can see, however, clearly that **the transformation rules, which one usually calls logical rules of deduction, have the same formal, that is, syntactical character.** (And that is the reason why we call the combined system of rules syntax, in widening the terminology of linguists).[93]

Within a few years, Zellig Harris would heed this call. Carnap continued:

> Since Aristotle the efforts of logicians (more or less consciously) were directed toward formulating the deductive rules as formally as possible, i.e. possibly so that with their help the conclusion could be "calculated" mechanically from the premisses. This was attained first in a strict manner only in modern symbolic logic; the traditional logic was too much hindered by the defect of the language of words.
>
> For a certain part of the language of science we already know a strictly formal theory, namely Hilbert's mathematics. It considers the symbols and formulas of mathematics without reference to meaning, in order to investigate relations of deducibility, sufficiency, consistency, etc. This mathematics is hence (in our manner of expression) the logical syntax of mathematical language. The logical syntax of the language of science meant here is an analogous extension with reference to the language of all of science.

NONSENSE AND COUNTER-SENSE

We discussed earlier the distinction made by Husserl between nonsense and counter-sense. As Ajdukiewicz reformulated it, nonsense consists of words that cannot be connected up with the rules of the language. In Carnap's view, this would become reformulated as a matter of *syntax*—at least as far as he wanted to use the term syntax.

Rules of syntax are quite independent of meaning, Carnap noted; even if we do not know the meanings of the words in the sentence "Pirots karulize elatically," we understand the first word as a plural subject, the second word as a verb, and the third as an adverb. The suffixes in this example indicate the lexical category, though Carnap notes with regret that not all words in natural languages display their category clearly; still, with this knowledge, the speaker of a language can draw various inferences even when she does not know the stems *pirot* and *karulize* (or is the stem *karul-*?).

Carnap invites us to think about formal systems which consist of strings of symbols, where only some strings "belong to a certain category of expression"—belong to the language, in this case. Think of the strings of symbols as purely formal, having no meaning, but having a formal syntax. There are many ways to study language, and this formal perspective is just one of them:

> In the widest sense, the science of language investigates languages from every one of these standpoints: from the syntactical (in our sense, the formal),

> from the semasiological [semantic], from the psychological, and from the sociological.[94]

But Carnap is interested only in the first, the formal or syntactical, and by this he means no more and no less than this: the structure of possible serial orders of symbols. In fact, he is interested in a pure version of such studies, rather than a descriptive version. By pure, he means one that only defines whether any given string is possible, regardless of whether it has ever appeared in the real world not—whether it has or not is the domain of descriptive syntax, and that is not his interest. He is interested only in pure syntax, which is nothing more than combinatorial analysis, in his words. "Descriptive syntax is related to pure syntax as physical geometry to pure mathematical geometry."[95] Later he made a similar point: "the synthetic physical sentences of descriptive syntax, which are concerned with the forms of the linguistic expressions as physical structures (analogous to the synthetic empirical sentences of physical geometry . . .). *Thus syntax is exactly formulable in the same way as geometry is.*"[96]

While Carnap had his eyes set primarily on the syntax of logically created languages, he does acknowledge that the "method of syntax which will be developed in the following pages will . . . also help in the logical analysis" of natural languages, because the syntactic concepts and rules will surely be applicable "to the analysis of the incredibly complicated word-languages" (that is, natural languages): "the syntactical property of [natural languages] is best represented and investigated by comparison with a constructed language which serves as a system of reference. Such a task, however, lies beyond the scope of this book."[97] But others will take up this project in the years to come.

THE LOGICAL STRUCTURE OF EVERYONE'S THEORY

Carnap outlined a program for philosophy in which the systematic study of language would play a major role. He forecast a program in which formal means can be established to specify well-formed sentences (by means of what he called formation rules) and in which relations of valid inference can be specified (by means of what he called transformation rules). The syntax of a language is the sum, then, of the rules of formation and of transformation.

The Carnapian project was one of outreach to the sciences of the world, or even more generally to the systems of knowledge that exist. Each such science, or system, had to subject its own development to a rigorous deconstruction and consequent reconstruction, with the democratic promise that however painful this might be for some disciplines, the standards that were set were the same for everyone. The deconstruction involved drilling into each concept as deeply as possible, with the intent of reconstructing it with help from two sources: an austere set of empirical terms which enter into clear and explicit relations with observations, and a rich language of logic that would show the exact ways in which observations are put together by a science to arrive at its scientific statements. This "rich language of logic" would be at his heart a common language shared by all sciences, and it was in this promise—that the same logic can be used for all sciences—that the democratic and egalitarian aspect of logical positivism could be seen.

In practice, this meant a formalization of the most important terms in a science, more often than not with the goal of understanding exactly which concepts depend on which other concepts. If we were to try to carry out such a program for linguistics, we would have to face *right from the start* the challenge of providing an explicit account for including meaning, if meaning is to be allowed to play a role at all.

This project was self-consciously the descendant of the efforts of mathematicians in the nineteenth century that we have discussed in which every effort was made to see whether the parallel postulate needed to be included among the axioms in order to develop the familiar Euclidean geometry. It was not enough to know that something was true; it was also necessary to understand what that true statement rested upon, both logically and empirically. Sticking with the example of meaning and linguistics, the linguist who wants to engage in the Carnapian project will want to know right from the start whether our reconstruction of the scientific terms of linguistics will rest upon any terms that in turn involve meaning. Do we want to be able to talk about "words" when we write a grammar of Swahili? Then we had better have a definition of that notion before we start writing. Will that definition involve meaning? The underlying austerity of the project makes it clear that the answer to *that* question is: we will include meaning in our definition of the term "word" only if including it allows us a satisfactory definition and there is no way to get there *without* including meaning.[98]

Linguists' response to logical positivism

LEONARD BLOOMFIELD

Logical positivism, as we have sketched it, corresponded remarkably well to the descriptive linguist's understanding of what he did at work. Protocol sentences? These take the form: "I heard [informant's name] say: [phonetic representation]." They were what filled his notebooks, and they formed the basis for the generalizations that he built up from the basic data.

Bloomfield was utterly delighted with the results of the Vienna Circle—we have already seen that he devoted an article in 1936 to their work, and to how it related to his own and that of other American descriptivists. "Within the last years a group of philosophers and logicians, known as the Vienna Circle, has arrived at the same conclusion concerning language" as Bloomfield had: to wit, that the terminology of mentalism and animism would have to be eliminated from the science of language, including such terms as consciousness, mind, perception, and ideas. But alas, Bloomfield appears to have misunderstood what Carnap and Neurath were trying to do. Bloomfield is perfectly clear in his desire to eliminate all talk of mind and ideas—what he calls "the terminology of mentalism and animism"—from linguistics, as from other sciences, including psychology, of course.[99]

As we have seen, what was important for Carnap was the possibility of reducing higher-level descriptions into a lower level language which could serve as a lowest level of epistemological support, and there were two worthy candidates to play this role, and only one of them was what Bloomfield wanted: that is, a language that talked about basic objects in the world, tagged with location and time. The *other* language which could serve as the basis for what statements about the world mean was the highly subjective report of my perceptions: we have followed this tradition (which can hardly be said to have a beginning) through Ernst Mach and Bertrand Russell.

JOSEPH GREENBERG

Joseph Greenberg was born in Brooklyn, New York, in 1915, of Central European Jewish parents. He went to Columbia University for his undergraduate degree and audited a course by Boas in his senior year. Alexander Lesser, his undergraduate advisor, suggested he go to Northwestern to study with Melville Herskovits, which he did.[100] Then at age 25, he went to Yale University on a post-doc which lasted just a year.

Greenberg remembered years later that this was the first time that he had encountered a community that was seriously engaged in developing what they viewed as a scientific account of linguistics. That meant getting rid of mentalistic assumptions and following the same path as the behaviorists in psychology.

Greenberg recalled that logical positivism was the dominant view at Yale when he arrived, and that Whitehead and Russell's *Principia Mathematica* was perceived as an important influence; Greenberg himself spent several years studying it, on and off. And "Bloomfield himself," Greenberg wrote, "recognized the kinship between structural linguistics and logical positivism by contributing a monograph on the linguistics aspects of science to the series *International Encyclopedia of Unified Science*, an endeavor planned and carried out by the leading exponents of logical positivism of the period."[101]

Bloomfield had Carnap on his shelf. "I remember that I was talking to Bloomfield in his office," Greenberg wrote many years later,

> when he walked over to his bookshelves and handed me a copy of Carnap's *Logische Syntax der Sprache* (at that time it was only available in the German original). He suggested that I read it. I can still remember him saying, "A fellow might get a lot out of reading this but, on the other hand, he might spend a lot of time and effort without it being of any real value to him as a linguist." He evidently felt that at his age it would be inadvisable to attempt it. I did read it and, of course, found it hard going because of an insufficient background in formal logic. Still, what it said about language was to me at that time both intriguing and stimulating.

In Chicago, when Greenberg was a graduate student at Northwestern University, he and a group of his fellow students went to hear Carnap give a public lecture—Carnap was a professor at the University of Chicago at that point. "For us it was like a pilgrimage."

> The logical positivists and the logicians of Russell and Whitehead's school were saying that natural languages were at once too complex and too irregular to be studied by exact methods. By this, as logicians, they meant not so much language itself as a subject of scientific study but the use of "ordinary language" in deductive reasoning. Hence it was necessary to devise artificial languages like that of the *Principia*, and moreover such "languages" must be completely formalized. By this they meant that it would be a calculus in which deduction

> proceeded by well-defined and purely mechanical rules which made no appeal to our treacherous intuitions about meaning. Only in this way could one avoid the fallacies of reasoning inevitably connected with the use of natural language. Hence the artificial languages devised by the logicians of this period took the form of axiomatic systems. They started with a set of primitive terms and relations and a primitive set of propositions (these latter might fittingly be called postulates). To begin with, these primitive concepts and propositions were purely formal symbols devoid of any semantic interpretation though later they might be provided with a semantic interpretation when they were applied to some empirical subject matter.

If the trip was a pilgrimage, the pilgrim came away the richer for it, but one visit was enough. Greenberg was set on being a linguist.

> At any rate, I felt that these were philosophical questions and that my own primary interest was in language. However, I did find that there were two valuable by-products from the logical positivist approach and particularly from the study of the *Principia Mathematica* that had preceded it historically. In an axiomatic system, each step in reasoning was required to be justified by exact methods of deduction, either from the primitive propositions of the system or from statements which already had been deduced from these by the same exact methods of deduction. Of course, the axiomatic method of geometry based on Euclid had a basic similarity to that of positivistic axiomatics, but my early exposure to Euclid did not have the same effect on me. This was probably because I had encountered it at too early an age for it to affect me strongly, but even more, I think, because the *Principia* and subsequent systems of the logicist school gave explicit rules for deduction, and by not giving the primitive propositions any meaning seemed to avoid the apparent vagueness and arbitrariness of the Euclidean postulates.
>
> In particular, the study of the *Principia* provided me with an intellectual discipline which has ever since stood me in good stead. Further, it forced one in any field to inquire concerning the logical relations of basic concepts to each other. Which ones could be defined in terms of others and which were truly primitive? Even the empiricist criterion of meaning, although it had to be rejected, still had a kernel of useful application. I believe that it is always relevant to ask regarding any statement what are the conceivable facts about the world which would decide its truth or falsity.
>
> Outside of my disagreement with the empiricist criterion of meaning, which even in its various revised forms seemed unsatisfactory in that it did not, to

use a favorite term of the positivists, really "explicate" what we mean by meaning, there remained other differences which derived from my anthropological and linguistic background. Natural language did not seem to me to be so irregular and complicated that it could not be described or even used for scientific purposes.

ZELLIG HARRIS

Zellig Harris noted the importance of Carnap's work on his thinking:

> It is widely recognized that forbidding complexities would attend any attempt to construct in one science a detailed description and investigation of all the regularities of a language. Cf. Rudolf Carnap, *Logical Syntax of Language* 8: "Direct analysis of (languages) must fail just as a physicist would be frustrated were he from the outset to attempt to relate his laws to natural things—trees, etc. (He) relates his laws to the simplest of constructed forms—thin straight levers, punctiform mass, etc." Linguists meet this problem differently than do Carnap and his school. Where the logicians have avoided the analysis of existing language, linguists study them; but instead of taking parts of the actual speech occurrences as their elements, they set up very simple elements which are merely associated with features of speech occurrences.[102]

A few years later, in what was probably Harris's most successful general statement of his view of linguistic analysis—a paper entitled "Distributional Structure"—Harris wrote,

> Some logicians, for example, have considered that an exact distributional description of natural languages is impossible because of their inherent vagueness. This is not quite the case. All elements in a language can be grouped into classes whose relative occurrence can be stated exactly. However, for the occurrence of a particular member of one class relative to a particular member of another class it would be necessary to speak in terms of probability, based on the frequency of that occurrence in a sample.[103]

And Harris mentions conversations with "Carnap and his follower Y. Bar-Hillel" as significant during the period in which he was developing the notion of transformation.[104]

But Harris was a very complex figure. Students who worked with him did not get the impression that Carnap was someone whose work they should read to better understand Harris—far from it. The record shows that they were quite wrong, but their recollections tells us something about the message Harris was sending to his students.

One of Harris's students, Noam Chomsky, told an interviewer:

> Recall that in those days, one wasn't supposed to read anything before the late Carnap, and that was read only to refute. There were exceptions for Frege and Russell, but limited ones. And there had been guys named Hume and Locke, but one didn't read them, just quoted falsehoods one had learned in graduate school. For Harris, none of this had any interest either, as far as I know.[105]

When we asked Sydney Lamb if he read Carnap or Reichenbach during the 1950s, he wrote: "I wasn't [reading Carnap or Reichenbach] and I don't know of any others who were."[106]

Hilary Putnam wrote, "Zellig never mentioned Carnap in my hearing."[107]

Paul Mattick wrote,

> Harris was not particularly influenced, in my opinion, by Carnap. (He did not think much of philosophy in any case; the only philosopher I knew him to speak of with some admiration was Nelson Goodman.)[108]

Thomas A. Ryckman wrote that he did not think that

> Harris found inspiration in Carnap. . . . To be sure, the idea of formal systems and of purely formal analysis was prominent in the philosophical atmosphere in the 1930s (the main source being the Hilbert school, especially the Hilbert-Ackerman "Grundzüge der theoretischen Logik," 1928; Leśniewski in Warsaw was known only from his acolytes). For Carnap and logical empiricism generally, much of the attraction of purely formal (logical) analysis lay in the promise of an antimetaphysical logic of science that would replace philosophy altogether. . . . *Logical Syntax of Language* is in any case a deeply flawed book, due to Carnap's notion of L-consequence. As we know, though Carnap by 1935 was enthusiastic about Tarskian semantics, some logical empiricists (particularly Neurath, who had been Carnap's closest ally) never accepted it, viewing talk of truth as a return to metaphysics.
>
> Harris was surely aware of these developments, but as I learned, there was no significant influence. Harris adopted formal (distributional, combinational) methods because there did not appear to be any empirically controlled semantic or non-formal methods. And, Harris did not look to logical methods

(concerned with inference) as a model for investigation of the regularities of language (see note 17 to ch. 2 of (Methods of) SL).

Nor do I think Harris was attracted to logical empiricism's view of philosophy. He did have a great respect for Dewey, whom he regarded as "more intelligent" than other philosophers. I recall that once, when discussing Carnap with Harris in the early 1980s, Harris referred to Carnap as a "wooden logician." The accuracy of that remark has always made an impression on me, even as I have come to know, and respect, a great deal about Carnap.[109]

In his obituary of Harris, Henry Hiż (1994) discussed Harris's view that a natural language could be its own metalanguage, without leading us into the kind of problem that Russell had discussed, the antinomy of the liar. Hiż says this about Harris's approach:

> Harris took the word *report* from Bloomfield, who used it as a translation of the German *Protokollsatz*, a term used by philosophers of the Vienna Circle, mainly Carnap and Neurath, for descriptions of supposed sensory perceptions by the person who has experienced them. It is not observed and not testable by others. Neurath argued that it is difficult to decide which statements are phenomenally primitive, and he questioned the supposed non-testability of some sentences. Bloomfield changed it to a sociological statement. Everybody who understands the report statement will react to it in approximately the same way. If people react in two different ways, the statement is ambiguous.[110]

YEHOSHUA BAR-HILLEL

Yehoshua Bar-Hillel was one of the people who formed the essential link between philosophy and the revolutionary changes that would come to linguistics in the mid-twentieth century. Born in Vienna in 1915, he moved to British Palestine in the early 1930s, living in a kibbutz, getting ready to go to Hebrew University, and planning to study either mathematics or philosophy. Soon after arriving on campus, he stumbled on the first three years of *Erkenntnis*, the journal that Carnap and Reichenbach were editing, and it was to be the blinding moment on his own personal journey to Damascus. The effect of this reading

> was nothing short of a revelation. Never before had I come across such an unrelenting strife [*sic*] toward clarity and testability in matters philosophical as in the articles of Carnap in these volumes; never before did I see such a powerful denunciation of metaphysical obscurantism combined with a thorough

> understanding and analysis of its seductive appeal and with the techniques of combating this appeal as in the contributions of Carnap, Neurath, Schlick and Reichenbach published there. My future was clear.[111]

Shortly thereafter, Bar-Hillel was given a copy of Carnap's *Logische Syntax der Sprache*,[112] and there would be no turning back. He was now a convert to whatever it was that Carnap was trying to accomplish. His friends called the book "Bar-Hillel's Bible"; "it was," Bar-Hillel wrote later, "doubtless the most influential book I read in my life."[113]

There were two more enormous influences on Bar-Hillel that were to come soon. The first was reading Ajdukiewicz's "Die syntaktische Konnexität," in 1938, an article published in 1935 (and much later published in a form translated in English, in 1967, under the title "On Syntactical Coherence"). We will examine this in detail, when we turn to the work of the Polish logicians in the 1930s in chapter 8.

The second influence was the shift to the study of *natural* languages, rather than artificial languages. He wrote,

> After four years of [military] service . . . I returned to . . . my thesis. . . . The question of the amount of philosophical insight one can obtain from direct analysis of natural languages and ordinary speech, with common sense and linguistic sensitivity serving as the main tools of investigation, in comparison with what can be done by an indirect approach through logically rigorous constructed language systems, the approach favored by Carnap and the "logical reconstructionists" in general, came to the fore and remained since then in the center of my interest. Strange to say, at that time it never seriously occurred to me that there could be a third approach, namely the one attacking natural languages and ordinary speech with the best methods of theoretical and statistical linguistics, respectively. But then both these disciplines were still in a rather poor state in the late forties.[114]

Packing it in at the end of the 1930s

By the end of the 1930s, logical positivism was in disarray. Much of that could be attributed to the destruction of academic life in Austria and Germany brought on by the rise of Nazism, and it is true that most of the positivist philosophers and scientists who survived the period—Carnap and Reichenbach, to be sure, but many others as well—continued to be active thinkers. But logical positivism is a youthful sort of philosophy, a

Beaujolais nouveau sort of philosophy, one that is most appealing when it is fresh, and it loses its sheen quickly. By its nature it is opposed to obscurantism, but that same nature makes it ill-equipped to come to grips with questions and issues that refuse to be dismissed by the strictures imposed by logical positivism.

In 1939, the philosopher Roy Wood Sellars published in *The Sociological Review* a thoughtful account of where things stood regarding logical positivism at the end of the 1930s. While it was thoughtful and largely respectful, it was also dismissive, and certainly could be read as a trashing. Towards the end he wrote,

> This new effort to work out a philosophy of science, which seems to be filled with a phobia for philosophic terms and distinctions, has sought a new path in terms of logic and linguistics. The magical word is now logical syntax.[115]

He did not think the movement—what at one point he called "this new and virulent form of positivism"—would go anywhere (he also called it "a virile challenge to the effete traditions of academic philosophy," but his tongue was clearly in his cheek when he wrote that):

> I fear . . . that the members of the movement so despised philosophy that they had to find out things for themselves that philosophers had long recognized.

They just didn't get it, was Sellars' point. They did not understand the philosophical questions that they were taking on, and in part it was because they didn't have an adequate background for it. Most of the Vienna Circle were scientists, not philosophers. "Why is it," Sellars wondered, "that scientists so often go off on a tangent when they undertake an excursion into philosophy?" The scientist can often bring something to the discussion; but "there has been a lack of balance, an unawareness of the complexity of the problem."

"In the early days of Wittgensteinian atomism and subjectivism," Sellars wrote, "the Vienna Circle made merry with metaphysics, not realizing that they were merely asserting a solipsistic metaphysics of their own, so they were accustomed to speaking of epistemology as meaningless."

Sellars ended his essay with a sensible note urging balance between the approaches of the disciplines, and an interesting prediction:

> Quite certainly, of course, the theory of perception worked out by a realistic philosophy must harmonize with biological, psychological, sociological, and linguistic facts. That, we can take for granted; but I am persuaded that no combination of formal logic and linguistics can be a substitute. . . . Formal logic is now a specialty. I prophesy that this [Viennese logic] . . . will undergo analysis and dispersion. Perhaps it will break up into logic and linguistics, perhaps it will merge with pragmatism and realism.

Quine

Willard Van Orman Quine was one of the most influential philosophers in the United States over the course of the twentieth century. He was born in 1908, in Akron, Ohio, and went to Oberlin College, where he read John B. Watson on behaviorism—which was at its peak at that point—and found it to his liking. He read Bertrand Russell on philosophy, which he was very much attracted to. In 1930, he entered graduate school in philosophy at Harvard. There he would study with Whitehead, and with C. I. Lewis (who had been a student of William James, Josiah Royce, and Ralph Barton Perry).[116]

In 1932, after finishing his dissertation, he went to Vienna to study what was happening there. Schlick invited Quine to join the Vienna Circle. Carnap had just left for Prague, but the two were able to meet when Carnap came back to Vienna, as he often would, and in February of 1933, Quine went up to Prague to spend more time absorbing Carnap's ideas, and then further north to Poland, to study with the logicians in Tarski's group.

There was no doubt that this was a life-changing experience for Quine. "I was then an unknown young foreigner of 23," he wrote. "It was extraordinary of anyone, and characteristic of Carnap, to have been so generous of his time and energy. It was a handsome gift. It was my first experience of sustained intellectual engagement with anyone of an older generation, let alone a great man."[117] Quine felt something much more extraordinary when he worked with Carnap than he had in Cambridge. "It was my first really considerable experience of being intellectually fired by a living teacher rather than by a dead book. I had not been aware of the lack. One goes on listening respectfully to one's elders, learning things, hearing things with varying degrees of approval, and expecting as a matter of course to have to fall back on one's own resources and those of the library for the main motive power. One recognizes that his professor has his own work to do, and that the problems and the approaches that appeal to him need not coincide in any very fruitful way with those that are exercising oneself."[118]

When Quine went back to Cambridge in May of 1933 as a Harvard junior fellow (1933–36, along with B. F. Skinner and Garrett Birkhoff), he was ready to present Carnap's point of view, which he did in November of 1934, to a group that included David Prall, Henry Leonard, Charles Stevenson, Nelson Goodman, and John Cooley. He told his audience that Carnap had "shown conclusively that the bulk of what we relegate to philosophy can be handled rigorously and clearly within syntax. . . . Whether or not he has really slain the metaphysical wolf, he has shown us how to keep him from our door."[119]

Carnap himself visited in December of 1935, on his way to the University of Chicago (he began teaching there in January of 1936, and was given a full position the following fall).

Nelson Goodman was a fellow graduate student with Quine, and the two of them shared many convictions and an interest in the philosophical approaches being developed in Vienna. Their joint work evolved into Goodman's dissertation, submitted in 1941, which later in turn evolved into *The Structure of Appearance*, Goodman's influential book.[120]

Quine's book, *Mathematical Logic*, was published in 1940, and it was an important step in the developments we have been following. Quine very clearly offered it as a step in the process of extending the reach of the algorithmic method. In the first few pages, Quine explained to the reader that mathematics can be, and indeed has already been, reduced to logic. We now have a formal language for expressing truths of logic and of mathematics. All that is fine, but can we find an entirely mechanical method which will determine which expressions in this formal language are in fact true, and which ones are not? We will call this a device of the First Sort. Here is what Quine wrote:

> This is a good deal to hope, particularly in view of the reducibility of mathematics in general to logic. Every mathematical problem would become soluble by a mechanical procedure—even the celebrated Fermat problem, which has resisted solution for three centuries. Publication of proofs in mathematics would never again be necessary; results would simply be stated subject to mechanical check on the part of the reader.[121]

In other words, is there a device that can take in a statement of formal logic, and output an answer—Yes, or No?—to the question as to whether that statement can be proven. Instead of answering this question (as we will see in the next chapter, it was already known there *is no device* that can do this), Quine stepped back and said:

> Diffident of so bold a project, we might try to formulate some less powerful notational criterion of logical and mathematical truth: a criterion whose fulfillment by any given statement is discernable only by luck rather than by an infallible routine test.[122]

The first hope was of a device that could provide a proof of any given statement, or its negation, and now this new, toned-down hope is a device (we will call it a device of the Second Sort) that can take an entire proof as its input and declare whether it is indeed a valid proof.

> Such, indeed, is the character of mathematical proof; a proof once discovered can be mechanically checked, but the actual discovery of the proof is a hit or miss matter. Our present more modest objective, then, can take the form of an explicit formulation of the notion of proof, or theorem, as will involve reference only to the notational patterns of statements.[123]

So that's it: please be satisfied with the not inconsiderable goal of being able to determine automatically whether a given proof is correct—because (as Quine now fills in) we know that we cannot have a device of the First Sort, no matter how much we might like to. Mechanical procedures do not find all propositions that human minds can know are true.

This conceptual organization would reappear in Chomsky's early work on linguistic theory, where he discusses and dismisses two conceptions of linguistic theory, one in which the theory computes the correct grammar directly from the data, and one in which the theory accepts the grammar along with the input and decides whether the grammar is the correct one or not.[124]

The last chapter of *Mathematical Logic* was called simply "Syntax." While Quine diffidently suggests that not all his readers may find it comprehensible, he obviously thought it was important. Almost three hundred pages had been devoted to developing a language for logic and in making statements, in English, about that language. Now Quine wanted to go one more step and develop a purely formal language which can be used to make statements about the language of logic. As he puts it himself, "Our medium of syntactical discussion hitherto has been ordinary language. . . . But now we have a syntactical notation which is just as strict and systematic as the logical notation whereof it treats." And Quine noted that "this kind of approach, whereby the medium of discourse about a formalism receives strict formalization in turn, dates from Gödel (1931) and Tarski (1933)."[125]

Here is the promise:

> To the scientist longing for non-quantitative techniques, then, mathematical logic brings hope. It provides explicit techniques for manipulating the most basic ingredients of discourse. Its yield for science may be expected to consist also in a contribution of rigor and clarity—a sharpening of the concepts of science. Such sharpening of concepts should serve both to disclose hitherto hidden consequences of given scientific hypotheses, and to obviate subtle errors which may stand in the way of scientific progress.[126]

Conclusions

About 25 years ago, Michael Friedman suggested a reconsideration of logical positivism from a perspective not that different from the one we take in this book. He wrote that the discipline of philosophy has given rise to "a large number of seriously misleading ideas about the origins, motivations, and true philosophical aims of the positivist movement. (One can hardly expect philosophical critics, concerned largely with their own agendas rather than with historical fidelity, to generate anything other than stereotypes and misconceptions.)"[127] Perhaps one might disagree with that; one could indeed expect something better, and perhaps Friedman expected something better too. What we can say with some confidence is that many of the reasons that logical positivism attracted so many young people are still alive today, and an understanding of what went well and what did not in the program of the logical positivists can still be of considerable interest today.

The scientific philosophy of Vienna and Berlin between the two world wars left an indelible mark on the thought of its era. Its earnest espousal of a way to resolve all philosophical issues into formal syntactic and semantic questions, and its impatience with historical and social context, make it as dated as a Ford Model T is to our eyes today. But it is not so distant that we cannot feel the excitement that nearly oozed out of the first years of its journal *Erkenntnis*. Its utter self-confidence, its utter certainty that things were going to get better if we only get our science right was terribly appealing, and it energized young people who then worked on some very hot topics, as we will continue to see in the chapters to come. Richard Rorty put it well:

> The sort of optimistic faith which Russell and Carnap shared with Kant—that philosophy, its essence and right method discovered at last, had finally been placed upon the secure path of a science—is not something to be mocked or deplored. Such optimism is possible only for men of high imagination and daring, the heroes of their times.[128]

CHAPTER EIGHT

Logic, 1900–1940

Logic is the science of what makes thought possible. In the four decades that we will examine in this chapter, there were important connections made that would tie formal or mathematical logic to mathematics and philosophy in new ways, and these connections were soon to have a major impact on linguistics as well. That these ideas would be so consequential for linguistics was undoubtedly far from the thoughts of the vast majority of linguists, but when we go back and read this material, we see that there were quite a few logicians who saw with varying degrees of clarity that there were consequences in their work for how we should understand language, and grammar as well.[1]

This chapter may pose more of a challenge for some readers, and we would encourage those for whom that is so to reflect a bit on why. We have just said that the ideas presented here are going to have an unexpected degree of importance for our view of the mind later in the twentieth century, and so you may quite reasonably expect to see a continuity of ideas across the barriers that define disciplines. But the intellectual values and assumptions of the world of the logician and the mathematician are not the same as those of the linguist, or the psychologist, or the philosopher, and it is very easy to look at what someone in a neighboring discipline is doing and wonder why they could possibly be interested in the questions that they seem to obsess over day and night. And when an idea *does* manage to reappear, from *our* discipline, in a neighboring discipline, it is very easy to look at what those folks are saying and tell oneself, "They really don't get it." Yes, it is important for us to understand where ideas come from and what their context was, but at the same time, the continuity of ideas does not always go hand in hand with a uniform interpretation (or understanding, for that matter) of the ideas.

We will focus our attention on three principal areas in this chapter: first, the development of three different ways of understanding the nature of mathematics and mathematical statements; second, the development by Post and Turing of the abstract *machine* as a model—perhaps a metaphor, perhaps a new way to think about mathematics—for understanding inference; and third, the development of logics, or something very much like logics, that generate not only proofs, but sentences of languages.

There is an important idea, however, which cuts across all three of these areas: the relation of things that are finite to things that are infinite. We human beings are finite creatures, with only finite time and resources at our disposal, and yet we know some things that are true and that go beyond our own limitations and finitude. The mathematician will remind us that we know about infinite sets, like the set of whole numbers. Those more interested in language than in mathematics might just as well note that we have a linguistic grammar at our disposal as speakers of a language, and we feel quite justified in making general claims about an infinite number of sentences (such as which sentences are grammatical, and which are not). This emphasis on the unlimited character of what we know about our native language will become a central concern in generative syntax. And philosophers have worried a good deal since David Hume about how it is that the finite and limited experience that constitutes the laboratory results of all of science can serve as the grounds for an infinite number of predictions about the future trajectory of the universe.

But of all of the fields that concern themselves with the relation of the finite and the infinite, it is probably logic that can best justify its claim to be the place where thoughtful people must go to understand the relationship between the finite and the infinite. And we will have the opportunity in this chapter to see what logic has to say for itself along these lines. Logicians in this period were already proposing that there was a connection between the infinite number of true and provable statements that arise in a mathematical system and the infinite number of sentences that are grammatical in a language.

Let's pursue the relation of the finite and the infinite a bit more before we continue. Take the positive integers as our simplest example of an infinite set. These integers form a sequence, in that every positive integer has a unique successor, and every number—except the first, which is 1—has a unique predecessor. There are some things that we know with certainty about all positive integers, some with absolute certainty. To make the point clear, we would do well to pick a concrete example. We know how to add and multiply positive integers, and we know that there are some integers which are the product of two smaller integers—as 15 is the product of 3 and 5—and others are not the product of any two smaller integers; we call these second kind prime numbers, such as 17 and 19. Every number can be expressed as the product of prime numbers in exactly one way: this is something that we are certain of, even though we cannot write out every individual case that the generalization covers. How is this certainty possible?

The most fundamental way of arriving at that certainty is a style of argumentation that focuses on properties that can be asserted of integers, and there is a very basic style of argumentation (called the *principle of induction*) that allows one to pass from a finite number of steps of argumentation to an infinite number of conclusions: or rather, it permits us to conclude that a property holds not just of one integer, but of all integers. To show that all integers have a particular property P, we show, first of all, that it is true for the number 1 (that is, $P(1)$ is true), and second of all, if it is true for any number n (that is, $P(n)$ is true), then it is also true for $n+1$ (that is, $P(n+1)$ is true). While something like this principle was employed by some mathematicians as early as the seventeenth century, an explicit awareness of the nature of this principle came only with the developments in logic that we discussed in chapter 3 in the work of George Boole, Augustus de Morgan, Giuseppe Peano, and a few others.

The recognition that a principle of induction could be built into models of grammar came later on. If a language has a finite number of words in it and an infinite number sentences that can be built upon them, then the early transformational grammarians (Zellig Harris and Noam Chomsky, as we will see in volume 2) established a first grammar that generated a finite set of kernel sentences, and then a second, transformational grammar whose rules had the property that if you pass in two grammatical sentences, it would produce a new grammatical sentence bigger than either of them. While the details of the style of induction are slightly different, the insight remains the same.

We will turn in a moment to the mathematical anxieties that contributed to the interest in rethinking the foundations of mathematics, and over the course of this chapter we will see how the developments that grew out of this concern led to new ideas about the nature of proof and the nature of valid inference. Everyone recognized that mathematics existed, that it was a large castle built up over centuries, and while some elements in it might need to be tossed out as foundations were refurbished, what was called for was a way of better understanding what was already known. Part of that better understanding would (on most people's views) derive from a more careful consideration of the individual steps leading from original principles to final conclusions.

The final section of this chapter turns to the ideas that logicians were exploring that would result in new ideas about language and grammar in the 1950s and 1960s. But the meaning of the formalisms today are in many ways no longer the same. The syntactic derivations that are explored in generative grammar—let us say, in a minimalist framework of the sort developed in the 2010s—involve analyses in which the well-formedness of many-stage representations are computed.

Three Approaches to the Philosophy of Mathematics

In the early decades of the twentieth century, three conflicting views emerged of how mathematics is to be understood: *logicism* first, and then in its aftermath *intuitionism* and *formalism*.[2] One would have thought that mathematicians would form a happy and contented family. Of course there are always issues of priority and credit, but what could lead mathematicians into the sort of furious combat that we will see emerge between Hilbert and Brouwer? Who could have known that they would become so rambunctious, so ready to come to blows over what we should and shouldn't do when we pursue mathematical theorems? The heart of the dispute we will examine is closely tied to the notions of *soft logic* and *hard logic* that we have encountered from time to time in this book—the way in which a conscious subjectivity is directly engaged in the discovery of truth.

The intuitionists were strongly committed to the centrality of soft logic in this sense, and the formalists were equally strongly committed to the elimination of soft logic from the true heart of mathematics. Curiously enough, the rupture between these two perspectives ended up rather sedately, and nowadays there is hardly a mathematician anywhere who worries about the niceties that the intuitionists and the formalists disagreed about so strongly. It would not be too far from the truth to say that in the end, the intuitionists lost the battle, at least within mathematics proper.[3]

But in a certain fashion the intuitionists' *project* was not completely forgotten. Their project was to question and to put limits on what could be impressed into mathematical servitude when a mathematician needed to drive a proof home. And in response to the intuitionists' deep misgivings about what they saw as excesses of mathematicians' unbounded presumptions, a new mathematics of computational complexity arose, one which lies at the heart of the theory of computational complexity today.[4]

We have seen a number of times that the mind sciences can go through periods—sometimes quite *long* periods—of anxiety about whether they are really scientific enough. But even mathematics has its periods of anxiety, and though the anxiety is not precisely about whether all practitioners are being *scientific* enough (the concern is more that they are being *rigorous* enough), the anxiety can be just as palpable when it comes to establishing the rigor and certainty of the conclusions that scientists and their scientific methods establish. And anxiety of this intellectual sort is one of the greatest causes of generational rupture, because the rising generation senses—we almost want to say, it *smells*—the rot and corruption of what is being handed to it by the preceding generation. Like a younger generation that is being handed the keys to the family chateau just when the roof has started to cave in and the basement has started to flood, a new, young group of intellectuals may hold it to be their progenitors' fault that things were not maintained the way they really should have been. If this seems to you like an unlikely metaphor for the growth of mathematics, read on.

In the first half of the nineteenth century, as we have seen, the foundations of geometry began to feel shaky. One might well wonder what could cause a field—a field whose rigor had been established by Euclid and his school of ancient Greeks and which had been used both to illustrate and to teach the very character of mathematical proof—to become anxious and uncertain of its own rigor. Ironically, what caused the anxiety was the lack of success that mathematicians had in establishing the exact and precise nature of the premises and postulates needed to establish Euclidean geometry. (There is a deeper truth lurking here: deep insights that lead to revolutionary changes in our understanding very often arise out of questioning the premises that a discipline is established upon.)

At the moment that this anxiety set in, there were two major questions that mathematicians were asking themselves: just how many completely independent postulates did Euclid really need to make in order to prove everything he set out to prove? That was the first, and the second was about the character of these postulates. Did any of them really require assumptions that go beyond fundamental logic? If we understand the world through a spatial faculty of "intuition" (using the term as Kantians were wont to), exactly where has the character of this spatial faculty crept into Euclidean postulates? The first of these questions might seem like more of a workaday sort of a problem: just how many independent postulates did Euclid really need? But there was a fly in the ointment. Geometers were not at all sure what to say about the premise that was needed which said that if on a plane, we have a line and a point that does not lie on the line, then there is exactly one line parallel to our first line that goes through the point. Many people tried to show that this postulate, known as Euclid's fifth postulate, or the parallel postulate, was *not* independent of the other postulates and could be proven from them, but no convincing proofs emerged despite their moral certainties that a proof was just around the corner. The natural way for mathematicians to proceed was to undertake some crafty reasoning based on the assumption that the parallel postulate was *false*, with the hope that such an assumption would lead to a contradiction: if such a contradiction arose, then they could appeal to the law of the excluded middle and feel comfortable with the conclusion that the parallel postulate does indeed follow from the other postulates of geometry.

Mathematicians started exploring what an anti-geometry, one that included the negation of the parallel postulate, would look like, and the further they explored, the less it looked like they were exploring a crazy geometry that was a rotten hull of contradiction. Quite the opposite: rejecting the Euclidian parallel postulate led mathematicians to interesting geometries with interesting properties: who knew? And so emerged non-Euclidian geometry.

In a sense, this was a step forward for geometers, but it raised serious foundational issues for mathematicians more generally, and for philosophers who worried deeply about mathematics (and as we have seen, Husserl was just such a philosopher, as was Carnap). Once again, we must bear in mind that there was general agreement on an essentially Kantian view: knowledge can be divided into certain knowledge and uncertain knowledge. Certain knowledge, in turn, is of two sorts, logic and everything else that we know through the mental categories of space, time, and causality. We know other things, with less certainty, through our interactions with the empirical world. Few were willing to consider the thought that mathematics rests on experience, and if Euclid's geometry has a certainty that rests on a mental category of space, what does this new non-Euclidean geometry tell us? Is the awareness of this new kind of space, with positive or negative curvature: is all of this accessible to us because we are able to go beyond the restrictions of our category of space? Or was our category of space really more flexible than we had thought it before?

No firm answer was generally agreed upon, but anxiety in the domain of space naturally led to a secondary anxiety among mathematicians. Just as geometry had been taken to be based on the category of space, numbers and arithmetic had been taken by many as the product of our category of *time*. Why time? Because numbers emerge originally from their nature as marking order. We count sheep by running them in front of us, and starting a count: one, two, three, and so on. If space, or our understanding of it, is subject to unexpected revision, could the same thing happen to arithmetic? The answer to this question was *yes*, as some mathematicians would come to put limits on the intuition we have that leads to our knowledge of numbers. This was the movement known as *intuitionism*. But earlier on, there was another interpretation of mathematics that arose at the end of the nineteenth century that deserves our attention. This was the *logicist* point of view, developed by logicians such as Gottlob Frege and Bertrand Russell, and for a while by David Hilbert, who would later become a central mover in the world of formalism.

Logicism

Bertrand Russell and Gottlob Frege were two of the most famous defenders of the notion that the mathematics of number was a specific subcase of logic, and their position came to be known as *logicism*. Logicism aimed to show that all the truths of arithmetic (which is to say, arithmetic as we know it, plus number theory as mathematicians style it) were ultimately reducible to logic, and hence were necessary just as they are not only in our universe, but in any universe at all.[5]

For some, what made logicism hard to swallow was the conclusion which seemed to follow from its basic premise: if all mathematical truths are part of logic, why is it there are *surprises* in mathematics? Any student of mathematics knows there are indeed surprises; who isn't astonished to learn that $e^{2\pi i} = 1$? Or that a group with a prime number of elements has no proper subgroups? It is very hard to study mathematics without getting the strong impression that there is a mathematical world whose denizens really have structure that can be explored and discovered. So the question can certainly be asked: is there any way to establish a doctrine of *logic*, containing all the incontrovertible statements about incorrigible inference, which does *not* also contain the notions of numbers?

The logicist position was that the answer is *no*: the principles of logic naturally and inevitably lead to the heart of the notion of number. If the domain of numbers contains surprises in it for the human learner, then the logicist was going to accept that for what it is: a fact about human limitations.

One of the challenges for the logicist position is how to deal with statements that claim to be about all numbers: for example, the statement that all numbers can be factored into prime numbers in exactly one way. What kind of principle of logic can allow us finite creatures to sensibly make statements which apply to an infinite domain, as the whole numbers are? One answer to that question—the most important one, in fact—is the principle of induction, which we mentioned just above: if we can prove a statement to be true of the number 1, and if we can show that whenever it is true of the number n, then it is also true of the number $n+1$, then we can draw the conclusion that the property holds of all numbers. But some sharp thorns remain wrapped around this question: is this "principle of induction" a principle of logic, or does it go beyond the central domain of logic?[6] There was no consensus on this question, but it was very important.[7]

Another serious challenge for this logicist point of view would emerge in the late 19th century when mathematicians began to talk rigorously about *sets*, and that discussion was also motivated by concerns about the nature of the infinite. The German mathematician Georg Cantor was the most influential in this development, but Gottlob Frege was already engaged late in the century in a great project to found arithmetic upon set theory, all starting from the belief that set theory could be taken to be part of the very most stable and reliable catechism: logic.

We saw in chapter 3 how Frege's project came tumbling down in 1902, when Bertrand Russell wrote to Frege and pointed out to him that there was a logical contradiction in the way he had developed the theory of sets, and set theory could not possibly then be a part of logic.

To understand how the problem arose, we must bear in mind two points. The first is that set theory rests on the fundamental recognition of the difference between being an *element* (or member) of a set, on the one hand, and being a *subset*, on the other. If x and A are both sets, we can still indicate that x is an element of A, written $x \in A$, *x is a member of A*. For example, x might be the set of all even integers (an infinite set), y the set of all odd integers (also an infinite set), and A a set with just two elements in it: x and y. In this case, x is an infinite set, and it is a member (but not a subset) of A, which is a finite set with two members. If we write $x \subset B$, we are saying something quite different: we are saying that every element of x is also an element of B. If x is infinite, then B is also. Being an element of a set is very different from being a subset of a set.

The second point is this: when we talk about finite sets, we can define a set either by listing all of its members or by stating a property that all and only the members share. We can define a set containing the numbers 2, 3, 5, 7, 11, and 13, or we can define it as all of the prime numbers less than 15; either way is fine. When we talk about infinite sets, we can only choose the second method of defining them, since we cannot list all the members in any finite amount of time. So this second, *intensional* way of defining a set is based on the principle that a set can always be defined once we make explicit a property that all and only the members of the set possess. This seemed to be such a simple and natural assumption: if we can provide an explicit criterion for membership in a set, then we may assume that the set exists.

But it occurred to a few people, and Bertrand Russell most famously, that serious problems could arise if we combine that last statement and we *also* allow a set to be a member of itself. If sets are things that could be members of themselves, then nothing is stopping us from defining the set of all sets that do not contain themselves, those sets that we called *normal*: $N = \{x \mid x \notin x\}$. As Bertrand Russell pointed out, if we ask whether $N \in N$, a contradiction arises. If $N \in N$ then by definition of N, $N \notin N$; that is what it means to be a member of N. If $N \notin N$, then $N \in N$ because that is the price of entry into the set.

The conclusion was either that there was no logically coherent formulation of the theory of sets, or else that something unexpected was preventing this question from arising, by virtue of N's definition violating something yet to be discovered. This uncertainty quite naturally led to considerable disciplinary anxiety to anyone who looked up from their daily toil.

RUSSELL AND WHITEHEAD, 1910–1913

Bertrand Russell published *The Principles of Mathematics* in 1903, shortly after his discovery of this serious challenge to a rigorous definition of set theory, and he continued to work on the foundations of mathematics. He joined forces with Alfred North Whitehead, and together they continued the logicist program in the face of the *antinomy* that Russell had presented Frege with, producing their monumental *Principia Mathematica* over the years 1910–13 and trying to show that with proper restrictions on what we say, we can avoid the antinomy, and we can establish a proper notion of arithmetic based on purely logical notions. Russell and Whitehead created something that they called the theory of types, allowing an infinite hierarchy of what a layman might refer to generically as sets. Each set would be assigned a level in an overarching hierarchy of levels, and a set could only be an element of a set on the next higher level, so concerns about self-containing sets could be eliminated. Onlookers who were paying attention to what was happening in mathematical logic early in the twentieth century could draw two quite different conclusions from this. For those prone to be anxious, all of this was cause for concern that even the most abstract of mathematical disciplines could discover that there were fundamental logical cracks in the foundations. But there was a different conclusion that could be drawn, which was that this was an excellent time to go ahead and explore the logical foundations of all the sciences. That kind of exploration is not easy, but it can be done, and there is a good chance that if it is done properly, we will learn something quite surprising along the way. And many scientists, including psychologists and linguists, found this second conclusion attractive. We have already met the scientists in the Vienna Circle: they were scientists dedicated to just this enterprise.

Intuitionism

There were two major alternatives to logicism that developed in the early twentieth century: *intuitionism* and *formalism*. The term *intuitionism* suggests the Kantian use of the word *intuition*, in which context "intuition" refers to a kind of knowledge which is immediate, and therefore indubitable, but based on something more than logic.[8] For the intuitionist, it is important to emphasize that whatever is based on intuition is based on an *act* of a mind (and in our case, that means an act of a human mind)—a mind that is paying attention to what it is doing. Intuitionism is thus the prototypical example of a system of thought that is founded on what we have called *soft logic*: the acknowledgment of a truth based on direct intuition. There was a natural connection between intuitionism, in this sense, and Husserl's phenomenology.

Intuitionism in mathematics was most radically and strenuously defended by the Dutch mathematician, L. E. J. Brouwer, and in a certain sense he pushed the idea of soft logic further than any other serious mathematician of his age: in his view, the comprehending mathematical mind not only *saw* mathematical truths—he actually built and constructed them, he *made* something that was not *already there*. In developing this intuitionist point of view, Brouwer rejected the emphasis on language that had been growing stronger in philosophy during this period, a trend that continued. Formulating mathematical statements in language, regardless of whether it was in formal language, or more or less everyday language, was of little mathematical interest to Brouwer. Expressing an idea in language was just the packaging and the marketing that had to be done after the real mathematical work was completed, and we should not confuse the mathematical act with its linguistic description. It was the active, creative, self-aware subjectivity of a mind that made real mathematics possible, for the intuitionists. The fuss about formalisms and logic is misplaced, says the intuitionist, and all that fuss is unreasonable in the same way that a person is who buys a bottle of wine because he likes the shape of the bottle rather than the wine inside.[9]

We have perhaps not emphasized enough how easy it is for a certain kind of soul to slip back and forth between the domains of mathematics and philosophy, and if ever there was a soul given to these wanderings, it was Brouwer (Kurt Gödel, a logician we will encounter shortly, was another such soul). There is no doubt that Brouwer was a great mathematician—his was the hand that created the modern field of topology, and his fixed point theorem is still studied by all students of topology. The greatest chairs of mathematics of his day were offered to him (though he had no interest in moving out of his native Netherlands, so he turned them down). But his concerns about the fundamental nature of existence were deeply rooted in his life from very early days. We can learn a great deal about his thoughts by reading his *Life, Art, and Mysticism*, which he wrote in 1905 at the age of 24 (Brouwer 1905). It expresses the yearnings of a soul who is sad and world-weary, and who is well acquainted with the writings of Meister Eckhart and Jakob Böhme, two of the most important mystics from the fourteenth and sixteenth centuries who placed importance on turning one's mind to an inner world through meditation and prayer. Eckhart and Böhme were part of an important tradition of *hermeticism* (of which Hegel's work can certainly be viewed as a development as well).[10]

Once we understand the central idea of hermeticism, we are in a much better position to understand Brouwer's intuitionism. Hermeticism acknowledges the existence of a God who created the universe, but who is not as distant and independent of the created world as he is according to the Platonic and mainstream Christian views. The creation of the world is not *merely* an act of God; it is what God *must* do. The world that he created is not one that he can move away from and observe from a distance; the creatures of his world (which includes hermeticists and all the readers of this very book) must in turn lead to a completion of God by their contemplation of him, and that contemplation is achieved through meditation and prayer.

To the modern ear, this certainly sounds like theology, and it might even be philosophy, and furthermore, we might be forgiven for not immediately seeing its connection to mathematics. But the connection is direct. The activity of the mathematician is an important part of the creative process of the universe, on this view. Hermeticism is a rejection of the simple Platonic view, the view which is held by virtually all working mathematicians, the view that the world of mathematics exists independently of the human mind and that it can be discovered and explored by the trained mind. The Platonist does not really know who created that world, but he does not need to care. The hermeticist, on the other hand, is directly engaged in the development of the universe, which includes the world of mathematics. He works *with* God, though not as his equal, but only as part of God's creation, to continue the development of the universe.

And that was exactly Brouwer's idea of intuitionism: he rejected the Platonic view of mathematics, the view that it is all there ahead of time and outside of time. The human mind creates the mathematical world, as a part of God's creation. The reader will recall our discussion in chapter 2 about two views of teleology, the transcendent view that sees God as the external entity who can define a final state that He wishes the universe to head towards, and the immanent view which sees goals as an integral part of the universe itself. Hermeticism is a philosophical and religious view that is committed to an immanent view of divinity.

For Brouwer, the world that we inhabit when we turn to the inner world of meditation and of mathematics is one which is not subject to the laws of cause and effect. The chains created by patterns of cause and effect in the vale of tears in which we live was what saddened Brouwer the most, and it was what he most wanted to escape from. We have encountered before the *concern* that the world is subject to laws of cause and effect—this has most clearly emerged in the context of anxiety that the reach of physical cause and effect is so great that there is no room left for anything else, such as inference and reasoning. But there are other great themes that may be threatened by cause and effect being too powerful and all-inclusive, and the greatest of them is teleology of one sort or another: that is, the notion that the world is tending in a coherent direction. We will see in the next chapter how important it was to Russian intellectuals such as Trubetzkoy and Jakobson that the decline of Western thought could be measured by how little room was left in the modern, Western scientific worldview for teleology and for an appreciation that systems both physical and human are moving in a coherent direction.[11]

What was undoubtedly the most contentious issue that separated intuitionists like Brouwer from all the mathematicians who rejected intuitionism was the status of *the law of the excluded middle*. If we consider two propositions—A, and the negation of A—then this law tells us that one of the two is true, because there is no value that A can take on other than being true or false. If we can prove that assuming A leads to a false consequence, then the law of the excluded middle tells us that we are totally justified in inferring that the negation of A is *true*. But that way of drawing conclusions was rejected by the intuitionists. Unless the methods of mathematics brings us directly to the conclusion that A is false or that A is true, we simply do not know its status. The philosophical position we have discussed would say that A has not yet reached the status *either* of being true *or* of being false, until such time as someone gets it there.

This was too much for most of the mathematical mainstream to accept, and David Hilbert, the leading mathematician of his day, the elder statesman of his discipline, and once an admiring colleague of Brouwer's, found the rejection of the excluded middle to be simply unacceptable. Too much excellent mathematics would be thrown out if any proof that used the law of the excluded middle were thrown out of the books of certain mathematical knowledge. This conflict in the nature of acceptable mathematical proof led to what was undoubtedly the most heated meltdown in the history of modern mathematics, which climaxed with David Hilbert's summary dismissal of Brouwer from the editorial board of the mathematics journal *Mathematische Annalen* in 1928. "Henceforth we will forgo your cooperating in the editing of the *Annalen*," Hilbert wrote him in a letter, "and thus delete your name from the title page."[12] The conflict involved not just Brouwer's philosophy of mathematics, to be sure; during the decade that followed World War I, there were countless opportunities for nationalist grudges left over from the war to grow and fester, and there is no doubt that such concerns played a role in this story. Hilbert knew that his own career would soon be over, and he did not want his journal to be at risk of falling under the influence of intuitionism after his departure. Cooler heads on the editorial board, like Albert Einstein, tried to smooth matters over, but to no avail. It was Einstein who gave this mathematical conflict the name by which it has come to be known: the war of the frogs and the mice![13]

Back to the mathematics. There was a strong link between the intuitionists' rejection of the law of the excluded middle and their insistence on *constructive* proofs: a proof is constructive if it provides a way to actually calculate the value of any numbers that it needs to use to prove its point. By setting the bar a good deal higher for proofs that are acceptable to the intuitionist, Brouwer was able to make the point that it is perfectly possible and reasonable to be confronted by a hypothesis which cannot be proven and whose negation cannot be proven either.

For some mathematicians, and we could place Hilbert among them, being armed with a particular set of proof-tools and being confronted with a statement which can be neither proved nor disproved could be a serious nightmare. If that particular statement is, in addition, *true*, then we speak of a situation in which that set of proof-tools (axioms, legitimate methods of inference) is *incomplete*: it is incomplete if there is a statement which we know to be true that can be expressed, but cannot be proved, within the system. In order to prove that a set of proof-tools is incomplete, we have to be able to make rigorous mathematical proofs about what can and what cannot be proven in that system. Creating a new field of mathematics in which proofs can be made regarding what can be proven in particular systems became an area of major concern during the 1920s and the 1930s, and it would have its greatest moment with Kurt Gödel's incompleteness theorems in 1931. Gödel would take the intuitionist's summary rejection of the law of the excluded middle and prove that in virtually any formal system you might care to construct, there will be statements which can neither be proven true nor proven false.

Formalism

The view that came to be known as *formalism* was based on the idea that to do mathematics right, what is required is its final formulation in a language that is so explicit that all of its assumptions are spelled out as axioms, and each step in a proof is justified by a particular explicit rule of inference, from a set of rules so carefully described ahead of time that a

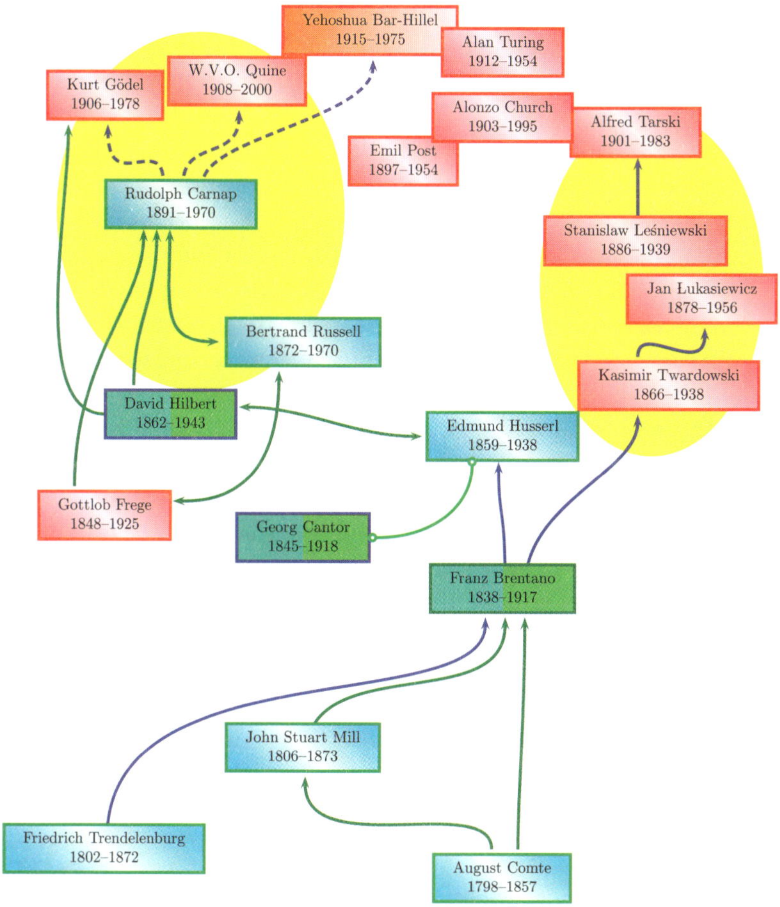

FIGURE 8.1. Logicians

mechanical method, one involving neither subjectivity nor insight, can check and determine whether the rules have been legitimately applied.[14]

This perspective could be justified in more than one way. On the one hand, as a method it would allow some to feel more secure in the conclusions that mathematics draws, a security that derives both from being sure that mistakes are not being made and from making explicit all the assumptions that one needs to make the argument go through. On the other hand, it was also a direction that was based on the idea that mathematics does not need to understand the *objects* of its study; it needs only to understand the relations between the objects. We could make up axioms that contain technical terms like *point*, *line*, and *plane*, but we would not be allowed to exploit our knowledge of what those technical terms really mean. The only steps permitted would be those justified by the particular axioms set up for the system.

Does the formalist believe that this is all there is to mathematics? The person who is skeptical of formalism, or outright opposed to it, is certain to say that if formalism is interested in nothing but mechanical operations, then it really misses the point of mathematics. But the mathematician who pursues a formalist approach does not think that the formalism is all there is: David Hilbert, the great German mathematician who was the patron saint of the formalist enterprise and who came to near blows with Brouwer the intuitionist over these issues as we have seen, knew that mathematics was more than an *arbitrary* set of axioms and rules of inference. So the problem is this, as we try to understand just what the formalist position was: once we have defined the rules of the formalist "game," how much more interpretation are we allowed to layer on top of the mathematical work that has been done? The skeptic will resist, and say that if you are a formalist, then be a formalist, and acknowledge that you really believe that mathematics is about sequences of symbols that you make on a piece of paper. But the formalists were quite sure they were doing something important for the discipline, and this accomplishment should not be misunderstood. And as we will see shortly, it turned out, to almost everyone's surprise, that the mathematical study of simply putting symbols on a very long piece of paper was not without important consequences![15]

The formalist was also very mindful of the fact that there is a wide gap between guaranteeing that a proof of a theorem is rigorous, on the one hand, and the task of coming up with the proof in the first place. *Checking* a proof to ensure its legitimacy as a proof is in most cases vastly easier than discovering the proof. Searching for the ability to do both of these things will be undertaken over the course of the century, but we cannot lose sight of the fact that establishing the validity of a proof is much easier than coming up with the proof itself. We touched on this at the end of the last chapter, and it will come back in other guises. There are circumstances under which *learning* can be thought of as discovery, or as a successful bit of exploration that arrives at someplace which is somehow better than where we started off. Learning can be a kind of creativity—but that makes it all the harder to understand.

Let's take a look at Hilbert's program, which was part of an effort that had considerable impact on the mind sciences during the twentieth century.

We have made reference before to David Hilbert's project of axiomatizing the sciences, which was an inspiration to the psychologists, linguists, and philosophers who devoted enormous efforts to proposing and exploring sets of axioms or postulates for their field. We saw this in A. P. Weiss's axioms for psychology, an effort which Leonard Bloomfield recognized, and for which Bloomfield tried to find an equivalent for linguistics. We have seen that this conception lay at the heart of much of the efforts of the Vienna Circle, and the hope that they had that they could provide something that was ultimately of use and value to the sciences.[16] The desire to develop an axiomatization for linguistics was part of the project of Zellig Harris and that of the early Noam Chomsky.

Hilbert himself initially supported the logicist program, as developed by Whitehead and Russell. He wrote in 1918,

> It seems necessary to axiomatize logic itself and then to establish that number theory as well as set theory is only a part of logic. This road, prepared for a long time—not in the least through the profound investigations by Frege—has finally been pursued by the ingenious mathematician and logician, Russell, with great success. In the completion of this extensive enterprise by Russell for the axiomatization of logic one can behold the crowning of the work of axiomatization in general.[17]

But Hilbert's views developed and changed, and a decade later, he would reject that earlier view and write,

> No more than any other science can mathematics be founded by logic alone; rather, as a condition for the use of logical inferences and the performance of logical operations, something must already be given to us in our faculty of representation, certain extralogical concrete objects that are intuitively present as immediate experience prior to all thought. If logical inference is to be reliable, it must be possible to survey these objects completely in all their parts, and the fact that they occur, that they differ from one another, and that they follow each other, or are concatenated, is immediately given intuitively, together with the objects, as something that neither can be reduced to anything else nor requires reduction. This is the basic philosophical position that I

> regard as requisite for mathematics and, in general, for all scientific thinking, understanding, and communication. And in mathematics, in particular, what we consider is the concrete signs themselves, whose shape, according to the conception we have adopted, is immediately clear and recognizable. This is the very least that must be presupposed; no scientific thinker can dispense with it, and therefore everyone must maintain it, consciously or not.[18]

Part of the formalist enterprise was similar in certain important ways to structuralism as it would come to be known in linguistics and anthropology. We have briefly alluded to that in our discussion of de Saussure, and it will be a significant part of our story in the next chapter. The formalist mathematician's enterprise includes careful attention to the linguistic form of the postulates with which a system begins, and the words of the postulates are understood as falling into two sorts. One is relational, and the other is substantive. If we were thinking about geometry, a postulate such as "any two points determines a straight line" takes "point" and "straight line" as substantive words—but the formalist then goes ahead and affirms that from the point of view of formalism, we need pay no attention to what those words really mean: we could as well replace "point" by "table" and "straight line" by "coffee cup." All that is important is that they refer to different sorts of things. Just as Saussure said about the individual sounds in his phonemic inventory: it does not matter so much which sound any particular symbol points to—what matters is that different symbols represent different sounds, and these symbols enter into relations with one other.

The axiomatic method for the construction of a domain did not start in the nineteenth century: it started two thousand years earlier, with the development that we associate with one particular ancient Greek, Euclid, who proposed a small set of axioms from which all of geometry could be deduced logically. But as we have seen, the nineteenth century saw the special status of Euclidean geometry taken away from it. Euclidean geometry was not the *a priori* truth that Kant had thought that it was; it was only one of three kinds of perfectly reasonable geometries, the geometry of flat space. But by the end of the nineteenth century, mathematicians knew that space could be curved, either positively or negatively. Space (and therefore geometry, its study) was more complex than we had ever suspected.

David Hilbert decided to rethink Euclid's axioms and make everything explicit that needed to be said to make the theorems of geometry really follow from the assumptions.[19] Hilbert worked on this during the

1890s and published the influential *Grundlagen der Geometrie* (*Foundations of Geometry*) in 1899.[20] He attempted to reduce geometry to a set of theorems about points, lines, and planes, using 21 axioms and nothing more. In retrospect, it is possible to see that he added to his conceptual tools various *sets* of things, like *sets of points*, but it was not until Russell and Whitehead's work more than 10 years later that people saw clearly that talking about sets of things was not at all as straightforward as it had seemed, even if you know well what the things in question are: the concept of *set* is not at all simple and comes with some complexities that were not at all obvious. You will recall the problem that arose from allowing a set to be a member of itself. What was deeply troubling about this case was that Frege's system allowed us to state a simple condition that an object had to have to be a member of pathological set. The condition was "*x* is a member of *x*," and there seems nothing pathological about that property, even if there is no guarantee that anything fits that description. But Russell's antinomy showed that the simple logical passage from *a condition being meaningful* to *the set defined by that condition* was indeed treacherous, perhaps irreparably so. Who knew? Who could have imagined?

Once Hilbert had accomplished this, he wanted to refurbish the foundations of other disciplines, both inside of mathematics and outside of it (and since his reorganization of geometry was squarely based on the mathematics of numbers, any serious and deep conclusions about the logical health of geometry would have to await a careful and rigorous review of the state of health of mathematics). Hilbert's general method began with a formal alphabet of symbols and a set of rules that expressed how these symbols could be combined in a well-formed way. Two more things were necessary: a set of axioms, and a set of rules for valid inference. (Again, we have seen this already, most notably in Carnap's exposition—but the project had come from Hilbert.)

Hilbert began with elementary objects, the natural numbers, with elementary properties, based on recursive functions and quantifiers, and with elementary proofs, ones which use only these elementary objects and properties, along with basic principles such as the principle of induction. The goal was to show that any statement that could be made with these elementary objects and properties, and that could be proven with mathematics of any sort, could be proven with only elementary proofs. Mathematics has an enormous set of proofs at its disposal, created by the ingenuity of mathematicians over the centuries; Hilbert aimed to show that slowly but surely, we could show that all of this was a linguistic expansion of the elementary objects, properties, and proofs, and that this expansion was for the benefit of the limited creatures that we are, not for the benefit of the mathematics in and of itself.[21]

Hilbert's program is fundamentally a parsimonious reconstruction of mathematics. We must find not only the truths of mathematics, but also the simplest way of arriving at each mathematical truth, where one part of what it means to be simple—one important part—is the set of assumptions and the set of inference tools we have available in order to draw a conclusion. We might as well say it, too: the sense of *parsimony* that is involved here is the one that is closer to *elegant* than it is to *simple*; there is no mathematics where there is no sense of elegance and, yes, of beauty.

Even in his younger days, when he was sympathetic to the logicist perspective, Hilbert emphasized the importance of the Decision Problem, the problem which has come to be known in many circles through its original name in the German, the *Entscheidungsproblem* (though that term appears to be due to one of Hilbert's students, Heinrich Behmann): when can we establish a proof that a class of mathematical propositions must be susceptible to mathematical proof or disproof? This kind of metaquestion was already present in his famous 1900 lecture in which he established the set of questions he bequeathed to twentieth-century mathematics. Twenty years later, those ideas remained important, though they were now reconceptualized from a formalist perspective. Behmann, Hilbert's student, lectured on this subject in 1921,[22] making very clear the connection between the kind of metamathematics they were developing and the nature of the "machine," in an abstract, not a concrete, sense. Behmann wrote,

> As is well known, symbolic logic can be *axiomatized*, that is, it can be reduced to a system of relatively few basic formulas and basic rules, so that *proving theorems* appears as a *mere [bloss] calculating procedure*. One has merely to write new formulas next to given ones, where the rules specify what can be written in every case. Proving has assumed the *character of a game*, so to speak.[23]

What does that mean, "a game"? A game is defined by the rules it has, by the rules that define what is a legitimate move and what is not.

> It is just like in chess [Behmann continued], where one transforms *the given position into a new one* by moving one of one's own pieces and removing one

> of the opponent's pieces if appropriate, and where moving and removing must be allowed by the rules of the game.[24]

The formalist is invested in establishing what the legitimate rules are, in mathematics or whatever discipline is under scrutiny. But is that all there is to it? Of course not:

> This comparison makes it blatantly clear that the *standpoint of symbolic logic* just outlined *cannot suffice* for our problem. For that standpoint tells us, like the rules of chess, *only what one may do but not what one should do*. In both cases, what one should do remains a matter of *inventive* thinking [*Nachdenkens*] and fortunate [or *lucky*: glücklich] *combination*. We require much more: not just every single allowed operation, but the *process of calculation* itself is specified by rules, in other words, we require the *elimination of thinking in favor of mechanical calculation*.[25]

It is a fine distinction, but we might argue with Behmann and insist that we do not require the elimination of thinking, but rather the elimination only of thinking that cannot be understood as mechanical calculation.

> When a logical mathematical proposition is presented, the procedure we ask for must give complete instructions on how *to ascertain by a deterministic computation after finitely many steps* whether the given proposition is true or false. I would like to call the problem formulated above the *general decision problem*. It is of fundamental importance for the character of this problem that *only mechanical calculations* according to given instructions, without any thought activity [*Denktätigkeit*] in the stricter sense. . . .[26]

But what is this "stricter" sense? Perhaps a "looser" sense?

> . . . are admitted as tools for the proof. One could, if one wanted to, speak of *mechanical* or *machinelike* thought[.] (Perhaps one could later let the procedure be carried out by a machine.)[27]

Hilbert's program was immensely influential, as we have already seen—it influenced the logical positivists in the Vienna Circle, and it influenced the methodological program of Bloomfield, as it would later influence the linguist Zellig Harris, and his student Noam Chomsky's early work.

Chomsky will ask linguists to agree that a grammar does not explain the grammaticality of a sentence unless the process that generates the sentence is one which could be carried out by a machine, just as Hilbert's group insisted that a proof needed to be amenable to a formulation that could be stepped through by a machine.

Let us close our discussion of Hilbert's program by emphasizing that Hilbert was in no way saying that mathematics was *just* a formal game of manipulating symbols (though it has been easy to find people who interpret his intent in that way). Doing mathematics involves smart people who think about mathematical questions, and then they develop proofs of the mathematical truths that they have uncovered. The fundamental, the real work of mathematicians is producing those proofs: that is the substance of the field of mathematics at any particular moment.

And then it was over

In 1930, Hilbert retired, and Brouwer stopped participating in the ongoing professional debate that his views on the nature of mathematics had started. The war of the frogs and the mice was over, and none of the younger mathematicians had any interest in seeing it continue. Hilbert's and Brouwer's voices were no longer to be heard, and to anyone who was paying attention, an angel seemed to pass; it was the end of a great generation. But now a new generation of equally brilliant young mathematicians was ready to take their place. These mathematicians included John von Neumann, whose contributions ranged from finding a sophisticated mathematical foundation for the newly discovered quantum mechanics all the way to the development of the modern computer, as we will see in volume 2, and Kurt Gödel, who astonished the world in 1931 by proving mathematically that Hilbert's program was *impossible*. Brouwer's decision to stop participating in the polemics was poorly timed, it seems: he was not there to welcome into his group any disillusioned formalists. Gödel's result was enormous and changed almost everything. It didn't follow that Brouwer was right, but the Hilbertian certainty that we would eventually know of every statement whether it is true or false seemed increasingly illegitimate.

Kurt Gödel was born in 1906 in Brno, which is today the second largest city in the Czech Republic (and it will be important in the next chapter as well, when we meet Roman Jakobson, who will spend many years in Brno).[28] Gödel left his home town at the age of 18, in 1924, and traveled the 75 miles to Vienna to became a student at the university there at an extraordinarily exciting time.[29] Among his teachers were the philosophers of the Vienna Circle that we have met: Rudolf Carnap, Moritz Schlick, and Hans Hahn, who would be his doctoral thesis advisor.[30] He participated in meetings of the Circle, but there was a great deal about the central ideas that the Circle was developing that Gödel had no sympathy at all for. He did not at all agree that the certainty of mathematical truths lies within their linguistic character, and this aspect of logical empiricism, which was perhaps its single most important part, did not appeal to him at all—quite the opposite, in fact.

Gödel's view of mathematics was much closer to Brouwer's intuitionist views.[31] Gödel believed that mathematical objects exist in a timeless immaterial world. What's more, just as humans are creative beings in a material world, there are beings in the timeless immaterial world, beings which we call angels and demons. And Gödel believed that God was responsible for the intricate structure of the universe as we know it.[32] "Every thing was created by God with a determinate goal. Nothing was created without a goal in mind."[33]

> Of course, Gödel had an interesting axiom by which he looked at the world; namely, that nothing that happens in it is due to accident or stupidity. If you really take that axiom seriously all the strange theories that Gödel believed in become absolutely necessary.[34]

More generally, Gödel thought that there was structure to the world on many levels, and that potentially, at least, every event could be meaningfully linked to many others. This was God's work, in his view. Cassou-Noguès described Gödel's position:

> Reason [presupposes] a sort of eye. There is an organ in the brain to perceive abstract concepts, mathematical objects for example, as the eye perceives objects . . . [there is] a mathematical eye . . . linked to cerebral centers of sensory perception and of language (in some fashion attached to both).[35]

> Some physical organ is necessary to make the handling of abstract impressions possible. Nobody is able to deal effectively with them, except in comparison with or on the occasion of sense impressions. This sensory organ must be closely related to the center for language.[36]

> We could possess for example a supplementary sense which shows us a space that is completely separate from space and time . . . and which is so regular that it can be described a finite number of laws. . . . I think that is the real situation, except that reason is not counted with the senses because its objects are quite different than those of the other senses.[37]

Gödel was thoroughly committed to the notion that the deepest attainable knowledge could be directly apprehended by a person: indeed, that was what his "supplementary sense" could provide. It was presumably for this reason that he was very interested in the lives of philosophers, and in particular specific moments in which something happened that changed them and their view of the world forever. Descartes had such a moment of crisis in 1619, when he felt his internal world was turned utterly upside down, and something happened with Husserl, too. Gödel said:

> At some time between 1906 and 1910 Husserl had a psychological crisis. He doubted whether he had accomplished anything, and his wife was very sick. At some point in this period, everything suddenly became clear to Husserl, and he did arrive at some absolute knowledge. But one cannot transfer absolute knowledge to somebody else; therefore, one cannot publish it.[38]

> Husserl reached the end, arrived at the science of metaphysics. Husserl had to conceal his great discovery. Philosophy is a persecuted science. Without concealment, the structure of the world might have killed him.

Gödel became deeply interested in Husserl's phenomenology at the end of the 1950s.[39] He perceived the closeness of spirit that linked intuitionism and phenomenology; it is this affinity that we have called *soft logic*, the view of logic that emphasizes the role of an immediately apprehending subjectivity. But Gödel did not believe that Husserl had accomplished what he had set out to do. He wrote about Husserl:

> Husserl could not communicate his ideas. He knew much more. This is not surprising: generally in psychoanalysis and other fields, many things—drives, will, decisions, and so on—are hidden. But we can only judge on the basis of what has been communicated.[40]

Gödel could not have been clearer about his deep agreement with the intuitionist position that the mathematician *sees* the objects that she studies:

> We have something like a perception of the objects of the theory of sets. I do not see any reason to have less confidence in this sort of perception, that is to say mathematical intuition, than in sensory perception.[41]

GÖDEL AND HIS PROOFS

Gödel had the great advantage over both Brouwer and Hilbert that he felt the intellectual pull of the views developed by both of them, and he hoped to find a better truth that would succeed at integrating the best from both. The profundity of Hilbert's program, we have seen, derived from a desire to develop mathematical tools that would not only talk about mathematical objects of the sort that mathematicians had studied up to that point, but would talk about the nature of mathematical proof. It is easy and tempting to think that the knower and the known can always be kept apart, but Hilbert wanted to use the same tools to analyze both.

Gödel used the distinction that had become coin of the realm in the Vienna Circle, the distinction between an object language and a metalanguage—and here, the object language is arithmetic and basic logical operators, and the metalanguage is the language in which we make our statements about valid proofs, and provability in general. He used the two levels of language, and he found a way to push meta-statements down into the language of the object language. The object language he focused on was an elementary language that included the basic notions of arithmetic, so of course it contained integers among its "objects." Gödel devised a way to translate meta-statements about proofs into numbers, employing a method that today we call *Gödel numbering*. Being a bit casual, if *g* is an integer, we can say "*g*'s proof," which just means the proof we would get by turning the integer *g* into the unique proof it encodes.

Gödel's proof consists of a construction whereby a single object can be simultaneously looked at in three different ways: as an integer, as a formal expression in a formal logic, and as a statement that has meaning to us as human beings looking down on the formal logic. That sounds difficult to arrange, and it is, and Gödel's brilliant idea was to make a system and a statement in it that would actually mean, "This very statement cannot be proven within the logical system in which it has been formulated." A slightly more technically worded paraphrase would be, "There is no sequence of logical steps in this system that constitutes a proof of this very sentence within the logical system in which it has been formulated." Let's call that statement *G*.

If the logical system itself is consistent, then it cannot provide the proof of both a statement and its negation—and in particular, it cannot provide the proof of both *G* and *not-G*. But if *G* can be proved, then *not-G* can be proved, and if *not-G* can be proved, *G* can be proved.[42] Hence if the system is consistent, neither can be proved, and there are statements that can neither be proven nor disproven within any particular logical system. Gödel's argument employed Whitehead and Russell's *Principia Mathematica*, but his exposition was so clear that it was hardly necessary to make the point that a parallel construction would work for any similar formalization of arithmetic.

And so ended one period of anxiety in mathematics. It ended in part because Hilbert retired from intellectual and personal engagement with the dispute between formalism and intuitionism, but more importantly, it ended because Gödel showed a precise, technical, and explicit method for dealing with a question that had seemed to many to be beyond the scope of technical argumentation. At the same time, Gödel's result was not entirely unexpected; several other mathematicians of Gödel's cohort felt that they were just about to achieve a similar result, and experienced to one degree or another a pang that it was Gödel who had succeeded first in proving the theorems (we will encounter one of them, Emil Post, just below).

There was another consequence of all this for the working mathematician. There was now a new game, a new kind of proof that a mathematician could work on, and that was the proof that a certain statement was not *provable*. Some of the great discoveries in mathematics over the last 80 years have been of that sort, a kind of proof that before Gödel was unimaginable.

There is also much to be learned from how logicians interacted with each other in this period, not the least of which is how they viewed their beliefs about logic and mathematics were rooted in their understanding of such traditionally philosophical notions as the nature of mind and reality. The views that were laid out and clarified during this period are broad and classic enough that we can use our understanding of the perspectives of the logicist, the intuitionist, and the formalist to better understand ideas about how language should be studied in years to come. The logicist aims to reduce a scientific domain to logic, which is to say, to concepts which are fundamental to any kind of thought, in our universe or any other. This perspective arises often in the study of language, and a slightly weakened version of it lay at the heart of generative semantics, a part of a battle in linguistics in the 1960s which we will explore at the end of volume 2. The intuitionist focuses on the fact that a legitimate understanding of a phenomenon is based on a subjective intuition of the object in question. We see echoes of this view in two later styles of linguistic analysis: the notion of native speaker intuitions, and some of the anti-formalist perspectives that are often called *functionalist* in current linguistic terminology. And finally, the formalist in linguistics is one who believes that the real heart of successful linguistic theory lies in the description of a formal object that will algorithmically generate the sentences of a language, the view that lies at the heart of generative grammar, as Chomsky originally proposed the idea in the 1950s.

The younger generation, that of Gödel and von Neumann, refused to be divided over the intuitionism/formalism controversy. They saw strengths to both sides of the debate, and integrated into their systems what was most useful. The rest of the world, looking at what was happening in mathematics, could draw two different conclusions. For those who tended towards being anxious, the moral could be drawn that even the best-run disciplines, like mathematics, are inevitably going to find cracks in their foundations, and trying to be logically perfect is a fool's game. The other conclusion, though, was that spending some serious time making explicit the assumptions of one's field was almost certainly time well spent, and from that effort would come a much deeper understanding of what one was doing.

The Chrome Machine of Logic

We discussed the notion of *machine* in chapter 4. The period between 1920 and 1940 was one in which the notion of the machine took on a new meaning. (The word *machine* is etymologically and conceptually linked to the word *mechanical*, and we will use the terms as meaning essentially the same thing, with the only difference that *machine* is a noun, and *mechanical* its corresponding adjective.) Machines until this time were firmly grounded in the physical: in the seventeenth century, when humans were thought to consist of a body, a mind, and a soul, the body participated in a physical world as an object, which interacted with other material objects just like a table or a stone would—and machines were of that same world. For some, like La Mettrie, man's body was just a machine (or just like a machine). As Western man grew to view the physical world as governed by a small set of laws regarding things in space and time, more and more of the world fell into the sphere of the mechanical.[43]

There are several distinct senses linked to the notion of machines and mechanical explanations, and even thoughtful people do not always keep them properly clear and distinct. It will serve us well to reflect a bit on the matter. One aspect of mechanism is grounded in a certain kind of materialism, and a belief that mechanical explanations derive from solid objects interacting locally in space and in time. The clearest examples of this sort of interaction are billiard balls and molecules of gas, which collide with each other and then move off to collide with other things. Their interactions are brief. An equally important example is that of two gears meshed with each other, each turning when the other does because one cog in a gear exerts a force on a cog in the other gear. In the most mechanical of cases, we expect a deterministic outcome from the interactions as well, something which we are certain to find with gears embedded in clockwork, but which is not obviously the case for billiard balls, or molecules, careening off of one another. Balls and molecules typically have round surfaces, and the direction of rebound is computable only to a certain degree of tolerance.

A different notion, *determinism*, lies uncomfortably close to the core of mechanism in its usage. Determinism is the name we give to the view that if we can get a full and detailed account of a situation, its future can be predicted unambiguously. There are two ways, typically, that accounts of events can fall outside of a deterministic frame: either by including the effects of a creature to whom we attribute some sort of will, or by including an irresolvable probabilistic element in the account. A creature with a will can be either a human being, or an imaginary creature: if we ask why the door opened when we approached, and the answer is that someone who was watching opened it (either an actual human being, or an imaginary imp), then the explanation is not at all mechanical. Judgments are a bit less certain in the case of irresolvable probabilities: when we learn that classical quantum theory teaches us that an observation of an electron forces it into one of several possible states, but the choice of which state it goes to is probabilistic and cannot, even in principle, be decided ahead of time, then we are not at all happy with the statement that the change undergone by the electron happens mechanically. Yes, it happens without the intercession of anyone's mind or judgment, but if the outcome cannot be predicted, then it is not at all clear that we would be content saying that the interactions are fully mechanical.

We saw earlier as well that another aspect that the locality of interaction is often taken to be a condition for a mechanical account, a condition which made it impossible for many philosophers to take Newton's theory of gravity seriously for a good long time. Newton's theory had the great advantage that it was quantitative and allowed predictions that left no margin for spirits or humans to be called upon to account for why an apple, or the Moon, falls to Earth. But the effects of Newton's gravity were effects at a distance, and Newton's theory provided no *mechanism* for the influence of an object, like the Earth, on another object at a distance, like an apple, if we insist that a mechanism only allows for local interactions.

The mechanistic view is sometimes associated with an insistence that what is most real of all is what exists at the smallest end of the great ladder of existence, the one that stretches from the smallest atoms all the way up to the greatest galaxies, and the most real forms of interaction are those at the smallest level; all interactions among larger objects are the result of summing up larger and larger numbers of interactions among component pieces. This view might even pass for scientific common sense today, in some circles: physics treats the smallest particles in the universe, and chemistry is a sort of glorified physics in which the general laws are the resultant of astronomical numbers of atoms and molecules interacting in keeping with the laws worked out by physicists of atoms and molecules, and the laws of biology are the resultant of equally large numbers of complex molecules and organelles and tissues built from those molecules, and so on. Large-scale systems are convenient simplifications of the additive properties of effects, and the realest of the real are those that are operative at the smallest level of reality, and that lowest level is composed of the most mechanical of interactions, largely reduced to particles pushing and shoving each other.

This mechanically minded view of reality is hell-bent on making sure that there is no level of analysis on the way from atoms (or subatomic things) all the way up to galaxies and beyond in which something new or different might intervene, which is to say, might enter into the picture as a new type of interaction among pieces, something that was not just a convenient summation of effects at the next lower level of analysis. It's physics all the way down, and there is no level of organization in which new sorts of forces or influences come into play that overwhelm the forces that physics has identified. So it's also physics all the way up, and physics can make sure that it leaves no room for new sorts of forces to appear as we look at higher and higher levels of structure or organization in the universe.

This view has been defended and attacked in various ways as long as it has been around, most notably since the time of Descartes. But there was a new kind of attack on it that happened in the 1930s, and it has a good deal to do with our story. What happened in the 1930s was a radical change in the way the term *machine* would be used. Even for those who still embraced the duality of two spheres of the physical and the mental, a major change was taking place, a change by which the notion of the machine climbed out of the physical world and right into the mental.

Ironically enough, the motivation for this unlikely happening was to be found in our discussion in the preceding section. Mathematicians and logicians were trying to come up with a rigorous definition of an inference that was so free of any subjectivity that no insight at all was necessary for taking the step from one proposition to another. In Hilbert's work, this meant finding rules of inference that even a mechanical process could carry out: even a machine. The intent was not to build a *real* machine; the search was on for a machine in an artificial world in which theorems exist. It was a world in which there was no temperature, no air conditioning, no batteries, no height and width, and no daylight savings time. It was a world that is only a bit like the real world in which you and I exist.

The search to find the right characterization of an inference that requires no subjectivity was one that was undertaken by a number of mathematicians and logicians during this period. It would be an error (though an easy one to make) to think that any of them were trying to restrict what counts as legitimate thinking. They were rather trying to come up with a satisfactory definition of some subpart of legitimate thinking, the subpart which can continue regardless of whether there is any subjectivity or not. What part of thinking could we with full confidence pass off to a dumb intelligence, an intelligence that had no insight and no creativity? Suppose we could write down all of the mathematical proofs that anyone had come up with as of January 1, 1935. Could we identify a subpart of them that consisted only of steps that could be followed by a mind that understood nothing and that only followed formal rules? Alan Turing's idea, which we will discuss in a moment, was that a natural way to provide an explicit account of what it means to only follow formal rules is to imagine a machine with a very limited range of options.

Up to this point, *mechanism* was the sort of thing that would be invoked in the context of showing that something could be accomplished without recourse to intelligence, or that it could be accomplished entirely in terms of physical, material things: things you could hit with your fist and make a thud. But this new sense of machine was not like that. In the first place, it was not trying to show that intelligence did not need to be invoked; it was trying to show that intelligence *was* being invoked, and intelligence turned out to be this machine. In the second place, the machine was absolutely not a physical, material thing: it was an abstract thing that lived in an abstract world. As we will see quite shortly, this abstract machine took on more characteristics of things that we are familiar with in the physical world (like having a "one" or a "zero" stamped on it), but it was an abstract object nonetheless.

Still, the architecture that Turing laid out for his machine, the one that we have taken to calling the Turing machine, was a threat to that simple exclusionary tactic. It was a threat because it pointed out that we could consciously and intentionally take the physical properties of the world, and with the very best engineering mind-set, build a machine (which means, by definition, a device doing nothing but obeying the laws of physics) whose behavior followed the principles of mindless intelligence. It wasn't that the machine obeyed laws of logic rather than laws of physics; we, the builders of the machine, ensured that as long as the machine remained in operation within the operating parameters established up front (appropriate operating temperature well below 150 degrees Fahrenheit, not too much corrosive salt in the air, and the like), the behavior would follow the principles of mindless intelligence. Physics and mindless intelligence were not in opposition to each other, and if the physical conditions departed too far from what the engineer had established as baseline conditions, the machine would no longer be performing logically even though it would be following the established laws of physics.

That was the revolutionary character of the Turing machine, visible to far-sighted thinkers even before the machine had been made metal. It opened the door to a vision of the world in which the laws of logic and thought are not opposed to the laws of physics, but in which the laws of physics can be used to embody the laws of logic, those laws of mindless intelligence. In its initial implementation, this feat was possible because human intelligence designed the machine, but the question could now be asked whether the process of Darwinian evolution could in some similar fashion create a device which ran entirely on physical principles, but which had been built (with no hand of an engineer) so that it had an architecture that would implement intelligence as long as it operated within the device's normal margins of temperature, energy, and the like.

In retrospect, this was a magical moment, or at least a pivotal one, and it marked the beginning of a new understanding of mind and machine, and ultimately language.

1936

1936 was the year in which the great breakthroughs were published, and we will look at three of them. The first of them, chronologically, came from Alonzo Church, with his lambda calculus, which would have great consequences for computer science, logic, and linguistics. The lambda calculus would be the basis for the programming language LISP, which drove so much of the early work in artificial intelligence, and it would also form the foundation of the work on syntax and semantics by Richard Montague in the 1960s, which eventually became central to work in formal linguistic semantics. The second paper, just months later, was written by a young undergraduate in England, Alan Turing, who presented an imaginary machine whose behavior lent insight into the nature of inference and proof. Turing's conception was perhaps the single biggest step on the way to the design of the modern digital computer. In certain respects, we can say that Church's view of computation gave rise to what are today called functional programming languages, like LISP or ML, and Turing's gave rise to imperative programming languages, like C or Python. The third paper was by Emil Post, who, like Turing, also created an imaginary worker with no imagination but the ability to carry out operations. Post was also one of the first to develop what we call today *string rewriting systems*, including the notion of *productions*. These notions were then developed by the young Noam Chomsky in the early 1950s, who built generative grammar as a way of looking at natural language on the basis of Post production systems.[44]

Church and the lambda calculus

Alonzo Church was born in 1903, in Washington, DC, and studied mathematics at Princeton University. His advisor there was Oswald Veblen, who was a very well-known mathematician involved in the developments of the foundation and the axiomatization of geometry (we mentioned earlier that David Hilbert had been instrumental in that development, though Veblen saw his work as closer to that of Peano than that of Hilbert). Just as we have seen over the course of this book, rising stars in mathematics had to spend time in Europe to become full-fledged citizens of their discipline, and Veblen made it possible for Church to go Göttingen—David Hilbert's department—in 1928, and to meet with Brouwer as well in Amsterdam.[45] Church's work after that shows the clear impact that these visits had on him.

Church began to develop a new mathematical way to specify any arbitrary computation, with the aim of providing a better response to the paradoxes that had been the source of turmoil in the world of logic and mathematics, one "in which we avoid use of the free, or real, variable, and in which we introduce a certain restriction on the law of excluded middle as a means of avoiding the paradoxes connected with the mathematics of the transfinite."[46] Part of Church's insight was that paradoxes arise because logicians allow expressions with free variables too much freedom, and he proposed an approach in which free variables were completely eliminated. When an expression contains a variable such as x, the variable must be bound. This means that $x^2 + 2x$ is not a legitimate expression, though it is fine to write down something that means "the function which takes a variable, call it x, and returns the sum of x's square and twice x's value," because outside of that expression, there is no free variable. Church used a notation with the Greek letter λ (lambda) to represent this, and so he would have written what we have just said as $\lambda x(x^2 + 2x)$. If you are going to have an expression like $x^2 + 2x$, then you must use something (typically a λ) to bind that free variable. That means that the "process" of replacing all of the xs in an expression by some value is *not* plugging in values to free variables; it is a process, a substitution, in which the set of occurrences of variables already bound by λ are replaced by the new value.

Part of that system left open the possibility that for some values of an independent variable x, a propositional function F might be neither true nor false (and in saying this, he acknowledged the influence of Brouwer, though his account was not Brouwerian at all), and in particular, Bertrand Russell's self-referring statements end up being neither true nor false, in a way that does not seem pathological at all.

What was especially innovative about his approach was his focus on the mathematical *function* as the most important entity in his conceptual universe, as opposed to, say, the set. His system gave a central position to the creation (or definition) of functions, and to their evaluations, each of which is a natural candidate for forming the fundamental unit of a logical analysis of mathematics. A function inside of another function could be bound by a λ. So just as there could be a function that maps a number to another number that is larger by 2 (we might write it: $\lambda x(x+2)$), there could be a function which takes a function and returns a function which generates a number which is 2 more than what the original function returns. We might write that as $\lambda f(\lambda x(f(x)+2))$.

Church went back to Princeton as a young professor in the Mathematics Department, where he was lucky enough to work with two outstanding young graduate students, Stephen Kleene and Barkley Rosser (though Church's good luck in this regard lasted his whole life: he continued to find outstanding graduate students to work with) who helped him to iron out bugs in his original lambda calculus. By 1936, he had developed a system that was robust enough to allow him to prove that Hilbert's decidability problem for first-order logic was not solvable. (This is similar to Gödel's result a few years earlier, but not the same: Gödel had shown that in certain formal systems, there were statements that could be neither proven nor disproven (and some of them were, in addition, true), while Church showed that there is no algorithm (of a certain sort) that can determine whether any and all statements in first-order logic are provable or not.)[47]

Church's lambda calculus would return to the development of linguistics in the 1960s, when Richard Montague, a student of Alfred Tarski's at the University of California, Los Angeles, began to use it to model English. Barbara Hall Partee, a young linguist there at that time, understood the value of such an approach to linguistics. Her important work and that of her students brought the lambda calculus into mainstream linguistics.[48]

RECURSION

The term *recursion* began to play an important role in logic during this period.[49] As early as 1888, mathematicians were beginning to offer definitions in a recursive style, which means offering a base case, and a method for extension which holds for either the base case or one reached through the method of extension. We can define an even number n in this way: first of all, 2 is an even number. Second of all, if n is an even number, then $n+2$ is an even number. No number that is not identified by that method is an even number. That's the way we would define the notion of an even number if we were to do it recursively, but nothing requires us to define it that way; we can always say, for example, that an even number is one that is divisible by 2 with no remainder. By the 1920s, this style of definition using recursion had become standard, but it was largely a *style* of defining a formal property.

Gödel made use of recursive functions in his work throughout the 1930s, borrowing heavily from the work of others in the 1920 (see Sieg 2008 for a careful study), in order to specify a new sort of function that soon came to be called "primitive recursive" (originally he just called it "recursive.") The notion of a recursive function is a bit complicated, but it would play an important role in the mathematics of computability. Its definition is itself recursive. After defining a few kinds of simple functions as "primitive recursive," including the *successor* function, which takes any number n as its argument and returns the value $n+1$, Gödel said the following. If we have two primitive recursive functions f and g, where f takes n arguments and g takes $n+2$ arguments, then we can define a new primitive recursive function h in two steps. First, if h's first argument is 0 (this is the base case), and this 0 is followed by n additional arguments, then h will produce the same output as f does when we pass it the same n additional arguments. If, however, the first argument of h is some number other than 0—if it is the successor of some number y, say—then h produces a more complex output. It turns to the function g, which takes $n+2$ arguments, and gives it y for its first argument, and for the second argument, it passes the output that it itself (that is, h) "had created" or "would have created" if its first argument were y (rather than the successor of y) and the other following arguments were the same, and then the rest of the arguments to g are the rest of the arguments passed to h. Yes, it is complex, but it is a compact way of saying that g provides a way of stepping through an operation in which h is called many times, and h serves to summarize, or keep track of, where we get after a certain number of such steps.

As we know well, the goal of this work was to understand what purely procedural and attention-free, creativity-free activity can accomplish. Alonzo Church wrote in 1935, "Following a suggestion of Herbrand, but modifying it in an important respect, Gödel has proposed . . . a definition of the term *recursive function*, in a very general sense. In this paper a definition of *recursive function of positive integers* which is essentially Gödel's is adopted. And it is maintained that the notion of an effectively calculable function of positive integers should be identified with that of a recursive function, since other plausible definitions of effective calculability turn out to yield notions that are either equivalent to or weaker than recursiveness."[50] Over the next six decades in mathematical logic, the term *recursive* came to be used to mean *computable*. Not everyone was happy with this terminological trend, and Robert Soare (1996) has been reasonably successful in persuading his fellow mathematicians to stop using the word *recursive* when they mean *computable*.[51]

So much for how mathematicians have used the term "recursion." The use of the term veered off in a *different* direction in the 1950s and 1960s in its use by "computer programmers who, like many of their colleagues, were not acquainted with recursive-function theory, let alone logic in general."[52] What was at issue was whether functions would be allowed to call themselves, just as they can call other functions. All computer languages allow for loops and iteration, but there was debate as to whether functions should be allowed to call themselves, and the question of whether this could be implemented was directly connected to whether memory-allocation could be done dynamically (i.e., at run-time, when the program is determining how deep the recursive, self-calling activity will be) or statically (i.e., when the program is being compiled). Church's lambda calculus and Gödel's proof are loaded with formal mechanisms that do the moral equivalent of calling themselves, but this is not at all the same as producing an infinite output: Turing machines, like any other sort of calculating machine, can get into loops, and the program (so to speak) can get out of those loops, or not, and one loop can be within another (so to speak) without recourse to anything resembling recursion. What identifies *recursion* is a computational process that keeps track of where it is in its computation, typically by some version of a push-down stack, and it keeps track of which functions' computations it has begun but not finished. We may say that recursion is permitted when a record of this sort allows for the possibility that one and the same function is marked as incomplete two or more times at a given moment of computational time, in what is a sequential mode of operation.[53]

Alan Turing and his machine

Church's paper appeared in the spring of 1936. The year before, Alan Turing, a 23-year-old college student at the University of Cambridge studying mathematics, took a course from Max Newman that included the work we have been discussing on the foundations of mathematics. The young Turing quickly developed the notion that today we call a "Turing machine," which he rather more modestly called an a-machine, an automatic machine, that could carry out purely formal mathematics, at least in Turing's mind. He wrote this material up in a paper in the spring of 1936, showing it to his teacher, who had just received an offprint from Alonzo Church of his paper, showing that Hilbert's decidability problem was unsolvable. Turing had just reached the same astonishing conclusion! While it was a disappointment for Turing to have been scooped by Church, Newman could see that the two papers described quite different approaches to the resolution of similar questions, and Turing's paper was accepted for publication later in the year in the *Proceedings of the London Mathematical Society*. Turing was able to transfer to Princeton, then, and work with Church over the next two years to earn a PhD at Princeton, before he went back home to England, soon to become engaged in the war effort, by developing in the real world the kind of computer he had imagined, a computer that could be used to crack the German code during the Second World War.

But long before Turing built a Turing machine in our world, he had built a machine which had some of the properties of an abstract, mathematical object, and some of the properties of a real machine that is made out of metal and plastic. It was really a chimera, an unworldly hybrid of metal and mental.[54]

A Turing machine has one part that is not entirely internal to it, an endless paper tape divided up into square boxes on which the machine is able to print a symbol from a finite alphabet, and which the machine can read. At any given moment, the machine can see exactly one box from the tape, and no others. It is also able to shift the tape one square at a time in either direction. These actions are all determined in a straightforward fashion by the machine's internal architecture, which has two parts to it. One is a very simple indicator which specifies what its current state is, a state chosen from a finite set of such states which has been determined when the machine was first built (so to speak—it is so easy to fall into using physical terms when discussing Turing machines), and the other is a large playbook which specifies exactly what the machine will do, with separate specifications for each state that it might be in, and for each symbol that might be appearing on the box from the tape that it currently sees.[55] That is all there is to a Turing machine.

What is remarkable is that Turing was able to show that his a-machine could compute any function that a mathematician could fully and explicitly formulate that was based on numbers. Turing's methods were very clever, and too clever for us to reproduce here, but when he was done, no one could read his paper and doubt that one of his machines could carry out any mechanical mathematical procedure that they might come up with.

One of the very central insights that Turing brought to the analysis of the abstract machine that he devised was the realization that time and finitude play a tremendously important role. You recall we began this chapter emphasizing the importance of the fact that we as humans are finite and limited, and yet in some ways we rise above our finitude to draw conclusions (many of which are certain conclusions) about an infinitude of objects such as numbers, or sentences. What Turing realized was that once he had designed one of his automatic machines, he could not always know whether, in carrying out its assigned operations on a tape that has a string of symbols printed on it, the machine would come to a halt. Now, in some of his analyses, Turing focused on machines that would continue to print sequences of 0s or 1s forever: if you are interested in computing numbers (not just whole numbers, but also numbers like π and $\sqrt{2}$), you cannot expect the machine to stop; it will never get to the end, since there is no end. So it was reasonable to declare that a machine that should compute numbers must continue to print out 0s and 1s on successive boxes, never ending (though it may eventually settle into printing 0s forever). For a machine whose purpose is to compute a number, Turing insisted that the machine always continue to print out 0s and 1s forever and ever, moving down the endless tape.

On the other hand, there are other tasks you might ask of a Turing machine which do not consist of asking it to compute a number, and cases where you want "success" to be associated with the machine coming to a halt, and in fact that is generally how the question is discussed nowadays: what are the conditions under which we can compute whether a Turing machine will halt, when we know the machine's internal architecture and the sequence of symbols that appear on its tape when it begins? (We can take the notion of "halting" to be associated with a particular state of the machine, one for which there are no actions that the machine is to make once it has entered into that state.)

Turing followed Gödel's clever idea that we discussed earlier, by which a single object could be viewed either as a number or as an expression in formal logic. Turing showed that his machine could be straightforwardly defined and specified by a single number, even if it is a large number, and even if the same machine could be described by many different numbers. (Today, this does not seem too surprising, perhaps, since we are accustomed to writing computer programs in a way that makes them look quite a bit like numbers, but all our experience today is built upon the insights that began with brilliant flashes like the ones in Turing's paper.) So in a very real sense, we can ask whether a Turing machine could answer questions about whether *another* Turing machine will halt when given this or that input tape, since that input tape contains a number, and the number could perfectly well be the number of a particular Turing machine.

You might wonder why the Turing machine that is fed the number of a different Turing machine should be able to reconstruct *anything* about that machine from its number, but that is a misplaced worry, because we have already been convinced that a Turing machine is potentially smart enough to do *anything* that can be spelled out in a complete algorithm. So despite the fact that the architecture of the Turing machine seems very simple, we should really think of it as a machine that can do anything that we can completely describe, when we start building it from the very beginning. So bear in mind that a Turing machine represents any kind of mechanical symbolic process that you or we could imagine.

Emil Post and his productions

Emil Post was born in the Russian Empire into a Jewish family in 1897, and like many others we have seen in our story, his family came to New York at the turn of the century, when Post was four years old. Post completed a PhD in mathematics at Columbia University, and then went to Princeton for postdoctoral studies in 1920. There he developed his ideas about the limits to mathematical proofs, some years ahead of others whose work would be more famous—a matter of profound frustration for Post, made hardly more bearable by the fact that he did not carry through his project to the point that others, like Gödel and Turing, did in later years. In retrospect, Post felt that he had come ever so close to achieving Gödel's results, about a decade earlier than Gödel, but he recognized that he did not get quite close enough to say that he had beaten Gödel to the prize.[56]

Quite independently of Turing, Post came up with the idea of an entirely unimaginative machine, which he called a "problem solver or worker" instead of a "machine." His article was received by the *Bulletin of the American Mathematical Society* on October 7, 1936. It is quite astonishing how similar Post's worker and Turing's machine are. Post's worker has a *symbol space*, just like Turing's machine had a tape, and each spot on Post's worker's symbol space is either marked or empty. The worker can mark a box, erase the box, move to a box to the left or right, and determine whether the box he is standing at is marked or not.[57] A set of directions for the unimaginative Post worker is made up of a numbered list of operations that tell him what particular action to take, and then which operation to proceed to next, or else it tells him what particular action to take, which will be followed by one of two different operations depending on the outcome of the prior action. In fact, when we read about the Turing machine today, what we generally read is a version of Turing's a-machine which has been modified to look more like Post's machine, due largely to Martin Davis's influential book in 1958. Davis had studied with Post in college, and later would edit Post's collected writings. Post's work lives on, as he might well have hoped for, in the way we think about the machine today that we *call* the Turing machine.[58]

GENERATIVE METHODS

Post published an important paper in 1944, based on an invited address he gave at the New York meeting of the American Mathematical Society. He began by explaining that he knew he was addressing mathematicians, and that as a group they had not found the work of people like Gödel and Turing easy to follow. He hoped to show his audience that there was an intuitive development that lay behind the forbidding formalism, and that if he could present that kernel, his audience would have learned something of considerable value.

His goal was to present a general characterization of certain kinds of functions of positive integers and certain kinds of sets of positive integers—those which are created in an explicit and step-by-step fashion, in a finite number of steps. He sneaked the word "generate" into the discussion quite early on. He wrote, "we rather imagine the positive integers 1,2,3, . . . generated in their natural order," and then a "corresponding process [is] set up which generates n^2" for each n, so that we *generate* the sequence of square integers.[59] And we could generate all sorts of sequences in this way, including fourth powers of each integer, and so on.

After briefly mentioning these examples, Post wrote,

> Several more examples would have to be given to convey the writer's [i.e., Post's] concept of a *generated set*, in the present instance of positive integers. Suffice it to say that each element of the set is at some time written down, and earmarked as belonging to the set, as a result of predetermined effective processes. It is understood that once an element is placed in the set, it stays there.[60]

Post was focused on explaining to the society of mathematicians why his notion of *generation* was useful and significant. He could have had no idea that it would revolutionize the world of linguistics, to the point where it was *generative* grammar that became the dominant view of language in the second half of Post's century. Emil Post played an important role in the larger development of the mind sciences, and he would never know it. What he saw of his own career was perhaps not enough, from his own point of view; he had been beat, he had been scooped by Alan Turing as far as the world was concerned, even if Post had also developed on his own the idea of a machine that was deterministic and unimaginative, and yet capable of computing complex algorithms. But the formalism that Turing used was different from Post's, and it was Post's notation that would be presented in detail in a well-received textbook just before his death, and which would attract the attention of a young undergraduate at the University of Pennsylvania, Noam Chomsky. We will return to that moment in volume 2. For now, let us simply bear in mind that Post's notion of *generation* was novel in the world of formal logic, and that it would be received enthusiastically by Noam Chomsky in the early 1950s to serve as the central notion of his generative grammar.

POST'S CANONICAL AND NORMAL LANGUAGES

Post's work was seminal in the development of what have come to be known as string rewriting systems, which is the mathematical study of systems which consists of sets of strings built up from a finite alphabet of symbols. It is not hard to see how the interest in such systems was very close to the interest in providing a concise and rigorous notion of mathematical proof. Post developed two important notions in his work: one of them was a *canonical system*, and the other was a *normal system*. They look and feel different, but Post was able to show that what could be done with the one could be done with the other.

A canonical system looks a bit daunting at first. We begin with an alphabet $\mathcal{A}$, and a finite set of initial words $\mathcal{L}$ which have been spelled out from letters in $\mathcal{A}$, and a set of production rules. Each production rule has an input (so to speak) and an output which is cobbled together from the input; each production rule is a recipe for that operation. The input to a production rule is a *set* of patterns (though Post does not use that homely term), each pattern consisting of a string that is made up of an alternating pattern of constants and variables. If we use capital letters like *M* and *N* to mark string variables, a pattern might look like this: "the *M* of *N* was *P*." This pattern would then match correctly a string like "the king of England was bald."

The way Post described it, a pattern took the general form like this: $g_{11} P_{i'_1} g_{12} P_{i'_2} \cdots g_{1m_1} P_{i'_{m_1}} g_{1(m_1+1)}$. That may not be easy to read, but it is intended to be a general form from which something like "the *M* of *N* was *P*" could be expressed. A production rule may have as many patterns as we like (and they can be of any length) to form its input. And then its output is *another* pattern, which naturally Post writes like this: $g_1 P_{i_1} g_2 P_{i_2} \cdots g_m P_{i_m} g_{m+1}$. This is just a way of saying that the output is going to be another pattern, where some particular symbols will definitely occur—those are the $g_1, g_2, \ldots g_{m+1}$, interspersed with some other strings—and those other strings, the various P_is in the output, must all come from one or more of the variables in the input patterns.

Post also defined a notion of a *normal* language, one generated by productions which are all of a certain very simple form: $\sigma_1\alpha \rightarrow \alpha\sigma_2$, where the σ's are specific strings that expressed in the rule, and α is just a variable that can cover any stretch of letters. Post was able to show quite remarkably that such a simple template is rich enough to be able to generate any language which can be generated by the much more general sort of system that he called a *canonical* system.

We would not normally expect the reader to go through all of this formalism, but this time there is a reason to do so, because this is the notation that Noam Chomsky would pick up in the mid-1950s to define string transformations in linguistics, and it is best if you understand where it comes from. We cannot overemphasize the importance of looking at Post's work on string rewriting systems.

In an odd stroke of fate, Post's work came to be better known due to the writings of others. In volume 2, we will look in some detail at a book published in 1950 by a young mathematician named Paul Rosenbloom. The book was entitled *The Elements of Mathematical Logic*, and it introduced Noam Chomsky as a young student to Post's work and to Rosenbloom's interpretation of that work. Rosenbloom was clearly thinking about the implications of Post's formalisms for linguistics; at one point, he wrote,

> With this tool at our disposal we can explain simply and elegantly many important mathematical and logical notions. One might also expect that many concepts in linguistics which have resisted all attempts up to now at clear and general formulation may now be treated with the same lucidity and rigor which has made mathematics a model for other sciences. The wealth of detail

and the manifold irregularities of natural languages have often obfuscated the simple general principles underlying linguistic phenomena.[61]

The Logicians' Grammar

The third major theme of this chapter concerns categorial grammar, and it is deeply connected, both personally and intellectually, to the Vienna Circle and, intellectually, to Edmund Husserl. Categorial grammar was the creation of the Polish logicians known as the Lwów-Warsaw school, an active group of researchers in several areas during the first four decades of the twentieth century.[62]

Looking ahead just a bit, the development of Montague grammar in the United States was an integral part of this development—Montague was a student of Alfred Tarski at the University of California, Los Angeles, and Tarski was himself a member of the Polish logicians and who came to the United States with the rise of Hitler. Montague grammar has been at the center of the recrudescence of interest in formal semantics in linguistic theory, both in the United States and in Europe.[63]

It was Kazimierz Twardowski who founded this school. We have already encountered him, but only just barely; in chapter 3, we learned that he was a student of Franz Brentano in Vienna, and we also noted some of his interactions with Edmund Husserl in the previous chapter, but now is the right time to say more about him and his role in this story. Twardowski was born in Vienna in 1866, and grew up there, studying with Brentano at the university, and then studying with the psychologists Wundt and Stumpf, important figures in chapter 5. But his parents were both from the Polish nobility, and for most of his life he lived in Poland, publishing and teaching in Polish.[64]

Husserl's ideas about language, about its structure and its basis in meaning, were widely discussed by the Polish logicians. Husserl and Twardowski were both students of Brentano's, but their difference in age was enough that they did not really know each on a personal level. Husserl's ideas about the relationship between a part and a whole were developed in his own third *Logical Investigation*, and this was adopted and advanced by one of the leading Polish logicians, Leśniewski—first in 1916, and then again 15 years later, when he introduced the term *mereology*, from the Greek *μεροσ* "part" and still used today, to cover this area of logical investigation.[65] In 1922, Stanisław Leśniewski developed a notion of *semantische Kategorien*

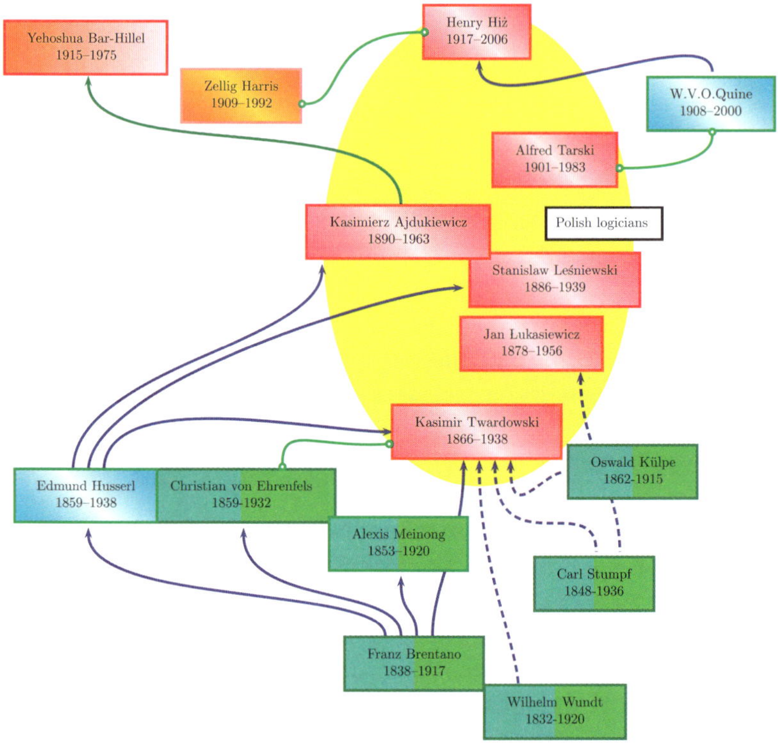

FIGURE 8.2. Polish logicians

(semantic categories),[66] and he viewed his work as based both on Aristotelian notions of categories and on Husserl's work. In 1929, he wrote:

> In 1922 I outlined a concept of semantical categories as a replacement for the hierarchy of types [of Bertrand Russell], which is quite unintuitive to me. Frankly, I would still today feel obliged to accept this concept even if there were no antinomies at all. From a formal point of view, my concept of semantical categories is closely related to the well-known type theories, especially with regard to their theoretical consequences. Intuitively, however, the concept is more easily related to the thread of tradition running through Aristotle's categories, the parts of speech of traditional grammar, and Husserl's meaning categories.[67]

One of Leśniewski's students, Henry Hiż, later came to the United States, where he eventually became a professor of linguistics at the University of

Pennsylvania. Looking back on his career, Henry Hiż later recalled his teacher who was "inspired by Edmund Husserl's *Logical Investigations* of 1901 and Russell's theory of types, as presented in . . . *Principia Mathematica* [and who] developed a grand theory of semantic categories. He used it in his system of foundations of mathematics, but he also claimed that no language can be in discord with his theory. His way of forming definitions was a precursor of Zellig Harris's linguistic transformations."[68]

We have just seen Hiż use the term "semantic" to characterize these Husserlian categories, but they were as much syntactic and distributional as they were semantic. Still, what is more important for us was the way that this approach made possible the project begun by Kazimierz Ajdukiewicz, and which eventually came to be known as *categorial grammar*. Ajdukiewicz was born in 1890, a member of the generation that was able to do their graduate studies just before the First World War broke out. He studied logic and mathematics in Poland (with Twardowski, Jan Lukasiewicz, and the famous mathematician Wacław Sierpiński), and then went to Göttingen University, where he could listen to David Hilbert lecture on mathematics and Edmund Husserl on philosophy.

Ajdukiewicz's idea of a categorial grammar was presented in a paper that he published in German in 1935.[69] The paper was called "Syntactic connectedness," and the way we read it today, it began with a bang. In the very first sentence, Ajdukiewicz wrote that "the problems of linguistic structure" had become "the most important problems of logic," and the main problem is that of syntactic coherence: how do we deal with the way in which verbs, which have a specific number of argument positions, become syntactically bound to their arguments in an actual sentence, in such a way that the meaning of the whole is the meaning of the parts properly integrated into a whole? "Such an arrangement of expressions is syntactically coherent." Sentences do *not* consist of words strung together; any given sentence is much more than a record of a sequence of choices made by a speaker, it is a product of a large number of relations between words that allow for only a small number of ways to constitute a coherent syntactic whole.

Here is the example that he gave to explain what he meant:

> "*John loves Anne*" is syntactically built up in a coherent way out of terms that make sense in the English language and is itself an expression that makes sense in the English language. On the other hand, "*Perhaps a horse if to shine however*" is indeed an arrangement of words that make sense in the English

language, but it lacks syntactical coherence and is not itself an expression that makes sense in the English language.[70]

Ajdukiewicz attempted to bring Husserl's ideas of categories closer to a treatment of natural language, and it was with categorial grammars that he proposed to accomplish this.[71] Ajdukiewicz alluded to Bertrand Russell's system of types, but only in passing, and he opted for Stanisław Leśniewski's theory of semantic categories.

Ajdukiewicz traced the term "semantical category" back to Edmund Husserl; we have seen that Husserl develops this notion in his *Logical Investigations*. Husserl had noted that we can categorize expressions in a given language in such a way that if two expressions (*S* and *T*, for example, such as "the chair" and "the table") are from the same class, when we replace an occurrence of *S* in an expression by *T* (so that "I sat on the chair" becomes "I sat on the table"), the result is just as coherent and meaningful as what we started with. These are Husserl's *logical categories*.

Ajdukiewicz divided these semantic categories into two sorts: *basic categories* and *functorial categories*.[72] Ajdukiewicz offered just a rough characterization of a functional category: it is one that is incomplete, that awaits something to be added to it to become completed (that is, these were Husserl's syncategorematic categories).

Basic categories of types, following Husserl (and then Leśniewski), consist of sentences and general names. There may be more, but Ajdukiewicz proposed that we can get started with just these two. Of the functorial categories, there is no limit; these can be characterized by the number of arguments that they require, the semantic category that they expect each of the arguments to be, and the semantic category of the functorial category once it has been associated with all of the arguments that it requires.

Ajdukiewicz laid out a fractional notation for his functorial categories, placing on top, in the numerator, the category that emerges when the category in question is associated with all of its required arguments, and on the bottom, in the denominator, the list of the arguments the category requires and the type of each. A functional category that requires two "names" as its arguments and with them produces something in the category of sentence will be written this way: $\frac{s}{(nn)}$. The logical connective of inference, →, has the syntactic category $\frac{s}{(ss)}$; Ajdukiewicz notes the categories of a more complex sentence:

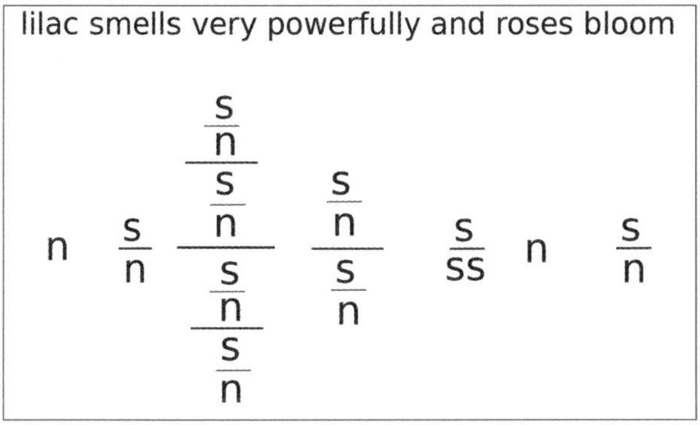

Fine: now how does a language express the linkage between an argument and the corresponding functor? In artificial language, we may use parentheses or linear order, or both, but in ordinary language, this is "determined by means of the order of terms, their inflectional forms, prepositions, and punctuation marks."[73] In fact, the linkage between functors and their arguments is the primary way in which syntactic coherence is achieved, and each principal part of an expression has its own *main functor*, again a notion going back to Leśniewski.

Any expression can now be analyzed as a descending hierarchy of expressions, each with a main functor; if the top functor is well-formed, by virtue of being analyzed as a functor with the appropriate number and kind of arguments, and if each argument is either a simple term of the right sort, or itself is well-formed by virtue of a recursive analysis of what it contains inside it, then the whole expression is *well-formed*; when there is an appropriate matching between the number and type of arguments that a functor requires and that it has in a given expression, then the expression is said to be *syntactically coherent.*

Ajdukiewicz noted that there are a number of distinct ways in which subparts of a sentence can be connected. He wrote, in a sentence that does not sound too dated, "To bring out the multiple mutual attachments of parts of an expression, symbolic languages resort to conventions regarding the 'binding power' of various operators, to the use of brackets, and to the order of terms."[74]

Ajdukiewicz pointed out that syntactic deletion may have to be reconstructed in order to correctly characterize a well-formed expression:

> We must notice that in ordinary language there often occur elliptical expressions; so in such a language it sometimes happens that an expression that

makes sense is not well-formed throughout, if we attend only to the terms explicitly contained in it. But we can easily get an expression well-formed throughout by adding the understood words that are omitted. Greater difficulties arise for a language like German that contains separable words; in this case we cannot give a criterion for a single word in a purely structural way.[75]

Ajdukiewicz then proposed what has come to be known as Polish notation for his expressions: each functor should be written down before each of its arguments. Instead of writing "$(p \vee p) \rightarrow p$," we write something like "$\rightarrow \vee ppp$." If we want to speak of the categories of these expressions, we write something slightly more complex to read "$\frac{s}{ss}\frac{s}{ss}sss$."

If syntactic well-formedness is preserved whenever we replace one expression by another from the same category, then the task of determining whether an expression is syntactically well-formed can be done just by looking at the categories of each word; we do not need to take into consideration the word itself. Ajdukiewicz pointed out how easy it is to determine if an expression is well-formed, once we have translated it into Polish notation: scanning from left to right, we take the first functor that we find that is followed by the indices that it is looking for, as expressed in its denominator. In the case of "lilac smells very powerfully and roses bloom," we begin with the sequence we had above, but rearrange it into Polish notation, so that a functor always precedes its arguments:[76]

and very powerfully smells lilac bloom roses

$$\frac{s}{ss} \quad \frac{\frac{\frac{s}{n}}{\frac{s}{n}}}{\frac{\frac{s}{n}}{\frac{s}{n}}} \quad \frac{\frac{s}{n}}{\frac{s}{n}} \quad \frac{s}{n} \quad n \quad \frac{s}{n} \quad n$$

We see that the first simplification that can be derived is this, which is gotten by reducing the second and the third element:

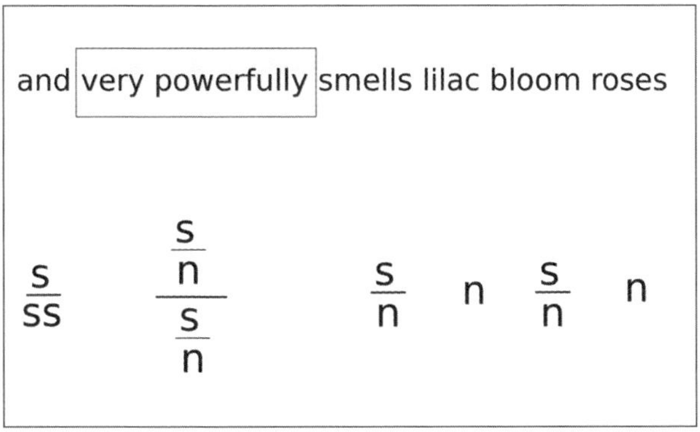

The next “derivative” is this:

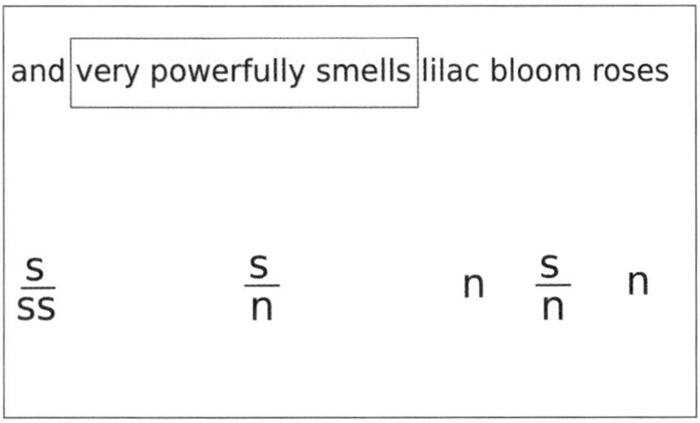

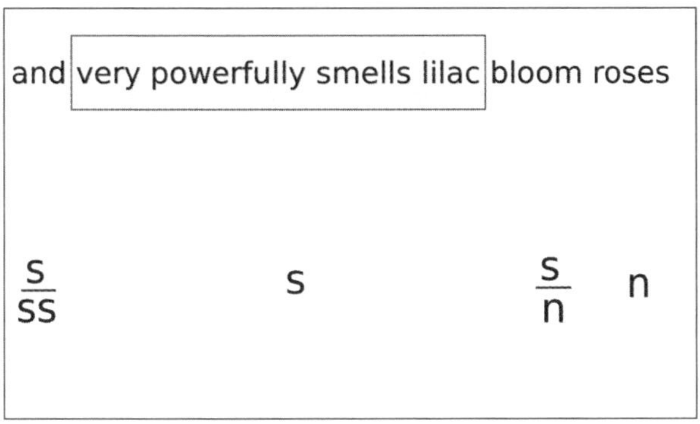

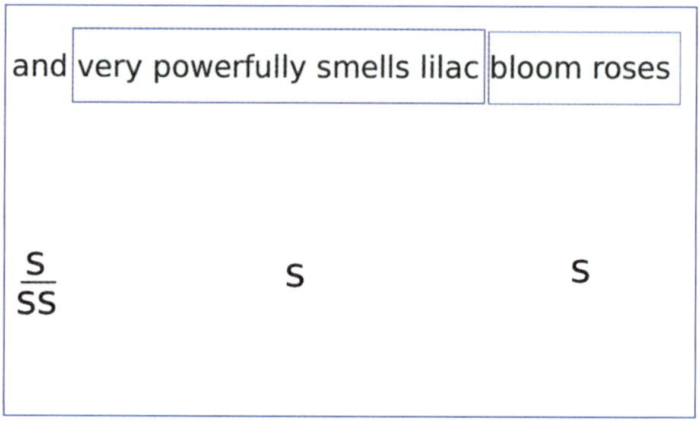

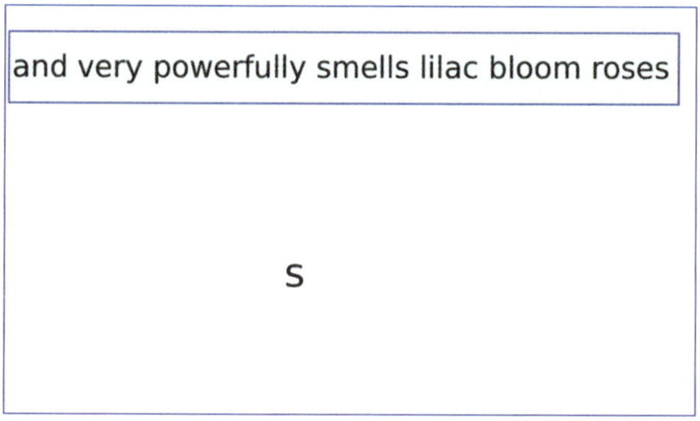

We now have an answer to the question: when is an expression syntactically coherent? It is coherent if it can undergo a process of the sort we have just observed, with a representation that is well-formed throughout, and it finally ends up as a single index, such as *s* in these examples, though it need not be simple, as in this case.

The linguist Emmon Bach later identified the guiding ideas of categorial grammar as three in number: first, the Fregean idea that the functor/argument relation in mathematics is present in language as well, in such relations as that between a verb and its arguments; second, a tight fit between syntactic and semantic generalizations; and third, a commitment to a form of logical monotonicity, leading one away from analyses based on syntactic movement (though in saying that, we are getting a bit ahead of ourselves.)[77]

To the United States

There were a number of distinctly personal ways in which this work came to the United States. We have already noted that Alfred Tarski moved to the United States with the rise of Nazism, and Quine, who was close to Tarski and more generally to the work of the Polish logicians, taught at Harvard.[78] More directly connected to the development of American linguistics was Henry Hiż.

We observed earlier that Hiż was a student in Warsaw, where he worked with Kotarbiński, Leśniewski, Lukasiewciz, and Tarski. In his seminar with Kotarbiński, "we read Karl Bühler's *Sprachtheorie* and contemporary Polish grammarians."[79]

He started studying there in 1937, but World War II interfered with that. In 1946, after the war was over, he went to Harvard to study philosophy. Quine recommended Bloomfield's *Language*, and Hiż found both Bloomfield's formalism and his behaviorism appealing.

Hiż was one of the very most important intellectual conduits between the Polish logicians and American linguistics. He began teaching at the University of Pennsylvania in 1951, where his students included a young Noam Chomsky. "They could take a heavy load; therefore I gave them a lot of Leśniewski, Tarski and the so-called Post production rules, in addition to some historical background."[80]

This development of categorial grammar by the Polish logicians was the first model of formal syntax worthy of the name. In volume 2, we will see it reappear in the context of American linguistics in the early 1950s when Yehoshua Bar-Hillel brought it to the pages of *Language*, after which it was developed further by Lambek, and it remains a major player in the way we think about formal syntax today.

Conclusions

This chapter completes our survey of four disciplines in the period from 1900 to 1940: psychology, linguistics, philosophy, and logic. It was not without difficulty that we pulled them apart and laid them out sequentially. They all grew from the sources that we described in the preceding chapters involving the nineteenth century, but the four disciplines also interacted constantly. Some figures, like Franz Brentano and Edmund Husserl, showed up frequently in unexpected places. Over the course of these four decades, the balance of influence between Europe and the United States did not remain stable. At the beginning, there was a strong presumption that an American had to study in Europe after his PhD, and most likely in Germany; German was the language of scholarship. By the 1930s, the tide was changing, as Europeans like Albert Einstein and John von Neumann were to be found in Princeton at the Institute for Advanced Studies, and an Englishman like Alan Turing had to come to the United States to write his dissertation.

The greatest change to be ushered in from the work we have been reviewing would be the invention of the computer during the Second World War. Alan Turing's a-machine became something real, and in volume 2 we will see how the incarnation of a computing machine, a machine that schemes, would change our understanding of how we view ourselves. But we would like to emphasize that already by the 1930s, there was a noticeable change in the way that abstract reality was playing out as well. We emphasized earlier that Turing's and Post's machines were abstract objects with provable mathematical properties, but they were built with a number of characteristics of the physical world, most notably the integration of a certain kind of *temporality* into their universe. Classical mathematics eschews time, as these things go. If a mathematician says that a function f goes from X to Y, and his words are brought to his attention, he will brush them aside and say that this is just a convenient way of talking about things. A function is a static pairing of objects, pairs made up of one object from one set and the other from another, and time plays no role.

In this new mathematical world of people like Post and Turing, abstract objects took on certain temporal and spatial aspects, and nothing illustrated this better than the architecture of Turing's machine. Things happen to the machine and the machine does things; most notably, it moves its tape along to the left or to the right, and as it does that, what is written on the tape remains the same. Even the notion of "remaining the same" contains temporality within it, of course. How new was this for mathematics? While there were hints of it in the past, it was nonetheless a big shift. When we learn to do long division on a piece of paper, we never mistake our algorithm for a deep fact about division or subtraction; the temporality of the long division algorithm derives from our own human limitations, not the nature of arithmetic.[81]

We began this chapter with the thought that stretching across everything we would see in this chapter was the shadow of the nature of the infinite, and we hope that you sense that now. There have been two aspects of this in the material we have reviewed. The first involves the infinity of things that can be counted, like the integers, and during this period there arose a better understanding of the difference between those things that are countably infinite and tightly under our intellectual control, like the certainty that there is no end to the set of prime numbers, and those things that are countably infinite and yet remain outside of our grasp, involving questions for which a mechanical seeker could spend an eternity looking and never, ever come to a conclusion. The other aspect of infinity that played a major role was the difference in size and character of the countable infinity of numbers, on the one hand, and the uncountable infinity of real numbers, that is, those numbers whose decimal expansion goes on forever without any expectation that there is a repeating pattern in the digits (a repeating pattern guarantees that the number is a rational number, hence countable). It was Georg Cantor who showed how we could determine (often, though not always) whether a set was countably infinite, or larger, more infinite, than countably infinite. That method was then used by the logicians we have studied in this chapter to explore the limits of mathematical activity, when it is viewed as a sequence of discrete steps. The key to that argument was always this: if you try to establish a list of all the elements of your set (which is just a way of counting those elements), I can immediately come up with an element in your set that can never be on your list.[82] Taking a step back from all of this, it is hard not to be astonished by the accomplishment of the finite human mind: it can sense and measure the difference in size between two sets, both of which are infinite, though one is infinitely larger than the other.

In the 1950s, the passage of ideas across the barriers separating these fields would become an enormous, riotous flood. The complex and arcane ideas of this chapter would be carried over to linguistics and psychology by people like Noam Chomsky and George Miller. That is the world we live in today.

CHAPTER NINE

European Structuralism, 1920–1940

Les années vingt et trente voient se dérouler les soubresauts de la lente et difficile transformation de la métaphore organiciste en structuralisme. —Patrick Sériot, *Structure et totalité*

The conception of linguistics developed in Central Europe during the 1920s and 1930s by Nikolai Trubetzkoy, Roman Jakobson, and their colleagues is an important and influential one for contemporary linguistics, but the precise nature of this development and of its influence is more controversial than it was even 30 years ago.[1]

Certain basic facts are hardly open to interpretation: Trubetzkoy was born in 1890 in an aristocratic Russian family, and Jakobson was born six years later in a bourgeois Jewish family in Moscow. Both were outstanding students and young scholars before the revolutions of 1917, and both spent the next 20 years in Central Europe, working together and with others in Prague and in Vienna. But the larger picture which includes the philosophical and political views that motivated them is only now coming into focus.

In some measure this is due to a pair of contingent and adventitious occurrences: Trubetzkoy died young, in 1938 at the age of 48, and Jakobson not only survived the tumultuous 1930s, but he arrived in the New World during the Second World War, and—not without difficulty and even opposition—succeeded in establishing an outstanding professional career in the United States after the war, where he had close and warm personal connections with the two most influential American linguists in the postwar period, Noam Chomsky and Morris Halle.

The story of the nearly 20 years during which Trubetzkoy and Jakobson worked together is full of contrasts of rupture and continuity, providing a veritable embarrassment of riches from which to choose in recounting their story. But at the same time we have been continually struck by a sort of false or imaginary continuity in the way in which the work of the early Prague Circle remains part of today's consciousness. Trubetzkoy's *Principles of Phonology* (*Grundzüge der Phonologie* in the original) is read today as a source of linguistic insight more often than Bloomfield's *Language* or Sapir's *Language*, yet *Principles* is read not in the way that Trubetzkoy would have expected it to be, but rather as a text that if read properly can offer us insights that can be appreciated by generative phonologists.[2] This is fine—but one-sided and incomplete. It comes as a shock to the reader today to see how deeply Trubetzkoy rejected the cultural values of Western Europe, and while we have enough historical hindsight to understand the cultural and social trauma caused by the Bolshevik Revolution and the consequent dissolution of the Russian Empire, we cannot reject out of hand Trubetzkoy's conception of his linguistics as a part of a much larger enterprise that included the study of culture, of geography, of social identity, and the very nature of the bonds that hold society together.[3]

We will explore first some aspects of the lives of Nikolai Trubetzkoy, and Roman Jakobson, and more briefly the Prague Linguistic Circle. Artificial though it may be, we will try to hold off on the discussion of the ideas that they shared, and the ideas on which they differed, till later in the chapter. In order to provide at least a sense of the dynamic of their careers and the world around them, we have to put their ideas on hold for a moment while we consider their interactions. After discussing both Trubetzkoy and Jakobson, we will turn to the positions that the Prague Linguistic Circle stood for, and then explore just a few of the central ideas of Trubetzkoy and Jakobson concerning phonemes, oppositions, and features.

But we can give you a warning right now that Trubetzkoy and Jakobson disagreed about a good deal, and not just as a matter of detail or implementation. Trubetzkoy saw logic—especially as developed by contemporaries such as Husserl—as a tool to understand complex systems. The phonemic system of a language was precisely such a complex structure, and providing an analysis of a phonological system of a particular language would bring to light dependencies between various properties. Just as Husserl had pointed to the fact that color, for example, was a property that only extended objects have—musical notes do not have color, there was a logic to the properties that things possess, and to understand the total logic of the human experience required understanding what depends on what. And he was sure that the right answer was not a simple binary system.

Jakobson did not share this understanding, though, as we will see, the chronology is complicated. Jakobson's interest in phonological features predated Trubetzkoy's, and once Trubetzkoy began working with great energy on the material that would become his *Principles of Phonology*, Jakobson's interests were elsewhere. But Jakobson never gave any reason to think that he took seriously a dependency logic of features, and the fact that he continued to press for a single set of features and a strict binary logic to deal both with vowels and consonants is the best evidence we could hope for that he failed to see the interest of the Husserl-Trubetzkoy perspective.

Let us turn first to the history, after which we return to the ideas. We will meet first Nikolai Trubetzkoy, then Roman Jakobson, and then look at the development of the early years of the Prague Linguistic Circle.

Nikolai Trubetzkoy

Nikolai Trubetzkoy was born in 1890 in Moscow—a wunderkind, a child genius. He was born into the family of a prince, a family that was part of the high Russian aristocracy. The greatness of his family began when an ancestor of his refused to convert to Roman Catholicism, and went instead to Moscow to meet with the Orthodox Czar Vasili III, father of the future Ivan the Terrible. A novelist could not have invented a better story of his family's origins, because this rejection of the Western church, the church both of Rome and of the German peoples, in alliance with the Eastern Russian Orthodox church, *is* the story of Trubetzkoy's inner life. He came from a family of intellectuals and philosophers who were deeply nationalistic, and he remained a part of that familial tradition even when the Russian Empire existed no longer. The events of his lifetime, the fall of the Russian Empire, all seemed to Trubetzkoy to be part of a long wave of Russian history, beginning with Peter the Great's opening to the West, which was a tragic betrayal of Russian tradition, and ending with the death of the last and final czar.

Nikolai's father Sergei was a well-respected professor of philosophy at the University of Moscow, a Plato scholar, and the first rector elected at the university—though he died at age 43, just after taking the position. Nikolai inherited his father's weak heart, and he spent a considerable time in the hospital because of it in the final years of his life. His family was noble, and as part of the intelligentsia, they participated in discussions of art, philosophy, and history on a regular basis.[4]

Within his family, Trubetzkoy grew up being exposed to a range of political views. His father was considered a liberal, but many others were profoundly marked by the "orthodox revival" that arose during this period.[5] His uncle Eugenij, a historian of art, was also a philosopher, like Nikolai's father, and he defended a traditional sort of orthodox Russian thought. Another uncle, Grigorij, was a political scientist.[6] Nikolai did not need to leave home for his intellectual stimulation, and in fact he was educated there entirely, by tutors. It was a small world in which the Russian elite lived: one of Trubetzkoy's tutors was the distinguished ethnographer Vladimir Bogdanov, who also taught Russian in high school and who taught Roman Jakobson.[7]

Young Nikolai's delight was in the non-Indo-European languages of the Russian territories—the Finno-Ugric languages, and the languages of the Caucasus, and the folklore and traditions of the far reaches of the Russian Empire. Indo-European languages and cultures did not attract him in the same way, but he early on recognized (as did all linguistics students of the time) that it was Indo-European that he would have to study in order to learn the theory and the technique of the field. Linguistics was, he recognized, the only one of the social sciences which actually satisfied the requirements of being a science.[8] And like his father, he studied philosophy.

Yet the essential theme of his studies was an opposition to Western thought and to Western science, as he found it in the Germanic and the Latin (or Romance) traditions. These powerful Western European traditions were not, he thought, the cutting edge of history: they were fast moving objects which threatened the health and welfare of Eternal Russia. What called to him were the languages of the Caucuses, and the myths and traditions that were specifically Russian, the endless territory that reached eastward. Here was a conflict that he would never succeed in resolving over the course of his life: German was the language of the greatest university system of Europe, and Central Europe was where the best linguistics was being written. Against his innermost inclinations, he decided to enter into the program of studies in linguistics and to learn comparative Indo-European studies and the languages which that enterprise was based on, like Sanskrit, in order to master the the most important linguistic framework of the day.

Trubetzkoy wrote about those days, later on:

> I first enrolled in the philosophy-psychology section of the History-Philology Department (Humanities), since I intended to study primarily ethnopsychology, the philosophy of history, and methodological problems. But I soon recognized that the philosophy-psychology section had little relation to my special sphere of interest. In my third semester I transferred to the linguistic section. . . . The scope and direction of instruction in the linguistic section left me dissatisfied. My main interest lay outside Indo-European languages. I nevertheless decided in favor of this section for the following reasons. First, I had already become convinced that linguistic science was the only branch of "anthropology" (ethnology, history of religion, culture history, etc.) that could pass from the "alchemic" stages of development to a higher stage only if, with regard to method, they would follow the example of linguistics. Second, I knew that Indo-European linguistics was the only thoroughly studied area of linguistics, that through it one could study the correct linguistic method. Accordingly I concentrated diligently on the studies prescribed by the program of the linguistics section.[9]

At the age of 23, just before the outbreak of World War I, he received a fellowship from the Russian government and spent a year in Leipzig, studying with Brugmann, Leskien, and the other Neogrammarians there, rubbing shoulders with Leonard Bloomfield as well as the French student Lucien Tesnière.[10]

But he did not consider himself a member of any Leipzig-oriented group of linguists. He viewed himself as self-taught, just as Saussure and Baudouin de Courtenay had viewed themselves, even though they, like Trubetzkoy, had spent time at the center of cutting-edge research in linguistics during their student days. The reader should take note of this; it is a curious view that many influential thinkers have developed about their own education, and we will encounter it again. Of course we all are self-educated in some respects; we read papers and books (after having decided, ourselves, what we are going to read), and we spend time lost in our own thoughts. But sometimes we find scholars emphasizing their own ignorance, or rather, protesting their own ignorance, and their theoretical innocence. It is in this light that we read with interest something that Trubetzkoy wrote some years later:

> I do not know Baudouin de Courtenay's work at all well, and of Shcherba's work I have only read his two theses ("Russian vowels" and "The Sorabe languages of the east"); I have never been the student of either Baudouin or Shcherba and I have created the essential aspects of my theory autonomously. . . . I have considered it to be more judicious to simply mention the names of the two great precursors of phonology in the domain of Slavistics, specifying that my theory is not directly linked to theirs and passing directly to laying out my conception in its totality, without going into detail on the origins of certain of its parts. In "Polabian Studies," there aren't any fundamental disagreements with Baudouin and especially Shcherba; there are things that one will not find in their work, but if my memory serves me well, there is nothing which contradicts their points of view. In my more recent work, I have more and more distanced myself from Baudouin's theory, which was, ultimately, inevitable. Nonetheless it seems to me that if we leave aside some of Baudouin's and Shcherba's sometimes unfortunate and inadequate formulations, and if we consider the foundations of their systems . . . our current points of view (those of Jakobson and my own) hardly contradict these theories and are only developments of them.[11]

Back in Moscow afterwards, he became an instructor at the University of Moscow, where he taught Sanskrit and began to develop his own views on the nature of language. He criticized the analysis of Russian syntax and morphology that A. A. Shakhmatov, a leading linguist, had just published, and he found the work of F. F. Fortunatov, his own professor and the founder of the Moscow School of Linguistics, too Western and too decadent for his taste, and not what Russian scholars needed.[12] He adopted the outlook of the movement that was known as "slavophile linguistics," which championed the notion that Russian had an internal structure that was unique to it, and not meant to fit neatly into Western typological categories.[13]

Trubetzkoy spent two years after that teaching at the University of Moscow, but then in 1917, he turned away from studying the Western Indo-European languages and focused his attention on Common Slavic and on the Caucasian languages. His interest in ethnology, folklore, ethnolinguistics, and the culture of the Caucasus led him to organize an ambitious program of fieldwork. He proposed a cartographic analysis of the eastern Slavic languages, undertaken from a holistic point of view where all of the various elements were shown to be intimately tied to one another, forming an organic unity. In Trubetzkoy's view, this project was very much a *Russian* sort of approach, just like the Russian Mendeleev had accomplished when he created the periodic table of the elements.[14]

Trubetzkoy was studying in the Caucasus at the time of the Bolshevik Revolution, and was never able to return home—he was, after all, a full-fledged member of the Russian aristocracy. His life, like that of other Russian émigrés at that point, was in chaos for the first few years after the Revolution. He moved from Kislovodsk, where he was at the time of the November Revolution, further south to Baku, in Azerbaijan. He wrote in a letter in December 1920 to Jakobson,

> During my wanderings in the Caucasus I came to Baku in March 1918, just in time for the "rebellion of the Muslims against Soviet Power," or, to be more exact, during that short time when the Armenians were slaughtering Tatars. I was alone there, had no means of subsistence, caught typhus, and after hospitalization got a permit to leave with great difficulty. I did not have a single acquaintance there.[15]

Trubetzkoy headed north towards Russia (or the Soviet Union), then to Rostov, and traveled west to Sofia, in the very Orthodox Bulgaria, where he was able to obtain a position as a professor in 1920. He stayed there until offered a professorship in Vienna in 1922, and Vienna remained his home until his death. The Vienna years were to be 16 years of intense activity.

Trubetzkoy was soon recognized in other centers of linguistics as an up-and-coming scholar.[16] In Paris, the Linguistic Society had begun meeting again after the war in 1919, reestablished by Antoine Meillet. In 1921, Trubetzkoy asked for membership in the Society, and he was received on June 18, with endorsements by Meillet and Vendryes, and over the years that followed, Meillet and Trubetzkoy maintained a correspondence marked by evidence of great mutual respect. Tesnière wrote that Meillet was "one of the few in France to favorably welcome the famous 'phonological' theory of Prince Trubetzkoy"; Meillet had "immediately seen the interest and the scope."[17]

In Vienna, Trubetzkoy's life was filled with work as a linguist, as a phonologist, and as a writer laying out the foundations of Eurasianism, and the substance of his work during those years will be the subject of the next two sections. In 1930, he began a fruitful conversation with his colleague Karl Bühler, the professor of psychology at the University of Vienna who we have already met. In 1935, Trubetzkoy published an essay "On Racism," where he attacked the biological conception of race attributed to the Nazis, while defending a specificity of each human community, a sort of cultural, psychological, social racism. This was properly viewed as pro-Russian and anti-German, and it led to a growing hostility towards Trubetzkoy on the part of Austrian Nazis, most notably the archeologist and future minister of education in the Nazi government that was to be, Oswald Menghin. Menghin began urging the Gestapo to investigate Trubetzkoy from the early days of the Anschluss in March 1938.[18]

Trubetzkoy's health grew markedly worse during 1937. His heart condition deteriorated, and he spent several periods hospitalized. After one of these serious episodes he was interrogated by the Gestapo, and his wife shared in a letter to Jakobson how upset it had left him. She let Jakobson know, too, that he was her husband's "only real friend." Trubetzkoy reached out to the young American linguist W. Freeman Twaddell about the possibility of leaving Austria and finding a position in the United States. A few days later, just after returning home, the Gestapo searched his house, and spirited away documents they found, including the papers of the International Phonology Association. Trubetzkoy's condition deteriorated rapidly after that, and he entered the hospital once again, but this would be the very last time. His heart grew steadily weaker, and he passed away on June 26, 1938. His magisterial opus, *Principles of Phonology*, was published the next year and would be translated into French in 1949, and into English in 1969. It has maintained the interest and fascination of phonologists over the decades since it was published, and earned for itself the position of the most influential book on phonology ever written.

Eurasianism

> The national substrate of the State which was previously known as the Russian Empire and which now is called the USSR can only be the set of the peoples who inhabit this state, conceived of as a particular nation, composed of several peoples and who, as such, possess a nationalism of their own. We call this country Eurasian, its territory Eurasia, and its nationalism Eurasianism. —Nikolai Trubetzkoy, 1927 (in Sériot 1993)

As we have seen, Trubetzkoy was a member of the elite of the Russian Empire in two respects: he came from a family of aristocrats, and he was a rising academic star, eager to challenge the linguistic establishment too influenced, he thought, by Western linguistic thought. And then the Revolution happened, and he was all of a sudden a refugee with no place to call home. The first years after the Revolution were a struggle to stay alive for many in Russia, and many intellectuals were simply deported. As the dust began to settle, centers of émigré Russians emerged in cities such as Prague, Paris, and Vienna. And those who were young enough to feel that their lives were still before them began to build a new understanding of who they were, as expatriate Russians. Trubetzkoy was a leader in a group that developed what they called Eurasianism, whose goal was to create a vision of the world, founded on both the Greco-Byzantine heritage and (lest we forget where we Russians came from) the Mongol conquest. Eurasianists rejected what came from the West: the Western tradition of the Enlightenment, and all those other trends that emerged from it, like socialism and communism in the nineteenth century. The Eurasianist view of Europe was that Europe was the home of the culture that had *colonized* Russia; the Eurasianist view of the Caucasus and of Central Asia was that there was a natural, organic connection between the peoples there and the peoples of Russia.[19]

THE FIRST STEPS OF THE EURASIANISTS

On June 3, 1921, Trubetzkoy gave a lecture at the Religious Philosophical Circle of Sofia which marked the first public presentation of Eurasianism as a new political and philosophical movement. Trubetzkoy was a relatively young man, just turned 31, and he was one of four Russians who would be the leaders of the Eurasian movement in its first few years, as it developed its overarching view of the world. In addition to Trubetzkoy, there was the musicologist Pyotr Suvchinsky, the theologian Georgi Florovsky, and the geographer Peter Savitsky.

Trubetzkoy was passionately interested in the foundations of nationalism, and he engaged as a philosopher and as an ethnopsychologist in the question of the meaning and sources of the psychology of peoples, questions which, as he wrote to Jakobson, "apparently interest you the most, and which is for me also the most important thing."[20] At this point, he was working on a book to appear in three volumes, whose title says it all: *Justification of Nationalism*.[21] The second volume was to be entitled

On Real and False Nationalism, and the third, *On the Russian Element*, but they were never published, nor finished. The first volume did appear, however, under the name *Europe and Humanity*.[22]

In its starkest form, Eurasianism was founded on the belief that Russia, the Caucasus, and Central Asia formed an organic unity. The Germano-Roman world to its west was the enemy of Russia—not only its enemy, but its worst enemy;[23] the Germano-Roman world had an impressive science, but at its core was a spiritual void and a stagnant pool of moral decadence. Part of this decadence was the universalism of the Enlightenment: "in a universal homogeneous culture, logic, rationalist science and material technique will always dominate religion, ethics, and the aesthetic."[24] And that was not what Trubetzkoy wanted: he wanted a culture that was particular to a nation in order to allow for the development of "the specific moral and spiritual features of each people."[25] Peter the Great was not the moral icon of the Russian people; henceforth that moral icon would be Genghis Khan.

As a movement, Eurasianism began not with a bomb but with a book: *Exodus to the East*, written by our four horsemen: Trubetzkoy, Savitsky, Suvchinsky, and Florovsky.[26] The four of them wrote a common introduction to this book which expresses the shock that they were feeling. "The essays that make up the present volume were put together in an atmosphere of a consciousness of catastrophe." That much is clear. "The segment of time within which our lives pass, beginning with the coming of the war, is felt by us as a pivotal, not merely a transitional, time. In what happened and in what is happening we see not just shock, but crisis, and in the future we anticipate a profound change in the customary countenance of the world." They wrote that they "honor the past and the present of West European culture, but it is not Western Europe we see in the future." It is, instead, Russia. And they said that they knew that "the world cataclysm, separating one epoch of world history from the next, has already begun. We do not doubt that the replacement of the West European world will come from the East."[27] They were living in one of the greatest ages of rupture in human history.

Trubetzkoy wrote about the cultural orientation of Russia. "It is widely believed in educated Russian society that the unique characteristics of this life can be described as Slavic." That, of course, lay at the heart of pan-Slavism. "This is incorrect," he went on. "From an ethnographic point of view, the culture . . . of the Russian people is an absolutely singular entity that cannot be accurately identified with any broader cultural zone or grouping of culture." So much for pan-Slavism. But then Trubetzkoy suggested that really there *is* a broader cultural zone. "Generally speaking, this culture comprises its own special zone and includes, besides the Russians, the Ugro-Finnic peoples and the Turkic peoples of the Volga Basin. Moving to the east and southeast, this culture merges almost imperceptibly with the Turko-Monglian culture of the steppes, which links it in turn with the cultures of Asia." The links to the west, however, "are not very strong": those are the links to the other Slavs, and to the Roman and Germanic civilizations. If Russians and south Slavs sometimes feel a kindred spirit, it is because they "both have experienced strong Turkic influences."[28] Eurasianism contained themes that brought it closer to the totalitarian ideologies of the twentieth century, such as its deep distrust and disdain for hedonism, materialism, and personal comforts, those well-known characteristics of the Romano-Germanic ideology, as the Eurasianists saw it.

The Eurasianists' ideas were rooted in conservative political discourse that began as a response to the French Revolution of 1789 and which found a voice in such writers as Joseph de Maistre, who fled France at the time of the Revolution and became the Prince of Savoy's ambassador in St. Petersburg in Russia, and Edmund Burke, the noted member of the English Parliament during the times of the American and French revolutions.[29]

With Burke, and even more with de Maistre, the principal political and philosophical themes which inspired Trubetzkoy were in place: inequality among men is a fact of nature, and only their equal submission to the monarchy would guarantee their integration into a harmonious social body, for which, in any event, the essential mortar would always be religion and, more generally, ancestral traditions. Thus for these four Eurasianists there could be no more important social study than ethnography.

The flip side of the coin, the inevitable conclusion, was that individualism, free will, personal liberties, the rights of men, and all of the values espoused by modern liberalism in the West were the causal factors that had led to the rupture of the fabric of modern society and its values. They were the cause of the decadence and the ruin of the Christian West. Their roots were to be found in the Enlightenment and in the very notion of progress which played such an important part in forging the ideological instruments of this decadence. What was needed now was a conservative counter-revolution which would be able to reconstitute a social organism which was rooted in traditional religious values. This conservative revolution resonated with a number of intellectuals in Germany during the 1920s, such as Ernst Jünger, Carl Schmitt, and Martin Heidegger.

The reader should take a moment to let this sink in. We are at present in a time and place in which universalism is so taken for granted that some may have trouble reading Trubetzkoy's anti-universalism and believing that he meant what he wrote—but he did. Trubetzkoy's Eurasianism was not a pan-Slavism of any sort: the more Europeanized Slavs were not welcome in his view of Russia, while the Turkic ethnicities, those speaking what we call Ural-Altaic languages, were integrally part of the Eurasian conception of a greater Russia. In fact, one could object to including Eurasianism under the banner of nationalism, in that the identity that it championed was far larger than the groups that had inspired nationalist movements up to that point. We must not forget that Trubetzkoy lived in exile outside of a country that had undergone one of the greatest upheavals that any political revolution had ever unleashed. Lenin and Stalin were the powerful leaders of a political force that claimed to represent the interests of Russian workers and workers of the entire world, and this Bolshevik force had been able to take over Russia because of the chaos that was the Great World War. The Western democracies had gotten themselves into an insane war that proved, in the eyes of many, the inadequacies of Western notions of democracy, capitalism, and individualism. The war that began in 1914 was felt by many, including Trubetzkoy, as a traumatic moment at a national and even international level, one which effectively condemned the moral stature of Western ideas of progress and the Enlightenment.

Eurasianism thus rejected the Enlightenment notion of progress, and Darwin's concept of evolution, and it offered a strategy for how non-Western peoples (including the Russians) should move out from under the domination of the Germano-Roman cultural domination.[30]

Eurasianists championed the view that the unique Russian essence had been rejected and denied for more than 200 years, ever since the reign of Peter the Great. This Russian essence was denied by a monarchy that gazed west towards Europe, an attitude that deprived the Russians of an understanding of their uniqueness. Darius Adamski states this clearly:

> This westernization affected Russians' way of dress; they were not permitted to wear beards, at the price of going to jail. The upper classes assimilated Western culture to the point that they adopted their language: the characters

> of Tolstoy conversed in French. And the poets who wrote in Russian often learned their language not from their parents or their governesses, but from their wetnurses. In a publication in 1921 [Trubetzkoy 1921], Trubetzkoy takes the position that only the Orthodox church, the bastion of Byzantine traditions, and the people, untouched by Europeanization, would be able to preserve Russian originality. [31]

Trubetzkoy wrote,

> If the Russian people are bound to a Turanian world essentially by certain features of its psychological profile, only its language links it to the Slavic world. Indeed, the "Turanian" world . . . is neither a racial nor a linguistic unity, strictly speaking, but rather an ethnopsychological unity. The Slavic world, by contrast, is nothing but a linguistic notion. It is through the intermediary of a language that a person opens up his interior world; language is the fundamental means of communication between individuals, and it is through the process of communication that personal collectivities are formed. This alone indicates the importance of studying the life of language from the point of view of personology. The history and the specific properties of proper Russian are extremely important in order to characterize the Russian national person, as important as the position of Russian among other languages.[32]

How would that awakening happen? Through a science of personology, which brought together the results of a family of related fields, including "historiosophy," "ethnosophy," and "geosophy," and, of course, linguistics. Personology would aim to determine the ideal conditions for the development of a society, and would be the "ideological system" of the Eurasian movement, allowing the Russian people to move beyond the left and the right of European construction. Personology proposed an analysis of psychology not at the level of individuals, but of the organic totalities that are nations. Against the philosophical, political, and social individualism of the Enlightenment and more generally of the Romano-Germans, the Eurasianists defended the organic unity—"symphonic," as they put it—that constituted Russia and that unified these different human faces in a harmonious whole. This integration was at the heart of the traditional teachings of the Orthodox church.[33] These notions are rooted in the Herderian notion of the nation, which itself had arisen in opposition to the Enlightenment views (recall our discussion in chapter 2): nations are cultural, geographical, and historical totalities, and each is particular unto itself. For the Eurasianists there is no universal culture; there are simply specific national totalities.

In the Eurasianist view, along with the rejection of individualism and of the universality of individual human rights came a profound rejection of democracy, and especially any tendencies which could be viewed as a dissolution of the nation in favor of the initiatives or preferences of individuals. Trubetzkoy argued for an "ideocratic" state, one directed by a single party constituted of morally superior citizens who would represent the "idea." If the government was not democratic, it was *demotic*: it should be fully supported by the people, and it should take actions for the best interests of the people, but it was not to be democratic. Factors which worked against the unity of the people were to be banished, factors like a free press or private capital. Liberalism and democracy are "the worst enemies of the ideocracy and the economy must function in a perfect national autarcy."[34]

WHO ARE THE TURANIANS? THE ORIGINS OF THE TERM

The Eurasianists called upon Russians to recognize the fact that they were organically a part of a greater *Turanian* unity. But who were the Turanians?

At the beginning of the nineteenth century a number of writers, notably in Germany, developed a strong anti-Russian polemic, part of which identified the languages, the ethnicities, and what they called the "races" of the Russian Empire as not being full-fledged members of the "noble Aryans" who had descended from the Indo-Europeans. We noted in chapter 2 that Max Müller used the term *Turanian* to refer to a group of languages that included the East Slavic languages, the Ural-Altaic languages, and he used the term to apply to the myths and the popular culture as well.[35] The term came from the name of the mythical kingdom Turan that had once been the rival to Persia, spreading from the Caspian Sea eastward to China, northward to Siberia, and southward to Persia. The Eurasianists took this term and made it their own, trying to establish a pride in a cultural unity and a new take on Russian identity.

The Russian psyche, said Trubetzkoy, still contained a "Turanian" or Mongolian aspect, but it was unconscious and needed to be awakened. Trubetzkoy observed that there was a characteristic that Russians shared with the people of the steppe: a *hardiness* of the soul (*udal'*): "a virtue that is typical of the steppe," Trubetzkoy wrote, "that the Turkish people understand but which is incomprehensible to the Romano-Germans or the [western] Slavs."[36] The Eurasians' focus on defining the Turanians as a group had as a consequence that the linguistic groupings motivated by nineteenth-century principles of language evolution were no longer providing the right answers to the question of who was closely related to whom. Instead, Trubetzkoy's focus on the existence of a *Sprachbund* brought to the fore the notion that languages could share many characteristics without being genetically related. It was a small step, then, for Trubetzkoy to offer the hypothesis that the grouping of the Indo-European languages was not a genetic classification, but rather the name we have given to languages that formed a *Sprachbund*, and which share many features and morphemes due to long-standing contact. He wrote,

> There is thus no reason at all to suppose that the existence of a protolanguage from which all of the Indo-European languages have sprung. We could just as well imagine the situation the other way around, that is where the ancestors of the Indo-European branches were originally dissimilar, and that it is only with time and with constant contact, as well as reciprocal contact and borrowing, that they have come to become more similar, without become totally identical. The history of languages shows both divergent and convergent evolution. It is even sometimes difficult to draw a line between those two things.[37]

In this way (though not only in this way) we see how the presence of teleological development and organic cultural totality in the Eurasianist worldview leads to a new position with direct implications for politics, culture, ethnology, and religion.

THE TOWER OF BABEL

Trubetzkoy published in 1923 an original interpretation of the effect God had on the world when He imposed multiple languages on it, after the disastrous Tower of Babel debacle described at the beginning of Genesis 11. In the science that allowed men to build the Tower, Trubetzkoy saw a "spiritual emptiness and moral decadence": theirs was a universalist and homogeneous world. Such a culture can possess only "the psychic features common to all human beings."[38] Today, we might think that this would be an excellent thing to discover, but Trubetzkoy took this to be limited to too small a domain, that of logic and material needs. This is most certainly not enough, and in Trubetzkoy's view one needs cultures that are specific to a nation, which allow "the specific moral and spiritual features of each people" to develop. Trubetzkoy's linguistic work supported his larger Eurasian view of the Russian Empire. No one could say it better than Trubetzkoy himself:

> Certainly each people who comprise Eurasia could, by considering certain features, be included in another group of peoples which was not exclusively Eurasian. Thus, if we take the criterion of language, Russians are part of the Slavic peoples, Tatars, Chuvash, and Cheremis and others are part of the group of peoples called Turanian; if we take [the criterion of] religion, Tatars, Bashkirs, Sarts, etc., are part of Moslem peoples. But these ties must be for them less strong than those which unite the Eurasian family: it is not pan-Slavism for the Russians, nor pan-Turanism for the Turanians of Eurasia, nor the pan-Islamism for the Moslems of Eurasia which must be found in first place, but rather Eurasianism. All these "pan-isms", which intensify the centrifugal forces of particular ethnic nationalisms, emphasize the unilateral link between one people and other peoples by a single set of criteria; this is why they are incapable of making of these peoples a true living multi-ethnic nation: a personal individuality. But in the Eurasian brotherhood the peoples are bound together not by a set of unilateral criteria, but by their community of historical destiny. Eurasia is a geographic, economic and historical totality. The destinies of Eurasian people are interwoven, they form an immense web which one cannot untangle, to the point where the removal of a people from it could only be accomplished by an act of violence against nature, which can only bring suffering. Nothing similar could be said of the groups of people who form the basis of pan-Slavism, of pan-Turanism, or of pan-Islamism. None of these groups is united to such a degree by the unity of the historical destiny of the people who comprise it.[39]

TRUBETZKOY'S RUPTURE WITH THE EURASIANISTS

Trubetzkoy broke with the Eurasianism movement in 1928, citing his disagreements with what he perceived as the new ideologies in the movement: Marxism and fiodorovism. By this time, it had become clear that a split had arisen among the Eurasianists, who became divided into a left wing and a right wing, separating those who thought that a *rapprochement* of sorts with the Soviets could be achieved from those who rejected that notion. The left wing was based in France (in Clamart, a small town just southwest of Paris), and it fell hook, line, and sinker to Soviet intelligence efforts to infiltrate and destroy it.[40]

"ON RACISM"

The question of *who we are* was the question that lay at the heart of Eurasianism, as it lies at the heart of any political movement and ultimately of any social group. For émigrés, most especially political refugees, the question is central to their life and is often overwhelming. Eurasianism appealed to a rejuvenated conception of geography and a new style of linguistic analysis to provide a new answer to the question on a political level. Who *we* are is also important in the intellectual context of linguists and philosophers, and as we will see below, both Trubetzkoy and Jakobson were deeply concerned with who allied with them among phonologists and who did not. And in this perilous moment between the two world wars, no question of identity was more laden with significance than the question of the identity of the Jewish people. On the one hand, the Russian Empire did not have a political tradition of tolerance for Jews, but on the other, by 1933, Jews were being forcibly disenfranchised from mainstream German society, and of the two linguists who had played a significant role in developing the ideas behind Eurasianism, one was a prince and the other was born a Jew.[41]

Trubetzkoy published an essay in 1935 which bore the name "On Racism" in which he addressed the status of Jews in European society.[42] The Eurasian rejection of the Darwinian model of evolution, and of the branching family tree model of linguistic evolution that had come with it, did not lead Eurasianists to embrace the biological racism that flourished in Nazi Germany at that point, and more generally in Western Europe. And regardless of what one might think of the Turanian spirit, it could hardly be forgotten that Eurasia was a conglomerate of ethnicities and languages from all over: biological purity did not make much sense for the Eurasianists, and the racially based biologism in Germany was being used to demote the status of Slavic groups more generally, a point that was hardly missed on the Eurasianists.

The Eurasianists' anti-Darwinism was thus dead set against biological racism, but that did not lead it to defend the equality of races, or that of individuals, nor the universality of human rights, that idea so closely tied to the Enlightenment. Between 1900 and 1920, a new movement appeared in art and philosophy which came to be known as the Futurist movement. Jakobson was one of the leaders and theoreticians of this new aesthetic. According to Trubetzkoy, Eurasianists should side with those futurists who feel the tragedy of modern culture, which the Futurist methods helped to reveal, but he tended to perceive Futurist art (and modernist art in general) as yet another sign of the destructive work of "Romano-Germanic" modernity. It was related to the socialists and the left as well as to the Jews. Here, Trubetzkoy came very close to the ideas professed by German Nazis. Warning Suvchinsky against getting too much involved with the Futurists, Trubetzkoy wrote:

> I am writing all this to you because I am terribly afraid that you, while communicating with this entire futurist gang, will lose the sense of measure and will give up your romantic position. You have a leftist temptation and love to the last outcry the field of art. Do not forget that leftism always leads to the devil, to Antichrist, and it is not for nothing that Yids are always with the left.[43]

For Trubetzkoy, what defined the Jew was his separate culture, his separate religion, his separate psychology, his separate history, his separate traditions. The Jew did not belong to Eurasia; he was an Other and he thus always stood in the way of the symphonic, organic totality that the Eurasianists privileged. For Trubetzkoy, Judaism was a communitarian "neurosis." But if a Jew rejected this separation, left his community and renounced it, and assimilated to the larger social reality, he could become a perfect Eurasian—just as fine as any other Eurasian.

Bear in mind that we have already seen that Trubetzkoy's wife told Jakobson that he was Trubetzkoy's only real friend: Jakobson embodied the true Turanian spirit. Jakobson would himself convert to the Eastern Orthodox Church in 1938, a few months after Trubetzkoy's death. Trubetzkoy's position against biological racism went hand in hand with a cultural and social racism. He wrote,

> What is unhealthy should be treated, and the treatment depends upon a correct diagnosis. When you treat neuroses, it often happens that the diagnosis itself is enough to allow the patient to achieve a full awareness of the cause of his state and to conceive of a true desire to combat it. The destructive consciousness of the Jews is a neurosis, a neurosis of a particular sort, which draws its origin from the feeling that there is an abnormal relation between Jews and goyim, a feeling reinforced by the influence of the Jewish milieu, which suffers from the same neurosis.[44]

The upshot, then, was that the appropriate way to treat the neurosis that is destructive of social harmony was to uproot its very cause, which was the existence of Jews *as Jews*, and to urge a policy of assimilation.

Trubetzkoy's Eurasianism shared with fascism and communism the principle that the individual was, and must be, understood as part of a larger organic whole, a totality, and this organic whole transcends any and all of the individuals within it.[45] This kind of organic whole could be identified by its destiny, which was the necessary working out of its history. In his brochure entitled *The Legacy of Genghis Khan*, Trubetzkoy wrote:

> The historical unification of Eurasia was from its beginning a historical necessity. And, at the same time, the very nature of Eurasia pointed to the ways in which this unification [would be achieved].[46]

Teleology, anti-Darwinism, and the rejection of Western mechanism

Trubetzkoy's Eurasianism was of a piece with his rejection of a mechanistic style of thought that he saw in the Western world, and this rejection would play a role both in his political writings and in his work on phonology with Roman Jakobson. In both spheres, the notion of evolution and change was a central concern, as we have already noted along our journey to this point. Our focus in this section will be on the side of Trubetzkoy's and Jakobson's emphatic rejection of Western mechanism, and some of the issues that are raised here will reemerge when we discuss the ideas at the center of the Prague school's notion of structuralism.

Both Trubetzkoy and Jakobson whole-heartedly rejected the mechanistic interpretation of Darwinian evolution, which we have described as the interpretation according to which the large-scale changes we see in the biological world are the agglomeration of a vast number of small, random changes. We noted earlier that the appearance of *The Origin of Species* in 1859 constituted a profound rupture in the way we looked at the biological world, its evolution, and the laws that may rule it. The very notions of evolution, adaptation, and especially of variation of species due to chance were bombshells in a garden of thought still largely defined by the views set forth in the Christian Bible, and many thinkers were quite dissatisfied with a mechanistic account of the development of species.[47]

Trubetzkoy and Jakobson saw a common thread running through the ideas of Darwin, of the Neogrammarians, and of Saussure, and they were convinced that this common thread of mechanistic thinking had to be identified, excised, and destroyed in favor of a new understanding of change. The new views would share some remarkably deep aspects of what was happening in Gestalt psychology at that moment, as we have already discussed, and we will return to this in connection with Trubetzkoy's interaction with Karl Bühler in Vienna. At the same time, we have already seen that the concern for establishing a place for teleology had the potential for considerable conflict: recall that Leonard Bloomfield was as dead set against teleology as Jakobson and Trubetzkoy were for it, at just the very same moment.

Trubetzkoy and Jakobson rejected the image of blind chance which they perceived behind the Western interpretation of Darwinian evolution. They could not accept the idea that linguistic change is like the walk of a drunkard who strolls down the street with no goal in mind. Was that a reasonable characterization of the Darwinian view? Surely not; the controlling mechanism for Darwin was the selective advantage that some of the variations found and others did not within their ecological niche. But reasonable or not, Trubetzkoy and Jakobson defended a *teleological* view. Teleological points of view were certainly not well received during this time, and much the same is true today. It is interesting to hear how Jakobson recalled his first discussions of this question with Trubetzkoy in the 1920s.

In October, 1926, Jakobson wrote a "long excited letter" to Trubetzkoy in Vienna in which he discussed an idea he could not dismiss, the idea that "linguistic changes were systematic and goal-oriented, and that the evolution of language shares its purposefulness with the development of other sociocultural systems."[48] Jakobson understood that this was not what most people thought, and he waited with trepidation for an answer from Trubetzkoy. It came, two months later. "I am in perfect agreement with your general considerations," the prince wrote back. "Many elements in the history of language seem fortuitous, but history does not have the right to be satisfied with this explanation." As long as we recognize that *language is a system*, then there must be a rationality to linguistic change. Trubetzkoy turned this into a criticism of Saussure:

> If Saussure did not dare to draw the logical conclusion from his own thesis that *language is a system*, this was due to a great extent to the fact that such a conclusion would have contradicted the widely accepted notion of this history of language, and of history in general. For the only accepted sense of history is

> the notorious one of "progress," that queer concept which as a consequence reduces "sense" to "nonsense."

It is important for us to see clearly both what Trubetzkoy and Jakobson rejected in the theory of evolution—in Darwin, in the Neogrammarians, and in Saussure—as well as what they wanted to see take its place. Jakobson captured his own perception of Darwin well; the vision of Darwin's which Jakobson rejected was this:

> Changes occur without there being any intention behind them; they are fortuitous and involuntary . . . language does not premeditate anything and its pieces move fortuitously . . . these disordered actions are only unfortunate burglaries that work to no end whatsoever.[49]

Jakobson said it even more sharply in 1927 when he alluded to work going on in biology in the Soviet Union that continued down the anti-Darwinian road. Some of this work was known under the name of "nomogenesis":

> Mechanical accumulation due to the play of chance or heterogeneous factors—that is the image that European ideology, the ideology that dominated in the second half of the nineteenth century, prefers to put forward over all others. Contemporary ideology, on the other hand, is bringing ever-more distinctly to light a functional system instead of a mechanically attained sum; structural laws instead of a wholly bureaucratic reference to neighboring cases; and goal-oriented evolution instead of evolution due to blind chance.[50]

Jakobson wrote,

> According to Darwin, evolution is the sum of the divergences resulting from accidental variations undergone by individuals which produce slow, ceaseless, and barely perceptible changes; there are countless hereditary variations, and they move in all directions. To this doctrine, contemporary biology, particularly Russian biology, more and more opposes nomogenesis: to a large degree, evolution is convergent, responding to internal laws that include enormous numbers of individuals over a vast territory, in leaps, saltations, paroxysms, by sudden mutations; the number of hereditary variations is limited, and they go in determinate directions. The problem of a structure of an internal law is also explicitly posed by modern geography (especially in Russia) with its notion of geographic individuality or *Landschaft*.[51]

It was the Neogrammarians who embodied this perverse mechanical perspective more than any others, in Jakobson's view: the Neogrammarians found the source of linguistic change to be in the random character of individual speech, and their notion of law had neither cause nor goal in it, and worse yet, they were *happy* about that. As we saw in chapter 2, there were good reasons for this: they were working hard to eliminate the Romantic and Hegelian tendencies in linguistics which saw language developing in a positive sense before history began, and then declining in the centuries since, those centuries of decline being what constituted *history*. It is striking that Jakobson seems to have no awareness of this; he did not see that the Neogrammarians were solving their own problem, which was to get rid of the myth of language development followed by language decline. The only thing that comes through in Jakobson's account of Neogrammarian thinking is an image of linguists whose vision was narrow:

> We are forced to get out of the Neogrammarian rut in diachronic linguistics too. The conception according to which phonetic changes are fortuitous and involuntary, and that in language nothing is premeditated made us present the phonetic history of a language as a sequence of troubles and blind destruction caused by factors that are extrinsic from the phonological point of view; on that view, these haphazard would be nothing but unfortunate, aimless vandalism.[52]

And Saussure was not a source of comfort either; he had drunk the Neogrammarian Kool-Aid.[53] We will return to how Trubetzkoy and Jakobson viewed their relationship to Saussure, but with regard to the most fundamental philosophical issue, they were disappointed at what Saussure left them:

> The neogrammarian conception of linguistic history is essentially the absence of a theory. The theory of historical processes is possible only if the object which undergoes the changes should be considered as a structure governed by internal laws, and not like an adventitious agglomeration. Saussure's doctrine viewing language as a system establishes necessary premises for a theory of language as a synchronic fact, but it continues to attribute to this synchronic system a fortuitous origin, it continues to view diachrony as an agglomeration of changes with an accidental provenance. A theory of diachrony is only possible if we conceive of it as part of the study of changes of a structure, and the structure of changes.[54]

And he added:

> For Saussure, changes occur without there being any intention behind them, they are fortuitous and involuntary. . . . Schleicher reconciled recognition of the internal, functional meaning of the language system as furnished by direct experience with the idea that language evolution was meaningless and was driven by blind chance by interpreting internal, functional meaning as a residue of the original perfection of the linguistic system. From that perspective, evolution is nothing more than disintegration and destruction.[55]

Once again we find the question of teleology arising as one of the most central issues. In its place Jakobson affirmed that "modern science (and principally Russian science) has defined itself by replacing the question '*why?* (i.e., *warum?*)' by the question '*to what end* (*wozu?*)'" and he proposed to replace "the mechanical views (of evolution) by a teleological approach." Jakobson wrote in 1928: "The overlapping between territorially, socially or functionally distinct linguistic patterns can be fully comprehended only from a teleological point of view, since every transition from one system to another necessarily bears a linguistic function."[56]

> In his interpretation of diachrony, Saussure stands firmly in the scientific traditions of the 19th century. For him, changes take place outside of all intention, they are fortuitous and involuntary, certain elements are altered with no regard for the solidarity which links them to the whole [of the system] and consequently can only be studied outside of the system; the movement of a system occurs under the influence of events that are not only foreign to it, but which are isolated [individually] and do not form a system in themselves. This leads to a profound rupture between diachronic linguistics and synchronic linguistics; Saussure's brilliant comparison of language and a game of chess loses its persuasive force if we adopt Saussure's opinion affirming that language never thinks ahead and that its pieces move fortuitously. That point of view leads us to present the history of the sounds of a language as a series of troubles and blind steps of deterioration caused by extrinsic factors. From the point of view of the phonological system, these chaotic actions would be nothing but a series of unfortunate accidents without any goal at all.[57]

The emphasis on the goal-directedness and functionality of language also led them to focus on acoustic phonetics more than on articulatory phonetics, in the belief that a speaker's goal is best understood as the creation of a sound, rather than as an articulatory act.

> It is the role of the goal of a phonetic event that is more and more the concern of the linguist, instead of the traditional one of causes. It is not by [merely] renouncing the notion of "phonetic law" that one is able to go beyond the tradition of the Neogrammarians, but rather by interpreting this notion teleologically and abandoning the mechanistic concept.[58]

While most of the passages we have cited here are from Jakobson, the same point is just as clear in Trubetzkoy's writing. He wrote in *Principles*:

> If, at each moment, a language is a system in which "everything is connected," the passage from one state of a language to another cannot be effected by means of changes which are stripped of all meaning. Since a phonological system is not the mechanical sum of isolated phonemes, but rather an organic whole in which the phonemes are the gears and the structure is subject to laws, "historical phonology" cannot be limited to the history of isolated phonemes, but must rather envisage the phonological system as an organic entity which is in the process of development. From this point of view, phonological and phonetic changes receive a meaning, a *raison d'être*. While being to a certain degree determined by general laws of structure—which exclude certain combinations and favor others—the evolution of a phonological system is at each moment directed by the tendency towards a goal. If one does not introduce this teleological element, it is impossible to explain phonological evolution. This evolution has thus a direction, an internal logic, that historical phonology is called upon to bring to light. Contemporary phonology makes the same point. This is perhaps the most striking difference between contemporary phonology and the theories of F. de Saussure.[59]

If we were to summarize the shift in perspective that we see in Trubetzkoy and Jakobson's overall picture of language and culture, and how it differed from that of the Neogrammarians whose work served as the heart of what they had learned as students, it would be that history is no longer the central character, because history has been replaced by *geography*. We must bear in mind that in the nineteenth century, history was what *provided an explanation*: we understood European languages now for the first time because we saw how languages descended from earlier languages. Trubetzkoy's skepticism about Indo-European languages forming a genealogical tree did not derive from a study of errors in the argumentation of the Neogrammarians: it was really a way of saying that geography presents a different kind of explanation of generalizations that we observe today.

Trubetzkoy's shift to phonological theory

During the 1920s, Trubetzkoy pursued his ethnographic passion, working on the description of a large number of Slavic languages. His grand project was to write an "Outline of a Prehistory of Slavic Languages," which would deal with the displacement of Common Slavic.[60] For these purposes he gathered the description of the vowel systems of these languages which he hoped could be put together in an atlas.

It was in 1927 and 1928 that Trubetzkoy's thinking and interests began to shift.[61] In October 1928, he wrote to Jakobson about something new:

> I have not worked very much this summer, I mainly took walks, because we have had splendid weather. I have made considerable progress on my "Polabische Studien," but I have still not finished them. One of the things that I have started is a project that I find fascinating: I have put together the phonological systems of the vowels of all the languages that I know by heart (34 altogether) and I have tried to compare them. Here, in Vienna, I have pursued this work and I have now 46 systems. I will keep on working on it from time to time until I have dealt with a hundred or so.[62]

Like a chess master who can play 20 games simultaneously in his head, Trubetzkoy took now to looking at all the vowel systems he knew, seeking a higher plan for them. He

> got some extremely curious results. For example, I have as yet found no language with an asymmetrical vowel system. All of the systems correspond to a small number of types, and can always be represented by symmetric schemas (in the form of a triangle, or parallel series, etc.). One can easily see various laws of the "formation of systems" (like, for example, the law that says that if a given language has front rounded vowels, their number can never be greater than the number of front non-round vowels, etc.) . . . I estimate that these empirical laws that can obtained in this way may be of capital importance, particularly for the history of a language and for its reconstruction.

Trubetzkoy quickly went from the amassing of data to the development of larger laws and principles, displaying the taxonomic genius which, as we saw above, Bühler compared to that of Mendeleev, creator of the periodic table of elements.

> In connection with my course on phonology, I have reconsidered several things. Among others, I am more and more convinced that we have distinguished too sharply between correlation and disjunction. In reality, there exists a whole series of nuances. Most clearly perceptible are the correlations which are limited to certain positions (because then in other positions, we find archiphonemes due to which we become aware of the marked and unmarked character in a particularly clear way).[63]

Trubetzkoy presented an early formulation of this new model in "Zur allgemeinen Theorie der phonologischen Vokalsysteme,"[64] and in a series of papers during the 1930s. Its most complete form would only be found in his magnum opus, published after his death.

The period that began around 1928 was thus Prince Trubetzkoy's great decade of phonology, in which he could focus his thinking on phonology and the systems that he found in his languages. His interests moved a bit from the ethnography and the geolinguistics, and certainly from the militant Eurasianism, that had held captive his attention to that point. He stayed with the structure of phonological systems throughout the 1930s right up until the end of his life, and he devoted himself to thinking, writing, and organizing whatever would help in the development of his view of phonology, one that he shared with Roman Jakobson.

We are struck by the fact that Jakobson seems to have stepped back from phonology during the Trubetzkoy's active 1930s, though the correspondence between them on matters phonological did not diminish. We will discuss this in more detail below, as we turn our attention to Jakobson.

Karl Bühler

Towards the beginning of 1930, Trubetzkoy found a source of energy and of encouragement right at home in Vienna.[65] In May, he wrote to Jakobson that he had been in contact with Karl Bühler, and in that letter we can sense that he has been energized by this.[66] In fact, he went so far as to say that "Vienna is showing interest in phonology," with evident pleasure.

Trubetzkoy had sent Bühler an offprint of a paper he had written on vowel systems, aware that Bühler was an important figure in psychology, and one who was particularly interested in language, and after that, Bühler had talked about Trubetzkoy's ideas more than once in public. Trubetzkoy seemed pleased to tell Jakobson that Bühler had warned him that he would have to change his understanding of where psychology was, that his understanding was out of date and restricted to associative psychology, which was simply passé. Trubetzkoy wrote that he was reading Bühler's *The Crisis of Psychology*. Jakobson wrote back the very next day to say that he agreed, that Bühler was indeed an interesting theoretician of language, and that "what he says about your psychological premises is certainly correct."

Bühler and Trubetzkoy influenced each other during the first half of the 1930s, and anyone who wants to read Bühler's *Theory of Language* ought to have Trubetzkoy's *Principles of Phonology* in front of him as well—and vice versa. It is often just the little effects that get noted: Trubetzkoy (1936: 8) indicated, for example, that it was Bühler's suggestion to replace German *zweiseitig* (bilateral) and *mehrseitig* (multilateral) with *eindimensionaler* and *mehrdimensionaler*.[67]

Trubetzkoy felt strongly that his theory of systems had a natural scope far beyond phonology as such. In 1936, he was asked to contribute to a special issue of *Journal de psychologie normale et pathologique*, and he wrote,

> The opposition is not exclusively a phonological notion. It is a logical notion, and the role that it plays in phonology is very similar to its role in psychology. It is impossible to study phonological oppositions (of which phonemes are merely the terms) without analyzing the notion of opposition from a logical and a psychological point of view.

We underscore today the importance of taking seriously this phrase "logical and psychological"; Trubetzkoy really did mean it when he said that it was as important for linguists to get the insights of logicians as it was to get the insights of psychologists. He went on to ask psychologists,

> In an article on "Contemporary phonology" (1933) . . . we offered our readers in the *Journal de psychologie* some of the results of our research on phonological oppositions. Today we propose to discuss this in greater detail, with the idea of attracting the attention of psychologists. We allow that in discussing

> this question in these pages of the *Journal de psychologie* we are not entirely innocent of self-interest. It is not merely to share with psychologists some results that may interest them, but even more to ask them to help us in this research, that we offer these remarks.

Bühler was generous in his open-armed welcome offered to Trubetzkoy's conception of phonology. In his *Theory of Language*, published in 1934, he wrote,

> As far as the signs of language are concerned, even before my contact with phonology I had some intimation of the of the axiom of the significative nature of language. . . . It was only phonetics as a block that seemed not to comply with the insight that the object of the sciences of language completely pertains to sematology in the same way as the object of physics pertains to mathematics. . . . The philosophical (and epistemological) wonder at that proved to be fruitful and was settled when I read Nicolai Trubetzkoy's programmatic treatise "Zur allgemeinen Theorie der phonologischen Vokalsysteme." . . . There I found a well-founded contribution to the theory of sounds which had the horizon of a well-rounded new discipline of linguistics, a contribution that no longer had the character of phonetics; this was just what I was looking for. It was thus shown to be possible and necessary to split up the scientific treatment of the sounds of language as logical insight demands.[68]

Working with Jakobson as Europe goes up in flames

Many years later, at the end of the introduction he had written for the collection of letters Trubetzkoy had sent to him, Jakobson wrote very movingly about how their phonological collaboration had come to its final moment.[69] Jakobson wrote that divergences had emerged between Trubetzkoy and himself over the role and distribution of correlations in phonological systems, and he said that Karl Bühler had great reservations from both a psychological and a philosophical point of view regarding Trubetzkoy's multilevel classification of phonological oppositions. In fact, he continued, Trubetzkoy's conception was both complex and heterogeneous, in a way that ignored the logical character of oppositions. Jakobson's view was that distinctive features permitted a systematic analysis of oppositions; he did not recognize that goal in Trubetzkoy's conception. Jakobson perceived what amounted to a struggle between them regarding the role of correlations in phonological analysis.[70]

Jakobson wrote that Trubetzkoy interpreted his procrastination as a lack of interest, or enthusiasm, for phonology.[71] But in retrospect Jakobson felt that Trubetzkoy had misinterpreted things. It was rather the *struggle* between them over the role of correlations in phonological analysis that had kept him from finishing his "General Phonology of the Word," which would have formed the first half of a larger project to include Trubetzkoy's work on the morphophonology of Russian. But with such a divergence between them, how could Jakobson write a major theoretical statement which would form the theoretical scaffold that the two of them could share? How could he find a bridge to link the growing differences between him and his mentor?

During this final period, Jakobson and Trubetzkoy met once in Brno, where Jakobson lived, and twice in Trubetzkoy's home in Vienna, hoping to better understand their divergences—once in 1936, once again in 1937, and finally in February of 1938. At that final meeting, Trubetzkoy's wife interrupted their discussion with news that she had just heard on the radio: Kurt Schuschnigg, the Austrian chancellor, had been invited to meet with Adolf Hitler in Berchtesgaden. Under great pressure from Germany, Schuschnigg announced he had appointed a Nazi sympathizer as the Minister of Public Security in Austria, a stopgap measure in his bid to forestall Hitler's annexation of Austria. Trubetzkoy felt that completing his *Principles of Phonology* while there was still time was his highest priority, and Jakobson left Vienna to go back to Czechoslovakia with the understanding that when that great manuscript was completed, the two of them would work out their differences.

The final three weeks of February 1938 saw day-by-day descent into political chaos in Austria, and on March 12, German soldiers entered the country and the Nazis took control. Three months later, Prince Trubetzkoy was dead.

Roman Jakobson

Roman Jakobson the Russian (1896–1920)

> Linguista sum, linguistici nihil a me alienum puto. —Roman Jakobson[72]

Roman Jakobson: a brilliant mind and a passionate curiosity that raced over every corner of human knowledge and art. An intimate of the intellectual and the artistic avant-garde. He would be a philosopher, a poet;

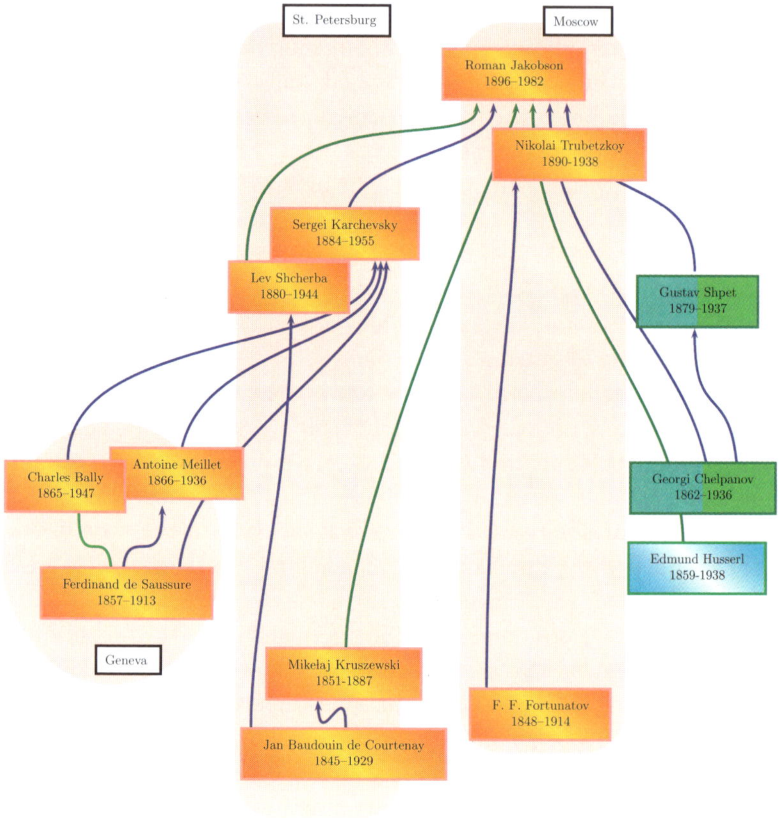

FIGURE 9.1. Trubetzkoy and Jakobson: early days

he was a linguist, a literary critic, a communicator, a popularizer, a bon vivant, a charmer—the polyglot who liked to say that he spoke Russian in 19 languages. Where did his enormous intellectual energy come from? Whatever its source, it seemed to have no limits; it fired on all cylinders, all the time. He is one of the most extraordinary of characters of this story, and he is the only one to play a major role in both this volume and the next. "Roman Jakobson, Russian by origin, was one of the most gifted Slavists and Bohemists, an unusual man both in appearance and in nature. A powerful man, with a rather large head, thick blond hair, and the face of a Roman god, he squinted in one eye. But he was not one to be bothered by such a troublesome defect. He overflowed with vitality, spoke with passion, and gestured with spirit."[73]

Roman Jakobson was born in Moscow, Russia, in 1896—he was six years younger than Trubetzkoy. Unlike Trubetzkoy, his background was not royal; his family was Jewish.[74] His family was well off. They had a four-story home in a posh part of town in Moscow, and other property in St. Petersburg as well. Roman's father was a chemical engineer, which provided them with a well-respected position in society. The Jakobson home was cultivated, and Roman grew up in an richly artistic and cultural milieu.[75]

The Jakobson family was close to the Brik family, who had a son Ossip, and to the Kagan family, who had two daughters, Elsa and Lili.[76] Roman, the two Kagan sisters, and Ossip Brik formed a tight group of friends who would play an important historical role in the development of the avant-garde, in Russia and in Europe more generally. Elsa was close to Roman, but (to Roman's chagrin) closer still to Vladimir Mayakovsky, a brilliant young poet who was at the leading edge of the Futurist movement in poetry.[77] When Roman's parents left Moscow after 1917, with a safe conduct permit that Roman obtained for services rendered, the Jakobson-Kagan-Mayakovsky group stayed in their Moscow home.[78]

Jakobson's gift and passion for learning languages would play a role throughout his life. He learned French and German at the same time as he learned Russian, and at a tender age he began to translate French poetry into Russian. He himself wrote poetry, until a moment in 1918 when he came down with typhus and renounced his effort to become a poet. Early on he thought seriously of pursuing his interest in science. But it was literature in the end that won out, though he knew that he wanted to approach it with the methods of science, which is to say, of linguistics. And thus Jakobson entered the Lazarev Institute, where he would remain until he went to university.

Founded in 1815 by an Armenian family, the Lazarev Institute quickly became a prestigious institution for students of the languages of the far-flung Russian Empire, taking on the role of preparing translators, administrators, and diplomats for the Russian government. Jakobson's years at Lazarev provide a natural context in which to understand Jakobson's interactions with diplomatic circles, and other governmental, and more or less secret, offices of Russian intelligence, first czarist and later Bolshevik.

At the beginning of the twentieth century, the Lazarev Institute was an important place where young members of the intellectual and artist scene in Moscow would meet. Beyond literature, poetry, and more generally all of the avant-garde movement in the arts, Jakobson's great passion revolved around folklore, ethnology, and the ancient languages and dialects of the Russian Empire. Like Trubetzkoy and like so many other young people in Europe through the century, Jakobson's connection to linguistics was deeply rooted in an appreciation of ethnography and the study of folklore and dialects. The kind of language that interested them was not that of the academic grammarians, but the living languages and dialects which they could go and study in real life among living people. This was precisely how Jakobson—like Trubetzkoy, like Boas, like Sapir, like Bloomfield, even like Humboldt and the brothers Grimm—began his initiation into linguistics.

UNIVERSITY OF MOSCOW

In 1914, Roman Jakobson was 18 years old, and he completed his studies at the Lazarev Institute of Oriental Languages. He enrolled in the Historical Philological Faculty at the University of Moscow, where he began his study of ethnology and Russian folklore, languages, and the development of a linguistics that would allow him to understand the underlying mechanisms of literary and poetic art, both ancient and contemporary. Years later, Jakobson remembered how he began his studies in linguistics. He sat down with the very first issue of a major linguistic journal, *Russkij Filologischeskij Vestnik*, and started reading. There he encountered an article by Kruszewski, who had been forgotten, and he was very struck by it—so much so that he bought a copy of a paper that Kruszewski had self-published, not finding a journal that would print it.

The education that Jakobson got at the University of Moscow during the World War I years was both broad and deep, and it put him in direct contact with the important thinkers we have discussed, most notably the psychologists Kurt Koffka, Carl Stumpf, and Wilhelm Wundt, and the philosopher Edmund Husserl.[79] Quite a number of academics who had studied with these leading European lights spent time at the University of Moscow during this period, and Jakobson drew much from his contact with them.

Georgi Chelpanov was a leading Russian psychologist at this point who had studied with both Wilhelm Wundt and with Carl Stumpf, and he taught at the Psychological Institute at the University.[80] Jakobson was a member of two of his seminars in 1914 and 1915, where they studied Kurt Koffka's 1912 publication "The Analysis of Ideas and Their Laws," and

Husserl's *Logical Investigations*; this latter Jakobson found especially striking and instructive.

Two visitors had a special impact on these discussions. Gustav Shpet was among the participants in this seminar; he had been a student of Husserl before the war, and he became a member of the Moscow Linguistic Circle, which Jakobson established. He also recommended Marty to Jakobson.[81] Another guest of the Circle was Serge Karchevsky, who had studied with Saussure in Geneva, and who was able to provide an account of Saussure's views even before Saussure's posthumous book was available.

Thinking back on his years at the university, Jakobson later wrote, "It was in poetics that the vital relations of the parts and the whole were most clearly apparent, and this stimulated us to think through and verify the teachings of Edmund Husserl and of the Gestalt psychologists by applying their principles to this fundamental cycle of questions."[82] Reflecting on his early interests, Jakobson wrote that "it seemed imperative to me, in the modest role of an apprentice, to master in depth the rudiments of the sciences that I was studying, particularly those of historical linguistics and dialectology, as well as logic and psychology." We will see how influential this work outside of linguistics was on him. Readers of this present book should find themselves unusually well-positioned to understand what drove Jakobson at this point: we have been following the ideas that Husserl and the Gestalt psychologists were developing at this point. Perhaps it does not need to be said, but it is clear that Husserl and the Gestalt psychologists were providing ideas and conversations that Jakobson and his young friends found both exciting and *liberating*.

THE MOSCOW LINGUISTIC CIRCLE

Roman Jakobson was the major figure behind the founding of the Moscow Linguistic Circle in 1915, though he was only a sophomore in college. He gathered together a group of students interested in poetry, linguistics, and popular, or folkloristic, traditions—ethnology, essentially. For Jakobson, the social circle in which he did his work was terrifically important, and he had many opportunities throughout his life to reflect on the opportunities he had to work together with colleagues in linguistic work. In the 1960s, half a century after the founding of the Moscow Linguistic Circle, he set some of his thoughts to paper.[83] He explained that the intellectual and cultural circles of his century were rooted in Russia in the groups that were organized in the 1820s and 1830s by young Russians who were aware of the "intensive culture life" of their peers in Western Europe. Jakobson wrote of the "seething, varied Russian intellectual life" of the period, and the "mutual disputes" of the various circles. Such a circle was called in Russian a *kruzhok*. The intellectual (and, to some degree, social) lives of these young intellectuals took on a fixed pattern: people would host their circles at their homes, and informal discussions were encouraged. Later in the century—in the 1870s and 1880s—Baudouin de Courtenay's circle of students and colleagues met in members' homes, forming a self-conscious *kruzhok* in Kazan, known in English as the Kazan Circle. There was a danger associated with being part of one of these circles, because you might come to the attention of the police. When the linguistics students in Moscow set up their circle, Jakobson (introduced by his very close friend and partner in ethnographic expeditions, Peter Bogatyrev) approached the head of the Moscow Dialectological Commission to ask whether they could gain some official recognition by associating themselves with the Commission, itself part of the Russian Academy of Sciences. The Academy approved this, and the Circle began its meetings in 1915.[84]

The Moscow Linguistic Circle engaged in spirited discussions with Gustav Shpet, Husserl's student, in the early 1920s, and Jakobson continued to be interested in Husserl's work during his period in Prague. Shpet had recommended Husserl to Jakobson in Moscow, just before he left for Prague, and not just Husserl, but also Marty, and the Gestalt psychologists.[85] Jakobson also noted that Viggo Brøndal, in the first issue of *Acta Linguistica*, the journal of the Linguistic Circle of Copenhagen, referred to the "Husserl's penetrating meditations on phenomenology," sure to become "a source of inspiration for all logicians of language."[86]

Jakobson also pointed to the influence of Hendrik Pos, a Dutch student of Husserl's, two years younger than Jakobson, especially in connection with a distancing from behaviorism—a behaviorism that intentionally broke the links between the subjectivity of the linguist and that of the person whose speech was being analyzed. This was not the phenomenologist's point of view, nor was it Jakobson's. Jakobson would insist that the linguist's job was possible only as long as he remembered that he shared an essential consciousness with the speaker. Pos made this clear, as Jakobson notes: "The distance between the originating consciousness and science is not without limits: the linguist is linguist thanks to the fact that he is a speaking subject and not despite that fact."

JAKOBSON AND TRUBETZKOY

Jakobson first met Trubetzkoy in the fall of 1914, when Trubetzkoy commented on a paper at the Moscow Folklore Commission and when, sometimes afterwards, Trubetzkoy discussed Jakobson's own presentation. Their mutual interest in folklore studies and Russian ethnology was what brought them together in the first place.[87]

Jakobson wrote,

> It was in autumn, 1914, at the debates of the Moscow Folklore Commission, after a paper by N.F. Jakovlev on the historical imagery of a Russian lyrical folk song, that I first met NT, heard his brief and pointed remarks, and intuitively grasped the vital vigor of the young prince's mind. A few months later, at the debates provoked in the same association by my primary, one could say probe lecture, devoted to the influence of folklore on an eighteenth-century Russian poet, NT raised questions of verse and its sound texture; from then on our long and frequent reencounters during and after the animated gatherings of the commissions for folklore and dialectology paved the way for an ever wider exchange of questions and answers between two inquisitive offspring, one younger and one older, of the fermenting decade of the 1890s.[88]

In the avant-gardist circles of the early twentieth century, the huge advances that were felt in mathematics, as well as in physics and in chemistry, were recognized and followed with keen interest, with the understanding that science and art were both revolutionizing the world. From the avant-gardist point of view, it made perfect sense that linguists should undertake the effort to become a science. Looking back on this period 50 years later, Jakobson would write:

> When I think back on it, when I reread the various things that show how the artistic, literary, and scientific avant-garde during the Twenties and Thirties in Moscow were tightly interwoven, I realize how important and rich were the effects of the fascinated awareness they had of Einstein's writings and those of his colleagues. And the Moscow Linguistic Circle, a young experimental association which tried so hard to find a renewed theory of language and poetry, and the later branch of this same tendency which we called the Praguean structural school, explicitly referred to the methodological efforts made by Einstein to reinstate the central problems of relativity and invariance.[89]

The historian of science Gerald Holton has made a similar point about the interwoven character of science and avant-garde art.[90] Jakobson shared the sensibility of the Futurists towards all that came from the newest science and philosophy, and it accompanied him his whole life. His extraordinary personality and his limitless intellectual curiosity grew into an ability to juxtapose and synthesize notions from the natural sciences with those of philosophy, linguistics, and more generally, the sciences of humankind. But he was never just a man of ideas; he formed close bonds with people, from Mayakovsky to Claude Lévi-Strauss, from Niels Bohr to Jacques Lacan. There was something magical about his intellectual strengths that would leave a mark on the development of the social sciences, as we will see in greater detail in volume 2. He was the broker of ideas, the smuggler who carried the precious treasure of one science to its neighbor, the intermediary who insisted on the importance of each field to all the others.[91]

It was at this point that History abruptly interfered with the Moscow Linguistic Circle. The Russian Revolution toppled the government, the czar was deposed, the socialists took the reins of government, Lenin's party overthrew Kerensky's socialist regime. The world shuddered, and then it turned upside down.

We noted above that Trubetzkoy's political orientation was on the right, but the young intellectuals and the brilliant artists in the avant-garde in Moscow and Petrograd were largely on the left side of the political spectrum. Mayakovsky and Jakobson were in Petrograd on the historic day when Lenin arrived on the train which brought him back to Russia from exile in Switzerland, and the two of them were in the crowd that gathered.[92] In the tumultuous days of 1917, Mayakovsky and Jakobson would attend meetings with Lenin or Zinoviev, but they did not join the Bolshevik Party. Their friend Ossip Brik did, and in 1920 he joined the Cheka, the political police. Years later, in the memoirs he wrote on his Futurist days, he wrote that Brik's decision would "ruin his [Brik's] life."[93]

Until the end of 1919 and the beginning of 1920, the internal situation in the Soviet Union was thoroughly chaotic, and a whole range of various political groups and forces coexisted. Jakobson adopted moderate positions. Indeed, he was always discreet about his political engagement during this period, about his contacts with the Russian diplomatic services, but 35 years later he noted that he was a member of the Democratic Constitutional Party (the "KD"), which formed the first government after the February revolution, and which was swept away by the Soviet Revolution in October. The KD Party was a moderate group that brought together elites and the urban bourgeoisie. V. D. Nabokov, father of the author Vladimir Nabokov, was a prominent leading voice in the KD. It defended a constitutional monarchy along the lines found in England, and Roman Jakobson was a member of the presidium of the student organization at the University of Moscow. During the period that followed the October 1917 revolution, the different intellectual, artistic, and political networks around Jakobson remained active, and it was only beginning in 1920 that political repression became widespread. In 1918, the Soviet government lacked mid-level management. Jakobson was a known quantity in Moscow and in St. Petersburg, and he was offered several different diplomatic or political missions.

The most astonishing came his way in 1918, when the Russian Soviet Federative Socialist Republic was negotiating a peace treaty with Ukraine, and the question of the precise placement of the boundary between the two countries needed to be resolved. (The reader will agree with us that the past often remains with us today; as we write these words, the boundary between Russia and Ukraine is being fought over with live ammunition.) The proposal was made to place the boundaries on the basis of the linguistic characteristics of the various dialect regions. It was summertime, and the only member of the Commission on Dialectology who could be located was Roman Jakobson. We can appreciate the irony that in just a few years, he would be the one to show the validity of the notion of a Eurasian Sprachbund, as his colleague Trubetzkoy argued; in 1918, the political boundary between Russia and Ukraine would be based on linguistic criteria. It was this work, incidentally, which Jakobson was able to parlay into the safe conduct which allowed his parents to leave for Riga, in Latvia.

1920 was a year of great moment for Jakobson. As the year began he was 23 years old, managing to survive in the post-revolutionary fervor that had seized Moscow. Jakobson's family had emigrated to Germany, but he himself had remained in Soviet Russia. He was not to remain much longer, and the end of the year would find him in Czechoslovakia.

In 1920, Soviet Russia was engaged in military actions both at home (what remained of the civil war that had first overthrown the czar, and that had then overthrown the Kerensky government and installed the Bolsheviks) and abroad. Russia was battling with both Poland and Ukraine. Czechoslovakia, a new country established by the Versailles conference at the end of World War I, was of great strategic importance to the Soviet leaders. They were seeking political recognition by another country, any country. They hoped that Czechoslovakia, created after the war through the work of Tomáš Masaryk (the philosopher we have already discussed) and colleagues, would be the first to grant them recognition. Czechoslovakia was important for other reasons to the Russians, too: not only did it contain a large and important community of émigré Russians, it was adjacent to both Poland and Ukraine, and cover for espionage and sabotage missions targeting those two countries could potentially be run out of Prague.[94]

Overtures from the Foreign Ministry in Moscow in early 1920 were greeted with a tepid response from its counterpart in Prague. Prague proposed dealing with humanitarian issues, such as the repatriation of Czechoslovakians stranded inside Russia. Diplomatic recognition might follow if the first steps were successful. Late in the spring of 1920 the Soviets created a bogus Red Cross mission, headed by S. I. Hillerson, and sent it to Prague.

It may sound odd to us today to think of the Red Cross as providing cover for a covert mission sent by one country to another, but history says otherwise. The Soviets had already sent a Red Cross mission to Poland for covert political ends, and there was at the same time a Red Cross mission in Moscow that came from the United States that was staffed more by Wall Street bankers than by doctors and nurses.[95]

Roman Jakobson was a member of the Soviet Red Cross mission, headed by Hillerson, that arrived in Prague on July 8, 1920. In his memoirs about the time, he explained that the mission was sent to explore diplomatic recognition and repatriation of prisoners, and that he left the mission in September, just a few weeks after he arrived with the mission. Judging from materials that we have today, Jakobson's characterization of the mission was disingenuous.[96]

Roman Jakobson the Czech (1920–40)

LIFE IN PRAGUE

> Modern phenomenology is exposing one linguistic fiction after another. It has skillfully demonstrated the prime importance of the distinction between the sign and designated object, between the meaning of a word and the content at which the meaning is directed. There is an analogous phenomenon in the sociopolitical field: the heated opposition to muddled, empty, harmfully abstract cant and phrase-mongering, the ideocratic struggle against "humbug words," to use the picturesque expression. —Roman Jakobson, "What Is Poetry?"

When Jakobson arrived in Prague in the spring of 1920, he found himself torn between the Russian mission which had brought him there and the academic world which he wanted to be part of. He remembered Nikolai Trubetzkoy, who he had met at the Academy of Sciences in Moscow, and he looked to see where Trubetzkoy was. He found Trubetzkoy's address, and wrote him a letter, a step which was the beginning of a long correspondence and then a friendship, an intense intellectual exchange which would last for 19 years and which would have a profound impact on the intellectual landscape of the international linguistic community. This correspondence had a considerable influence on Jakobson's choice of linguistic research topics, and the debates and divergences found in these letters structured Jakobson's thought and his writing.[97]

Jakobson was at this point a brilliant young intellectual—he was 24 years old—and drawn to a thousand new ideas, and it is not hard to sense a young man who is not quite sure of himself. He took nearly 10 years to convince himself that he could defend a dissertation to obtain the PhD that he both wanted and needed.[98] A brilliant young intellectual, a conservative in matters of politics, but in the arts, in literature, and in poetry, a fervent and enthusiastic defender of the avant-garde as he found it developing throughout Europe. If ever there was someone who incarnated both continuity and change, it was this young Roman Jakobson. All of the anxiety of the age about modernity—what it was, what it meant for the arts and for society as a whole—found him like a storm finds a lightning rod. As far as Jakobson was concerned, he would be modern or he would be nothing at all.

We have already looked in detail at what it meant to be modern in Central Europe in the 1920s, especially the movements in philosophy and psychology. The world was coming to see how Albert Einstein had revolutionized physics, and quantum mechanics was about to provide yet another revolution. To be modern meant to be a scientist and to believe in science: the mathematical sciences first of all, but also the sciences of texts, of poetry, and painting; the sciences of peoples; and the science of their languages. Jakobson knew that science was not done in isolation, and he had already put that knowledge into practice in Moscow: science consists of programs, of manifestos, of groups, circles, organizations, and official affiliations. His life in Prague would illustrate and illuminate this understanding.

Such are the principal themes that occupied Roman Jakobson at this point in his Praguean career. All of them are found in the epistolary interchange he shared with Trubetzkoy. The two men were built differently, to be sure. The austere Trubetzkoy was the man behind the great phonological project, focused on the writing of his magnum opus throughout his life in Vienna. Jakobson's life in Prague was always on the run, a bit more like Saussure's life had been in Paris. Jakobson never encountered an idea that did not captivate him, if only for a moment, in the arts, in all that was Slavic, in journalism, in cinema, even in religion. As Morris Halle would write, "Among the 134 texts that Jakobson published between 1920 and 1939, there are studies on all of the great themes that interested him throughout his career."[99]

All the great themes! Already in Moscow he set his sights on everything worth talking about, and all of that came with him to Prague. His job at the Mission brought some money with it, and he published his thoughts on poetry, and even more on verse in Czech, which would influence Trubetzkoy's thought. And on ethnography, and with his friend Peter Bogatyrev on diachronic phonology. He was interested in moras and intonation as well ("Die Betonung und ihre Rolle in der Wort- und Syntagmaphonologie" [1931a]) and in morphology ("Zur Struktur des russischen Verbums" [1932b]), both works in which Husserl's influence and the phenomenology of perception are again to be found. At the same time, Jakobson was active in the avant-garde of the artistic circles, but he was also taking courses in diachronic phonology and Czech grammar at the University of Prague.[100] Anywhere there were Slavists, he was to be found, editing a number of medieval Czech texts, proposing a reinterpretation of Czech historiography that would tie Bohemia not to the west, with its Germano-Roman past, but to the east, with its historical, linguistic, and most of all religious roots, tied to Saints Cyril and Methodius.[101]

The founding of the Prague Linguistic Circle

VILÉM MATHESIUS

Vilém Mathesius was a Czech linguist who became by the mid-1920s the intellectual leader of a group of young linguists. He met Jakobson in 1921, shortly after the young Russian had arrived in Prague. Mathesius was nearly 15 years older, an established academic, a professor of English. He had been a student of Marty, and of Masaryk, both students of Brentano.[102] Mathesius was particularly interested in Gestalt theory, which was the cutting edge of psychology in the German-speaking part of Europe.

Bear in mind that both Ehrenfels and Carl Stumpf were locals, since they were professors at the German university in Prague.[103]

Mathesius and Jakobson talked about the possibility of establishing a discussion circle on the model of the *kruzhok* to which Jakobson had given the name of the Moscow Linguistic Circle. Jakobson would later write that Mathesius "was a man of great talent who grasped perfectly what needed to be done in linguistics . . . a born organizer."[104]

By 1925, Mathesius thought that the time was ripe for a linguistic circle in Prague. He was not unaware that a Linguistic Society had just been created in the United States as well, where the names of Leonard Bloomfield and Edward Sapir were receiving a good deal of attention.[105] During its first year of meetings, the Circle just included a half dozen local young men interested in language, but in 1926, the meetings grew to include out-of-town visitors who would make presentations.

The Circle would meet about a dozen times per year and invite speakers to present material for extended discussion. A major focus lay with finding a new way to think about linguistics in which history, and positivistic notions of causality, could be left aside in favor of another way of thinking.

Mathesius was undoubtedly speaking for all the members of the Circle when he spoke warmly about the function, both intellectual and personal, that was played by the constant intellectual dialogue that the Circle stimulated. He found nothing comparable at the university: there were no opportunities fostered by the university to encourage young and not so young scholars to get together regularly and discuss new ideas, both their own and those of scholars around Europe. This was precisely what the Circle did, though, and Mathesius felt that there were two factors that were critical to its success: the Circle was comprised of a small group of people who came to know each other well, and they shared an unusually large number of intellectual interests.

THE FIRST MANIFESTO

Let us pause and reflect upon the nature of life and work for Jakobson and the other linguists in Prague and in Europe at this moment. Technology had not yet produce the internet, of course, and if the telephone existed, it was too expensive to have an impact on intellectual discussions. Communication depended heavily on the mail, and researchers invested in correspondence, and in the absence of copying machines, manuscript copies were not at all uncommon. Travel was by train and by boat. But the academic world was smaller and denser; the academic universe as a whole was smaller and younger than it is now. Intellectual thunderbolts were striking at an unheard of pace. We have already talked about Gödel's theorem, but we have barely mentioned Einstein's theory of relativity and the still more disorienting shift in the image of the physical universe caused by the development of quantum mechanics during the 1920s—a shift which raised very serious questions for anyone who thought that they could appeal to modern science to defend the belief that the word was a rational place in which all events were the causal response to preceding causes.[106]

This was a time of *manifestos*: we have seen the 1929 manifesto of the Vienna Circle, and there were many manifestos in the arts, including ones from the Dadaists, from the surrealists, from the Futurists, and many others. Of course writing a manifesto is a political act, and even more a social act. It is *more* than a statement of *credo*, of "what I believe": it is a statement of what *we* believe, it is a call for others to join a movement. The political zeitgeist of this historical moment is here particularly transparent.

To be a time of manifestos is to be a moment of rising self-consciousness as a group, and that impulse was as evident as it was in the United States, where linguists gathered together to form the Linguistic Society of America. In Europe, linguists had met in seminars and in laboratories, but the need was felt for a larger and less ad hoc organization. In 1927, the linguists at the University of Nijmegen in the Netherlands (which had only been founded four years earlier) took it upon themselves to call for an international congress for linguists the world over. This effort was led by Catholic academics in the Netherlands who were involved in the development of a network of Catholic universities which was officially recognized by the pope only 20 years later.

Two of the organizers, Christian Uhlenbeck and Monseigneur Joseph Schrijen, sent out a questionnaire to a large group of scholars interested in linguistics in preparation for the first international congress of linguistics, which was to be held in the Hague in the spring of 1928. They began by noting that linguistics had become a science (yet again!), an autonomous science, a science that was internationally recognized and which had as its goal not just the Indo-European languages, but all the languages of the world. The letter included a set of questions to which they invited responses: "What are the most appropriate methods for a complete and practical presentation of the phonology of any language?"[107]

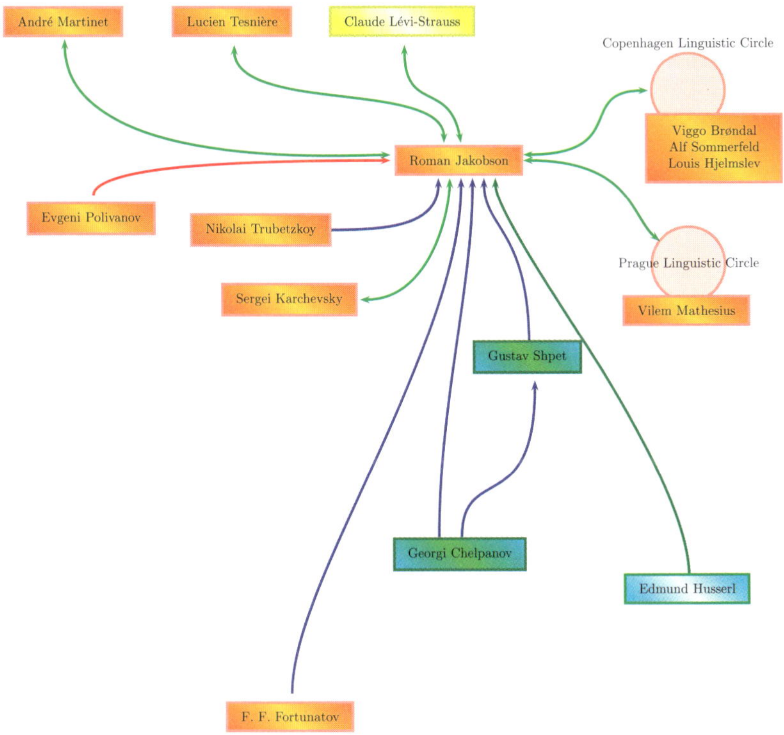

FIGURE 9.2. Roman Jakobson

Jakobson wrote a brief answer to this question in the form of a manifesto. He was, we must recall, 34 years old and a student, with neither a dissertation nor a university post to his name. But he had an idea which he could formulate, and he was proud of it. He sent it both to Trubetzkoy and to Serge Karchevsky, the former student of Saussure and private docent professor in Geneva, whom Jakobson had met in Moscow. Both agreed to attach their names to it, though Trubetzkoy had some advice for Jakobson: "Put yourself in the place of someone who had never heard any discussion of these questions. . . . Don't forget that linguists, on the whole, are truckers who are hardly used to abstract material."[108] Jakobson then sent it off to the organizers of the upcoming meeting in the Hague.

In this statement, Jakobson wrote:

> Any scientific description of the phonology of a language must above all include the characteristic of its phonological system, that is to say, the characteristic of

> the inventory, specific to each language, of the significant differences between acoustico-motor images. A more detailed specification of the types of such differences would be very welcome. It is especially useful to envisage phonological correlations as a class separate from significant differences. A phonological correlation is constituted by a series of binary oppositions defined by a common principle that can be thought of independently of each pair of opposed terms. Comparative phonology must formulate general laws that govern the relations between the correlations found in any given phonological system. . . . However this delimitation itself will not suffice either: it would be necessary to specify the types of phonological differences that are significant. There are two fundamental types of differences between acoustico-motor images. These are—to employ terms drawn from logic—differences between disjoint images and differences between correlative differences. . . . Under these conditions, linguistic thought is capable of abstracting the third term (or the term of comparison) from concrete pairs; in addition, the substrate common to the two terms of each of these pairs can also be abstracted, and thus forms a real entity in any given phonological system.[109]

Jakobson recalled that they all thought of their ideas as "deviations from traditional dogma," and likely to generate "severe opposition." So he was amazed at the positive feedback he got from the organizers, including linguists, like Meyer-Lübke, from an older generation. "We were especially happy with the unanimous response—behind the scenes—of the international avant-garde of our science, to these proposals."[110]

Looking back, Jakobson realized that the ideas that were brewing in Prague were not different in spirit from those being considered by young people across Europe (he mentioned Holland, France, Norway, Poland, Rumania, Hungary and Russia). There was a "drift," he wrote years later, "which unites the work of all these explorers and strictly distinguishes them both from older tradition and from some different doctrines which found their outspoken expression likewise in the '30's."[111]

What Jakobson saw as the central idea that had captured the imagination of the young Prague linguists, and so many of the other young Europeans, was the idea that language was a tool of communication, and therefore the central idea in linguistic analysis should be how language served as a means to communicative ends. It was necessary, they thought, to combat the "inveterate fear of problems connected with goal-directedness."[112] He was quite right; Western thought was dominated by positivism, and the strong belief that to be scientific required a belief in a system where the only causal real factors were from the past to the future in time.

THE CIRCLE, REORGANIZED

The next year, 1929, the Prague Circle was reorganized, and a new set of theses was drafted which would represent the official position of the Circle. It was no longer a social forum for open discussion; it had theses now to defend and to promulgate.

Jakobson wrote a letter to Trubetzkoy on April 16, 1929, in which he explained what was happening:

> The initiative core of the circle has now concluded that the circle in its function as a parliament of opinions, as a platform for a free discussion, is a relic, and that it has to be transformed into a group, a party, which is tightly interlocked as far as scientific ideology is concerned. This process is taking place at present with much success. An initiative committee of sorts has established itself in the circle.[113]

As Keith Percival has noted,[114] this marked the beginning of a conception of a scholarly group striving to organize itself along the lines of a "monolithic political party."[115] This was a new twist on how a scholarly group could construct its own picture of itself, a picture in which it was virtually a political party. That, at least, was a kind of group familiar to all its members. On the other hand, both political parties and scholarly groups could function as salons for the discussion and evaluation of new ideas (or even not so new ideas). But while scholarly groups generally exist precisely for that function, political parties will not always see their goals best served by endless debate and argument over goals and methods. Both a political party and a scholarly group are likely to have a central organizing committee, but only the political party will offer the board the power to specify the group's official position on the questions of most interest to the members. And only a political party would dream of removing a member from its rolls if he failed to take the right position.

In 1930, the Circle's bylaws came to include the following passage:

> He who is excluded from the Circle ceases to be its member. Each member of the Circle has the right of proposing to the executive committee the exclusion of a member whose conduct is at variance with the purpose of the Circle.[116]

This language in the bylaws of a linguistics association is shocking, even offensive, today. Toman notes that Jakobson described the Circle in 1933 in these words:

> Already in 1929, during the Congress of Slavists in Prague, [the Circle] presented itself as a militant and disciplined organization, with precise programmatic theses. The novelty of the structure of this Circle . . . appears in the fact that the Circle renounces carrying out the task of a parliament containing diverse currents and proclaims openly in its statutes that it aims at collaborating in the progress of linguistic research on the basis of the structural function method, and that the activity of any member of the Circle which shows itself in opposition to this program will result in his expulsion.[117]

Many years later, looking back on the history of the Prague Circle, Hajičová remarked, "The strength of the Circle was in its spirit of dialogue, which kept the Circle receptive to new ideas, rather than in any set of postulates common professed."[118] Times must have changed.

Jakobson and phonology

In 1930, Jakobson defended a dissertation in Prague, and the next year he left Prague for Brno, where he began to teach linguistics, Russian philology, and Old Czech literature. He became professor there in 1933.[119]

We saw above that Jakobson felt in the mid-1930s that he had to defend himself against Trubetzkoy's charge that he had lost interest in phonology. That reproof had come in a letter that Trubetzkoy sent to Jakobson, a letter that was both supportive and critical, and that showed that he understood how easy it was for Jakobson to allow his energies to be dissipated in too many directions. Jakobson's life was too bohemian from Trubetzkoy's point of view (and it was too hard to resist pointing out to Jakobson that he really did live in Bohemia), when Trubetzkoy wrote this to him in 1935:

> Bohemian journalism leads to bohemian intellectualism and kills scientific thought. You have always been attracted by bohemianism. There is no danger in this when you are young. But sooner or later we all arrive at the point where it is time to settle down. You say in your letters that you do not have any more scientific ideas, that you are exhausted and that you absolutely have to "break with your subject." And with this pretext you let yourself be taken

> up by the concerns of *Slovo o slovesnost*, by journalism, by your connections to the Czech literary bohemians, by the disputes among various small groups of Czechs and other trifles. That, I think, is what is blocking your scientific creativity. I do not believe in this scientific sterility that you speak of. I suppose that *mutatis mutandis* you are living the same thing as me, that is to say the transition from a scientific youth that went on too long to intellectual maturity. Maturity is not yet old age, it is not the sign of sterility . . . if under the pretext of stopping your scientific activity you throw yourself into Czech journalism, you will indeed throw away your talent, you will simply let yourself go, and go downhill. Attempts to make one's youth last forever are always in vain.[120]

Jakobson was blessed to have a mentor who would be so frank.

The reader will recall that Jakobson defended himself of the charge by saying that it was a fundamental disagreement with Trubetzkoy about the binary nature of features that had slowed down the progress he had hoped to make on his larger statement about phonology. We did not question Jakobson's remark at that point, but let us return to it now. The thoughtful reader might even have remarked at that point that Jakobson's rejoinder did not really contradict Trubetzkoy; there are few things that can slow down a student's research than an advisor who tells him that his theoretical principles are unconvincing.

Whatever the reasons, when one looks at the collected writings of Jakobson published many decades later, Trubetzkoy does not appear to have been wrong about Jakobson's engagement in phonology in the early 1930s. Jakobson's phonological production in the period before 1931 was extensive.[121] But between 1931 and 1938, the year in which Trubetzkoy died and Jakobson was forced to leave Czechoslovakia, there is surprisingly little on general questions of phonology. Jakobson was active and busy, but to judge from his publications, his interests in film, the arts and criticism were greater than his interest in phonology. His principal linguistic publication from this period involves the nature of case in Russian, one which would inspire many linguists in the following decades.[122]

Jakobson and Eurasianism

We looked earlier at Trubetzkoy's engagement with the ideas and the political movement of Eurasianism. Jakobson shared Trubetzkoy's devotion to it. In 1931, he wrote:

> Is the human community of the Eurasian geographical world different (and if so, in what respects?) from the neighboring worlds, above all that of Europe and Asia? The economic geography, in correlation with the facts of physical geography, observe the totalizing character of the Eurasian world. The historical destiny of Eurasia confirms its indissolvable unity. The study of the racial coefficient of blood types establishes the essential anthropological difference between the people of Eurasia and those of Europe and Asia. And finally ethnology, rid of its long dependence on the genealogical table of languages, can establish the characteristics particular to the Eurasian cultural circle.[123]

As Sériot noted, "Jakobson spent virtually all of 1931 defending the idea of the existence of Eurasia on a linguistic level . . . the idea was taken up again at the end of his life in the chapter on 'The factor of space.'"[124]

Perhaps it is possible to read Trubetzkoy's and Jakobson's work on phonology without any understanding of their broader interests or the questions that motivated them; we have, after all, been reading *Principles* that way for decades. But it is a thin reading which misses much of what Trubetzkoy valued in his own work. His goal was the development of a movement allied with Russian cultural nationalism, and his interests were in philosophy, ethnology, culture, and ethnolinguistics.

What is more, Trubetzkoy and Jakobson saw a substantive link between their linguistics and their Eurasianism during this period. Jakobson provided a scientific grounding for the definition of the Eurasians as a language community: the Eurasian languages all shared a back/front phonological distinction that pervades their phonologies, and a lack of tonal distinction in their phonologies as well. There are two ways of interpreting their views: on the one hand, they drew linguists' attention to surprising effects that arise in a context where a group of genetically unrelated languages are spoken by a multilingual community, an effect that we call "languages in contact" today. On the other hand, it is hard not to see their discovery as a bit of analytic overkill, in the sense that they knew the conclusion that they wanted to reach, and kept looking until they found linguistic characteristics that drew boundaries exactly where they wanted them to be.

Fleeing the Nazis

But European history continued to march on. Trubetzkoy, his mentor and friend, passed away in 1938, suffering from both a hereditary heart condition and Nazi interrogations. In March of 1939, after the Sudetenland crisis in which the Western allies capitulated to Hitler's demands, the German army marched into Czechoslovakia, and the country was scuttled. At that moment, Jakobson was in Prague and warned by friends. He hid himself in the city, looking for a means to obtain papers to leave the country.

With the help of the Danish ambassador in Prague, Jakobson was able to leave, and on April 21, he arrived in Copenhagen, where Brøndal and Hjelmslev had arranged for an invitation for Jakobson to give a series of lectures. With this trip, Jakobson and his wife began a series of dangerous moves around Europe to keep a step ahead of the Nazis, with the assistance of Jakobson's colleagues and friends, many of whom he knew through his efforts to make phonology (and linguistics more generally) an integrated scientific community in Europe. A personal and professional warmth seems to have developed between the Scandinavian linguists and those of Prague during the 1920s and 1930s. In Copenhagen, he reached out to Twaddell in the United States to see if there was a chance he could obtain legal status there, but the chaos that reigned at the time ruled that out. He sought out visas from Denmark, from Norway, from Sweden, and from France.

Astonishingly, though, this period during which he and his wife fled the advancing German army was intellectually productive for Jakobson. The five months he spent in Copenhagen were spent pursuing a project dear to the Prague Linguistic Circle: with Brøndal he worked on the *Atlas phonologique des langues du monde* and the *Projet de terminologie normalisée pour la phonologie*, two important elements of the Prague Circle's program, the first motivated by ethnological interest, and the second inspired by Husserlian thought. In September of 1939, it was becoming clear that his position in Copenhagen would soon no longer be safe, and the Jakobsons were able to get to Oslo, at the invitation of Alf Sommerfelt, before the Germans invaded Denmark in April of 1940.

Jakobson stayed in Oslo until April of 1940 as an invited professor at the Institute for Comparative Research in Human Culture. This institute was founded in 1922 by Fredrik Stang (whose son would years later become a famous linguist) and Alf Sommerfelt, the leading Norwegian linguist of his day, with the aim of facilitating communication among scholars in a world where political boundaries were making that difficult.[125] The institute was based on a premise that all people shared a character that led them to respond to challenges in similar ways, and that what was different across the cultures of the world were the results of the needs of peoples who responded differentially to the particular challenges that they encountered. The institute played an important role in the development of ethnographic and sociological study in Europe, and it adopted from its very beginning an orientation aligned with the sociological approach developed in France by Durkheim and his associates.

But Jakobson's activities in Oslo were not focused on ethnology. With the distinguished Norwegian Semiticist Harris Birkeland, he worked on religious history and biblical philology, another facet of his restless intellect, and one that remained with him his entire life, leading him to the history of Slavic religion, medieval philology, the philosophy and the theology of Saint Constantin (Saint Cyril),[126] even working on medieval Hebrew.[127] As Jindrich Toman underscores, his theological position is remarkably constant.[128] Long before Martin Luther's Reformation and what Jakobson called the "Germanic pretension," the evangelization of the Slavs (a word for which he associates an etymology linked to medieval Latin *sclavus*) was a magnificent story of linguistic, religious, cultural, and popular emancipation. In 1939, he wrote:

> The right to conduct the liturgy in the national language represents the right of the nation to the highest cultural value in the medieval hierarchy of values, and thus ranges over the totality of values: the entire culture, especially the entire literacy, assumes national traits. From its very beginnings, Cyrillo-Methodianism inseparably connects the national element with an element that is uniquely democratic. The right to the highest spiritual values is made accessible to each nation and to the entire nation.[129]

Thirty years later, in a toast that he raised at a conference in Prague in 1969, he declared:

> The Holy German Empire of the German Nation would cease to exist, as would that of the Habsburg monarchy and Hitler's "Thousand-year Reich"—all of them tried to deny Constantine's [i.e., Cyril's] ideas, but these ideas are still here, Constantine's name is still alive while the names of the emperors, monarchs and Führers are now hardly remembered. The fate of Constantine's teaching revealed the power of the Word, the permanence of his idea. No one spoiled it, nor could it be spoiled, because it was a value of a different order. "Wisdom speaks in it", as Constantine's deep words say, and wisdom can only be overcome by still deeper wisdom, by nothing else.[130]

But on a continent that was falling country by country to the advancing armies of the Nazis, Jakobson's long journey was not yet at an end. On April 9, 1940, Germany invaded Norway, and the Jakobsons tried to cross the border into Sweden at Särna. They were held there until they were able to obtain "alien" status, which would prevent them from being deported. They went on to Stockholm and then to the University of Uppsala, where Jakobson would work until May 1941, when he was able to obtain French, Swedish, and American visas—and Norwegian nationality as well, in addition to the Czech nationality he already had.[131]

It was during this year in Sweden that Jakobson wrote *Kindersprache, Aphasie und allgemeine Lautgesetze*. Though he was an expert on neither language acquisition nor aphasiology, Jakobson had already taken some first steps in these directions in 1939 at the 5th International Congress of Linguists, following in the path that had been blazed by Karl Bühler.

In this work, in 1941, Jakobson defended a model of the acquisition of phonological systems based on the progressive acquisition of oppositions that are specific to a given language. At the beginning, the child has much content and little structure. She has, we may say, access to all the possible sounds of language, but little or no structure. As she learns, the possible articulations become more limited along the lines that derive from the markedness relations in the language that is being learned. The progressive acquisition of markedness[132] begins with the universal triangles of [*ptk*] for consonants, [*iau*] for vowels, and the universal unmarked CV syllable. Aphasic pathologies follow a path that is the reversal of the path of acquisition, progressively destroying the oppositions and the marked elements, in the extreme all the way back to the most elementary triangles and CV structures. (The reader may well remember that this insight *actually* goes back to the French psychologist Théodule Ribot [see chapter 4].) These elements concerning the acquisition of markedness systems and linguistic pathologies will come back in a number of publications in the 1950s, as we will see later on.

On May 20, 1941, the Jakobsons were able to get two tickets on the ship *Remmaren*, which left Göteborg for New York, a two-week trip. On the boat was another couple that the Jakobsons knew: Ernst and Toni Cassirer, who were at that point Swedish citizens and, like the Jakobsons, fleeing the advances of the Nazis across northern Europe. Cassirer was the most brilliant post-Kantian of his generation, a widely respected philosopher who was forced to leave Germany in 1933, and who had just spent six years in Sweden. Cassirer's wife later recalled, "The conversation lasted . . . nearly the whole fortnight of the passage and was extremely exciting and rewarding for both scholars. Whether it was stormy or not, whether the mines were dancing before us or not, whether the war news were positive or not—the two scholars were discussing their linguistic problems with the greatest enthusiasm."[133]

In June of 1941, Jakobson arrived in the United States, and a third chapter of his life would begin. Two years later, Jakobson would become integrated into the American academic system and become eventually the doyen of American Slavic studies. We will return to this in volume 2, when we encounter a new Jakobson—essentially the third Jakobson—who will have traded Eurasianism for universalism, and Edmund Husserl for the American Charles Sanders Peirce.

But the Jakobson whose career flourished in the United States is not entirely the one that Europeans remember. V. V. Ivanov was a Russian linguist 30 years younger than Jakobson, who had both a brilliant and a troubled academic career. When his thoughts went back to Jakobson in 1982, just after Jakobson's passing, he recalled that "Roman Osipovič often talked about the generation of the 1890s."[134] Trubetzkoy was born in 1890, and Jakobson in 1896; from our point of view, that *is* Jakobson's generation.

> These people seemed to him to have been able to accomplish so much because their formative period fell in a time which preceded great catastrophes of our age, such as the First World War. Although Jakobson was half a decade younger than they, the people of the '90s (about whom he wrote so much . . .) were the artists, poets and composers who contributed most to his own formation. Perhaps as a prodigy, perhaps as a man who was always ahead of his own time, he became part of the generation that preceded his own.

Ivanov meant that Jakobson had the opportunity to breathe deeply of the air that preceded World War I:

> It was before the First World War that Jakobson received from Germany and first read Husserl's *Logische Untersuchungen*. So he began to work in this relatively peaceful pre-war decade, in which he saw the beginning of everything that was to initiate the unprecedented flowering of art, literature and science of the twentieth century. It is to these times that he belongs, and it is through him that we were in touch directly with an epoch that otherwise would have remained for us pure mythology.

Ivanov was born in 1929, so the prewar and the interwar years were ones that he did not know directly.[135]

> He was the living embodiment of this mythology, a personality of a different dimension, both in his scholarly activities and in his life, on a scale that our age, the generation that followed him, never achieved.

Ivanov was of generation born in the 1920s; he was born just a bit after the dissident physicist Andrei Sakharov, a bit before the émigré linguist Igor Melčuk. Judging simply by the calendar, this was the same generation as the American linguists Noam Chomsky and Morris Halle, but the New and the Old World were very different for this generation.

Ivanov wrote of Jakobson's "not fitting in" with the epoch that followed Jakobson's, and he ventured the opinion that Jakobson

> belonged indeed to the epoch of the Renaissance that began the century, but that had no continuation; and Jakobson himself was a man of that half-realized Renaissance. He was that gigantic promise that was held out by the beginning of the twentieth century. The extent of his enormous achievement equaled the almost inconceivable situations in which he worked.

Is it accurate to say that Jakobson's generation had no continuation? Over the course of this book, we have studied intellectual continuity, and the continuity that is grounded in teacher-student relationships, and as we will see in volume 2, Jakobson was part of a continuity that reached into American scholarship from the end of the 1940s through the 1960s, though with very few exceptions, that continuity was embodied in the work of Slavists.[136]

Ivanov continued:

> Moreover, he was not frightened by the catastrophe that followed . . . he recalled his last talk with Trubetzkoy in 1938 just before the Anschluss of Austria. He told me that exactly at that point, the idea of distinctive features came to his mind, an idea that he later developed over the course of many years, and that resulted in the concept known to us as the universal system of distinctive features. He said that the idea occurred to him in that tragic moment when both Trubetzkoy and he knew that they were meeting for the last time, when the occupation of Czechoslovakia that forced Jakobson to change his entire life was close at hand. He remarked that this was very much like the

> rest of his scholarly experience, and he emphasized that the most important ideas come in moments of catastrophe; nor did he see any contradiction in this.

This larger-than-life sketch of Jakobson seems spot on—*except* for the timing of Jakobson's insight regarding features. Jakobson's account is part of a historical image in which Trubetzkoy passes a baton on to Jakobson at the very last possible moment, a baton that Jakobson took with him and developed over the next 25 years in the New World. This is an appealing story—an appealing myth, if you will—but it is not what we have read in the published work of Jakobson and Trubetzkoy and not what we have recounted here.

Structuralism and the Prague Linguistic Circle

Structuralism

We have already encountered the term *structuralism* in the context of the development of psychology, and in the realm of linguistics, we have encountered it in discussions of the work of the Kazan school, the Geneva school of Saussure, and among the American linguists. Now we encounter it in the writings of the Prague Circle in the 1920s. In 1929, Jakobson wrote,

> Were we to comprise the leading idea of present-day science in its most various manifestations, we could hardly find a more appropriate designation than **structuralism**. Any set of phenomena examined by contemporary science is treated not as a mechanical agglomeration but as a structural whole, and the basic task is to reveal the inner, whether static or developmental, laws of this system. What appears to be the focus of scientific preoccupations is no longer the external stimulus, but the internal premises of the development; now the mechanical conception of processes yields to the question of their functions.[137]

With these words, Jakobson sounded a call with the central themes of structuralism as he understood it. But what exactly is, or was, structuralism, structuralism as it developed from the Prague Circle's understanding of language, and perhaps as it evolved later in anthropology just after World War II, and then in a range of social sciences and humanistic endeavors? We *owe* you an answer to this question, difficult as it may be and impossible as it may be to provide an answer that satisfies a common curiosity. Jakobson was not clear about what the structuralist project was, because clarity was surely not his aim and because he cared more about creating fertile soil than planting seeds in tight, narrow rows. He undoubtedly perceived things about the movements in arts, letters, and sciences which he felt but did not always understand, and his magic included the gift of saying things worth hearing even when he did not completely understand what he was saying. Perhaps that is too harsh; perhaps we do not all share the same view of what it means *to understand* something. Our sense is that Jakobson did not understand relativity, or quantum mechanics, or information theory, even though he obviously enjoyed seeing connections between linguistics and those other disciplines. Nonetheless, he had an ability that allowed him to say things that in some primal fashion surpassed his own understanding. No one is obliged to treat his pronouncements as charitably as a believer might treat the expressions of a Greek oracle.

We will therefore embark on an account that we believe to be both faithful to Jakobson's general view, and as clear as possible, in a domain where clarity was about as helpful as a flashlight would be in a planetarium (which is to say, not very).

SYSTEMS AND STRUCTURES, CHANGE AND EQUILIBRIUM

The important themes of this structuralism are these: First, there is the notion of structure as an organic whole, in contrast, somehow, to a mechanical agglomeration. There is a suggestion lurking that a living object (which is something *organic*) has ways of being put together that clockwork does not. It would not be unreasonable to sense a chord resonating with German Romanticism in such a position. Each part of the whole interacts with a number of other parts, in a way that is potentially complex, and in a way that the parts do not interact with whatever is outside of the system.[138] There is thus an *inside* to the system, and it is this inside that is of more interest than the outside. Just as important is the idea that the system as a whole, and the parts that form the system, have functions: without function, there is no system. And as they interpreted Darwin's theory of evolution to assume a mechanical view of the world with no room for organicism, they rejected Darwin.

Having rejected a mechanistic interpretation of Darwinian change as they understood it, what could Trubetzkoy and Jakobson offer as an alternative way to understand language change? The answer was *structures* and *systems* in phonology—but what exactly did they mean by that? They were not very clear on this point, even though it was central to their enterprise. From our vantage point today, we can see three points that they were trying to make. The first was that if we look at the phonemic inventories of a large number of languages, we find order and system in virtually all of them, and there must be a reason for that; it could not have happened accidentally. The second was that there are generalizations that we can make about phoneme inventories, such as the existence of a contrast between two sounds is possible only if a certain other contrast is found (we will consider an example of this below). Again, there must be a reason for these generalizations, and chance alone does not provide a reason. And third, a point that is made much less clearly, but which we can see when we step back from the linguistics and see their effort in a larger context: a system, such as a phonemic inventory in a particular language, is not just a set of markings on a piece of paper; there are specific *relations* between pairs of sounds, and no doubt relations among larger sets of sounds (like among all the consonants, or among all the vowels). The crucial notion here is *opposition*, and sometimes it seems as if oppositions are more fundamental than phonemes themselves for Trubetzkoy—but even if they are not more fundamental and more real than phonemes, they are not *less* real than the phonemes; for Trubetzkoy, we do not, and languages do not, start with a set of phonemes and then explore what kind of oppositions can be discovered between the pairs of phonemes. The oppositions are as ontologically real as the phonemes. To draw an analogy: when the solar system evolved out of a ball of gas and dust, it evolved into a massive star at the center, and a small number of larger agglomerations which eventually became planets. To understand the evolution, we have to focus not just on the mass of each separate piece, but on the forces of gravity that held the proto-planets to the stars, and which pulled each planet towards all the other planets as well, in a fashion that led to all of the planets lying on the same plane: a regularity that is too important to ignore. When a system evolves, Trubetzkoy and Jakobson might have said, under the constant influence of forces linking the individual elements, the resulting system has characteristics of structure that a random collection of items never will.[139]

Here is another way to think of it. Trubetzkoy and Jakobson saw *systems*, organizations of elements in which we could imagine that every element is attached to scores of other elements by invisible linkages—springs, if you will. If we observe the system and see it at rest, it is not because there are no forces acting on the elements, but rather because the system has found a dynamic equilibrium, one in which all the forces on the elements are balanced. To make a simple physical analogy: build a network of springs that connect a set of rings, like the set of coils that might lie underneath a bed mattress, and see what happens if we grab a ring and move it: it will move back to where it was once we let go. Cut through one of the springs, and the rings to which it was attached will move, just enough to find a new equilibrium state in which some of the springs are now tighter than they were before, and others looser. This intuition was held to be true for the phonemes that comprise the phonological system of a language just as much as they hold for the study of the natural world.

Beyond the internal interactions of the pieces was another notion: language is a functional system, which means that it has aims. Language exists because the person who speaks has something that she is intending and trying to do. At the same time, language forms a system, and we cannot understand anything about a language if we do not bear that in mind—a synchronic system, a state of a language at a given moment, such as today, or even right now.

This was a rejection of some ways of interpreting mechanism, in which the past influences the present by virtue of forces that are acting on the present, while the future does not influence the past, nor does it influence the present. But for Jakobson, this was not enough: language was a system in which there was a goal, and that very fact meant that it was trying to get somewhere (so to speak) that it had not yet arrived at. This was true both on the level of the individual who speaks, and on the level of the social organism that constitutes a language. Trubetzkoy wrote in 1932, "the evolution of a phonological system is at any given moment directed towards a goal."[140]

It is important to understand—what perhaps, from our vantage point today, is not at all expected—that there was a deep connection among three ideas: the idea that linguistic explanation could include the notion of a *goal*; the idea that languages could be related without being genetically related; and the idea of Eurasianism, which is to say, the reality of a the Russian (and later Soviet) Empire as a legitimate social organism. Perhaps the clearest statement of this was offered by Jakobson:[141]

> In the current ranking of values the question of where we are going is considered more important than the question of where we are coming from. These are no longer questions of historical origin, but rather of the direction in which we choose to go in order to define a nationality; the idea of class has replaced the idea of caste; in social life, as in scientific currents, the issue of community

> of origin takes a back seat to that of the community of functions, and indeed disappears entirely in light of the considerations of unity *de tendance*. The goal, which yesterday was an idea that was neglected and ignored is becoming rehabilitated everywhere. . . . Next to the traditional conception of family relations of languages is appearing the notion of community of linguistic trends.

This emphasis on *goals* always played a role in Jakobson's thought, and after his arrival in the United States, there were two movements that focused on goals that Jakobson espoused. The first was Peirce's pragmatism, and the second was the cybernetics movement, of which Jakobson would be a strong supporter. It was, in a sense, the idea that Norbert Wiener, the future inventor of cybernetics, viewed as the very central idea of the cybernetics perspective, and so it should not be surprising that Jakobson will find himself aligned, intellectually and politically, with the cybernetics movement, as we will see in volume 2.

There is more than one way in which the notion of a goal can be understood, and in light of the fact that the notion of synchronic linguistic analysis was still in the process of gaining a respectable status in the 1920s, it is easy for some confusion to arise about what a *goal* could be. The simplest situation is like the one that Wiener would be worried about during World War II: if a gunner is trying to hit an airplane, the place where the bullet should be going is not where the airplane is now, but where it will be in a few seconds (the bullet and the plane forming an ad hoc structure, in this case). A goal, then, can be understood as an expectation of a future occurrence which has an impact on what happens right now. That impact is not causal in the usual sense of the term, in the sense that a future occurrence cannot cause something in the past, but present belief about the future might be able to be the cause of some effect in the future.

Jakobson's sense of goal was linked to the notion of *function*. In his brief 1928 statement, he wrote that the Neogrammarian position had not been given a theoretical foundation, and such a foundation cannot be given unless the *purpose* of the system is taken into account. "Every transition," he wrote, "from one system to another necessarily bears a linguistic function."[142] The evolution of a linguistic system bears little resemblance to an anti-aircraft gun trying to model where its target will be in three seconds. The goal that Jakobson had in mind was a synchronic state that had characteristics that could explain why the system as a whole had shifted to the newer state. For him, this meant "a reinforcement of [the notion of regular sound-change] by the substitution of a teleological approach for the mechanical approach." If each language forms a system, then comparison of systems is a sensible activity even if the languages being compared are not genetically related. While linguists had engaged in such comparisons before, most comparative work was done in the context of trying to reconstruct earlier forms. For Jakobson and Trubetzkoy, it was equally interesting to see how similarities could be found among languages that were spoken in the same area that were *not* the result of inheritance from a common, earlier language.

STRUCTURALISM AND GESTALT PSYCHOLOGY

Some of the views of Praguean structuralism have much in common with what Koffka was developing at the very same time in the 1920s for Gestalt psychology, as we have seen in chapter 5. It shares certain sensibilities with the organicism that fit so naturally with the German Romantic movement of the late eighteenth and early nineteenth centuries, but the most striking difference is that the prospect is open now for a mathematical formulation of its principles, much as the atomistic billiard ball model encouraged a mathematical model. In accounting for the billiard ball collision, the mathematics was based on finding a post-collision trajectory in which both momentum and energy were conserved; in the case of dynamic equilibria, the mathematics leads us to look for a quantity akin to the potential energy inherent in a stretched spring—or rather, the sum of such energies over a large number of springs—and then look for a solution in which the first derivative of that energy was zero, providing us with an optimal balancing of forces. Do we think that either Trubetzkoy or Jakobson were thinking of computing first derivatives and setting them equal to zero? Certainly not; there is no reason at all to think so. But across the sciences methods were being developed, and had in fact already been developed by the 1920s, in which large systems could be quantitatively understood even when there were spring-like interactions between large numbers of pairs of elements. And at a certain level, that was exactly what Trubetzkoy and Jakobson saw in the structure of phonemes of a language.

As soon as a physical system attains any level of complexity, it often turns out that the appropriate mathematical model to use to understand its dynamics is not one that resembles Newton's laws of motion in its familiar formulation, but one in which a system with a large number of elements has achieved a lowest overall energy level as a result both of internal preferences of the individual items and of the forces that link each item to its near neighbors. Gestalt psychologists emphasized the relevance of the notion of a *field* to psychology, and a field is an expression of the sum of the effects at a particular point that arise from points both near and far. In such physical systems, the overall energy level of the system will be lowest when the elements are aligned in parallel fashion, each element coming into alignment with what it perceives as the dominant trend in its neighborhood—you may think of the magnetic orientation of an atom when it finds itself in a magnetic field, for example. Cases like this, which were at the forefront of science in the 1920s, lent credence to the notion that we could better understand the dynamics of a physical system by seeing how the system behaved as if it were trying to maximize or minimize certain physical quantities.[143]

BACK TO HUSSERL

In addition, nothing could have been clearer in Jakobson's mind than the direct connection between Husserl's philosophy, Gestalt psychology, and Praguean structuralism. We have in this book been traveling with the evolution of these ideas, and the reader has been introduced to Husserl and to the controversies among the German Gestaltists precisely so that the reader might understand that for someone in Jakobson's day who was well-read (we might even say, for someone who was cultured in his day), these links were perfectly clear. But the reader today might not see this.

As we see it, there were three themes in Husserl's work which struck a chord with Jakobson. The first was Husserl's commitment to accepting consciousness as a reality that deserves our attention and our analysis. The second was the understanding that admitting consciousness led immediately to recognizing goals as an important part of the human psyche, and the third was the philosophical problem that Husserl had inherited from his teacher Brentano, the logical analysis of the part-whole relationship. Jakobson was very clear and explicit about the impact of Husserl's ideas regarding the part-whole relationship, but it is hard for us today to get a clear sense of what he is talking about—primarily because we do not think of turning to philosophers for answers to questions that seem real to us, and which seem more technical than philosophical (though he did write a paper precisely on this, called simply "Parts and Wholes in Language").[144] Let us think about the treatment of syntax, of grammar. While it is not the whole story to suggest that in days gone by, people thought about speech as being composed of one word simply following another, there is some truth to that, and it was an important part of the work on syntax during this period, all the way up through the 1950s, to document and catalog the ways in which grammar was *not* composed of a simple chain of words. But people did not have the logical tools to move ahead full steam. It took new ways of thinking about things to allow linguists to speak of phrases and syntactic constituents (we saw some of this in the last chapter). And we are still in the throes of those discussions: there are many linguists who dispute that words could be composed of subunits in a fashion parallel to the way in which a sentence is built up of words. Those whole-word morphologists are struggling with exactly the questions of what it means to say that a linguistic artifact, such as a word, is "made of parts." Jakobson and Trubetzkoy were concerned with phonology, and they objected to Saussure's statement that spoken language consisted of a string of phonemes. Jakobson's objection to Saussure's view of a string of segments was that the segments were themselves logically decomposable in a fashion that led him to distinctive features, but from the point of view of phonology today, Jakobson *did* rather casually accept Saussure's assumption that there was just a single string of segments present in a phonological representation, when today phonologists would beg to differ, noting that phonological representations consist of two or more parallel autosegmental tiers of phonological segments. The reinterpretation of phonological representations as it is done in autosegmental and metrical phonology is exactly an example of the kind of questioning of the logic of parts and wholes that Husserl urged us to consider.

It is worthwhile to listen to Jakobson explain to linguists several decades later what he thought was happening during the 1920s and 1930s. In *Main Trends in the Science of Language*, he writes:

> The legend of a "militant anti-psychologism," allegedly proper to this [structuralist] movement, is based on several misunderstandings. When phenomenologically oriented linguists resorted to the slogans of anti-psychologism, they used this term in the same way as Husserl did when he opposed a model of a new, phenomenological psychology with its fundamental concept of intentionality to the orthodox behaviorism and to other varieties of stimuli-responses psychology.[145]

Even if that is not quite correct (and it is not), it gives us an understanding of what Jakobson thought. Husserl's rejection of psychologism, as we have already seen, was not a rejection of psychology, or a rejection of one school or other of psychology, but it was a rejection of the view that logic, or mathematics, or philosophy, was logically dependent on psychology, a rejection of a view that is known today as scientific realism.

> This Husserlian model and kindred psychological orientations met with vivid interest among linguists and with their readiness to cooperate as well. One may recollect the contact and convergences between the research of F. de Saussure and E. Claparède, the explorer of parts and wholes, as well as N. S. Trubetzkoy's fruitful discussions with Karl Bühler and the assiduous attention which linguists of the two hemispheres paid to the progress of Gestalt psychology. What seems to remain particularly instructive, are warnings of both American experts in the relationship between language and mind, E. Sapir and B. L. Whorf, to the Gestaltists who, as far as language is concerned, rather "let the matter drop" since they "have neither the time nor the linguistic training required to penetrate this field" and since "their ideas and terminology inherited from the old laboratory psychology are a liability rather than an asset" (Whorf et al. 1956).

It was the Berlin school of Gestalt psychology—the ones who became known, eventually, in the United States—who let the matter of language drop; Karl Bühler was also a Gestalt psychologist, but one from whom both Trubetzkoy and Jakobson learned a great deal, as Jakobson noted just below.

> In a similar way, Sapir, although aware that linguistics was destined to have a special value for configurative psychology, suspected that "a really fruitful integration of linguistic and psychological study lies still in the future," because linguistics is one of the most intricate fields of inquiry for psychologists (Sapir 1929). Finally, our links with the so-called Prague school of psychology and with its initiator, C. von Ehrenfels, the first propounder of the focal concept and label Gestalt, certainly left their imprint on the upgrowth of the Prague Linguistic Circle. The only offshoot of modern linguistics which the allegations of antiphilosophical, antimentalist, and antisemantic bent really suit has been the linguistic activity of the so-called mechanists (as labeled in L. Bloomfield's *Language*, 1933, Ch. 9), a group of American linguists influential mainly in the forties but now nearly vanishing.

This is the central point; nothing matters more.

> The rigorously restrictive problem of mechanistic investigation may be interpreted, however, as a set of useful reductionist experiments, irrespective of

> the philosophical credo of the experimentalist. At any rate, despite all the particularities of this regional team which separate it from all the other groups of linguists in the present-day world, the analysis of linguistic structures is the common denominator of all the contemporary scientific currents; and this persisting trait sharply distinguishes the linguistic research of the last four or five decades from the main routes and targets of the anterior period. Ernst Cassirer's view of "Structuralism in Modern Linguistics" deployed before the Linguistic Circle of New York on February 10, 1945, raised the adequate slogan "structuralism versus mechanism" and interpreted structuralism as "the expression of a general tendency of thought that, in these last decades, has become more and more prominent in almost all fields of scientific research."

We will return to Cassirer's paper in volume 2, where it can be seen as one of the first statements by the European academic refugees installed on the new shores of the United States (Cassirer had been able to obtain a position in philosophy at Yale shortly after his arrival in New York).

We have by no means plumbed the depth of Husserlian thought in Jakobson's linguistics. Holenstein was a German philosopher who published a book in 1976 on Jakobson's relationship to phenomenology, and he spent a good deal of time in conversations with Jakobson as he was writing the book *Roman Jakobson's Approach to Language*. The reader gets the sense that if there could be an authorized biography of Jakobson the phenomenologist, this book would be it.

Jakobson saw a connection between Husserl's notion of *foundation* (*Fundierung* in the original German) and implicational universals. Linguists today associate the term *linguistic universals* with Joseph Greenberg's work (though Greenberg's direction seems to have been influenced by Jakobson's work at this point; Greenberg was co-editor with Martinet of *Word*, the European-oriented linguistics journal based in New York after World War II).[146] But Husserl used the notion of *foundation* in order to be able to talk about logical (that is, not psychological) relationship between (any kind of) properties: for example, if something is red, then it must be extended in space, and if a number is a prime, it must be an integer.[147] Understanding Husserl on this point is certainly not easy, but it is reasonably clear that he has in mind a dependence which is anything *but* empirical—it is, we might say, as far from Hume's situation as one could possibly get, if by "Hume's situation" we refer to the case of two objects or events being frequently present together, and an observer who decides that the two therefore enter into a sort of causal relationship. Husserl is not interested in empirical dependence here at all. He does indicate that he is concerned with situations where one can say:

1. If *a* exists, then *b* also exists.
2. If *a* exists, then *b* does not exist.
3. If *a* does not exist, then *b* does not exist either.

And this is exactly what Jakobson declared that he would study in his very early paper on Slavic phonology.[148]

Holenstein's view of Jakobson's interpretation of Husserl seems to us more than plausible (although it does not strike us as a good interpretation of what Husserl in fact had in mind—that is quite another matter), and, as we have noted, it is quite easy to read Holenstein as an authorized account of Jakobson's views. The most interesting of Jakobson's features were "relational," in his view, such as the opposition between *compact* and *diffuse*, or the opposition between *acute* and *grave*. These were oppositions that Jakobson saw operative at a rather high level, in the sense that he believed that the *same* opposition could be present in two different instances of oppositions in which the differences between the two pairs are quite different to the untrained mind. Jakobson wrote, along these lines, "in phonemics the combinability of distinctive features into bundles or sequences is restricted and determined by a considerable number of universal implicational rules. For instance, the concurrence of nasality with the vocalic feature implies its concurrence with the consonantal feature. A compact nasal consonant (/ɲ/ or /ŋ/) implies the presence of two diffuse consonants, one acute (/n/) and the other grave (/m/)."[149]

Holenstein remarked,

> Jakobson models his formulation of the rules of implication on Husserl's doctrine of foundedness. His application of the rules to the problem of invariants and universals, however, goes beyond Husserl and the entire classical conception of the determination of essences. . . . In phonology certain features were found to appear in all systems, others only in some systems. Of significance here is the discovery that features are not randomly present or absent. The presence or absence of a feature is bound to the presence or absence of other features. This means that substantive elements are not universal in the absolute sense, but the laws of implication are.[150]

While we do not think that Husserl would have agreed that this move illustrates what he was trying to do, it does appear to us that Holenstein is quite correct in saying that Jakobson was at least inspired in this line of thought by the kind of questions that Husserl asked.

Neogrammarians, structuralism, and the disinterment of Saussure

> Depending on the context, Saussure serves the Prague linguists as a model to be propagated, as a companion in arms in the avant-garde, and as a rival against whom to measure their own conceptions. —Elmar Holenstein, *Roman Jakobson's Approach to Language*

The phonology of the Prague Circle is often described as a continuation of a structuralism that grew out of Saussure's ideas. And while this is certainly true, the fact is that it is as false as it is true. And how could it be otherwise? How could Jakobson and Trubetzkoy *not* view their own work as revolutionary, and based on insights that no one had ever had before? And how could we not take their point of view seriously in understanding the ruptures that were at stake? Certainly Trubetzkoy himself found little to recommend the idea that he was following in Saussure's footsteps. When a review of Jakobson's "Remarques sur l'évolution phonologique du russe" found Saussurian ideas within, Trubetzkoy found the notion "absolutely revolting." He felt that the English linguists did them an injustice when they identified Trubetzkoy and Jakobson as "purely and simply the school of Saussure."[151] In a letter to Jakobson, Trubetzkoy wrote that he had reread Saussure's *Course* and found nothing but a lot of old ideas.[152] Trubetzkoy wrote to his friend: "For inspiration [on a paper he had to write], I have reread de Saussure, but on a second reading he impressed me much less. There is comparatively little in the book that is of value; most of it is the old rubbish. And what is valuable is awfully abstract, without details."[153] Jakobson later wrote, in his preface to the letters that Trubetzkoy had sent to him, that Trubetzkoy "insisted on the profound difference of our views and approaches from the far-off doctrines of Baudouin de Courtenay and Ferdinand de Saussure. He rejected the fallacious endeavors 'to identify us with the Saussurian school.'"[154] Trubetzkoy also wrote:

> I have read the introduction to Baudouin's *Science of Language* (1909/1910 edition), the *Theorie phonetischer Alternanten* [*sic*] also by the same author, I am finishing the book by Stumpf that I was finally able to get a hold of. Reading

> Baudouin, I see more clearly what distinguishes him from "us". The path we have taken seems [to have been] much longer than we could have thought.[155]

Just how did Trubetzkoy and Jakobson themselves, then, view the relationship? Of course they were, in *some* ways, admirers of Saussure, Trubetzkoy's comments notwithstanding (none of us want to be held responsible after our death for everything we write in letters to our friends and colleagues). Jakobson had learned of Saussure through Karchevsky, who had studied with Saussure in Geneva, and both Jakobson and Trubetzkoy greatly appreciated his radical opposition to the Neogrammarians.[156]

Jakobson remembered:

> The first information that we got about the *Cours de linguistique générale* [*CLG*] was brought to us by the young Russian linguist S. I. Karchevsky, who left Geneva during the war and came to Moscow. He was one of Saussure's last students. Karchevsky spoke to us about his teacher, but he didn't have the book itself. . . . When I arrived in Prague in 1920, I contacted A. Sechehaye and I received several copies of *CLG* and other works from the Geneva school. I was the first to send these copies to my friends from Moscow. That was when our work on Saussure began.[157]

Today, Saussure is remembered in large measure for his great dichotomies, separating *langue* and *parole*, *diachrony* and *synchrony*, and so on. Jakobson and Trubetzkoy were critical of all of them. Synchrony and diachrony were *not* logically incompatible; the very first of the Prague theses of 1929 made that clear, and they took this as a big step away from the Geneva school.

The dichotomy that separated *langue* and *parole* fared no better. Jakobson wrote in the third of the Prague theses: "Each functional language has its system of conventions—the *langue* in a narrow sense; it is therefore an error to identify one functional language with *langue* and another with *parole* (in Saussure's terminology), for example intellectual language with *langue* and emotional language with *parole*."[158]

In hindsight, Jakobson underscored clearly the ambiguity of the Pragueans vis-à-vis Saussure:

> The point was to make clear what it was in Saussure's teaching that drew us towards it, sometimes linked us genetically to it, and what on the other hand radically separated us from it.[159]

The English edition of these discussions of Jakobson's were a bit different: Note how the vocabulary of rupture and continuity come to the fore in Jakobson's account. The English version has this:

> The question at hand was to specify those aspects of de Saussure's teaching that were shared by my views and those that separated us from each other. First and foremost there proved to be a considerable break with the Genevan precepts even in their two fundamentals, namely the arbitrariness of the linguistic sign and the rigid insistence on the linearity of the verbal form. Our entire phonological analysis, with its systematization of minimal phonological elements, clearly illustrated this fact. It was perfectly logical to base an exposition of the new approach precisely on an explication of these essential divergencies [*sic*]. One of the most fundamental and fruitful principles of Saussure was that of the "oppositions" upon which the entire system of a language is based. On this point, I followed Saussure with increasing insistence from the moment I became acquainted with his *Cours*, which Sechehaye sent to me in 1920 shortly after my arrival in Prague.[160]

Jakobson also wrote,

> The Neogrammarians worked on the history of linguistic facts, and not on the history of languages, because the notion of a language as a whole did not exist in Neogrammarian perspective. It was Saussure who reintroduced it, but he did not take it all the way. It was his great force to have understood that the description of language has to be oriented toward the system, towards the structural laws of this system. But as far as historical linguistics goes, Saussure remained within the Neogrammarian's rut. I would say that it is one of the symptoms and one of the cores of the tragedy of Saussure. He did not know how to find new paths in this area.[161]

What Trubetzkoy and Jakobson held against Saussure was his failure to break cleanly and definitively with the static, atomistic, and mechanistic linguistics that lay at the heart of the Neogrammarians' view of language. Jakobson wanted to see those characteristics replaced by a new emphasis on function, on dynamics, and on teleology guided by interpersonal communication. This orientation drew him to Sapir and Whorf, in fact, and it is what his aversion to Bloomfield was based on:

> In his theoretical considerations of the bases of language and linguistics, Bloomfield had followed a tortuous path that led from the dogma of psychologist

> Wilhelm Wundt (1832–1920) to the speculations of Albert Paul Weiss (1879–1931), a peculiar behaviorist mystic. Weiss's hypnotic influence on the anything but philosophical spirit of Bloomfield had the unfortunate result of narrowing the range of concrete questions addressed by Bloomfield in his work on different languages and on language in general. . . . Unfortunately, it was precisely these attempts by Bloomfield to construct a scientific theory that would eschew the semantic aspect of language that grew after his death to have such an important and sterilizing influence.[162]

For Jakobson, the perception side of language was always more important than the production side, and language was in the ear more than it was in the mouth, an emphasis that led him to focus on an acoustic basis for his distinctive features. This meant a return to Saussure's position. If, as Bergounioux has emphasized, the French have defended a more sociological conception of this collective consciousness while the Germans have defended a more ethnological conception,[163] linguists at the end of the nineteenth and the beginning of the twentieth centuries saw language as a collective psychological phenomenon, and Jakobson adopted Saussure's position as outlined in his *Cours*:[164]

> Many phonologists focus almost exclusively on the act of phonation, that is, the production of sounds by the organs (mouth, larynx) and pay no attention to the acoustic side. This method is not correct. Not only is the impression produced on the ear given to us as directly as the motor image of the organs but indeed it is that which is the natural base for any theory. The acoustic given already exists unconsciously when we approach phonological units; it is through the ear that we know what a *b* is, or a *t*, etc.[165]

NEOGRAMMARIANS

If it was galling to Trubetzkoy to have his phonology compared to Saussure's, we can only imagine how much worse it was to find himself put in the same box as the Neogrammarians, when he felt that his view was a huge step ahead of what had been accomplished by the Neogrammarians and phoneticians:

> M. N. van Wijk . . . claims that the notion of phonological system is already to be found in the work of the 19th century linguists, that of the neogrammarians (notably of K. Brugmann) and of H. Schuchardt. This is most unfortunate. One would have to be blind and deaf not to notice that the stops in Greek

> form three parallel series (τκπ, –δγβ, –θχφ) or that the stops in Sanskrit form four: (*k*, *c*, *t*, *ṭ*, *p*, –*g*, *j*, *d*, *ḍ*, *b*, –*kh*, *ch*, *th*, *ṭh*, *ph*, –*gh*, *jh*, *dh*, *ḍh*, *bh*), etc. Since neither the Neogrammarians nor H. Schuchardt were deaf or blind, they did not fail to recognize and even to remark upon this state of affairs. But between remarking upon it and the notion of a phonological system in the current sense of the term there is not merely a considerable distance, there is a fundamental difference. For the neogrammarians, the "phonic system" was nothing but a synthesis of isolated elements. Its regular structure was just fortuitous, it was unexpected and inexplicable, and it just got in the way. They noted the fact, but they tried to forget it as soon as possible, so as not to be obliged to study it in itself and to seek its causes.[166]

Let us interrupt Trubetzkoy for a moment, so that we can point out that what he is saying is very implausible, and surely he was aware of that at some level or other. It would not be unreasonable to say that Jakobson and he were making a special effort to better understand the consequences of the patterning that he found in inventories of phonemes across languages, better than the phonologists who had preceded him, phonologists who had focused on different problems. But Trubetzkoy wildly overstates his case.

> Taking the system as a starting point, going from the system to the isolated phoneme—that is what a Neogrammarian, so fearful of any hint of teleology, would have considered as a failure of method. . . . The only two linguists from the period before the Great War for whom the phonological system was not the more or less fortuitous result, unexpected (and hence illegitimate) result of a synthesis, but the starting point of the investigation and one of the fundamental principles of method, were F. de Saussure and J. Baudouin de Courtenay. Phonologists today, far from ignoring the origin of their ideas and their predecessors, as Mr. van Wijk claims, have always underscored the genealogical connections which link them to these two great masters. However, each impartial observer must agree that phonology today has fewer atomistic elements than was the case in the systems developed by F. de Saussure and by J. Baudouin de Courtenay, born at a time when atomism and individualism dominated all of science.

This is a refrain that we continue to hear: complaining that the previous generation was not very smart, because they did not worry about what my colleagues and I are worried about today.

SAUSSURE'S INFLUENCE

In light of all that Trubetzkoy and Jakobson wrote about Saussure, what can we conclude about their relationship to Saussure's work, and about how they saw that relationship? The first conclusion must be: it's complicated. Trubetzkoy certainly felt little inclined to give Saussure much credit for advancing the discipline of phonology, nor did he appreciate being viewed by others as following a path that had been blazed by Saussure. We have suggested that Trubetzkoy and Jakobson held it against Saussure that he did not undertake what they themselves wanted to do, which was to thoroughly disengage from the Neogrammarian project and its mechanical view of language. They did not seem to perceive that they were living and working in a world that had been made possible, at least in part, by the downstream effects created by Saussure's views and his writing.

Jakobson remained more circumspect than Trubetzkoy throughout his whole life regarding his own relationship to Saussure and to Saussure's impact on linguistics. It was after Jakobson left Czechoslovakia and arrived in the United States that he was able to offer a broader picture of his own thought and how it related to Saussure's. That comparison served a purpose—a rhetorical purpose, we may say, though meaning nothing more by that than that Jakobson could use the thought of a long deceased linguist to serve as a point of comparison of his own ideas, and to some degree, as a symbol of his intellectual heritage.

Françoise Gadet posed a similar question in her study of Jakobson a few years after Jakobson's death. She interviewed Jakobson's friend and editor, Cornelius van Schooneveld, who said that he had asked Jakobson why he was so attached to Saussure. Jakobson's answer was straightforward: "we needed a flag to cover the ship. . . . Il fallait bien un pavillon pour couvrir le navire."[167]

Phonology

> Disregarding all other considerations, a change in generation has taken place. Generations always advance by steps. In Copenhagen it was discovered for the first time that we are not alone in being active as an outpost. We are followed by the younger generation which has learned from our work and which can work independently. In any event, the Congress gave me wings. —Nikolai Trubetzkoy, to Roman Jakobson

We will look at four of the ideas that lay at the heart of Trubetzkoy's and Jakobson's views of phonology in the 1930s. There was, first and foremost, the notion of the *phoneme*, the closest thing to a substance that could be found in the world of phonology, an entity inherited from Baudouin, Saussure, and Sapir. Sound systems are made up of phonemes, and each language has a specific set of phonemes, which may change gradually over time. The second was the *opposition*, the relationship that holds between pairs of linguistic objects (typically, phonemes) that differ in an essential fashion. The third was the *feature*; features are the elements that arise in a phoneme out of the opposition which holds between pairs of phonemes *as a pair*. Finally, we will look briefly at the notion of *markedness*, a notion that arose out of the Trubetzkoy-Jakobson team's work, and which eventually became part of the everyday toolkit of linguists everywhere.

Between Trubetzkoy and Jakobson there were differences of perspective on how phonologists should treat different oppositions, a difference that lay at the heart of the disagreement about binarism. Of the two of them, it was Jakobson who was the binarist, and in practical terms, binarism amounted to the view that what was most important about oppositions was that they could be understood as based on a set of basic, purely binary oppositions. Trubetzkoy was not at all convinced of this, and believed that teasing out the differences between different types of oppositions was a big part of what phonological theory would be responsible for, once the smoke had cleared and it was clear just what phonology was.

There is much more, to be sure, that we will not explore here, including some crucial ideas, like positions of neutralization, and archiphonemes; about the very important distinction between phonology and "morphonology," we will say nothing at all, despite its real importance to the development of the field. But there is already enough for us to discuss.

The doctrine of phonemes

The phoneme is the smallest unit that carries the potential to mark a contrast between two words. Trubetzkoy recognized that in an important sense, his work on phonemes was pulling much the same load as that of a number of different linguists. "It is only in the last few years," he wrote, "the last fifteen years or so, that the idea of a fundamental distinction between sounds and phonemes has begun to spread around the world. Several linguists have reached this conclusion quite independently of one another. Among these linguists we must count Edward Sapir, whose

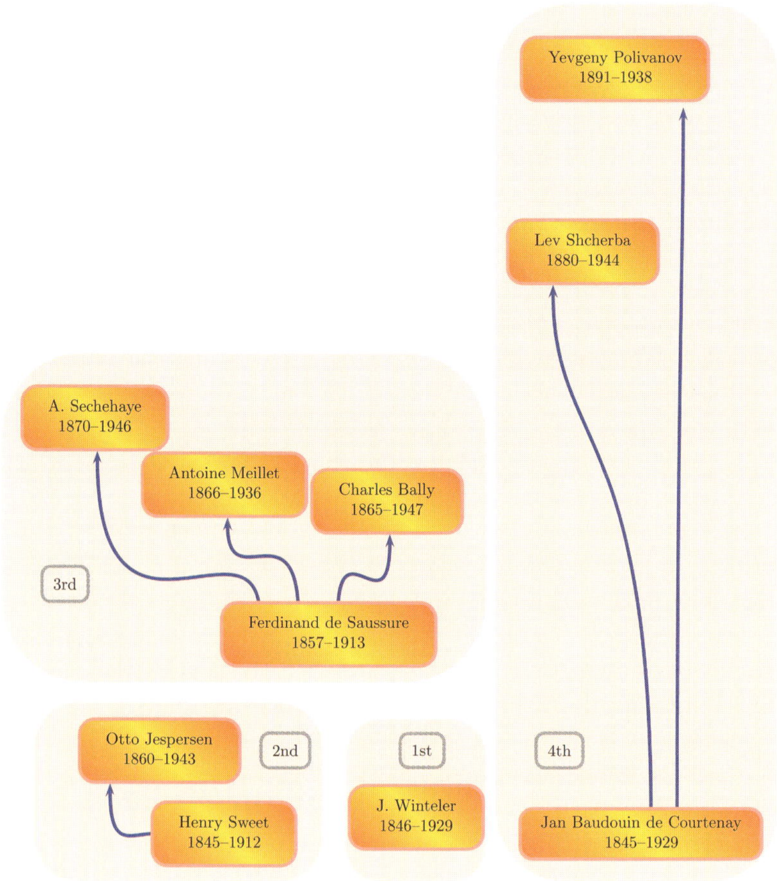

FIGURE 9.3. Trubetzkoy's versions of the phoneme

theory of 'sound patterns' was created independently of J. Baudouin de Courtenay and even of F. de Saussure."[168]

We ourselves could hardly disagree more with Trubetzkoy when he says that these advances in our understanding of the phoneme were all independent of one another (and we trust that the reader who has made it this far with us feels just the same), but we find it of great interest that he wrote it—and that he undoubtedly believed exactly what he wrote (though do recall the observation he made about genealogies just a few pages back.)

We will not find a linguist more sensitive, both politically and intellectually, to issues of rupture and continuity than Nikolai Trubetzkoy.

These issues come up both in his scholarly works and in his letters to colleagues, and his views are often quite surprising to us, given what we know today. Near the beginning of *Principles*, he explained that there had been five groups of linguists who had developed the notion of the phoneme, and it is less clear in his discussion there[169] how independent he thinks each group was of the others. Trubetzkoy's view is given schematically in Figure 9.3.

Trubetzkoy gave no indication that any of these groups affected any others—he implied that there was little or no contact, but he was not entirely clear on this, and we encourage the reader to compare this to our genealogy, which shows any number of real lines of intellectual connection that Trubetzkoy overlooks.

The doctrine of oppositions

At the core of Trubetzkoy's vision of phonology, along with the phonemes, was an image of the organization of these phonemes by means of a set of *oppositions* between pairs of phonemes. The notion of the phoneme was one that he adopted and adapted, but we need to better understand the meaning and the source of the notion of the opposition. First of all, we need to take them at their word when Trubetzkoy, Jakobson, and Bühler all tell us that the opposition is a notion that comes from logic. "*Opposition* is not exclusively a phonological concept, it is a logical one," Trubetzkoy wrote in 1936, and "it is impossible to study phonological oppositions (of which phonemès are only the terms) without analyzing the concept of the opposition from the points of view of psychology and logic."[170]

The logic to which Trubetzkoy was referring was not the sort of symbolic logic that we teach today in philosophy departments, which in all likelihood is the only sort of logic that the reader is familiar with. The logic that he was referring to would be found, for example, in Husserl's *Logical Investigations* and in Ernst Cassirer's *Substance and Function*. This kind of logic is rooted in Aristotle and in Kant. It enjoyed a long tradition going back more than two thousand years, and it predates the studies in logic that we explored in the preceding chapter. It is the study of concepts and relations in their purest possibility. It is quite different from the view of the Vienna Circle, for example, if by that we mean the view that a meaningful statement must either describe contingent facts about the world or else be a statement about language, and about how words and symbols are *used*. For Trubetzkoy, logic was a neighboring discipline which could shed light on the possible and the necessary relationships between objects of various sorts.[171]

The idea that phonology is centered on the analysis of oppositions has become so familiar that it is easy to forget that it is a doctrine with some teeth to it, so to speak. As a doctrine, it *excludes* many things from phonology, even things that might be of interest to some linguists. There are aspects of a language that have nothing, or nearly nothing, to do with marking contrasts in a language, such as the study of the rhythm and dynamics of speech, or much of what is studied under the rubric of variation today, like what marks the difference between socially identifiable styles of speech. Trubetzkoy called the study of such things *phonostylistics*, which was not a part of phonology proper. Saying that the focus of phonology is on those aspects of sound that are used to make contrasts in a language is not an anodyne observation, and it rules out many things that *some* linguists might very well think essential to the nature of language.[172]

Two phonemes are in opposition in a direct sense if they can appear in the same sound context and mark two distinct words—*bell* and *sell*, for example, as we have seen; *s* and *b* are in opposition, though two sounds can be in opposition in a more indirect sense if they never can be found in the same environment (something that is often true of a vowel and a consonant). But that is just the beginning of the story, for there are some kinds of oppositions (that is, relations of pairs of phonemes) that are more fundamental and perhaps more important than the others—the leading candidate for this status being a *correlation*, which we will return to. With some insight into the kinds of stronger and weaker, more and less important, relations between pairs of phonemes, we will find emerging a structure that bears some very rough similarities to the periodic table of the elements.[173] Neither Jakobson nor Trubetzkoy were the clearest of writers, though Trubetzkoy's *Principles of Phonology* shows a considerable effort to present a set of theses logically. But the reader of *Principles of Phonology* would really like a clear statement about what an opposition *is*; we would be grateful if Trubetzkoy had written something like this: An opposition always has three parts to it, as a matter of logic. There is, first of all, a pair of units which are juxtaposed; that is the first part. The second part is what they have in common. The third part is a pair, too, consisting of what it takes to reconstruct each of the original units from what they have in common, which is to say, what it is that makes each of the original pair of units different from what they have in common. If we consider the opposition of *t* and *d* in English, for example, we can ask what it is that the *t* and *d* have in common, which from an articulatory point of view is a gesture of the tongue to the top of the mouth, along with a closing of the air passage to the nose. The sounds differ in that the *d* has something—we call it "voicing"—that the *t* does not. If we combine this voicing with what they have in common, we have a *d*; if we combine what they have in common with nothing at all, we have a *t*. In this way, an opposition is formally speaking a triple, a set of three things.

Trubetzkoy's vision of a structuralist analysis of a vowel system (or more generally, a complete system of phonemes in a language) included understanding the different kinds of oppositions that can be found in a language. There are different *ways* in which two phonemes can form oppositions, and those different ways typically are reflections of how the larger phonemic system sits over them, encasing them in its totality. Trubetzkoy wrote:

> Each phoneme has a definable phonemic content only because the system of distinctive oppositions shows a definite order or structure. In order to understand this structure, various types of distinctive oppositions must be studied.[174]

Jakobson did not share this view, because he was more attracted to an elementary logic of opposition, in which there is just *one* kind of opposition. It seems to us that this was a great bone of contention between them. Let's read closely something that Jakobson wrote about this in his introduction to Trubetzkoy's letters:

> NT's effort to build up a multifarious classification of phonological oppositions met with Karl Bühler's philosophical and psychological doubts. . . . This complex and heterogeneous scheme seemed to overlook the logical essence of *oppositions* and to hamper the consistent dissociation of phonemes into distinctive features as their actually oppositive components. Two polar tendencies, one to restrict and the other to generalize the part of correlations in the phonological pattern, was the latent reason for my incessant postponements in completing my "Phonologie générale du mot," planned as the first section of the [book he planned to write with Trubetzkoy].[175]

There is a lot that is lurking behind this last sentence, and a reader today would have to be pardoned for reading it too quickly. Decoded, it says this: "NT builds up in this book a theory of oppositions, which includes a typology or classification of oppositions, different kinds of oppositions.

I myself was never convinced of this direction; it has seemed to me from the first that there are only two kinds of oppositions: the *real* ones, which I call 'correlations,' and the others, which I call 'disjunctions.' And I am not the only one who thinks so; Professor Karl Bühler agrees with me as well. In all honesty, it was our disagreement about this that stood in the way of NT and me finishing our major work on word phonology, and now it is simply too late."

We can see already that Trubetzkoy had begun to develop a typology of oppositions, and it is a typology based on logic, in the sense that Husserl explored in his *Logical Investigations*: the notion of opposition is abstract enough that it must be applicable to any study of a system of objects that are of the same sort, in some sense. The same logic could be applied to groups of people, or melodies, or any structured ensemble.[176]

Hendrik Pos was a philosopher who had been Husserl's student, and who interacted a good deal with the Prague Circle linguists during the 1920s and 1930s. His point of view was probably just a bit more Husserlian than those of Trubetzkoy or Jakobson; in 1938 he wrote,

> Recent research has revealed the important role that the opposition plays in all the layers of language, from phonology to syntax. The opposition is one of the principles that constitutes the system of language. The discovery of this fact has been a great stimulant to linguistic research; it has profoundly modified its method: instead of recording in great detail all sorts of isolated facts, we aim now to establish an order which permits us to see the structures. . . . Philosophy has an interest in making precise how the idea of the opposition contributes to knowledge of linguistic facts, because the notion of the opposition belongs to logic. The opposition is not an isolated fact: it is a principle of structure. It unites two distinct things which are linked in such a fashion that thought cannot posit the one with out the other.[177]

One of the crucial steps taken by Jakobson and Trubetzkoy was the notion that if phonemes can be put into contrastive pairs—for example, voiced and voiceless consonants like *d* and *t*—then a third object emerges from that pairing: the object that consists of what the two phonemes have in common. This is an idea that we suspect comes from a reading of Husserl: if two concepts A and B can be compared in such a way that we can determine what it is that A has that B does not, and what B has that A does not, then out of this mental operation emerge three things: what A and B have in common, what A has that is specific to A, and what B has that is specific to B. This logical operation led them directly to the notion of the archiphoneme: what A and B have in common is an archiphoneme.

MULTILATERALISM

Once an opposition was found in a system, Trubetzkoy would then ask whether there are other sounds in the language which also share that set of properties, the ones that *t* and *d* have in common. In the case of English and its *t* and *d*, there *is* another phoneme, *n*, which also is a consonant and which shares the gesture of the tongue tip, the gesture that *t* and *d* have in common. The *n* has something the *t* and *d* do not—the feature of nasality—and the *d* has something the *t* does not, which is voicing. Because there is a positive answer to the question as to whether there are any other sounds that contain what it is that *d* and *t* share—there is an *n*—the original opposition between *t* and *d* is called *multilateral*. If there had been no other sounds, it would have been called *bilateral*. The Pragueans had been developing these ideas; part of this was well expressed by Karchevsky, in a paper he wrote in 1927.[178] For Trubetzkoy, the difference between oppositions that are bilateral (such as voicing is in German and Russian) and multilateral (such as point of articulation is for consonants in virtually all languages) was highly significant; from Jakobson's point of view, it was shallow—it was shallow because when we look more closely, we will see (or so thought Jakobson) that what looks at first like multivalued features are really combinations of always-bilateral oppositions. In any event, for Trubetzkoy, an opposition between two phonemes A and B was bilateral just in case what A and B have in common is shared by no other phonemes in the system, and multilateral otherwise.

PROPORTIONALITY

The second important characteristic of an opposition for Trubetzkoy was whether it was *proportional*, and this property depends not on what the two phonemes share in common, but rather it depends on what the pair of characteristics that precisely defines how they differ. In the case of English *t* and *d*, the difference is that the *d* has something, voicing, which the *t* does not, so this *differand*, as we might call it, is the ordered pair (∅, *voicing*). If there are other pairs of sounds for which this is the differand (such as *p* and *b*), then the opposition is called *proportional*, since we can say "*t* is to *d* as *p* is to *b*." Otherwise, we say that the opposition is *isolated*.

CORRELATION

The correlation is a relation between phonemes that emerged early in Jakobson's thought about phonemic oppositions. A correlation is a relation between pairs of phonemes which is found in more than one pair of phonemes, and which is best understood as the presence versus the absence of some phonological feature. "Voicing" is a typical example, found in a language where there are two or more pairs of phonemes that differ only in voicing (such as *p* and *b* alongside *t* and *d*). If an opposition is not a correlation, then it finds itself in the more general category of *disjunctions*. The reader today will be forgiven for having no idea why this is called a "correlation," because the central meaning of that term today is a statistical one. We often say that two events are *correlated* if they tend to happen together, and we may wonder if there is a relation of causality behind the correlation.

But that is *not* the sense intended Jakobson (or Trubetzkoy), who were patently alluding to the Aristotelian use of the term, which is totally obscure today. In view of the fact that Aristotle's ideas about logic dominated the Western world for two thousand years, it is no surprise that "Aristotelian" may be simply replaced by "logical" in the usage of some, as is the case for both Trubetzkoy and Jakobson. In the Aristotelian tradition, two terms are correlatives if their meaning is possible only by viewing each as relative to one another, and one cannot be understood without the other: like *father* and *son*, *double* and *half*, *whole* and *part*—or *voiced* and *voiceless*, according to Jakobson and Trubetzkoy. Relatives of this sort are one of just four kinds of opposites that Aristotle analyzed: the other three were contraries, contradictories, and possession and privation (the last two being joined together).[179]

GESTALT

Trubetzkoy was well aware that understanding the *inventory* of phonemes of a language was not the whole story of the phonology, and the Gestalt psychologists had an important point to make. He wrote something that at first sounds quite surprising: "Of course, the matter should not be oversimplified. The phonemes should not be considered as building blocks out of which individual words are assembled." Despite Trubetzkoy's rejection of that statement, it is one that phonologists do, indeed, often make. He wrote that instead, "each word is a phonic entity, a *Gestalt*, and is also recognized as such by the hearer, just as an acquaintance is recognized on the street by his entire appearance. But the recognition of configurations presupposes that they are distinct. This is possible only if individual configurations are distinguished from each other by certain characteristics. The phonemes are then the *distinctive marks* of the configurations of words. Each word must contain as many phonemes, and in such a sequence, as to distinguish itself from any other word."[180]

It is not hard to hear the allusion to Bühler's *Gestalt* here. He continued: "As a Gestalt, each word always contains something more than the sum of its constituents (or phonemes), namely, the principle of unity that holds the phoneme sequence together and lends individuality to a word." He called upon the central analogy of the Gestaltists, that which compares a melody and a scale: it is entirely true that all of the notes of the melody are members of the scale, but they are not *just* members of the scale; they also include ordering and repetition.

Now, Trubetzkoy appealed to logic in his suggestion that there was a purely logical distinction regarding the way two elements may be related in an opposition. When we compare two things, and we abstract away what they share in common, we focus on what individuates each of the two from that which they share in common. That is a pair of things: one item for what the first element possesses in addition to what the two share, and another item for what the second element possesses in addition to what the two share. There are, Trubetzkoy suggested, only three possibilities now: first, there may be an opposition of something versus nothing; second, there may be an opposition of more and less of the same thing; third, there may be an opposition of two different things. There are no other possibilities.

The first one Trubetzkoy described as a contrast between the presence and the absence of a *mark*.[181] That mark could be many things: voicing, lack of voicing, nasalization, lack of nasalization, roundedness, non-roundedness, and so on. Of the two things in opposition, the one with the mark would be called *marked*; the other, *non-marked*. And this type of opposition was very important in phonology.[182] Secondly, there could be a difference in the amount of something that the two had: two objects that have exactly the same shape and composition, but differ in length, could be said to differ in this way, which he called "gradual." Gradual oppositions are oppositions in which the members are characterized by various degrees or gradations of the same property. Thirdly, equipollent oppositions are oppositions in which both members are logically equivalent, that is, they are neither considered as two degrees of one property nor as the absence or presence of a property. A glottalized *d* and an *n* have a *d* in common, and they differ by virtue of one of them possessing glottalization, and the other possessing nasalization.

The doctrine of markedness

The final topic we will look at is *markedness*. Jakobson recalled how he first encountered the concept of markedness. It was in a letter that Trubetzkoy had written to him in July 1930, in which he explained that they had been glossing over something important in their thinking about phonological distinctions and oppositions. Sometimes "the presence of some mark is opposed to its absence."[183] One value of the opposition is somehow understood in a more assertive, active, or positive sort of way, while the other is interpreted in a less assertive and more passive sort of way, as a sort of lack rather than something positive.

> One member of the correlation is **positive** or active while the other is **negative** or passive. At least this is the case if the opposition is binary . . . only one member of the correlation is perceived as actively modified, as carrying a positive features, while the other emerges merely as lacking this feature, as passively immutable. The two members are perceived as equally active, equally deviating from the norm in the opposite directions only when there is a third, absolutely passive or neutral member, that is, only when the correlation is ternary rather than binary. . . . So I believe that we should speak not about the principal and subsidiary variants of the archiphoneme but about the active and passive or positive and negative features of the correlation.

Jakobson was hugely enthusiastic about this idea, which he told Trubetzkoy he found to be one of Trubetzkoy's most important ideas ever, and he went on to explore this notion for the rest of his life. How are we to tell which value is marked and which unmarked? From Jakobson's point of view, the choice could depend on the context: one distinction might hold for both consonants and vowels, but the marked value might be different in the two cases.[184]

Jakobson wrote back to Trubetzkoy:

> I am increasingly convinced that your notion of correlation always being a relation of a marked and an unmarked series is one of your most remarkable and

> productive ideas. I think that it will become important not only for linguistics, but also for ethnology and the history of culture, and that correlations encountered in the history of culture, such as life/death, liberty/oppression, sin/virtue, holidays/workdays, etc., can always be reduced to the relation *a*/not *a*; the relevant thing is to establish what constitutes the marked set for each period, group, people, and so on. I am convinced that many ethnographic phenomena which seem at first glance to be identical, such as conceptions of the world, are actually distinguished by the fact that what is considered marked in one system is considered unmarked in the other.[185]

Jakobson perceived markedness in life outside of linguistics, as in social anthropology and in cognitive psychology, symbolic domains where the collective psyche is present. Looking back years later, he wrote about what he was thinking at the time of the letter that had come from Trubetzkoy, about how he was thinking about the suicide of his close friend Vladimir Mayakovsky, who we met briefly above. Mayakovsky was just a few years older than Jakobson, a brilliant poet larger than life in much the same way that Jakobson was, and in those intense years between the revolution and Jakobson's departure for Prague, Mayakovsky was an intimate of the same circle of friends as Romka, including the two Brik sisters. His love life was tumultuous, almost beyond measure, and his treatment by the Soviet government had its ups and downs as well. In 1930, he committed suicide. Jakobson's recollection of the moment when markedness came into his life was fused with his recollection of Mayakovsky's death:

> At that time, we were both deeply affected by Mayakovsky's suicide, which took place in April of 1930. We understood his lines about unmarked, "easy" death, and about the fact that "to make a life is markedly more difficult," and we realized that, according to this upside-down view of the world, not death but life "required motivation."

His wife commented, "For Mayakovsky, life was a marked category that could only be realized when there was a motivation for it; for him it was life, rather than death, which required motivation."[186]

Over the years that followed, there would be many suggestions made by many people as to what markedness means. From the very beginning, there were at least two competing visions: one involved the first kind of opposition mentioned just above, where there were two items in opposition, with one having something and the other nothing over and above what they share in common. A very different interpretation involves expectation against some background, or background knowledge: the marked member of an opposition would be taken by some later scholars as the one that was least unexpected, the most expected, especially when a specific context was being considered.[187]

Death, War, and Pestilence

We have lingered in the world of Prince Trubetzkoy and Roman Jakobson for quite a long time. At the beginning, we saw them for the brilliant young scholars and avant-gardists that they were, and we saw how they managed to navigate through the Soviet Revolution, each in his own way, and each knowing that they would be obliged to live out their lives outside of the Russian Empire and the Russian cultural sphere of influence. They were cosmopolitan, they were cultured, they studied philosophy. They not only studied history, they knew that they were living that history now, they were agents in the unfolding history of the Russian-speaking peoples.

The enormous object that stands before us, though, is Eurasianism. What needs to be said? First, Eurasianism was of the utmost importance to Trubetzkoy, and to some degree, no doubt lesser, to Jakobson. It also needs to be said that this is a side of these linguists that we have never heard about. Even now that we have heard about it, and learned a bit about Eurasianism's credo, it still seems like a lot to swallow.

We studied in considerable detail the fashion in which rethinking the meaning of history was central to so much that went on during the nineteenth century. Eurasianism was not a nineteenth-century movement. Instead, Eurasianism provided an *alternative* to history—just as positivism did, though in a very different way. Eurasianism replaced history with geography, progress with stability, and mechanism with teleology.

Undoubtedly you, like us, keep glancing back and forth between the Eurasian political agenda and its deep-seated distrust of Darwin and Western determinism. Is there really a strong connection between the politics and the philosophy of history, the philosophy of biology?

Yes, of course there is. We did not mention it at the time, but we encountered much the same phenomenon in the last chapter when we moved slowly through the career of Leonard Bloomfield. He underwent an epiphany of sorts during his years at Ohio State University, when he sloughed off his German culture and took on the mantle of a logical positivist, declaring that all of that puerile talk about minds and mentalism was *passé* and would be eliminated from both our grammars and our careful patterns of speech. Bloomfield with his positivism of the 1920s is as distant from us intellectually as Trubetzkoy and his Eurasianism of the 1920s is. And there is no hope of understanding what they thought and what they wrote if we do not understand the cultural milieu in which they lived.

What could we possibly derive from all of this—you the reader, or we the authors? Surely the fact that we can see how odd those worldviews are should give us confidence that we are not ourselves in the pall of anything similar, no? But as our theme is rupture and continuity, we have to face the question: how did we distance ourselves from those views, and have we come out of them naked, ready to face the universe without the comfort of a positivism, the comfort of an Eurasianism?

The answer is: it's complicated. We have brought the story up to the first shots of World War II, and the world that emerged out of that horror will be one that is radically changed. It will be changed enough for us to better understand positivism and Eurasianism, or at least to see it with some perspective that comes from having some distance. But we still have our greater philosophies being spun around us. And we will do our best to understand what it was that positivism, Eurasianism, and all of the other *isms* have turned to in the decades that followed the Second World War.

CHAPTER TEN

Conclusions and Prospects

Midnight in the Century

Our story has brought us now to a staging point of history, as the world prepares itself for the greatest moment of rupture it has ever seen. Even though we have kept our attention on the mind sciences, we have felt the cold draft of war on our back as we wrote the last 100 pages, and perhaps you can feel it as well. We have moved forward with linguistics, psychology, philosophy, and logic, and now the forward motion halts and war begins. Some of the root causes of the war have made their appearance in our account so far: we saw the university in Prague split into a Czech-speaking university and a German-speaking university, as claims of language morphed into claims of ethnic and cultural identity, feeding the dogs of war.

Europe exploded into war in 1939, and the United States followed two years later. Out of that war emerged a new world which is ours, yours and ours. The war had its horrors: no fewer than 50 million people dead, slaughtered, tortured, and starved. As many more were refugees, clinging to their rafts and their suitcases, while technology, industry, and bureaucracy organized an auto-da-fe for everyone.

To borrow Victor Serge's image: it was midnight in the century.

Let us hold everyone and everything in suspension, and let us take stock of what we have seen and what we have learned over the course of this book.

Guideposts

Stories and histories

We have walked with you through more than 100 years of thoughts and thinkers, and we have used time in a very careful and considered way, as you have no doubt noticed. We have tried to ride the current of time like a surfer on a breaking wave, staying in the narrative present, and rarely looking either backward or forward except as the past and the future emerge in the imaginations of the thinkers themselves—each of those thinkers holding tight to their surf boards as they ride their waves in.

One consequence of this way of presenting our material to you is that you may well think of all sorts of ways in which the issues we have discussed have great currency today. For example, we discussed Darwin's ideas about evolution in chapter 2, and Trubetzkoy's criticisms of those ideas in chapter 9, and there is an enormous literature today involving language change and variation which grows out of Darwin's views. Why did we not tell you all about the work going on today, work that proves how important and relevant Darwin's ideas were?

Because that is your job.

Our job is to show you that you cannot understand the present without understanding the past and how we got here. We do not believe that once you know the history, understanding the present is simple and that controversy fades to black. No: the controversy remains. If anything, we hope that our discussion makes it *harder*, not easier, to dismiss the views of the thinkers of the past that we have looked at, and harder to resolve the serious questions we face today.

We *test* theories by comparing their claims to reality, and we *criticize* theories by comparing them with alternative theories, and those are also the ways in which theories get better, by evolving in response to failures in tests and to criticisms that find vulnerable spots.

This is the benevolent side of the *battles in the mind fields* that we have been exploring: the battles between ideas are what makes our theories better. Those forces that are involved in these battles are not enemies; when we have done things right, the forces that disagree with us are the forces that allow us to make our own positions stronger.

In chapter 1, we emphasized the importance of distinguishing the social world of actors and the intellectual world of ideas. Battles may seem to play out in both spheres, but they do so in very different ways and by very different rules. Let's look into this a bit more.

From ideas to conversations

You will recall that we told you at the very beginning that the book you were about to read was not a history. Now that you have reached the final chapter, you may not agree with what we wrote; you may think that this is indeed a history, and you may wonder why we misinformed you.

We could have said that we had written an intellectual history, and that the subject of our study was the ideas, the arguments and the counterarguments, the proofs and the refutations of the proofs that have swirled about in the mind sciences. But intellectual history is not exactly what we have presented you with. There are many questions that an intellectual historian would have pursued more vigorously than we have. An intellectual historian would feel far more obliged than we do to give credit to the right people for coming up with brand new ideas. We have recused ourselves entirely from that particular endeavor.

The basic unit that we have been studying is not the idea, but rather the conversation. A conversation typically has two participants, each of whom has a point of view, and those points of view agree on some elements and disagree on others. A conversation can be viewed as a confrontation of two ideas, and it can also be viewed as a social interaction between two people, or groups of people. That is its richness: a conversation can be understood both in its human character, and as a way in which different points of view become expressed, become clearer, and develop. (And sometimes one of the speakers changes his mind, too—and his ideas may thus end up with one fewer proponent.)

Imagine with us, then, that over the course of this book we have been observing many, many thinkers mingling on a very large patio, clustering into conversations, with individuals moving from one conversation to another. It is a powerful image, because understanding it fully requires both understanding the content of what these people are saying to each other (that would be the intellectual side of things) as well as the reasons that people converse with each other as they do (this is the social side). Is there a sharp line between the two? Not always; the classical notion of rhetoric, for example, links the two, since rhetoric takes as its study the way that through language one is able to transmit conviction and belief, from speaker to hearer.[1]

Vocabulary and conversational style do change from era to era, and from discipline to discipline, though these differences are secondary once we recognize them for what they are. Sometimes our task has been to find ways to make differences in vocabulary and conversational style disappear by an effort of translation. We have learned that this work is well worth the effort, and a good part of our work in this book has consisted of us trying to convince you of the value of making such an effort.

Sometimes the conversations that we follow as we do our scholarship involve two people who live and work at the same time. The conversations that we are talking about often are played out not on a patio, but in articles and books, and when the conversations are between contemporaries, it may be easy to determine who it is that a particular scholar is criticizing—or it may not be. And we have noted cases where a scholar continues into his advanced years still fighting resolutely against the opponents who were active during his early formative years—recall what we said about Titchener responding to Herbart long past the time when anyone cared about, or remembered, Herbart.

So we allow ourselves the freedom to say that scholars are engaged in conversations with people who are no longer with us in body, but who are still with us because they have left their books and articles. We can still engage in conversation with Locke, with Descartes, and with Brentano because we have their books which we can read. To that extent, we keep them alive so that they can teach us, and help us make our ideas better.

Seen from afar, this multitude of conversations may seem like it produces a buzz of white noise, but there is a harmony of sorts that governs this megaconversation, and we would describe it not as a buzz but as a vast polyphony. We could even imagine a sort of geography to what we have described as a patio: psychologists may hang out on this side of the patio, and philosophers over there; linguists stay near the beer, but tend to move around a lot. What brings people together is sometimes agreement, but quite often it is disagreement and controversy. Agreement alone may make people feel comfortable with each other, but agreement is not enough to keep a conversation alive for very long. Conversations need differences to keep them going. The anthropologist Bronislaw Malinowski coined a term for conversations that have little content to them; he called them "phatic communions," ways in which people get together and use language to stroke the social bonds that link them. We are more interested in other sorts of conversations.

When you are a young scholar, just starting to learn, you select a small number of conversations with good reputations to pay attention to, and then you listen carefully. At first you probably do not understand too much, but eventually you do, and then some time later you begin to think you might even have something to contribute to the conversation (though you may be too shy to speak up at first). You learn a lot from the good questions that other people are asking, and you begin to get a sense of what questions feel proper and what questions simply are never asked. You learn what makes a good argument and how to develop an idea in a way that will gain others' attention.

These are conversations, but they are not anodyne. They are conversations with passionately engaged individuals. By speaking of a polyphonic conversation, we must not imagine people sitting down and singing Kumbaya, about how we are all sisters and brothers; these conversations can be sharp and pointed, and they nearly come to blows on occasion. The disagreements may become aggressive over time, and a loser might well leave a conversation in search of another, if he really does not like losing, or losing face.

Sociologists following in the steps of Erving Goffman have developed a notion of a person's "face," whose construction, development, and at times, destruction takes place in a purely social (that is, relational) context, and that aspect of human conversations carries over to the conversations that we study here. Each agent builds and defends his *persona*, his face, which he presents to others, and it will surprise no one that the wounds that a person's face suffers can be more violent and more painful than those inflicted on a real human's body.

Thus our core values tell us that the only time simplifying an idea is permitted is in order to explain it initially, to make it perhaps more comprehensible at first. But those same core values insist that we work as hard to harden and strengthen everyone's arguments and points of view, because the stronger the idea is that you are fighting against, the better will be your defense as you work to defeat it.

The social engagement of conversations

The megaconversation that we have been talking about is one that is engaged in by human beings. These humans' relationship to their ideas is complex: they adopt ideas, they cling to ideas, they develop ideas, they might even fight for them or attack their own ideas' rivals. We find that we cannot understand the great conversations without also understanding the speakers themselves, the debaters, the stentorians, the charming raconteurs. These are *people*, and everything we know about people can help us better understand how they behave when they are engaged in this great set of conversations.

We accept the autonomy of the people and the conversations that they are engaged in, the autonomy of each sphere with respect to the other. Studying how scientists interact within a laboratory or across a discipline seems to us fundamentally inadequate if the ideas that they are championing are left out of the analysis, and to that very same degree, an analysis of the ideas offered with no understanding of how the ideas' proponents engaged in their defense is missing something essential.

When we say this, we mean in particular that we do *not* embrace the metaphor that Daniel Dennett suggested when he wrote that a scholar is just a library's way of making another library.[2] That way of thinking about our subject reflects a need (which we do not share) to reduce one form of dynamic to another. Perhaps linguists are more adept at working with multiple autonomous systems than philosophers are, each system having its dynamic but sharing structural properties that links those different dynamics together.[3]

The people in our story have lives, as individuals living in their time and context, but more interesting for us is the realization that we can see as we look at these people, there are dynamics that are more social than individual, and which have greater explanatory power than the forces that are psychological or individual.

By "forces," we mean to refer to what it is that makes it more or less likely that an individual will try and will succeed in a particular effort, in a particular place and time. In that sense, forces may be helpful or they may hinder; they may be local or they may be generational. These forces do not *determine* univocally what an individual does, though they may have a mighty effect on that, and they do not determine what an individual believes or discovers. But when we look at the flow of ideas and people (precisely as we have done in this book), we find coherence in both of those streams, and to some degree (although perhaps a smaller degree) coherence between the flows, between particular ideas and the particular social context.

And so what we have done, then, in this first volume is to create a sort of social history of ideas and people, with the goal in mind of establishing a larger picture which can be applied to understand the development of the mind sciences over the decades to follow.

In the shadows

There is an insight that is often associated in psychology with Sigmund Freud, but this insight was pursued more recently with great energy in sociology by Pierre Bourdieu as well: there is much to be learned by uncovering what was hidden and repressed, and the process of uncovering is the single most important step that can be taken in pursuit of liberty because what is hidden or unknown continues to act in our lives as a source of alienation. (Bourdieu was famous for his characterization of his kind of sociology as "social psychoanalysis.") The two of us have felt this any number of times over the course of writing this book, and if we have succeeded, you the reader may have as well.

If Freud is associated with the term "repression," Bourdieu was more apt to speak of "amnesia" to refer to much the same phenomenon. *Amnesia* is a Greek word formed with a negative prefix *a-* from a stem related to *mneme* "memory," but what interested Bourdieu was what interests us as well, which is not *individual* amnesia, but rather social amnesia, which is very different in character. Bourdieu was also fond of referring to *anamnèse*, a word with no direct translation into English. We will coin a term for temporary use: *anamnesia*, with the same negative prefix *an-* (a variant on *a-*) attached to *amnesia*. The French word *anamnèse* has a medical use: it is used to refer to a statement summarizing the symptoms that the patient has reported. In a slightly extended sense, it also refers to the psychological material that has emerged over the course of a patient's psychoanalysis. It is this latter context which emphasizes the idea that the content of *anamnesia* was not merely forgotten, but also repressed.

Social amnesia is forgetting in a special way: it is forgetting by a social group, with no recollection that they have even forgotten something (an individual amnesiac may be very well aware that he has forgotten many things!). Bourdieu emphasized something peculiar about social amnesia: it goes hand in hand with mistaking things that are historically conditioned for things that are *natural*, that is, for things that were not created by humans at particular historical moments (and for particular historical reasons). His emphasis was on people, by and large, and he was interested in how the categorization of people in historically particular ways could be passed off and perceived in a society as merely a matter of objective nature. Our interest extends to abstractions as well as to people.

For Bourdieu, then, a good part of what he was trying to accomplish, in his work as a sociologist, was therapy for a society, which is why he was fond of describing it as *social psychoanalysis*, as we noted above.

At times we have called this book an effort of anamnesia as well.

At other times we have described this book in a different way, as an attempt to reappropriate history for another generation. At the end of the day, the two descriptions (one very Bourdieu-ian, the other more political) do not differ by much. To reappropriate history puts a bit more emphasis on the positive value of history, while the psychoanalytically tinged expression of anamnesia seems to suggest that all that has been forgotten and repressed was traumatic and negative in character, which is not at all the impression we have intended to share with you, the reader.

We do not mean to say that the forces become any the less strong by virtue of our being aware of them; perhaps they do, but we would rather see ourselves in the same situation as the strong swimmer who can get to where she wants to go by avoiding the current when it flows the wrong way, and by swimming into it when it flows the right way. Any sea-faring captain in the seventeenth century understood that idea.

Telling ourselves our own comfortable stories

It is not always *comfortable* for the people who are engaged in the sciences more generally to consider these forces. Indeed, it can be quite uncomfortable, on occasion. Take a very simple example: in chapter 1, we offered a list of more than 25 writers, going back to the early nineteenth century, who explained that linguistics was *finally* a science, and some of the recent examples were from our own esteemed colleagues. There is certainly no shame in being on that list; most of the citations come from the most revered of our intellectual forefathers, and we are certain that every single person was saying what they really believed to be true, and they could provide sound reasons to support their beliefs. But we could imagine that someone would think we were making fun of them by including them on the list, as if being part of a larger story about how the mind sciences evolve were tantamount to doing science badly in some sense or other. That is not so; these are, as we just observed, influential and important linguists, but no one enjoys being analyzed.

The business of reassuring ourselves that we are, finally, a science is a bit of a nervous tic, akin to a harmless neurotic symptom. No one wants to acknowledge they have such tics, and even less do they want to have them analyzed and cured. Freud was all too aware of this. He made one trip to the United States, when he was invited by G. Stanley Hall, who we recall from chapter 4. Hall was president of Clark University, and he persuaded Freud to come and give a lecture. The story goes that as Freud was arriving in New York in 1910, he turned to Carl Jung and said, "They don't know that we are bringing them the plague."[4]

So we tell ourselves that regardless of whether there are social and psychological forces that have had their effect on us, we still have the fortitude to treat the ideas fairly. And oftentimes that is exactly correct: we do.

But we construct stories for ourselves about who we are, and we largely do that by constructing stories about the groups to which we belong; we discussed this in chapter 1 in connection with each group's account of its own identity. These stories are not crutches or failings; they are part of being human. At worst, they may be simplistic and inaccurate in point of fact. But that does not mean that there should be no such stories at all: that isn't going to happen.

Roman Jakobson is one of the great characters of this story, and the one who most reminds us of a man riding a huge surfboard on an enormous wave in Honolulu. We discussed some of his views on phonological features in the preceding chapter, and we paid closer attention than most people do to the differences that separated him from Trubetzkoy. This view emerged from our reading of Jakobson's papers in the 1920s, and the letters that Trubetzkoy wrote to him over the course of their 18 years of collaboration. We perceive a Jakobson who was frustrated by Trubetzkoy's failing to be convinced by his understanding of features and the special status he felt belonged to binarity from a logical point of view. We also are very aware of the creative potential that can emerge from two linguists working together who do not always see eye to eye, which is not a bad description of Trubetzkoy and Jakobson trying to write a book together (and failing at it).

You will recall that V. V. Ivanov remembered Jakobson's account of how the notion of binary features occurred to him during his final meeting with his friend Trubetzkoy. We find Ivanov's account likely and plausible, but at the same time what Jakobson recalled decades later (and shared with Ivanov) does not align very well with the picture that we find from the documents in the 1920s and 1930s, where the picture that emerges is complex and riddled with disagreements. Disagreements between co-authors may be fruitful, but they typically get written out of the story, just as intellectual dead-ends often are.

It is this matter of *getting written out of the story* that captures our attention, over and over. There are occasions when disagreements are written out of the story (we have suggested that Jakobson and Trubetzkoy's disagreement over the nature of features is one such), and there are cases when agreement is written out of the story—for example, when one of the leading American behaviorists, Edward Tolman, suggested he did not see much of a difference between what he was saying and what Gestalt psychologists were saying (recall the discussion in chapter 5). In both cases, the strongest force leading us to forget those parts of our own history is the desire to keep history simple. The simpler the story, the more effective it can be, and the most important way to keep intellectual history simple is to make sure it seems like whatever questions arose were resolved, since resolved questions can be forgotten.

We mentioned another example in chapter 6 when we discussed Bloomfield's book *Language*, published in 1933. His students and his immediate successors took it as a manifesto of American independence in linguistics, but we commented there that the bibliography that Bloomfield included gives an entirely different impression. You will find very few Americans there, and not a single article from the journal *Language*, which Bloomfield had been so instrumental in setting up in 1924. With his bibliography, Bloomfield sends a message—how conscious this message was, we do not presume to say—that his work is grounded in nineteenth-century German linguistics, and that the current American linguistics scene is not where the real action is to be found.

So our *reason* for wanting to remember our past is not in order to be allowed to wallow in it, it is so that we can learn from it, and deal better with our own present. There are two great dangers that face us in our present moment: we can fail to understand the weaknesses of our own positions, and we can dismiss others' positions by failing to understand them. It is our belief that the kind of study that we have undertaken in this book is the best way to avoid both of those very serious hazards.

It may happen that a young person approaches a group with high hopes of finding an exciting dialogue going on, but finds no such thing at all—like what John B. Watson found at the University of Chicago, where he went to study psychology. He listened to the functionalists there, and he heard the disagreements that they were engaged in with the structuralists (recall chapter 4), and no matter how long he listened, he felt no sympathy, no resonance with either side. He walked away and found some other people who were more interested in animals and animal learning, and he served as a core of a new conversational group that had very little of interest to say to the functionalists or to the structuralists.

It is our faith in disagreement and controversy that most separates our account of the mind sciences from what we read in other people's books. Many decades ago, Thomas Kuhn drew the world's attention to what he called *paradigms*, which help to establish ground rules that workers in established fields all agreed upon, and which allow what Kuhn called "normal science" to play out without calling fundamental beliefs into question. That was *his* view, in any event, of normal science, though it hardly describes any of the moments that we have looked at in this book. You might take this as evidence that the mind sciences are "pre-paradigmatic," but that would be a mistake, in our opinion. It is controversy and disagreement that feeds all of the conversations that we have been listening to.

The study of the history of ideas is complex and difficult, in part because it is hard to make clear what the ground truth is that would serve as the empirical support of that enterprise. As we noted in chapter 1, we can sometimes identify ideas which move underground, like a mole, to pop up in a disciplinary context just beyond a fence that might have segregated off a neighboring field. There are other times when it is just about impossible to avoid saying that a certain idea was in the air, or part of the zeitgeist or the intellectual climate of a certain time or place. This is a very hard notion to pin down, because so often such notions are used without being overtly pointed to by a speaker, and speakers may well not be aware of the role the ideas play in their thoughts. We have seen in this volume a number of researchers try to develop a model of the mind as some sort of machine, and in the next volume, this urge will become much greater with the advent of the modern digital computer. While we can point to several writers whose publications brought computers to the attention of the public in the 1940s and 1950s, they were just some of the more visible and prominent voices among a much larger chorus of those who talked about the impact computers would have.

To repeat, then: our view of the evolution of the mind sciences is one in which the operative unit is an ongoing conversation between people. These conversations are not casual *how-are-you*s; they are discussions of real questions that matter. For there to be a discussion, there must be some questions of interest to at least two people, and there must be a willingness to engage in a back and forth conversation. Each person must listen to the other person and make an effort to understand, and be willing to change her ideas and positions as information or arguments come to the fore.

Geography

Languages and geography have played a very significant role in the picture we have seen. Our patio metaphor suggests quite rightly that the languages that you speak comfortably may have a significant impact on what you learn, say, and study. Through much of the nineteenth century, it was German that provided the best support for people who wanted to study psychology and philosophy. With the rise of the bourgeoisie and the relative fall of the aristocracy throughout this period, the intellectual no longer looked to find a job as a tutor to a member of the rich nobility. That was what Descartes had done, in the middle of the seventeenth century; his last job was as tutor to Queen Christina of Sweden, whose early morning tutorials (she set the schedule, and she must have been an early riser) led to Descartes's untimely demise. By the nineteenth century, as we have seen, the move was towards providing professional positions for people to teach and to earn some money, sometimes even enough to live on. This move was made first in Germany, as we saw in chapter 2, and by the end of the nineteenth century, some far-sighted members of the rising industrial class in the United States were engaged in establishing institutions which encouraged research and the development of scientific professions. As long as, and to the extent that, Germany was the leader in this movement, mastery of the German language was simply a necessity for anyone interested in engaging in the larger conversation, much as Latin had been centuries earlier in Europe—and Arabic had been, 500 years before that.

In each of the last five chapters, we have watched the beginning of a shift of balance from Europe to the United States. Some of that shift was due to the small steps in founding universities in the United States (Johns Hopkins, Stanford, University of Chicago, and the like), institutions that could attract good enough junior Europeans, and in a few cases some European stars. But with the rise of Nazism in 1933, the ground shifted ominously, and Jews and intellectuals became aware that they could not take their personal safety for granted in Europe. Many were able to come to the United States, and most of the immigrants shifted their scholarship to English from German.

As English has emerged as the most widely spoken language in international academia, Americans who are anglophone from an early age are not faced with the linguistic disadvantage they might have had in the nineteenth century (and would have had, if their parents did not speak German to them at home), but there remains today a linguistic barrier of a second order for today's anglophones. The search for our own roots that we have undertaken in in this book requires a good deal of reading of original sources that have not been translated into English.[5] Sometimes the work of a scholar during his European period and his American period feel quite different, and the earlier work may simply not have been translated into English. We were quite struck by this in reading the work of the Berlin Gestalt psychologists, whose work in German assumes a reader has a good deal more familiarity with contemporary physics than their later work in English does. We have discussed Roman Jakobson and his thought in Russia and in Czechoslovakia, and in volume 2 we will see a newer Jakobson. While his earlier work is published in his *Selected Writings*, much of that work is reprinted in the original languages without translation.

Geography has had its effects at levels other than the national. We saw in the nineteenth century a quickening of the pace of domination of a relatively small number of cities in which important universities were thriving. This was not the beginning of the trend; it had started as early as the thirteenth century, when European universities, modeled on Arabic schools, began to emerge in Spain, Italy, France, England, and elsewhere. But the pace grew faster, and we have inevitably seen the predominance of such cities as Leipzig, Berlin, Vienna, Paris, Oxford, and Prague, and smaller cities such as Göttingen. In the next volume, Cambridge, Massachusetts, will join this group of intellectually dense and potent locations.

Laws, mechanism, cause and effect, and teleology

We have had many occasions to talk about the notions of scientific laws, machines, mechanism, cause and effect, and, especially in the last chapter, teleology, and yet we have really only scratched the surface of the matter. We would like to summarize some of the most important points that we have seen and look ahead to what we will encounter in volume 2, *Dissent in the Mind Fields*.

We have seen two different ways of understanding talk about scientific laws. One way is solidly tied to nineteenth-century positivism, and it is most closely linked to the name of Ernst Mach, though much of it is already found in Comte. Scientific laws, on this view, are extremely compact expressions that can be expanded to create an infinite number of descriptions and predictions. The other view focuses not so much on the role of scientific *statements* within a theory, but rather emphasizes the belief that a scientific law is a statement regarding things that exist but are not seen, that is, regarding objects whose existence provide a simple and coherent account of what is observed. The clearest example of this is the atomic hypothesis and its various reformulations as physics has dug deeper and deeper into the depths of the atom.

The desire to describe the world in mechanical terms is a different matter, orthogonal to the first distinction we have just made, and it concerns what is *meant* by a mechanical account of something. We have seen that not everyone agrees on this point. At the time of Isaac Newton's formulation of his laws of motion and of gravity, many natural philosophers found his account of gravity unacceptable because it was *not mechanical enough*, and it was not mechanical enough primarily because it seemed to posit effects at a distance from their causes (in this case, the effects of gravity). Others found Newton's model mechanical enough, because they decided that being able to reduce many questions to quantifiable answers was absolutely good enough, and indeed, that was the dominant view for 300 years after Newton, until the development of the general theory of relativity.

In the case of the mind sciences, the search for a mechanical explanation has often taken the form of trying to describe a discrete sequence of events, each of which is the effect of the preceding event and the cause of the following event, just like when a set of billiard balls moves in straight lines on a table except when two collide and then rebound (and much the same image could be used to think of molecules forming a gas). The billiard balls (or molecules) move along straight lines when they are not colliding, and that is what Newton's first law of motion describes. Newton's laws do not do a good job of accounting for the dynamics of collisions, though the principles of conservation of momentum and conservation of energy (which came later) suffice to account for most of what happens after one of these collisions. It is not unreasonable to say that one of these collisions *causes* the change in velocity of each of the billiard balls.

And there is a natural way in which time can be used to organize a discrete set of events: the sequence of collisions that each ball is engaged in forms can be simply organized in time in such a way that a cause always precedes its effects.

But the combination of Newton's laws of motion and his law of gravitation gave rise to a very different world view, one in which looking for cause and effect is the wrong way to frame the problem. Newton had to invent the differential calculus to do his physics, and in doing so, he showed that the dynamics of the physical world had to be conceived of in terms of three quantities for each moving object: the object's location, the first (time) derivative of the location, and the second (time) derivative of the location; the second of these is called the *velocity*, and the third the *acceleration*. These three notions—location, and its first and second derivatives—are well defined if, but only if, they are defined over an interval of time (it does not matter how large or small the interval is, but it must be an interval, and not a moment in time). Newton's second law of motion says that there is a simple, linear relationship between the force at a given time acting on something and the second derivative of that thing's location.

There is no notion of cause and effect in this theoretical model that Newton proposed. Yes, we can say that each object undergoes a motion which is the "result" (but not in the sense of cause and effect) of its first and second derivatives, and the second derivative at each point is determined by the force found there and the mass of the object in question. There is no cause and there is no effect: there is only a dynamical system. There is *before* and there is *after*, but time is continuous, and it is not discrete except insofar as we provide simplified models of the reality.

At some level, Trubetzkoy and Jakobson must have understood this, or something not too far from it: they understood that the model of time in which the world moves forward by a sequence of causes and effects was a simplification imposed by the observer, and not an adequate account of reality. Newtonian reality connects immanent forces (for Newton, gravity) with the second derivative of an object's location. But there is no *before* and there is no *after* as far as a relation of causality is concerned. We do not say that "gravity *caused* the motion": gravity is the *name* that we give to a particular force, which has a value at each point in space and at each point in time.

In the last chapter, we saw psychologists and linguists who were struggling to extend this Newtonian dynamic system to the analysis of thought and language. The laws of the Berlin Gestalt psychologists took the rough form of Newtonian dynamics, and looking back on it now, it becomes clear that this is what Trubetzkoy and Jakobson were groping for when they talked about teleology. Their teleology was the linguistic equivalent of Newton's forces.

These questions will recur in volume 2, and indeed, are of great consequence for the cognitive sciences right now, as we write these words. We could summarize them in this way: if we can formulate a problem in a format that is based on spatial location and a continuous notion of time, then we can formulate a dynamical analysis of the evolution of the system.

But there is another, separate strand of thought that we have been exploring which involves only the discrete model of language and thought. When we say a discrete model, we mean a situation in which reality can be described as a (possibly infinite) sequence of objects or events, and we have already observed that in some respects this is an idealization. But for the analysis of language and of thought, this has seemed to almost everybody to be a reasonable idealization. People do utter one word after another, for the most part, and (perhaps) we say one sound after another.[6]

It seemed natural to most scientists to divide sequences of cause and effect into those sequences where the cause and effect was purely mechanical from another kind, where something blocked that purely mechanical character (but bear in mind that the notion of "purely mechanical" remains to be worked out). There were two kinds of thing that could block pure mechanism: one was randomness, and the other was an act of will or intent.[7]

Throughout much of the nineteenth century, there was an easy chain of inference rarely made explicit, but which was present nonetheless, based on a tacit view of words in a language: the view that to a first approximation, from the point of view of the speaker, the words of her language are a given, and it is only under unusual circumstances that the vocabulary of her language is affected by her will or intent. On the other hand, what a person *says* is her own choice, made of her own free will, while using the words that her language provides. (This is why Saussure took the study of syntax to be part of the study of *parole*; this was equally a fundamental premise in Whitney's linguistic worldview.) To the extent that an act is carried out of one's own free will, an act should not be viewed as part of a mechanical chain of inference—free will blocks the chain of mechanism, so to speak—and hence the production of sentences is not essentially mechanical in nature.[8] The same argument could be turned around: some aspects of language and language change are mechanical, most notably sound change; no room is left for free will or decisions in sound change; therefore some aspects of language do not involve personal free choice. These are the exceptionless laws of sound change.

To make matters worse, a notion of the unconscious was growing throughout the nineteenth century, with the realization that some parts of human behavior are best understood as the result of unconscious beliefs or desires. This step left open the question as to whether the unconscious could legitimately be treated as a mechanical process, which brings us to the question: does the unconscious *choose*, or does it just *react* in a fashion that includes neither randomness nor free will?

This concern regarding the proper domain of mechanism lay at the heart of the Neogrammarian's program, which consisted of a search for inviolable (and hence mechanical) laws of sound change, and for other principles (perhaps laws?) of analogical change, operative when laws of sound change were not the whole story.

Anna Morpurgo Davies quite rightly pointed out that the Neogrammarian's appeal to mechanism should be understood as a rejection of the view of language as an "organic" object or process, a term endorsed by the earlier Romantics.[9] These terms are vague and fuzzy precisely where we would like them to be clear and precise. The term "creative" is often encountered in connection with what is "organic," and rarely if ever in connection with what is "mechanical," but this is a connection that needs explanation. In dividing the principles of language change into sound change and analogy, it was natural for the Neogrammarians to see creativity as always a sign of analogy, and perhaps a necessary condition for analogy. But things are not simple: Hermann Paul objected to dividing words into morphemes, much preferring to talk about analogies. What is the plural of *fax*? He would say that *tax* is to *taxes* as *fax* is to *x*, so solve for *x* to get the plural of *fax*, which must therefore be *faxes*. Paul insisted that if working out these analogies leads the analyst to the discovery of stems and affixes, these *remain completely unconscious*. Now, that may or may not be *true*; but even if it is true, can we draw additional conclusions about these analogies not being creative, or mechanical? In Paul's case, it seems to be the case that "unconscious" was a hedge: it was a way to say that the analysis was not 100 percent real (this is, admittedly, an interpretation). But the significance of something being conscious or not faded from linguistic concerns over the next century, until the rise of generative grammar (in which native intuitions sometimes played a significant role) and views of mind that declared they were in league with generative grammar, such as the work of Jerry Fodor (1983).

When all is said and done, what we inherited from the Neogrammarians was the desire to see the complexities of language as a mechanical result, as the result in some fashion of a machine. The Neogrammarians were the first modern linguists to integrate the idea of mechanism into their model of language. They struggled to determine just what the boundaries were within which mechanism could be restrained, and beyond which the mind remained in control. But we can learn a good deal from paying attention to that struggle, because we are still engaged in it today. As we will see in the next volume, with the rise of a theory of syntax, it was no longer possible to view the sequence of words a person uttered as the result of her free will: the grammar is *responsible* for permitting only some sequences of words, and not others. How does the free will of the speaker interact, so to speak, with those places in the grammar where options are left open, places in the grammar where the speaker may make a choice? This is a question that even today is rarely even recognized as a question of mainstream linguistics. If a grammar allows a noun to appear in a particular utterance, the grammar will specify *noun* and leave it open which particular noun appears there. Today's computational linguist (who is not wearing the hat of a mainstream linguist) will object and say that there is a probability distribution over which nouns will appear in that position, and she may show that she can (really!) create a better speech-recognition algorithm by embedding that distribution into the probabilistic grammar. The mainstream linguist will demur, saying that the step of integrating a probability distribution over nouns is a mistake for a linguist; the linguist should leave the choice open to the speaker, for that is where the speaker's free will can take charge.

We will not pursue this point here, and we have raised it just to point out that we are still in the process of figuring out where the boundary lies that separates free will (or choice) from mechanism in the linguistic analysis, and that this question first arose with the Neogrammarian model. The rise of the computer model of mind has made the question more complex and more interesting, but it has not resolved the issue.

It had escaped no one's attention that when a Neogrammarian declared the existence of a sound law, that law was operative only in certain places (and certain times): someone who was skeptical of the Neogrammarians might have said that it works where and when it works, and only then. Is it fair and honest to call something like that a "law"? Not if Newton's law of universal gravitation is the kind of law that linguists are aspiring to. But to the extent that language is in the head of the speaker (as virtually all Neogrammarians claimed), it was not unreasonable to say that the law had lawful control over everything that transpired in a particular speaker's head from the moment that she learned it to the moment that she forgot it or she died.

A law, then, is exceptionless, and whatever its existence is exactly, it does not exist in some fashion that makes it depend on consciousness, awareness, or creativity.

Several things occurred that made this picture of mechanism as sequence-of-actions-without-will-or-intent less appealing. One was positivism, beginning in the 1820s, which was skeptical of science discovering invisibles; it emphasized that doing science was rather the construction of *statements*, especially compact generalizations that summarized a large number of observables. The positivist movement may have begun in the 1820s, but its impact continues to be felt even today, as part of the spirit of modern times. One strand of positivism is of *the-proof-of-the-pudding-is-in-the-tasting* sort: a scientific account is an information-handling device that takes some data as input. These facts play the role of boundary conditions. The scientific account then provides some predictions as output, predictions that are expressed in observable terms which can be matched up to reality in a noncontroversial way. That positivist style says, in effect, don't worry about the internal matters in my theory; I'm allowed to put there anything that allows me to make a better set of predictions.[10] In many respects, mainstream linguistics today implicitly holds to such a methodological premise: a new object in a linguistic theory is allowed in just in case it allows for better predictions with regard to grammaticality.[11]

This positivist view of science, then, emphasized that science is an advanced way of organizing knowledge and down-played the degree to which science is providing us with a fancy instrument to look at the hard-to-see construction of the world. This de-emphasis on using theory to tell us how the universe is constructed connected well with something the godfather of empiricism had emphasized: David Hume had cast out the legitimacy of using cause and effect as a real relation between existing events in the natural world.

The second thing that made this less appealing was something that lay at the heart of structuralism, as we saw it in the preceding chapter. Structuralists (such as Trubetzkoy and Jakobson) rejected the notion that mechanical cause and effect *was* the ultimate ground-level truth. An adequate account of language, they argued, had to include operative forces that lay behind individual observations, and these forces could only be defined by identifying what they were pushing *towards*.

Positivism said that science's discoveries are *not* necessarily in the objects being studied; structuralism said that to study language requires a vocabulary including *cause* and *effect*, and that mechanism (as the local impact of one object on another) was not the guiding metaphor of scientific explanation.

At the very same time, logic was creating things that seemed like little imaginary machines, ones which had no room in them at all for creativity, will, or self-awareness, and which would run according to a small number of strict rules, much like billiard balls do, or molecules in a perfect gas. These machines, like Turing's and like Church's, had a time-like element to them, like the physical universe and like language, but they were totally controlled by a small number of explicit axioms, just like Euclid's geometry was.

The time was ready for an explosion, like a room full of vapors would be. A behaviorist like Clark Hull would have no idea what hit him when the explosion came: he was so convinced that Newton's laws of motion were the right way to do science, so sure that discovering Newtonian laws for psychology was the ultimate condition for psychology to achieve the exalted position of a *science*, that he insisted that his laws of psychology should look just like Newton's, and no matter how often the Gestalt psychologists said that they were proud of Newtonian physics too, Hull never acknowledged that perhaps a new kind of mathematics might be right for the study of human conduct.

Prospects

The second volume of this book begins with the way in which the generation that came after Sapir and Bloomfield saw their task, with a discussion of Zellig Harris and Charles Hockett, two of the most influential linguists who followed in the traditions established by Edward Sapir and Leonard Bloomfield in the United States. Harris is interesting for several reasons, the most striking of which is that he is widely recognized as an important and influential linguist, but the views that he held and which he wrote about in his books were not at all like those attributed to him by linguists today. It is enough to make you wonder whether they actually read what Harris wrote.[12] A charismatic teacher, Harris worked closely with a number of students who became influential linguists, including Lila Gleitman and Noam Chomsky.

Charles Hockett studied with both Sapir and Bloomfield, and was very involved in developing mainstream linguistic theory in the 20 years that followed World War II. Both Harris and Hockett became engaged in the central example of theoretical rupture studied in volume 2, the rupture between mainstream American linguistics in the 1950s and the rise of Chomskyan linguistics. But this rupture was largely a ghostly nonevent, in a sense: Harris himself made no remark distancing himself from Chomsky, nor did Chomsky undertake serious criticism of Harris's theories. And yet the rupture and the schism occurred. Hockett's case was different in some ways from Harris's, and while he was actively supportive of Chomsky's early work, by the mid-1960s, he declared that he had been mistaken in thinking that Chomsky was pursuing the same ultimate goals that he was, and Hockett became disengaged from mainstream linguistics.

The entry of the United States into World War II was an enormous event in the United States as well as in Europe, and New York was the place where a good deal of the head-on encounter between European émigrés and American academics took place during the war years of 1941–45. We will follow the path taken by Roman Jakobson after he left Sweden and came to the United States. The New School for Social Research had already been engaged in providing a temporary academic home for German refugees as early 1933, and when Jakobson arrived, there were a number of francophone academic refugees who became part of a French university in exile forming part of the New School. Jakobson became part of that group.

New York was also the home of one of the War Department projects at this time which was charged with developing pedagogical materials for a large number of languages used in the different theaters of the American war effort. Many of the young linguists who had studied with Sapir and Bloomfield in the 1930s were drafted, or otherwise integrated in the project at 165 Broadway in Manhattan. Over the 1940s, a sometimes difficult and sometimes frigid relationship developed between some of the American linguists and some of the European émigrés. By no means was this kind of encounter unique to linguistics; similar stories have been told about the lives of other refugees fleeing Hitler and coming to America, many coming from established professional positions and many others coming from perfectly ordinary working-class lives. The United States struggled, in some ways, as it integrated these refugees, and if we are to understand the dynamic of postwar linguistics and psychology, we need to take this larger context into account.

Roman Jakobson had a very successful academic career after the war. He was a professor at Columbia University for several years, and then was offered a position at Harvard University, where he moved in 1949. That move was part of a larger picture of a growing academic dominance in Cambridge, Massachusetts, home both to Harvard University, the nation's oldest and most prestigious educational institution, but also to the Massachusetts Institute of Technology, which would be totally transformed by the new funding model for higher education that emerged in the postwar period.

MIT was one of the most important partners in the alliance established between industry, education, and the War Department during World War II. It was centrally engaged in the development of radar and to a lesser degree with the development of the electronic computer. Norbert Wiener, a colorful and brilliant mathematician at MIT, contributed to a new way of thinking about problems both scientific and human that were looming on the horizon. Wiener had his own ideas about information and about computers, and he coined the word *cybernetics* in a bestselling book that he published just after the war.

Wiener knew that he was not alone in his belief that the modern computer, which had become a physical reality during the war but which had yet to become part of the economy, would lead to great changes in the decades to come. Along with a number of other academics and intellectuals, he participated in a series of meetings in the New York area between the late 1940s and the mid-1950s that were known as the Macy Conferences, and these meetings give us a very good picture of how reflective individuals were trying to understand the changing world. All of them had been adults during the 1930s; those who had been in Europe had lived the horrors of the rise of Nazism, and those who had been in the United States had lived through the Great Depression, when nothing the bankers or the government did would breathe life back into the listless body of the American economy. With the war over, no one had any solid reason for being confident that the United States' economy would not slide back into a period of massive unemployment. Would computers help us avoid that, or would they help drive us back into a deeper depression? No one could answer this question with confidence.

Psychology and the other social sciences in academia were changing in similar ways. The dominance of behaviorism that was palpable in the 1930s could no longer be taken for granted. Though some behaviorists, notably B. F. Skinner, continued successful academic careers during this postwar period, more and more research was done which accepted the idea that humans had minds, desires, points of view, and cultures. The presence of the computer on the sidelines encouraged this; the computer seemed to provide a lower limit on how intelligent a social scientist's model could possibly be. If a computer could, even in principle, accomplish some task, it would not be breaking any canons of scientific method to assume that a human could also accomplish the task.

And so, over the first decade following World War II, three deeply entangled dynamics developed: the creation of the peacetime computer, the emergence of cybernetics as an intellectual venture, and the continued growth of the social sciences, all of this set against the background of global polarization.

The polarization of the postwar world set the foundation for seeing far more than just military and economic competition, balance, and challenge. European politics had for centuries been tied up in political and military alliances, and while a few countries had long histories of rivalry, most of the continent could look back and see how a country that was a rival in one decade could become an ally in the next. The nineteenth century was not just the century in which the modern nation state became stable, as we saw in chapter 2; it was also a period during which larger military and diplomatic alliances would emerge and collapse. World War I was essentially the direct result of a confrontation between the Entente Alliance and the Central Powers that could not be resolved diplomatically. But that confrontation was political, not ideological.

The world that emerged after World War II was one that in retrospect was *ready* to be polarized, which is to say that two extreme political and ideological positions emerged which demanded that everything find its proper place along the axis that stretched from the American sphere to the Soviet sphere. It was not the first time in history that something like this had happened; the Crusades had some of this character as well, as Christendom fought the Islamic world. But now the polarization was coming much closer to being truly global and universal than it had ever been in the past.

In the study of language, two trends emerged in the 1950s: the beginning of an American theory of syntax, and the development of automatic translation systems of the sort that could take Russian texts and produce English translations. The theories of syntax were largely developments of Leonard Bloomfield's steps towards a theory of syntax that was not Eurocentric, based on work he had done in the 1920s and 1930s. But even his most devoted followers recognized that Bloomfield had only scratched the surface of the problem, and that left a great deal for others to do. We will survey the work done through the mid-1950s on developing an American theory of natural language syntax.

As Europe rebuilt its infrastructure after the Second World War, linguistics came back to life there, and we will look at some of the striking continuities that emerged, focusing on Martinet and his version of functionalism, as well as his influence on dialectology and sociolinguistics, on the divergent paths laid out by Weinreich and by Labov, and on the Scandinavian schools.

We turn then to the work of Noam Chomsky, whose interest in linguistics was piqued not only by his father's work on Hebrew grammar but also by his work with Zellig Harris at the University of Pennsylvania, where he was a student. During four years as a Harvard junior fellow, he developed a powerful new approach to the study of language that was deeply informed by the work of his teacher Zellig Harris, by work in modern mathematical logic, and by the new ideas coming to the United States from what had been the Vienna Circle before the war.

We will explore Chomsky's first development of a theory of grammar that he developed in a manuscript entitled *The Logical Structure of Linguistic Theory*, an obvious allusion to both Carnap's *Logical Structure of the World* and his *Logical Syntax of Language*. It was a brilliant attempt to bring together a theory of syntax, a theory of phonology, and an answer to the problems of empiricist learning and generalization that the philosopher Nelson Goodman had been exploring for 20 years (as we mentioned in chapter 7, Goodman was also Chomsky's teacher at the University of Pennsylvania). Chomsky's book was not published until 20 years later, but it was widely distributed in *samizdat* form, and it formed the basis for a number of very influential publications, beginning with the short monograph *Syntactic Structures* in 1957.

Psychologists and computer scientists were at the same time developing models that allowed deeper insight into how intelligent behavior could exist in a world that is also fully determined by the laws of physics. Computers in the 1950s were available to a small number of young researchers, and the ideas that they implemented were available to even more. The style of breaking down problems that had first been described by Charles Babbage became the standard way of programming a digital computer, and it provided a metaphor, or a way of providing a story that appealed to psychologists, and at the same time it offered a kind of explanation of human behavior, and very possibly, of human thought.

One of the leaders of this new trend in psychology was George Miller, a psychologist at Harvard University, who interacted a good deal with the young Noam Chomsky. George Miller and Jerome Bruner got support from the Harvard administration to open a center for cognitive studies, and Cambridge became an important stopping point for anyone in the world working on problems of thought that were influenced by the formal models that had grown out of the logics that we explored in chapter 8 of this volume.

We end volume 2 with a study of two successive ruptures within linguistics: first, the rupture between Chomsky's generative linguistics and the mainstream American linguistics that engaged the kind of models that Harris and Hockett had helped to develop, and then the first internal rupture within generative linguistics, the strident period when Chomsky and several younger linguists disagreed about the relationship between transformational syntax and logical representation. Some years later, this period of four or five years came to be known as the "generative wars," pitting Chomsky's interpretive semantics against the model of the generative semanticists. In some ways it is reminiscent of the war of the frogs and the mice that we looked at in chapter 8, pitting David Hilbert against the inroads that L. E. J. Brouwer was making with his intuitionism. It was for some painful, and when it was over, there was no one present who could declare that the war was over, or even who had won. No soldiers came home from that war to return to their prewar jobs. It is the most complex case of rupture and continuity that we will explore.

At that point, we will end our historically driven view of linguistics and the mind sciences and explain what principles of ethical living and scientific progress we have drawn from these reflections.

Conclusions

In reliving the past that we have witnessed over these 10 chapters, our goal has been to show a way to reappropriate this history of ours. We believe that as we better understand where our ideas have come from, we emerge with a far deeper and more powerful understanding of what our current beliefs are, and what it will mean to make them better. *All* of the things we believe are solutions to problems, or rather were *once* solutions to problems. When we forget the problems to which the ideas were solutions, we lose the close relationship that others might once have had with them: we become now on a *respectful* relationship (we use "vous") with the ideas, rather than a familiar relationship (where we would use "tu"), and we give them a power over us which should not exist, and which those ideas do not deserve to have! The scientist, the thinker, the human has the ongoing responsibility to evaluate and improve the set of ideas in our science, and we therefore must understand precisely why we adopt the scientific analyses that we do.

That, then, is our first conclusion: that it is our option and our responsibility to reappropriate the history of our fields, in order to better evaluate it and in order to simply make it better.

Our second conclusion concerns the relationship between the social and the conceptual. On the social and the personal side, we have drawn your attention to many forces that have had an impact on the development of the mind sciences, and it is possible that you have at times wondered if there is no end to the distortions that can beset these fields at their worst moments.

We believe that the forces that exist in the social world of individuals and groups are inevitable, and that they sometimes have noxious effects on the development of the mind sciences. But we also believe that these unwanted effects result not only from existence of the social forces, but also from an ignorance and lack of awareness on the part of the actors in this drama. To put it simply, the human and social forces can be used for good ends or for bad, and it is up to the individual actors to choose those ends. To expand on the metaphor we introduced just above, social forces are like currents in the ocean and the air, and an effective captain of a sailboat uses those currents. She uses them in full awareness of where she intends to go, allowing her experience and her understanding to fill in the direction she needs to go before tacking, and to allow her to be sensitive to changing conditions either in the winds or the seas.

All of which is to say that acknowledging the existence of social forces in the mind fields is not a cause for despair or pessimism. The forces exist, but where they direct us to go is up to us. Our task is to appropriate the study of our own history so that it can most fully permit us to set and understand our goals, and then to accomplish them.

Notes

Preface

1. Archibald A. Hill once observed, with tongue in cheek, that there was a "strong tendency for differences among linguists very nearly to reach the proportions of early morning trips up-river with pistols for two and coffee for one" (Hill 1991: 78).

2. Morris sadly passed away as the finishing touches were going on this manuscript. He is greatly missed.

Chapter One

1. Bourdieu 1984, 2001.

2. Matthews makes a remark along these lines regarding Bloomfield's *Language*; see chapter 6 below, and also Matthews 1993.

3. In *The Rediscovery of the Mind*, Searle remarked, "The books I read in my philosophical childhood—books by Wittgenstein, Austin, Strawson, Ryle, Hare, etc.—contain few or no references to other authors. I think unconsciously I have come to believe that philosophical quality varies inversely with the number of bibliographical references, and that no great work of philosophy ever contained a lot of footnotes" (xiv).

4. Both Descartes and Leibniz are rationalists, by today's reckoning, but this difference in how they regarded mentalism separated them every bit as much as whatever differences there might have been between rationalists and empiricists.

5. We are sympathetic to the remark made by Pietarinen in a paper on the Dutch symbolics group:

> The limited recognition of this group is due not so much to individuals as it is to the philosophical community at large. During the

> last decades, philosophers have been guilty of a "crime against science" by withholding credit from developments in the history of linguistics, logic, and mathematics. The genetic origin of some of the key ideas routinely relegated to the analytic genre [of philosophy] may often be traced back to these largely-forgotten developments. (2009: 469)

We would simply add to this that the philosophers are no *more* guilty of this than any of the other disciplinary groups.

6. We use the academic present tense is to describe Bourdieu's sociological analysis.

7. Randall Collins (1998) has developed a sociological analysis of philosophy that shares some aspects of Bourdieu's work; we have found Collins's work helpful, as well as some of his aphorisms, such as the observation that "intellectuals are people who produce decontextualized ideas" (3). Our trees of intellectual genealogy were inspired by some of Collins's genealogies, and also by Boring (1948). Collins's detailed look at intellectual genealogies has left a strong impression on us, and we have been struck on more than one occasion by the justness of his remark that "as usual, intellectual energy is propagated down the wires of interpersonal contacts, while the content of ideas is rearranged by horizontal strains of opposition reconfiguring the attention space" (718). See Zuckerman 1977: chap. 5 for an interesting discussion of similar genealogies among Nobel laureates.

8. Chomsky 2002: 95.

9. Joos 1957: introduction.

10. Bourdieu 1980. A riff off of Marx, to be sure: the estate of a lord inherits his first-born son.

11. The references to what follows are Gibbs 1857: 5 (the passage is dated May 1838); Whewell 1858: 2:258; Dwight 1859: 215; Müller 1862: 3–4; Müller 1864: 1; Whitney 1867a: 277 (the passage is dated 1864); Schleicher 1874: 1 (translated by Winfred Lehmann in *A Reader in Nineteenth Century Historical Indo-European Linguistics)*; Whitney 1865: 1; Saussure 2002a: 115; Merz 1903: 539, 539n; Bloomfield 1924: 319; Collitz 1925: 1; Bloomfield 1925: 1; Trubetzkoy 1933: 245–46. Bloomfield 1933: 3; Zipf 1936: 3; Hockett 1942: 3; Bloomfield 1943: 198 (in *Language* 19, speaking of Franz Boas); Chomsky 1945: 289; Bloch 1949: 92; Hall 1949: 119; Fries 1952: 1; BBC Cambridge Language Research Unit 1952 (aired on BBC Radio, March 21, 1956); Benveniste 1954 [1976]: 6; Fries 1970: 198; Newmeyer 1986: 20; Piatelli-Palmarini in Uriagereka 1998: xxv; Culicover and Nowak 2003: 5; Boeckx and Piatelli-Palmarini 2005: 447.

12. For the impatient reader who wants an example to chew on, take the case of the theoretical object known as the syllable. During the 1890s, Saussure (2002b) developed a theory of what he called "sonants," which took high vowels to be not flesh and blood phonemes, but abstract voiced coefficients that were interpreted

by virtue of the phonotactics of the "sound chain." For some recent discussions, see Laks 2012, Tifrit 2005, and Goldsmith 2011.

13. For example, Gleason 1955, Hockett 1958, and Joos 1957.

14. Our discussion in this book is also at times a struggle against a popular, romantic, and often indefensible interpretation of Thomas Kuhn's view of the history of science, an interpretation that glorifies the rupture of a scientific paradigm with respect to its predecessors. We will return to Kuhn at the moment when his *Revolutions* fits into our story in volume 2, but we can remind the reader now, without revealing any secrets, that Kuhn's greatest impact came from the way in which he brought to life the sense that scientific research was not plodding and ahistorical. Quite a number of scholars and linguists have tried on Kuhn's model for size and found it wanting when they focus on linguistics. Our interest in Kuhn is elsewhere: we see him as part of the larger story, as someone providing intellectual tools in the very history that we are describing; see Matthews 1993.

15. There exists a relatively small literature that addresses the history of linguistics from a point of view informed by the sociology of knowledge and of science. Amsterdamska (1987) developed just such an account of nineteenth-century comparative linguistics, in a book that grew out of a dissertation that she wrote with Robert K. Merton and Harriet Zuckerman at Columbia University. She defined a school of thought—the focus of her analysis—as "a group of scholars or scientists united in their common divergence, both cognitive and social, from other schools in their discipline or specialty or from the discipline or specialty as a whole" (9).

16. "He who does not know the torments of the unknown cannot know the joys of discovery, which are the most powerful that a human mind can ever feel. But yet by a quirk of nature, this joy of discovery which is so sought for, so hoped for, vanishes as soon as it is found." Claude Bernard 1865: 387.

17. Berlin 1980: 175.

18. Boudon has written a good deal of sensible material along lines that are entirely congenial to our perspective, such as Boudon 1990: chap. 7.

19. Bourdieu 1975.

20. Mannheim 1923 [1952]: 299.

21. "A science which hesitates to forget its founders is lost." Whitehead 1916: 413.

22. Mannheim 1923 [1952]: 298–99.

23. Danziger 1979a: 206.

24. Jakobson 1973a: 12, the source of this and the two following quotations.

25. Cassirer and his wife were traveling on the same boat as Roman Jakobson and his wife. The two men spent the duration of the voyage discussing the same subjects that we are discussing here.

26. Quoted in Hendel 1949: 56.

27. Skidelsky 2011: 48, citing Charles W. Hendel, "Ernst Cassirer" in Schilpp 1949: 57. Cassirer had been a student of Hermann Cohen, who was not an easy

man to be a student of. Cassirer's wife described Cohen (after Cassirer's death): "His stormy temperament was combined with a burning desire to get his way—using all means at his disposal—in matters close and important to him. . . . Cohen was a zealot, who could not understand or tolerate opposition. . . . Cohen regarded himself with great pride as the head of the 'Marburg School' which he had founded. . . . [He] *would* animate his pupils with all the fervent love of truth that was in him. Every deviation from his point of view seemed to him treachery" (Skidelsky 2011: 47, who cites Cassirer 1981: 89).

28. Bjork 1983: 75.

29. *On the Origin of Species* (Darwin 1859: 2).

30. Tocqueville 1840: 2. Tocqueville was responding to a more familiar, but opposing, point of view, one which was expressed well by Thomas Hobbes in *The Leviathan*: "So also in Reasoning of all other things, he that takes up conclusions on the trust of Authors, and doth not fetch them from the first Items in every Reckoning, (which are the significations of names settled by definitions), loses his labour; and does not know any thing; but onely beleeveth." Chapter 5, from Thomas Hobbes, in *The Leviathan*, No publisher given, but the cover page reads "Printed for Andrew Crooke, at the Green Dragon in St. Pauls [*sic*] Church-yard."

31. See Boring 1963b: 21: "Mankind wants its heroes. . . . The history of science is spattered with the aggressive demands of great and lesser men for recognition and acceptance as leaders, but one hears less about the equally urgent need for followership, because the ranks of disciples are not limited by anyone's range of apperception."

32. Carnap 1928 [1967]: xvi.

33. Boring 1963a: 16, 18 for both quotations.

34. Ibid., 21.

35. Boring 1942: 312.

36. As with so many topics dealt with in this book, the subject of ideology is one with a vast literature. The term is analyzed largely by sociologists; Mannheim 1938 is an important classic. The authors have profited as well from the discussion in Boudon 1986 and in much found in Bourdieu. The use described here is quite close to that proposed by in Ricoeur 1974. A case can be made that Francis Bacon was the first in the Western tradition to explore what has come to be known as ideology. He was wearing his hat as a sociologist *avant la lettre* in his *Novum Organum* (Bacon 1620), where he talks about what he calls *idols*.

37. Feuer 1975.

38. Ibid., 1.

39. Ibid., 4.

40. Feuer 1969: viii.

41. *Guardian*, October 15, 2013, referring to Michael Grove.

42. The strong political attachment to the notion of ideology will return in chapter 9 when we discuss Eurasianism, a political philosophy that Trubetzkoy was instrumental in founding. One recent writer noted,

> According to one witty commentary, the Eurasianists had lost Mother Russia and also failed to find a Mother Europe. When Europe proved an alien world, there followed a fundamental reexamination of the self—what was Russian in a Russian. [Cf. Ladis K. D. Kristof, "The Russian Image of Russia: An Applied Study in Geopolitical Methodology," in *Essays in Political Geography*, ed. Charles A. Fisher, London: Methuen, 1968, p. 374.] The urge to sort out the profound identity crisis was thus one of the most potent driving forces of Eurasianism. As one perceptive observer, the Russian philosopher, V.V. Zenkovsky (himself an émigré who left Russia at the end of 1919), had noted quite a while ago, "Not ideology, but psychology, is essential and influential in Eurasianism." [V.V. Venkovsky, *Russian Thinkers and Europe*, Ann Arbor MI, ACLS: J.W. Edwards 1953, 106.] (Torbakov 2015, 125)

43. There was a complex political background to this development. The circle that came to be known as the Ideologues was a group that included such notables as Madame de Staël, Destutt de Tracy, Maine de Biran, and Benjamin Constant, who viewed themselves as continuing the concerns of the Encyclopédistes and the Enlightenment more generally in the decade following the French Revolution, notably in the period 1794 to 1803. They were committed to the fight against "metaphysicians" and others searching for first causes and essences, and they sought the origins of thoughts in sensation, with the belief that once this method was writ large, it would lead to a better and more stable vision of society. The origin of the term *positivism* also lies with this group. Madame de Staël, following Buffon before her, used the phrase *positive science* to refer to a quantitative method based on observation and measurement. Saint-Simon (who was close to Madame de Staël) adopted this usage as well, and Auguste Comte, who was a disciple of Saint-Simon early in his life, continued the use of the word; it is his usage of the term *positivism* that is generally acknowledged today. See Pickering 1993: 61ff.

44. Hoyningen-Huene 1995: 355.

45. Paul Ricoeur makes the case that people who tie ideology to social domination (as many marxists do) fail to see that what is interesting about ideology is something broader with greater reach. Too great a focus on domination, especially of one class by another, leaves us with nothing to say about a wide range of questions, and to the extent that ideology is understood as lying not very far from error and deception, it becomes all the more difficult to say anything useful about the story that one constructs for oneself as a member of a group. The

difficult part of all this is that one's interest—both *self*-interest and the interest of the groups to which one belongs—can play a role in making one's assumptions about the nature of the group's activities seem natural and hardly open to challenge of any sort.

> It seems to me that we must escape from the clutches of the fascination of the problem of domination in order to deal with a larger phenomenon, which is that of social integration, of which domination is one dimension, but neither the unique condition nor the essential one. However, if we take it for granted that ideology is a function of domination, it means that we also admit uncritically that ideology is an essentially negative phenomenon, not too far from error and lies, and even closer to illusion; in the contemporary literature on the subject, one no longer even subjects the all too familiar and natural idea to serious criticism that ideology is a false representation, whose function is to hide the allegiance of individuals, professed by an individual or a group, that they have an interest in not recognizing. Therefore if one does not want to come to grips with this complex of distorted and unconscious [self- or group-] interest, nor simply take it for granted, we must, it seems to me, untie the knot that binds the theory of ideology and the strategy of suspicion, even if it means showing by description and by analysis why the phenomenon of ideology brings out suspicion as the immediate response. (Ricoeur 1974: 329)

46. Book of Genesis, chaps. 6–9; Quran, Sura 71.

47. That this is a traditional strategy in philosophy is no news to philosophers, who have often remarked on it. John McCumber (2007), for example, refers to the "gesture of consigning all one's predecessors to the status of mere babblers," and he calls it "standard for modern philosophers from Descartes to Reichenbach" (105) and refers to "Descartes's dismissive note on the prejudices he had imbibed from his education" (243).

48. Royce 1892: 343, and also Barrett and Aiken 1962: 84.

49. Cohen 1960: 105.

50. Ibid., 105.

51. Cohen 1952: 506.

52. Allegre 1985: 21.

53. Koerner 1975: 725.

54. Ter Hark 2003: xii and xiv.

55. On these questions, see Gross 1998, who observes, "belief that a happening is a discovery (and therefore entitled to priority) arises in the relevant scientific communities when a set of normative requirements is satisfied as a consequence of reading scientific articles" (163).

56. Wright and Bechtel (2007: 44–54) discuss the connection between views of mind, machine, and modern ideas of mechanism, with a somewhat different emphasis than ours. They present a useful analysis of the notion of mechanism in this context, focusing on the importance of deciding whether the essence of a scientific account is one that provides a model that is in harmony with observations by virtue of generating or justifying statements about observables, or an account that provides a description of what the things themselves in the real world are actually doing, observable or not. More crudely put, the question for the scientist is whether her model should be understood as providing a structural account of a hidden process that exists in the real world—if not, then the model is justified by the fit between its predictions and the observables. Strong arguments have been made for both positions, but a scientist may feel torn in both directions. There are three natural responses to this dilemma: the first is that if the first view is weaker than the second, it is already a high bar for a theory of language, and we might as well aim for it since we still have a long way to go; the second is that the first view provides a target that is so low that even if we were to reach it, it would have little or no scientific value—only the second describes what real science is; and the third is that the second view is so riddled with untenable assumptions that it is silly, or unrealistic, to think that we can agree on a definition of what the real things are in the world that have even a hope of providing insight into language structure. To be sure, each of us can choose any of these positions, or any other.

57. The secularization of the Western worldview has a complicated history, but Isaac Newton was a strong defender of the *voluntarist* view that God continued to *act* in the world, and that he had not simply created the world and then contemplated it from afar. For a detailed account of this view, and how it related to Newton's conflict with Leibniz, see Shapin 1981. A different understanding of God's role was developed in the wake of this new view of physics that was called *deism*, and which thought God was maintaining a hand's-off attitude towards the universe, but that movement came later.

58. Descartes's position was less attractive to Christians as well in the years that followed. Blaise Pascal, just a generation younger than Descartes, would be more in the spirit of his times, because Pascal was more comfortable with a belief in God that was based on faith and on grace granted by God, rather than on rationalist argument, as Descartes had argued. If Descartes were alive today, he would wonder why so many people thought that belief in God was based on faith rather than reason. For twenty-first-century Christians who value faith above reason, Descartes is harder to understand than Pascal.

59. This is a difference that even a child understands today: a gadget can stop working because its battery is dead, but having an internal source of energy is not to be confused with intelligent (or intelligent-seeming) behavior.

Chapter Two

1. Monod 1876: 27.

2. There is a large literature on the history of science in the nineteenth century, and we have profited from it. See the collection of papers in Cahan 2003. As Cahan notes, three of the syntheses proposed in the twentieth century for the scientific perspectives in the nineteenth have had considerable impact, though they paint vastly different portraits of the age; these are the works of John Theodore Merz (1903), of J. D. Bernal, the influential marxist, and of the historical sociologist Joseph Ben-David.

More generally regarding the nineteenth century, we have been influenced by our reading of the monumental triptych of the British marxist historian, Eric Hobsbawm. See Hobsbawm 1962, 1975, 1987.

3. Samuel Butler wrote, in 1863, "The world begins to feel very small when one finds one can get half round it in three months." We all know what Jules Verne was able to do with that idea.

4. We have in mind a remark in Chomsky 1979: 57.

5. If Erasmus and Luther eventually opposed each other from a doctrinal point of view, they started off from quite similar positions. In his *Opera Omnia* (1523), Erasmus wrote, "why does it seem inappropriate if someone sounds forth the gospel in his native language, the language he understands—the French in French, the English in English, the German in German, the Indian in the language of India? It seems to me more out of place—even ridiculous, rather—that the uneducated and women, like parrots, mumble their psalms and the Lord's Prayer in Latin, although they do not understand what they themselves are uttering" (Erasmus 1992).

6. A particularly striking case is the case of Martin Luther and his German translation of the Bible; his usage there established, more than any other text or usage, what would become known as Modern German.

7. Bras 2008.

8. German Romanticism was a movement that began roughly at the time of the French Revolution and lasted until the middle of the nineteenth century. Its effects were most visible in the arts, where it emphasized the importance of feeling over thought and genius over effort; it had little sympathy for the values of the Enlightenment. Among philosophers, Fichte was a leading voice of romanticism in Germany. Friedrich Schlegel was one of a group of friends, most of them poets, who formulated the central ideas of romanticism; he was also arguably the first serious historical linguist of the nineteenth century.

9. Given the Greek word *demos*, one might be justified in referring to this view as the *democratic* view, but that word is charged with other conceptual freight nowadays.

10. Fichte 1808.

11. Forster 2010; Danziger 1983.

12. The English word *folklore* was coined during this period (1846) by William Thomas in a conscious calque on the German *Volk*.

13. See Bright 1997, and for more recent references (not all of which we have been able to obtain), see also the anonymous Wikipedia article entitled "A Language Is a Dialect with an Army and Navy," with its critical cyber-apparatus.

14. Smail 2008.

15. Needless to say, the story of this development is complex. For a fascinating recent account of a late eighteenth-century contributor to the question, see Harvey 2014.

16. Joseph 2012: 203.

17. Allegre 1985.

18. Coneybeare and Phillips 1822: ii.

19. Ibid., iii.

20. Oldroyd 2003.

21. Whewell 1839: 228.

22. Humboldt 1849: 111.

23. See Buffon 1749–1789; Jussieu 1824; Cuvier 1817; Saint-Hilaire 1818.

24. Linnaeus was also a great traveler, and he scoured Europe in search of specimens. Visitors can still see the botanical garden that he created for his collection at the University of Uppsala.

25. Later in the century, when structuralist methods of analyzing a corpus were getting underway, the work on taxonomies was not far from the thoughts of those putting together the new perspectives. Saussure wrote, in 1872 (when he was 15 years old), that "we have a sort of classification that unites not only species, but types (e.g., taxons). I do not claim that χλαδος for example is the same word as χαλαμος. I say simply that if we allow a primitive form *kal*, these two words can be derived from it. . . . We should not say that the dissimilarity between two words joined by a common root leaves the door open to allow just any word in: if the limits of a root are wide, they are also sharply marked." Saussure 1978 [1874]: 86.

26. It was over the course of the eighteenth century that the notion that matter was neither created nor destroyed became clear, a discovery often associated with the French scientist Antoine Lavoisier. The nineteenth century saw the development of an awareness of the conservation of energy in all its forms, and in light of that observed conservation, it was easy to reach the conclusion that energy had a metaphysical reality no less than matter did. (Sarton et al. 1929). Only a few short years into the twentieth century, the notion was introduced that neither matter nor energy were always conserved, because one could be converted into the other.

27. See notably Alter 2005; Richards 2002.

28. Darwin 1859: 422.

29. Labov 1981.

30. Darwin 1859; Labov 1981, 1994, 2001; Laks 2007, 2013; Bergounioux 2002.

31. Alter 2005: 4.

32. Many scientists were convinced of the existence of atoms in the nineteenth century, but it was not until 1905 and the article that earned Albert Einstein his Nobel prize that the controversy faded away, and the anti-atomists had to admit defeat.

33. Mendeleev 1889: 635–36.

34. Ibid., 634–35.

35. Ibid., 638.

36. Bühler 1934 [2011]: 319.

37. The term *synoptic* comes from the Greek "seen together" and refers to the practice of presenting the Gospels juxtaposed, by content, on a single page, so that it is clear what material is presented in more than one Gospel, and how the different presentations are similar or different. This graphical practice would have an important influence on the methods used to compare sister languages descending from a common ancestor.

38. Jones 1798: 422–23. Jones's creation of modern comparative linguistics is recounted, for example, in Bloomfield 1933: 12; Saussure 1995: 2; and Pedersen 1931: 18.

39. For a new view on this development, see Campbell and Poser 2008. See also Burridge 2013 for a recent overview, and Koerner 1975.

40. Before Jones: Robins 1990; Leibniz, from Metcalf 1974: 251.

41. Metcalf 1974: 251.

42. Quoted in ibid., 233; see also Campbell and Poser 2008.

43. See Godfrey 1967: 57–59, and especially Trautmann 2006: 18–20, which presents an important case for viewing the question of language origins against the background of the understanding of the genetic relations among peoples, and the difficulty with which European thought extricated itself from the background myths of Eden, Moses, and the Tower of Babel.

44. Jespersen 1922: 33–34. See Metcalf 1974 for extended discussion, and Godfrey 1967. Jespersen's comment in this paragraph reads a bit like an unprovoked criticism; it only makes sense if there lies behind it some expectation of work that should be done by anyone noted by history books. There will *always* be questions left for younger people to follow up on. That is the nature of knowledge, which begets both ignorance and questions. If Jones left work for those following him to accomplish, so did Jespersen; of whom can that not be said? The right answer to the credit problem takes on a special importance from the point of view of the linguists cited, because the discovery of Indo-European has an essential role to play in the account linguists give themselves as to their origin.

45. See Norman 1929, and more generally, Benes 2004.

46. Demoule 2003; and see Demoule 2014 for important developments of this point.

47. These two citations are from Renan 1862: 9–10.

48. Gobineau 1853; Chamberlain 1899.

49. Regarding Aryanism as an ideological response to Semiticism, see Olender 1992 and Demoule 2014; see also Grafton 2004.

50. For a study of the influences of Indo-Europeanism on the European (and especially the German) intellectuals, see Cowan 2010. The chapters of the book examine the intellectual struggles of such luminaries as Johann Gottfried Herder, Friedrich von Hardenberg (Novalis), Friedrich Schlegel, Friedrich Schelling, Schopenhauer, Hegel, and Nietzsche with the "mythical image of India" (5). See also Marchand 2009.

51. Schleicher 1863: 7.

52. It is not at all insignificant that they all began as Semiticists, before turning to Sanskrit.

53. See Honeybone 2005 on the relevance to generations in nineteenth-century linguistics.

54. Schlegel 1808; Koerner 1990, esp. 243.

55. The classic discussion of morphological systems in contemporary linguistics remains that of Greenberg 1960. It would take us far afield to enter into a discussion of Schlegel's deep concerns about the depth of the decline of modern European culture; see Gérard 1963 for a detailed discussion of Schlegel, Bopp, Humboldt, and other linguists of their generation, and the concerns about the decadence of society in the West. We have left out of our story so far any mention of Friedrich's brother August, who was as famous among German romanticists as Friedrich was, and who will appear briefly below.

56. Schlegel 1849: 449.

57. Schlegel 1808: 28. This translation is from Lehmann 1967. Note the use of *mechanical* here, in light of the ongoing discussion of the word *teleology*.

58. Benes 2004.

59. See Demoule 2014 for a discussion of these scenarios.

60. Example from Pedersen 1931: 57–58.

61. Verburg 1949.

62. This is a particularly striking aspect of Distributed Morphology.

63. Bopp 1820: 27.

64. Ibid., 34.

65. Bopp 1856 [1885]: xi.

66. Verburg 1949: 453. There is an enormous literature on the rise of nineteenth-century historical linguistics, of excellent quality, written by scholars who often feel that they disagree deeply with all the others. Pedersen 1931 is an excellent read; Koerner's more recent and scholarly "European Structuralism: Early

Beginnings" (Koerner 1975) is also useful in this context (even though it places its focus elsewhere, as the title indicates).

67. Sweet 1980: 2:406.

68. *Gesammelte Schriften* 3: 297, cited in Sweet 1980: 2:408.

69. The terms developed by the Schlegels and Humboldt would remain important over the nineteenth century, and Edward Sapir would use them as well in his book *Language*. On the range of opinions regarding the influence of Herder on Humboldt, see Koerner 1987.

70. See Shibatani and Bynon 1995; Humboldt 1999.

71. Humboldt wrote,

> The originally independent *significance of suffixes* is therefore no necessary obstacle to the purity of true inflection. Words formed with such inflectional syllables seem no less determinate than where internal change occurs, but to be merely simple concepts, cast in varying forms, and thus to fulfil exactly the aim of inflection. Such significance does indeed call for a greater strength of the inner sense of inflection and a more decided mental sound-mastery, which in this case must overcome a tendency for the grammatical form to degenerate into compounding. [!] A language like Sanscrit, which chiefly employs such original, independent, meaningful syllables for inflection, shows by that very fact the confidence it reposes in the power of its animating spirit. (1858: 106)

72. Amsterdamska 1987: 37.

73. Humboldt introduced a fourth category, polysynthetic languages, as well; see Humboldt 1999, which was originally published in 1836.

74. See especially Ruëgg 2004; Collins 1998; and Amsterdamska 1987.

75. In Napoleonic France, the imperial university was essentially an institution for the awarding of degrees. Teaching was very much a secondary function, and taking a degree did not require taking any courses, while research was essentially nonexistent. The peripheral institutions (the Ecole Normale Superieure, the EPHE, the Collège de France) took on the pedagogical and research functions that were in effect abandoned by the university system during the nineteenth century.

76. Amsterdamska 1987: 68–69. See also Ferry, Pesron, and Renaut 1979.

77. These American universities, the first to adopt the values and goals of the German universities, did not have an easy time in recruiting outstanding European scholars. It is notable that Johns Hopkins recruited the noted British mathematician James Joseph Sylvester in 1876, after a long career in England in which he was the target of anti-Semitism.

78. See Paul 1972.

79. Auroux, Bernard, and Boulle 2000.

80. The work of the Grimm brothers notably included the German dictionary, whose publication began in 1854 but would not be completed until 1961, with its thirty-second volume. It is an immense historical and etymological dictionary, presenting the history of each word, its meaning, and the range of its dialectal forms and uses. The question of the linguistic and cultural identity of the German people is central to the work of the Grimms. In his history of the German language (Grimm 1848), Jacob Grimm considered the geographical range in Europe over which German was spoken, and concluded, "Our language is also our history" (185). Jacob Grimm wrote, in 1846, in "What is a nation (Volk)?": "a nation is the embodiment of people who speak the same language. That is for us Germans the most innocent but also the proudest definition because it . . . turns our gaze to . . . near future when all barriers fall and the natural law will be acknowledged that neither rivers nor mountains divide nations but that language alone can set boundaries around a people that has pushed past mountains and streams." Cited in Benes 2008: 147.

81. Cited in Amsterdamska 1987: 36.

82. Koerner 1975: 759: "If there was a major 'breakthrough' in 19th-century linguistics, it is with the work of Schleicher in the 1850s and 1860s and not with the Neogrammarians from 1876 onwards, as has generally been asserted in the histories of linguistics available to the present."

83. Recall Labov's discussion of the 15 Darwinian criteria.

84. Schleicher 1863: 20.

85. Alter 2005: 129.

86. The discussion that follows is based on Müller's *On the Stratification of Language* (Müller 1868), notably from pages 7 to 20.

87. Alter 2005: 21.

88. Koerner 1980.

89. Whitney 1879; see its preface, p. v.

90. Whitney 1867b: 22–23.

91. Ibid., 22–23.

92. Ibid., 19–20.

93. Whitney 1875: xxi.

94. This, and the three following quotations, are from Whitney 1872: 342–43.

95. Whitney 1875: 310–11.

96. This and the next seven quotations are from Whitney 1875: 59–63.

97. This and the next quotation are from Whitney 1875: 67–68. The status of a phonetic alphabet was a subject of considerable moment in the International Phonetic Association.

98. Whitney 1867b: 47.

99. Ibid., 50–51.

100. The Neogrammarian challenge is analyzed in Amsterdamska 1987, which heavily informs our account here. See also Kemmer 2013, whose perspective is similar in some respects to the one developed here, and Davies 2006 and Blust 1996.

101. The term was due to Whewell; he is someone who should be read and cited today (Whewell 1832: 126). Uniformitarianism stands in opposition to catastrophism, the view that the important changes in our past have been due to sudden causes from outside the system, such as the arrival of a huge meteor falling from the heavens.

102. Osthoff and Brugmann 1874.

103. This translation we use here is from the Linguistics Research Center site of the University of Texas at Austin, https://liberalarts.utexas.edu/lrc/resources/books/reader/14-h-osthoff.php.

104. Ibid.

105. Ibid.

106. Kiparsky 1974.

107. We have already alluded to Labov's modern interpretation of this material: Labov 1981 and 1994.

108. This is from Brugmann's contribution to Appendix I of Lanman 1897: 80–81.

109. Joseph 2012: 175.

110. Amsterdamska 1987: 124.

111. Paul 1886. The three quotations that follow are from pages 1–3.

112. Amsterdamska 1987.

113. Mugdan 1984.

114. Saussure 1968: 43.

115. For a recent discussion, see Radwanska-Williams 1994.

116. The quotations in this paragraph and the next three are from Baudouin de Courtenay 1895: 150–52.

117. Bloomfield 1939.

118. Halle 1962.

119. This and the quotation in the following paragraph are from Baudouin de Courtenay 1895: 171.

120. This and the following quotations are from Baudouin de Courtenay 1895: 174–75.

121. Grout 1853: 441.

122. Gibbs 1853: 471.

123. Holmboe 1855: 429.

124. Baudouin de Courtenay 1895: 160.

125. Baudouin de Courtenay 1870: 54–56.

126. Baudouin de Courtenay 1889: 125–28, including the following quotation in the text.

127. Baudouin de Courtenay 1870: 59–60.

128. His complete bibliography (Buss, Ghiotti, and Jaeger 2003) includes two monographs and 60 articles over 34 years, plus in addition three unfinished books and just over a dozen posthumously published articles. Roman Jakobson was very influential in establishing Saussure's place in the history of general linguistics; other major linguists, including linguists as different as Bloomfield and Trubetzkoy, saw Saussure's take on general linguistics to be representative of ideas in the air at that point.

129. Of the many sources on Saussure, we have primarily benefited here from the critical edition due to De Mauro (Saussure 1972), with its appendices, and the monumental biography by John Joseph (Joseph 2012).

130. Saussure 1972: 321.

131. De Mauro 1972: 1.

132. Saussure 1972: 355.

133. Joseph 2012: 172.

134. Ibid., 178.

135. Taking Saussure at his word, the choice of Leipzig was of not great significance; he wrote: "I should add that I was going to Leipzig rather at random, simply because my friends from Geneva, Lucien Gautier, Raoul Gautier, Edmond Gautier and Edouard Favre, were studying there, in the schools of theology and law. My parents preferred, since I was only 18 and a half, a foreign city where I would be surrounded by fellow Swiss" (1960: 21).

The reader of Saussure's memoirs must choose whether to take this on its face, or to see it as a way of underplaying the significance to Saussure's relatively brief but in some respects tragic period in Leipzig—or perhaps better, any responsibility on Saussure's part for finding himself in the community of leading scholars in the world concerned with the questions that he cared about. The authors are of mixed minds on the matter.

136. Here is a recollection of Saussure's from years later. The reader may decide how seriously to take this anecdote. Saussure wrote,

> When I gave [a certain presentation] at Curtius's seminar in 1877 on the fact that *ā* is in alternation with *ă*, M. Brugmann was not present, but finding me the next day in the great court of the university, he came up to me and asked me in a friendly way, as something that interested him, (and this is literally from Brugmann), "ob noch weitere Beispiele als *stātor: stătus* und *māter: păter* wirklich für diesen Ablaut vorliegen." If we were to tell the story today that M. Brugmann asked if there were more than three examples for the ablaut *ă*: *ā*, whoever were to say that would seem to be telling a tall tale. But that is exactly just how little the current generation is capable of judging either the state

> of affairs in 1877, or the exact role that should be attributed to the researchers in question. Nothing could be simpler, than to open, for example, the *Grammaire* of Gustav Meyer, who was the first to ignore my name, *ă:ā:ō*; *ă:ē:ō*; and *ă:ō:ō*, to notice these facts that are so clear that nobody had even bothered to look up; and that is why it is, I repeat, very characteristic that in 1877 M. Brugmann himself did not realize that there were a lot of examples [pour un seul fragment] of ablaut, like *ă:ā* which seemed new to him in principle (and everything about *o* could certainly have been taken straight from my *Mémoire*. (Saussure 1960: 23)

137. In retrospect, he emphasized the feeling he had had at the time, the feeling of being an outsider, both because he was Swiss, and because German was not his mother language; these two things made it easier and more comfortable for Saussure to fall into spending time with the other students who were there from Geneva.

138. Translation from Joseph 2012: 132.

139. Saussure 1960: 21.

140. Benveniste 1964: 21.

141. Grammont 1912: 387.

142. Benveniste 1964: 24.

143. Saussure 1995b.

144. Saussure 2002b. The two manuscripts published a century later by Marchese (Saussure 2002b) show us a Saussure who was a phonologist well ahead of his time. His analysis of sonority, syllabification, and awareness of the role of phonotactics is thoroughly contemporary. Cf. Laks 2012. Consider, for example, the first appearance of the notion of distinctive feature, this notion which will become so important in the work of Jakobson. Emile Benveniste wrote,

> One notes in Saussure's writing the phrase "distinctive features," which gives it a curiously modern sound. The identity of a language is provided by the sum of its distinctive features, which is to say, by what distinguishes Gothic from the other dialects. This is most likely already the seed of what would be the fundamental principle of Saussurean linguistics, that of distinctive features and oppositions as the defining element of linguistic entities. And when, in this same report of 1881, Saussure says he has dealt with the phonetics: "graphical system, the vocalic system, the consonantal system," we can perceive the weight which is imbued in the term *system* by the author of the *Mémoire*. (1964: 29)

145. The tension between the study of *langue* and *parole* has been heightened in the last twenty-five years by the recent emphasis on doing corpus linguistics, a style of linguistic research that focuses on corpora. In an earlier day, it might

be thought that when we look at speech (or *parole*), we are merely look at what happened to have been said, with the unspoken view that in some sense the sentences found in a corpus are a random sample of sentences that could be generated from the correct grammar. The study of corpus linguistics rejects that view by bringing to light structure found in actual discourse that is not grammatical structure, but is rather discourse structure.

146. "L'essence double du language" in Saussure 2002a (or see Saussure 2006 in English). The appearance of these manuscripts has led to a major reconsideration of Saussure's thought.

147. Saussure 2002a: 273, fragment 3347.

148. Saussure 1995: 317.

149. De Mauro (in Saussure 1972: 476) takes the remark to come from the editors, not Saussure: "Nothing in the manuscript sources shows that Saussure uttered this famous phrase, and even less that it somehow represents the 'fundamental idea' of his teaching," and he calls it the seal, or imprint, of an "editorial manipulation." See also Godel 1957: 119, 181; Bouquet 1997; and, more recently, Gasparov 2013: 57. Bailly himself wrote *in 1913* that up until the 19th century, language had never been studied "for itself, in its true function." That does not mean that he did not hear it from Saussure, but it certainly loosens the bond linking the phrase to Saussure.

150. Meillet 1966.

151. Brugmann 1886 [1888]: 1:244.

152. Saussure 1879: 135.

153. Ibid., 47–48.

154. Alter 2005: 209.

155. Schuchardt 1885: 29-30. Translation by Theo Vennemann and Terence H. Wilbur.

156. Ibid., 30.

157. Bloomfield 1908.

158. Bloomfield 1884, 1888.

159. Cf. Stratton and Ewing 1920.

160. Edgerton 1928: 216–17.

161. Auroux 1994.

162. This is a much more important issue than these brief words would suggest. The treatment of liaison consonants in French, a subject which one of us (Laks) has worked on for many years, is one where the determination of what belongs to the spoken language and what to the written language is extremely illusive and difficult—in part, but only in part, because each can influence the other.

163. Helmholtz 1885: 104.

164. Rousselot 1897.

165. Sweet 1877, 1908.

Chapter Three

1. *Critique of Pure Reason*, Kant 1781.

2. A recent view of this history from a point of view similar to the one we describe here can be found in Chater et al. 2015, chap. 1.

3. The awkwardness of that translation reappears time and again; in the bilingual version of Frege's *Foundations of Arithmetic*, we find the following, in a discussion of the not so obvious equation 135664+37863=173527. Frege suggests that this is a provable, not obvious, statement. "Kant thinks he can call on our intuition of fingers or points for support," reads the English; "Kant will die Anschauung von Fingern oder Punkten zu Hilfe nehmen," reads the German original. What is this thing, this intuition of a finger?, we might ask. The English continues, "Moreover, the term 'intuition' seems hardly appropriate," which is the least we might say, though the German says, "Der Ausdruck 'Anschauung' scheint auch nicht recht zu passen," and why does Frege think that "intuition" (or *Anschauung*) is inappropriate? It is because "even 10 fingers can, in different arrangements, give rise to very different intuitions." It is clear that Frege has in mind different geometrical arrangements that make 10: two rows of 5 dots, or a triangle with rows of 4, 3, 2, and 1 dots, for example. This is not at all what the English word *intuition* means; the sense that Kant was trying to evoke was quite clearly more like that of "immediate presentation to consciousness."

4. See below, and chapter 8 as well. We have attempted to limit as best we can our references to developments in mathematics, and this limitation has had as a consequence that there is too little discussion of George Cantor's work. Ferreirós 2007 is an especially good work to complement this present book.

5. Müller 1887: 146.

6. T. H. Green, *Works*, 22: 224, cited in Müller 1887: 147.

7. Gauss, *Werke*, 1900: 177, cited in Gray 2008: 48.

8. Gauss, *Werke*, 1900: 200, cited by Gray 2008: 48–49.

9. We draw on the discussion in Coffa 1991.

10. Guillin 2004.

11. Comte 1996: 53.

12. Ibid., 59.

13. This is perhaps not obvious today. The traditional distinction between substance and accident served as an account of change: it is substance that remains while accidents of the substance change.

14. Entropy is a difficult notion to describe simply. It is a measure of the disorder of a system, but in a particular sense. It is based on the recognition of an important distinction between macrodescriptions at a human level, and microdescriptions, at a molecular level. Some ways of understanding the world, such as determining the temperature of something, only make sense at a macroscopic

level, and to such macrodescriptions correspond an astronomical number of different microscopic states. Entropy is a measure of the number of different microscopic states a system will run through while it seems to us, on a macroscopic scale, that we are in an unchanging and stable equilibrium. Coopersmith 2015 provides a good account of this material.

15. Comte 1996: 72.

16. These three quotations are from Mill 1865: 63, 65.

17. Gall 1810.

18. Compare with the current perspective of Fodor 1983, especially chapter 1.

19. Mill 1865: 51.

20. Ibid., 50.

21. Comte 1830: 79.

22. Alter 2005: 91–92; Whitney 1867a.

23. Whitney 1867a: 31.

24. Ibid.

25. Comte 1830: 15–17.

26. Wundt 1894: 95–96. Translated by Danziger, in Danziger 1979a: 207.

27. Mach 1897: 369, cited by Skidelsky 2011: 240n20.

28. It is important to bear in mind that Austria and Germany were two quite different cultural regions, despite the fact that they shared a common language. The rise of the German university system in the first half of the nineteenth century, for example, was an important event that was strictly German. The view that Kant and Hegel (and the German Romantic movement) were the central facts of recent intellectual life was taken for granted in Germany, but held little traction in Vienna, or in Prague (Janik and Toulmin 1973, Smith 1994) where English empiricism found the sort of sympathetic audience that it did not find in Germany.

29. Mach 1901: 278–319; Janik and Toulmin 1973.

30. This passage in Mach 1901 appears to have been added in the fourth edition. It is on page 50 of a German edition posted by Bernd Payson (https://bernd-paysan.de/mach.pdf), and on page 249 of the French edition *La méchanique* (Paris: Hermann, 1904).

31. Cassirer (1950: 105) described Mach's view:

> Because the capacity to comprehend and remember details is limited, the material must be arranged in order. For instance, if we knew for every period of time the space traversed by a falling body we might be well content. But what a prodigious memory it would take to carry the pertinent table of s and t in our heads! Instead of so doing we employ the formula $s = \frac{gt^2}{2}$, that is to say, the principle by which we can find the appropriate s for any given t, and this provides a complete, convenient, and compendious substitute for the table. This principle, this formula, this "law" has

> not one iota more of factual value than the isolated facts taken together, its worth lying merely in its convenience. It has utilitarian value.

This final statement was certainly going too far. Ernst Cassirer expressed why, in the context of criticizing Bertrand Russell: "The content of the concept," he wrote, "cannot be dissolved into the elements of its extension, because the two do not lie on the same plane. . . . The meaning of the *law* that connects the individual members is not to be exhausted by the enumeration of any number of instances of the law; for such enumeration lacks the generating *principle* that enables us to connect the individual members into a functional whole." See Cao 1997 for a general discussion.

32. Mulligan and Smith 1988.

33. Mach 1905; see also Winston 2001.

34. Mach 1905: 276–77.

35. Ibid., 277.

36. In twentieth-century behaviorism, this view would justify a large part of the program. B. F. Skinner was clear on this: "the terms 'cause' and 'effect' are no longer widely used in science. . . . The terms which replace them, however, refer to the same factual core. A 'cause' becomes a 'change in an independent variable,' and an 'effect' a 'change in a dependent variable.' The new terms do not suggest *how* a cause causes its effect: they merely assert that different events tend to occur together in a certain order." Skinner, *Science and Human Behavior* 1953: 23, cited in Winston 2001.

37. Novák 1988.

38. Such as, notably, Smith 1982 and 1994.

39. Stumpf 1919: 13.

40. See Blackmore 1998: 77.

41. Stumpf 1919: 16.

42. Mulligan and Smith 1985; several important points of the next few paragraphs are developed in this paper.

43. Cited in Mulligan and Smith 1985, from Brentano 1982: 1. In the original, Brentano used the term "Psychognosie" for what is here translated as "descriptive psychology."

44. Brentano 1874 [1995]: 13.

45. We do know that Brentano maintained prayer and meditation as an important part of his daily life; he wrote passionately about that to his student Stumpf. Concentration and meditation practices regularly include strengthening of the practice of developing an awareness that is not engaged in mental manipulations, and Brentano's writing about introspection made it clear that he recognized that some mental efforts at introspection were simply very hard, a recognition that suggests his conclusions were based more on a systematic ap-

proach to introspection, one that did not grow out of the needs of the psychology laboratory.

46. Husserl 1919 [1976]: 47–48.

47. Ibid., 48, including the next three quotations as well.

48. Ibid., 51, including the next three quotations as well.

49. Ibid., 53, for all quotations in this paragraph..

50. Ibid., 53–54, for all quotations in this paragraph.

51. Ibid., 54–55.

52. Ibid., 55.

53. Stumpf 1919: 10.

54. Ibid., 12.

55. Ibid., 44.

56. Brentano's relevance and importance can be described in another way: through the great influence of his lectures, and of his teaching more generally, both analytic philosophy and phenomenology can claim Brentano as a direct ancestor: analytic philosophy through Meinong (if one focuses of Meinong's influence on Wittgenstein and Russell) and phenomenology through Husserl, both of whom were students of Brentano.

Fabian 1996: 162 cites a letter from Ehrenfels to Brentano dated December 5, 1907:

> Especially with respect to our relationship I then developed the following directives: Brentano is an extremely intellectually productive personality who unfortunately, like most brilliant people, suffers from a characteristic concomitant disadvantage: from one-sidedness and biasedness which is a part of his particular, eminently developed character. Trying to convince him of some kind of result in a certain field of science or of even a general cultural field which was disagreeable to his nature, would turn out to be a completely futile effort which would lead me to becoming emotional and to falling out with my admired and highly deserving teacher to whom I am deeply indebted. So from then on I was much more determined to adopt the behavior of the pupil towards the teacher in my future conduct to Brentano (a style familiar to me anyway) and to accept gratefully all good and worthy things that he still would give me. But when dealing with him I consciously intend to exclude all intellectual and emotional reactions which to my sense of delicacy cannot be assimilated by him—and I shall not be affected by his underestimation of what I appreciate . . . or by his scornful and derisive treatment of what for me great and praiseworthy. . . . If he were an ordinary person, such behavior would altogether be too arduous and perhaps

> incompatible with self-esteem. But concerning Brentano, I do not feel humiliated at all in this role of a pupil.

57. Schuhmann 1988.

58. Masaryk wrote a dissertation on sociology, but it was not accepted, and a later dissertation was, though apparently it was not viewed as very strong academically.

59. Novák 1988: 4.

60. Demetz 1998. Jakobson discussed the influences that Masaryk speaks of in his early *Foundations of Concrete Logic*, notably William Dwight Whitney (*Language and the Study of Language*, and *Life and Growth of Language*), and the Neogrammarians: Paul's *Principien des Sprachgeschichte*, Delbrück, and Marty's *Ursprung der Sprache*. "The basic problem Masaryk poses in his *Foundations of Concrete Logic* is the place of linguistics in relation to other sciences," Jakobson wrote (1930: 65).

61. Roman Jakobson would later note that "our links with the so-called Prague school of psychology and with its initiator C. von Ehrenfels (1859–1932), the first propounder of the focal concept and label *Gestalt*, certainly left their imprint on the advancement of the Prague Linguistic Circle" (Jakobson 1973a: 17). Like so many of the philosophers and psychologists of this particular time, he was a serious musician; he had, in fact, studied harmony with Anton Bruckner.

In his later years, Ehrenfels wrote about his two teachers, Brentano and Meinong, with insight:

> So let me confess right away that I regard Brentano as the greater of the two as regards productive capacity. For keenness of intellect they were perhaps evenly balanced. But Brentano was, in my opinion, by far the more fortunately endowed scholar. He had an immediate instinct for that which was clear and essential and also for the admissibility, where appropriate, of abbreviated methods of thinking, whereas Meinong's mind seemed to be directly attracted to that which is intricate, minute and laborious. My impression was that Brentano also excelled more as regards economy of effort and the methodical influence exercised by the style of his verbal and written presentation. What we need is the brevity of clarity and not the prolixity of superfluous assurances. . . . I must here stress that I was brought most impressively into contact with that living quality which can best be described as scientific conscience or scholarly morality not by Brentano but by Meinong. And yet all the conditions ought to have been here more favourable for Brentano. Brentano was from the beginning for me the more imposing intellectual personality; he was by far

> the elder and more distinguished of the two (and in those days, as a lad coming up to Vienna from my native Waldviertel and the small town of Krems an der Donau, I still laid some store by outward distinction). Brentano held tutorials lasting several hours, and a private recommendation soon brought me into personal contact with him. Brentano was a charming interlocutor and an attractive figure in speech and appearance. None of this was true of Meinong. And yet it was through Meinong and not Brentano that I came to grips with moral seriousness of scholarship and a scientific sense of responsibility, the categorical imperative of the seeker and disciple of truth. (Fabian 1996: 164)

62. Smith, *Austrian Philosophy (*1994: chap. 8), points out similarities of this concept to Husserl's *Philosophy of Arithmetic* (1891: chap. 11), and then developed further in the *Logical Investigations* ten years later. Smith also refers to work by Meinong in work in the 1880s, in which Meinong cites Ehrenfels.

63. Friedman 2000: 47; Blackmore 1972: 47.

64. Friedman 2000: 47.

65. A good overview can be found in Peckhaus 1999.

66. Boole 1854: 3.

67. Ibid., 28–33.

68. Ibid., 159.

69. Boole 1997: 74 for this and the quotations in the following paragraph.

70. Ibid., 75.

71. Boole 1854: 24. The two following quotations are from this work, pp. 25 and 38.

72. Bertrand Russell made a point with an irony that we are sure he was aware of—though not all of his readers caught the irony, it would appear. Russell in his *Principles of Mathematics* (1903) remarked that he had read Frege's work (and he knew that Frege was attacking much the same problems that he was), but he did not understand Frege until he had independently arrived at a similar position. From Russell's internal point of view, reflecting on his own understanding of where his ideas stood, that makes perfect sense, but to anyone other than Russell, it is completely unreasonable. If you read my paper on a subject, and then come to the same conclusion, I would not be sympathetic to the view that you had arrived at the same position independently. I would say that I had at the very least helped you considerably on your way to your conclusion.

73. Reck and Awodey 2004, Frege 1997, Kreiser 2001. See also Grattan-Guinness 2000.

74. Flitner 1986: 126–27, cited in Reck and Awodey 2004.

75. Carnap 1963: 4.

76. Russell 1945: 784.

77. Russell wrote, near the beginning, "Professor Frege's work, which largely anticipates my own, was for the most part unknown to me when the printing of the present work began; I had seen his *Grundgesetze der Arithmetik*, but . . . I had failed to grasp its importance or to understand its contents" (Russell 1903: xvi).

78. On the goals of ideography, see Frege 1879 [1967]: 73.

79. Bear in mind, however, that at the time Frege was writing, mathematicians did not use the term "*function*" with the same kind of generality we do today; the term "function" was generally assumed to have as its domain the real numbers or the complex numbers (and not, for examples, the integers or some arbitrary set).

80. Frege 1879 [1967]: 23.

81. It is true that in a wide range of cases, the act described in the two cases is subtly different, a point considered in detail by Nicolas Ruwet. When we say *John saw himself*, we typically mean either he saw himself in a mirror, or else in his imagination—perhaps in another time or place, while nothing parallel is expected when the object and the subject are not the same person.

82. This discussion is closely linked to the property of *compositionality*, a notion that begins to emerge in the work of Boole and Frege. We say that the semantics of a system is compositional if the meaning of a composite expression E is a regular formation of the meaning of the components of E and of the way in which the components are put together in forming E. An idiom such as *kick the bucket* is the clearest case of a violation of compositionality: the meaning of the whole is more than the meaning of its parts and the meaning of how the parts are put together.

83. Frege 1879 [1967]: 20.

84. Cited in Schmit 1992: 31.

85. The term first appeared in German in 1866; see Kusch 1995: 101.

86. We will return to this in chapter 8.

87. Heijenoort 1967: 127.

Chapter Four

1. On the subject, there is again a large literature. A good overview of the reanalysis of the history of psychology in recent decades can be found in Arens 1989. Ash 1995 is an excellent account of the complex rise of psychology. The paper by Ben-David and Collins 1966 has been especially influential on what we write in this section. The artificial and self-conscious nature of the selection of 1879 as the starting point of Wundt's laboratory is discussed lucidly in Boring 1963b: 22–23. When Wundt arrived in Leipzig in 1875, he was given an old and unused auditorium. Some students, and some non-students, began conducting research there in 1879, some of it published in Wundt's journal, which was founded in 1883. Somewhere along the line, Wundt and his colleagues started to

refer to the space as the Psychologisches Institut. It was officially recognized as such in 1894.

2. Friedman 2000: 29.

3. Fechner 1860 [1912]: 566.

4. It is the Weber-Fechner law that accounts for the decision to include the logarithm in the definition of *bel* and *decibel*, familiar to students of phonetics today.

5. On Wundt, see Espagne 1998, and also Danziger 1979a, which is also useful. This paper by Danziger was responsible for opening up a discussion of the inadequacy of the anglophone misunderstanding of Wundt, which in turn had derived in large measure from E. G. Boring's description of Wundt's ideas. Danziger 2001 is a very useful statement of Wundt's perspective on psychology. See also Nerlich and Clarke 2001 and Rieber and Robinson 2001.

There has been interesting discussion of the role played by academic competition among German universities in the development of Wundt's laboratory (see Bringmann and Ungerer 1980, for example). It is interesting to see how academic careers in late nineteenth-century Germany continue to present-day America. See also Blumenthal 1970; Blumenthal 1975; Blumenthal 1979. Levelt 2014: chap. 6 is a good source on Wundt as well.

6. Danziger 1983.

7. One might disagree with our formulation here; one might believe that for Wundt, the individual mind was all that exists, and that a culture's belief system emerges out of the interaction of such individuals. We take these as being two different ways of saying much the same thing.

8. See Danziger 1979a, an important paper bringing out the underlying issues. See also Danziger 1980b.

9. Kusch 2005: 132ff.

10. The reader will recall our allusion above to the uncertainty regarding whether energy was a substance, like matter but distinct from it; to the extent that energy could be understood as the cause of movement (as potential gravitational energy causes a body to fall), this suggests that energy is a substance. We emphasize this not so much because it sheds light on what was happening in physics, but rather to emphasize that even within physics and chemistry, controversy could continue with regard to what was real and what constituted substance.

11. Kusch 2005: 133, citing Wundt 1908b: 260–62.

12. Danziger 1979b: 31.

13. Letter to Adolf Meyer, April 29, 1918, reprinted in Leys and Evans 1990: 215.

14. Danziger 1980a.

15. Wundt 1896: 11; English translation, paragraph 2.4.

16. Wundt's view is close to one that is described by twentieth-century philosophers as *neutral monism*, but that term is generally used to describe a position on what really exists—a metaphysical or ontological view. In this book, we are

more concerned with overarching connections among the mind sciences, and focus therefore on perspectives on human experience rather than ultimate claims on reality. Kusch 2005: chap. 4 has an excellent discussion of this side of Wundt and others of his generation.

17. Wettersten 1988.

18. Ash 1995.

19. Dewey 1910: 506.

20. James maintained a warm relationship of friendship with Ernst Mach throughout his life; he had met Mach in Europe when he was young, and they remained in contact by correspondence. James wrote very glowingly of Mach's work when it was published in the United States. See Blackmore 1972.

21. James 1890: 192.

22. This kind of wit could fly both east and west. When Whitney's book was published in 1867, Heymann Steinthal published a review of it, noting that it would likely be a popular book, because it was "easily accessible to the common mind," and [in Alter's words:] one should not expect from this sort of book the same depth of treatment one would find in a work written for a German audience! (cited in Alter 2005: 170).

23. James 1890: 1:549.

24. This quotation and those that follow are from James 1895.

25. James 1895: 122.

26. James 1890: 10.

27. Ibid., 308, for this and the following three quotations.

28. Sokal 1990; Ross 1972.

29. Ross 1972: 65.

30. Ibid., 81.

31. Ibid., 85.

32. Ibid., 104.

33. Buckley 1989.

34. Sokal 1990: 114.

35. We should emphasize that as far as we can see, there is really *no* connection between this early psychological usage of the term *structuralism* and the later uses of the term in a number of social sciences, most notably in linguistics. As we will see in chapter 9, the term *structuralism* took on a new life in the context of the Prague Circle of linguists, and many later scholars responded to the Praguean call to interpret Saussure as a leading figure in the establishment of structuralism. See Percival (no date), who directs us to Titchener 1898. See also Joseph 2002: 54n1, who points to Titchener 1898. Joseph says that Angell first used the term *structuralism*, and this was in 1907. See Koerner 1975: 721–22 as well. Koerner's note 4 reads, "Jakobson reports that he had used the expression 'structural method' at the Congress of Slavists in Prague on October 7, 1929. . . . Recently, O Szemerényi . . . has claimed that V. Mathesius was

the first to introduce the term 'linguistique structurale et fonctionnelle' *TCLP* 4:291. . . ."

36. Titchener 1898. Cf. also the critical-controversial papers of Caldwell (1899: 187) and Titchener (1899: 290). On Titchener, see Bjork 1983: chap. 4, and also Boring 1952 and Evans 1972. See also Titchener 1898; Hindeland 1971; and Leahey 1981.

37. Citation in Bjork 1983: 187, who in turn cites Murchison 1961: 2:340.

38. Quotations are from Boring 1952: 33. Titchener was an enormously influential academic in psychology, just as Whitney was in linguistics. Mary Henle (1984: 13) noted that Titchener was "perhaps the foremost experimental psychologist in America. For example, when the president of Harvard wanted advice about a major appointment, he consulted Titchener, even though Titchener had, a decade before, turned down what was then the major post in American psychology, the professorship at Harvard. Titchener's *Textbook of Psychology* was the first experimental work since Wilhelm Wundt's *Physiological Psychology* to be published in all three of the major languages. When jobs were available, Titchener moved his men around like pieces on a chessboard. For example, when Langfeld got the position at Princeton, Titchener wrote to Boring, 'I suppose that the path is now open for Pratt to stay with you at Cambridge. I think at the moment—though I may possibly change my mind—that I shall let Kimball Young get the Smith place and put Bishop into Pittsburgh.'"

39. Cited in Bjork 1983: 97–98.

40. Titchener 1925: 313–14. The quotations from Titchener in the next two paragraphs are also found here.

41. Bjork 1983: 87; Bjork provides citations.

42. Caldwell 1898: 408.

43. All of the following quotations from Titchener are from Titchener 1898: 449–52.

44. All three quotations from Boring 1952: 32.

45. Danziger 1979a: 206. The following quotations are from this page as well.

46. This functionalism had nothing to do with the functionalism that the linguist Martinet would propose several decades later.

47. Dewey 1884: 282.

48. Ibid., 279.

49. Ibid., 280, including the quotation in the next paragraph.

50. Angell 1936: 10. This is another example of the self-deprecating tone we observed in James, and will see again, an expression that is unimaginable in one of his German colleagues.

51. Angell 1907: 62–63, containing the next quotation as well.

52. Ibid., 69.

53. The history of psychology in France and England is an area in which considerable work is being done at present. We have profited from the work of Serge

Nicolas: https://sites.google.com/site/prsergenicolas/home/publications, and our information draws on his sources, notably Nicolas 2013.

54. See Brooks 1998, and after that, Guillin 2004. On Cousin, see also J. Goldstein 2005, 2013, and D. Goldstein 1968.

55. J. Goldstein 2013: 49.

56. See Nicolas and Murray 1999.

57. See Guillin 2004; Nicolas and Murray 1999.

58. Ribot 1870.

59. Ribot 1879 [1885]: 6, cited by Guillin 2004: 171.

60. Ribot 1877: 367, including the next quotation..

61. Nicolas and Murray 1999: 281.

62. Ribot 1881, 1883.

63. Wolf 1973 is the source of the information in the following paragraphs.

64. If the experimenter latches onto a behavior that he observes in the subject on the first experiment, and then continues to use that behavior as an indicator with some significance associated with it, then who has trained whom? The subject has trained the experimenter, with neither of them aware of it.

65. Nicolas and Ferrand 2002: 272.

66. Binet 1903.

67. Gould 1981: 181.

68. Ibid.,187.

69. Gould 1981; Demoule 1999.

70. Darwin 1871: 228.

71. Ibid., 226.

72. However, we would be oversimplifying if we limited the scientific revolution to a view of the world as a mechanism. While the view of the world as machine was vital, another view was every bit as important, the notion that mathematics could explain the world. We have become so accustomed today to viewing mechanisms and local collisions, on the one hand, and mathematical formulas that predict trajectories and gravitational forces, on the other, that we easily forget how far apart these two traditions are. In a number of studies, Frances Yates has shown the influence of early Renaissance hermeticism in the renewal of interest in studying the natural world with mathematics, and in exploring hidden, or "occult" forces, such as magnetism, reported by Gilbert in 1600 and gravity, by Newton in 1687; see, for example, Yates 1939, 1968. As Yates shows, this latter tradition was quite distinct from the one that emphasized local mechanical interactions.

73. The most striking instance in this book will emerge in chapter 8 as we look at the Turing machine. The impact of the computer on our thinking about the mind will be the central theme of volume 2.

74. Menabrea 1842. Menabrea wrote years later that "this translation was accompanied by notes of the greatest interest, which develop most clearly what I

was only able to express in an incomplete manner" (1884: 181). The article and Lovelace's notes were published by Richard Taylor (1843). Most of these materials are available online at http://www.fourmilab.ch/babbage/sketch.html (from which we draw our quotations), but without the author indicated, no doubt because it was a woman, and it would not be until 50 years later, and following Menabrea's later remark, that the work was attributed to Ada Lovelace, a fact that Babbage had never concealed from Menabrea.

Chapter Five

1. Angell 1907: 85.

2. Ibid., 81. Just slightly below this, Angell notes that the "mention of this classic target for psychological vituperation recalls the fact that when the critics of functionalism wish to be particularly unpleasant, they refer to it as a bastard offspring of the faculty psychology masquerading in biological plumage" (ibid.).

3. Cohen 1979. Angell later wrote of "behaviorism, which my student and assistant, John B. Watson, developed in such an extravagant manner" (1936: 26). See also Balkan 1966, Buckley 1989; Hackenberg 1995; and Wozniak 1997.

4. Cohen 1979.

5. Ibid., 55. See also Larson and Sullivan 1965, which provides much more interesting material on the Watson-Titchener relationship.

6. Hannush 1987: 137. Watson's son, John B. Watson, also agreed to be interviewed on his father's life, and his remarks are similar in tone. "He deeply believed that any expression of tenderness or affection would have a harmful affect upon us."

7. This and the next three quotations are from Watson 1913: 158–59.

8. This and the next four quotations are from Watson 1913: 162–63.

9. This and the next quotation are from Watson 1913: 164–65.

10. This and the following six quotations are from Watson 1913: 166–69.

11. E. B. Titchener, in a letter to R. M. Yerkes, April 2, 1914, cited in Larson and Sullivan 1965: 343.

12. On soul and mind in nineteenth-century psychology, see Reed 1997.

13. Schnaitter 1999: 213, based on Moore 1999.

14. Watson 1930: 11.

15. Ibid., 6, including the following quotation.

16. This and the following two quotations are from Watson 1929: 25–26.

17. Ibid., 33.

18. McDougall 1929: 41, including all of the quotations in this paragraph.

19. Boring 1929b: 79.

20. For example, see Gardner 1985.

21. In one of his last publications as something close to a behaviorist, Karl Lashley (1923) presented a thoughtful analysis of the range of different opinions within behaviorism. He wrote, "The history of the movement is still reflected in the tendency of its exponents to stress the experimental method rather than interpretation, in the lack of any specific formulation of the relation of the science to the specific problems of the older subjective psychology, and in a certain shifting of ground in behavioristic discussions which indicates that the behaviorists themselves are not yet quite certain of the philosophical implications of their system" (237–38). It is not hard to see that he is uncomfortable standing too close to these "behaviorists," who do not seem to be very reflective people, as far as Lashley could see, and in a few years he would distance himself even more. Lashley continued, "Too often a statement of an extreme position is followed by a partial retraction or qualification which leaves the reader in doubt as to the degree of heterodoxy expressed" (238).

22. Much of the information here is from a letter from E. A. Esper written to Charles Hockett, and quoted nearly in full in a concluding chapter of Bloomfield and Hockett 1970.

23. Weiss had immigrated to the United States from Germany shortly after his birth, but was raised speaking German at home. Esper 1968: 113–14.

24. Ibid.

25. Amsel and Rashotte suggest, "Were Hull alive today he might find it necessary to revive his crusade against the imprecision of a cognitive approach. On the other hand, he might think the battle irretrievably lost." Hull 1984: 63.

26. Cited in Smith 1986: 153.

27. This is cited in the introduction by Amset and Rashotte, in Hull 1984: 1, and the quotation that follows is from there as well. Hull came back to the analogy to Newton in Hull 1935.

28. Hull 1952: 155.

29. Hull 1984: 2–3.

30. Gengerelli 1976: 685.

31. Ibid., 686.

32. Krueger and Hull 1931, cited in Smith 1986: 161.

33. Hull 1930: 154. We encourage the reader to note the resemblance of this to a remark we encountered in chapter 1.

34. Morawski 1986.

35. Hull 1935: 495.

36. Ibid., 496.

37. Ibid., 511.

38. Hull, Gladstone, Felsinger, and Yamaguchi 1947: 234.

39. Ibid., 237. All of the immediately following quotations are also from here.

40. Ibid.

41. Heider 1989: 146–47.

42. Ritchie 1964.

43. Smith 1986.

44. Tolman 1932: 394. The next quotations are also from this page.

45. Ibid., 8.

46. Tolman (1932: 7) noted that he adopted the terms "molar" and "molecular" from the philosopher Donald C. Williams; Williams in turn adopted them from C. D. Broad. In a letter that Clark Hull wrote to a colleague on July 14, 1943, he indicated that Tolman felt strongly about the use of these terms:

> A couple of weeks ago I received a very strongly worded letter from Tolman, in which he told me in an astonishingly vigorous manner that he was thoroughly peeved at two things: first the fact that I had not credited him with the invention and introduction into psychology of the concepts of molar behavior and symbolic constructs [and another thing]. . . . I must confess that the violence of his reaction quite astonished me. I had always supposed him to be an exceedingly mild and amiable person. His letter was anything but that. (Hull 1984: 82–83)

47. Tolman 1932: 418.

48. Ibid., 47.

49. Ibid., 419.

50. Koffka 1935: 3.

51. Smith 1986: 86.

52. Boring 1929a, quoted in Smith 1986: 87.

53. Ritchie 1964.

54. Bruner 1983: 109.

55. This and the two following quotations are from Hull 1984: 55–56.

56. See Weidman 1999 for an account of various interesting aspects of Lashley's life and opinions, many of which violate our sense today of propriety, to put it mildly. The correspondence between Watson and Lashley is not pleasant to read.

57. Watson and Lashley 1913: 139.

58. There were those who saw him as a behaviorist, despite all of his protests to the contrary. McDougall (1929: 45) referred to Lashley in a footnote: "One of Watsons's most vigorous disciples, Dr. K. S. Lashley, taking his cue from his leader, has recently described me as 'bouncing back and forth between accurate description and the exhortations of a soap-box evangelist.'" Joel Isaac (2012: 101) describes Lashley as an "ardent neo-behaviorist," and cites Boring 1961: 55–56 as his source.

59. Lashley 1931: 14.

60. Ibid., 5.

61. Ibid., 14.

62. This and the following five quotations are from Bruce 1998: 74-76.

63. In a letter to Clark Hull (July 15, 1946), Lashley wrote, "I had hoped that you would also discuss the continuity theory . . . this point of view lays great emphasis on attention. . . . We behaviorists threw out attention in 1915 but it has come back for me as the central mystery of psychology." Bruce 1998: 73.

64. Mach, in turn, cited Loeb's work. There is a direct genealogical line—Mach to Loeb to Watson—in the anti-metaphysical strain that eventually, with Watson, sought to eliminate mind.

65. Whitman 1898: 328.

66. In 1912, Jacques Loeb discussed the implications of a mechanistic perspective on human action; this sounds indistinguishable from the view that Watson would pronounce over the next ten years under the banner of *behaviorism*. See Herrnstein 1972. For a popular account of Jacques Loeb, see Lemov 2005. From Loeb 1912, a rather bankrupt analysis of ethics:

> If our existence is based on the play of blind forces and only a matter of chance; if we ourselves are only chemical mechanisms—how can there be an ethics for us? The answer is, that our instincts are the root of our ethics and that the instincts are just as hereditary as is the form of our body. We eat, drink, and reproduce not because mankind has reached an agreement that this is desirable, but because, machine-like, we are compelled to do so. We are active, because we are compelled to be so by processes in our central nervous system; and as long as human beings are not economic slaves the instinct of successful work or of workmanship determines the direction of their action. The mother loves and cares for her children, not because metaphysicians had the idea that this was desirable, but because the instinct of taking care of the young is inherited just as distinctly as the morphological characters of the female body. We seek and enjoy the fellowship of human beings because hereditary conditions compel us to do so. We struggle for justice and truth since we are instinctively compelled to see our fellow beings happy. Economic, social, and political conditions or ignorance and superstition may warp and inhibit the inherited instincts and thus create a civilization with a faulty or low development of ethics. Individual mutants may arise in which one or the other desirable instinct is lost, just as individual mutants without pigment may arise in animals; and the offspring of such mutants may, if numerous enough, lower the ethical status of a community. Not only is the mechanistic conception of life compatible with ethics: it seems the only conception of life which can lead to an understanding of the source of ethics. (31)

67. This and the next two quotations are from Boring 1929b: 117.

68. Ibid., 118; Boring 1963b: 81.

69. Boring 1929b: 118. Boring put the same point even more pithily elsewhere. "Introspectionism," he wrote, "got its *ism* because the protesting new schools needed a clear and stable contrasting background against which to exhibit their novel features. No proponent of introspection as the basic method of psychology ever called himself an *introspectionist*." Boring 1953: 169.

70. Boring 1929b:118; Boring 1963b: 81, which includes the passages that follow in the text.

71. Heider 1989. We would like to emphasize how important this point was at the time and how important it would become in the future. Whether it makes sense to take the stimulus and the act of sensation as a process that precedes and operates independently of perception is a very important question, and the empiricist tradition took it largely for granted that this was possible. This view carried over into the Bloomfieldian view of phonology, for the categorization of sounds into allophones of a phoneme was assumed to be possible independent of the perceptual side of the system, which was the organization of phonemes.

72. But when the better known group of Gestalt psychologists—Wertheimer, Koffka, Köhler—began to publish, he "found himself forced out of the role of the progressive rebel into that of a defender of an established view" (Heider 1989: 134).

73. The best sources in English are the books written by Wertheimer, Koffka, and Köhler. A good recent review can be found in Wagemans et al. 2012, and an excellent historical overview can be found in Ash 1995. On the American side of Gestalt psychology, see Sokal 1994.

74. A good discussion of this split in Gestalt psychology by the second decade of the twentieth century can be found in Boudewijnse 1999.

75. Koffka, reply to Benussi, written in 1915, cited in King and Wertheimer 2005: 115.

76. Wertheimer 1922.

77. But some people did declare themselves commited to mechanism, such as Leonard Bloomfield, and we will see what he had to say in chapter 6

78. See also Riskin 2016.

79. Wertheimer, "Gestalt Theory," in Ellis 1938.

80. Ellis 1938: 8.

81. Köhler makes this point clearly, in an abbreviated summary of Kohler's 1920 book, given by Ellis in his *A Source Book of Gestalt Psychology*:

> Let us consider under what conditions a physical system attains a state which is independent of time (i.e., a state of equilibrium or a so-called stationary state). In general we can say that such a state is reached when a certain condition is satisfied for the system

> *as a whole.* The potential energy must have reached a minimum, the entropy a maximum, or the like. The solution of the problem demands not that forces or potentials assume particular values in individual regions, but that their total arrangement relative to one another in the whole system must be of a certain definite type. The state or process at any place therefore depends in principle on the conditions obtaining in all other parts of the system. (18–19)

82. Wertheimer 1922, partially translated by Ellis 1938: 12ff. The paper appeared in a festschrift for Carl Stumpf.

83. This is nearly a quotation from Mildred Focht (1935: 12), who cites Köhler 1929: 148ff.

84. *Psychologie, Lehrbuch der Philosophie*, 546, cited by Focht 1935.

85. Ash 1989; King and Wertheimer 2005.

86. Stumpf's work played a much more important role in the development of Brentano' s ideas of the relation between parts and the whole than this may suggest. We may put it this way: understanding parts is easy, but understanding the whole is always and everywhere a harder task. Stumpf and Husserl were two of Brentano's students who carried Brentano's questions into the 20th century.

87. Gundlach 2014:139.

88. Ash 1995: 133. Both quotations from unpublished letters to Ehrenfels.

89. Gundlach 2014.

90. King and Wertheimer 2005: 102, citing Köhler 1929: 110–11.

91. Köhler 1959: 728.

92. Cited in Ash 1995: 131.

93. Heider 1989: 137.

94. Ibid.

95. Ibid., 133. Heider goes on:

> [The Berlin group of Gestalt psychologists] saw G. E. Muller as the principal advocate of a theory built on elements, bits, and crumbs. I remember a session at the Psychological Congress that met in Bonn in 1927 when a discussion developed between Muller and Köhler. There was this white-haired old paladin of German psychology with a reputation almost equal to that of Wilhem Wundt himself, presenting his case in a dignified and courteous way. Köhler replied, and I felt embarrassed by his sharp sarcasm. And Bühler, professor in Vienna, who had written about a form of Gestalt psychology that different from that of the Berliners in some of his theoretical assumptions, was equally despised.

96. King and Wertheimer 2005: 171. This student later married Wertheimer.

97. Wertheimer 1945:106. The following quotation is found on p. 107.

98. On the emigration of German psychologists to the United States during this period, see Wellek 1968.

99. Swarthmore College was a hotbed of research for decades after Köhler's arrival, where the faculty included German immigrés trained in Germany such as Hans Wallach and Solomon Asch, and a younger immigré Henry Gleitman, who was one of Edward Tolman's last students, and students who became luminaries in postwar psychology, such as Ulric Neisser and Robert Rescorla. One of us (J.G.) learned a good deal about the history of psychology from listening to Henry Gleitman.

100. Koffka 1922: 531–32.

101. This and the next three quotations are from Koffka 1935: 63.

102. Ibid., 64. Heider pointed out that Koffka made it clear in *Principles* that his objections to the Graz approach to Gestalts was that it was vitalistic and "introduced a profound dualism into psychology" (Koffka 1922: 559). Köhler does much the same in Köhler 1929: 176, where he wrote that "even those who were particularly interested in the new topic found it extremely dfificult to recognize at once its radical consequences for psychological theory."

103. Boring 1961: 41.

104. This and the following quotation are from Helson 1925: 343.

105. Ibid., 345.

106. *Saturday Review*, 23 May 1925, 773.

107. This and the following two quotations are found in the introduction to Hull 1984: 22–23.

108. The quotations that follow are from Köhler 1929: 51–54.

109. This and the following three quotations are from ibid., 53.

110. We use the word "value" as a term to cover both scalars and vectors.

111. When we say that one thing affects the changes in another here, we mean that the second equals an integral of the first, but this is not a relationship of cause and effect.

112. Springs also create a force at certain locations, whose measure is linearly proportional to the degree that the spring is stretched.

113. This conceptual, and mathematical, metaphor emerged again in the 1980s in the context of neural networks, first due to the work of the physicist John Hopfield, which in turn led to the recrudescence of work on both recurrent and feedforward networks.

114. Sebeok 1987.

115. See Vonk 2004 for general comments on Bühler's program; see also Mulligan 1988, and Marthelot 2012.

116. There appear to be two editions of this book; the one online is quite different than the printed version that we have. But they are not described as such, at least in the United States. Kusch describes Wundt as a "purist," that is, as someone who "insisted that psychology be both conceptually and institutionally

independent of other disciplines," notably logic, physiology, and the applied sciences, while the Wurzburgers were "happy to allow permeable borders with respect to at least some other fields," and Kusch calls them "promiscuist" in this sense.

117. William Weimer (1974) has gone so far as to describe the obscurity of the story as a case of "suppression of psychological history": "As an illustration of the suppression of psychological history that can only be seen as such with the advantage of hindsight (granted by the revolution in cognitive psychology and psycholinguistics), consider the developmental and cognitive psychology of Karl Bühler. Bühler is virtually unknown among English-speaking theoretical psychologists" (248).Weimer discusses additional complexities that are quite germane to our discussion.

118. Nerlich and Clarke 2001. For a recent perspective on ellipsis, see Merchant 2001.

119. It appears that the support was based on backing for the work that Charlotte Bühler was engaged in, and on which she lectured in the United States during 1924–25.

120. Musolff 1997.

121. Sebeok 1987: 133.

122. Bühler and Moritz Schlick jointly supervised about forty doctoral dissertations.

123. Sebeok (1987) noted that Bühler's widow would later write that Jakobson "leaned extensively on Karl, (yet) did not befittingly acknowledge his debt to Karl," a perspective Sebeok argues is not true: "the opening sentence of Jakobson's celebrated *Kindersprache* [1941] . . . is a direct quotation from a 1935 paper of Bühler's" (133). Sebeok's account defends Jakobson from the charge of not citing Bühler, but we can learn from Charlotte Bühler's text of the sadness and, for Bühler at least, lack of professional recognition that entered the Bühlers' life when they came to the United States.

124. This and the following quotation in this paragraph are found in Ash 1995: 310–11.

125. See Mulligan 1988 for a discussion by precisely such a philosopher; we are indebted to his perspective in this and the next paragraph. Quotation from p. 204.

126. Ibid., 223.

127. See ibid., 209, who observes that this point was also made in Bühler 1919. In "Phonetik und Phonologie" (Bühler 1931), Bühler uses the term *Helligkeit* for the feature that would be called [acute].

128. Bruner 1983: 59.

129. Ibid.

130. Ibid. Allport's degree was from Harvard, where he had studied with Münsterberg and Herbert Langfeld, both students of Wundt, as we have seen. He followed a familiar model: he spent a year studying psychology in Germany,

where he was impressed by "the briliance of the Lewinian approach" and the Gestalt school more generally. But he was drawn to social psychology and the study of personality, not a booming field in the United States in the 1920s. He was a member of the Department of Psychology at Harvard from 1930 to 1967, and he will play a significant role in volume 2.

131. Ibid., 72.

132. Ibid., 73. Boring himself saw a rough side to his own character, and late in his life, he was willing to attribute it to his own insecurity. He wrote, "Here I may say a little about my ruthlessness, for I had that reputation when I was Director of the Harvard Psychological Laboratory in 1925–49. I would make tough judgments that hurt people, that interfered with their careers. I never felt, though, that I was cruel; I had too much desire for affection and admiration for that. Sometimes I was called legalistic, and that word shows me what was going on. I stuck to principle. . . . Indeed I was tough and ruthless. I did hurt people, but I always believed that I was refusing a smaller loyalty to maintain a larger one" (1961: 15–16). He discusses the nature of his insecurities just above this quotation.

133. These quotations are from Heidbreder 1933: 3.

134. Ibid., 4.

135. Ibid., 7.

136. Ibid., 14.

137. Ibid., 15.

Chapter Six

1. Atkins's paper is available at http://www.alaskool.org/native_ed/historicdocs/use_of_english/prucha.htm; see also Linn et al. 2002.

2. See Smithsonian Institution, Bureau of American Ethnology at https://archive.org.

3. Ratzel 1895; Goldschmidt 1959; Lewis 2001.

4. Whitfield 2010.

5. Haeckel 1879.

6. Boas 1888.

7. Emeneau 1943: 35.

8. Jakobson 1944.

9. Teeter 1964: 198.

10. Ibid.

11. Emeneau 1943: 37.

12. Lowie 1944: 311.

13. Benedict 1943.

14. Lowie 1944: 311.

15. Harris 1968: 252–53.

16. Ibid., 253.

17. Lowie 1947: 315.

18. Ibid., 318.

19. Boas 1912, 1934, 1940; Boas et al. 1934; Gossett 1963: 418 wrote, "It is possible that Boas did more to combat race prejudice than any other person in history."

20. Sapir 1949.

21. Darnell 1990.

22. Darnell and Hymes 1986: 210.

23. Sapir 2011: 121.

24. Kroeber 1984: 131.

25. See Darnell 1998, who says that Angell expected Sapir to set up a department of anthropology. The reader will recall that the Institute of Human Relations was the brainchild of the dean of the law school at Yale, Robert Maynard Hutchins, and Hutchins became the president of the University of Chicago just shortly before Sapir left to go to the Institute of Human Relations at Yale.

26. Darnell 1990: 260.

27. Lagemann 1989: 157, 166. Kluckhohn will return in the second volume of this story.

28. If we think about the range of sounds that exists in a particular language, we note first of all that no matter how many sounds may appear in that language, there are sounds that do *not* appear in it. English, for example, has the sound /z/, as in the word *zoo* or *fuzz*, while many European languages do not; the Scandinavian languages do not, for example, nor do most dialects of Spanish. So each language has a range of sounds that appear in its words, and for the most part, each language excludes other sounds from appearing in its words.

The more subtle side of the phoneme is this, however: we may find that in some languages, two sounds in the language *contrast*, while in another language, the very same sounds appear but do not *contrast*. Consider the sounds /s/ and /z/. In English, these two sounds contrast, by which we mean that we can find two words (such as *zoo* and *sue*) which are identical except that they differ just by the two sounds of the pair: take *zoo*, replace the /z/ in it with /s/, and we have *sue*. As this example brings out, the result has nothing to do with spelling, and everything to do with sound. Given the pair *zoo* and *sue*, we conclude that /z/ and /s/ contrast in modern English. The same point, of course, can be made on the basis of other pairs of words, like *maze* and *mace*, or *zip* and *sip*.

Mexican Spanish is one of the relatively few dialects of Spanish in which the sound [z] appears, as in a word such as *mismo* 'same', in which the sound that is spelled 's' is pronounced with a [z]. In fact, we do not find the sound [s] before an *m* or an *n* in Mexican Spanish, and that is where we *do* find the sound [z]. In short, [z] is the variant that we find instead of [s] in the position just before an *n* or an *m*. In this case, there is no contrast between [z] and [s], and the two sounds belong to one and the same phoneme.

The notion of the phoneme is used in other ways. For example, American dialects of English differ with regard to the status of the sounds that linguists identify with [a] and [ɔ]. In some dialects, especially on the East Coast, these two sounds are contrastive, and we find many pairs of words that are pronounced differently, based just on the difference between these two sounds: for example, *ahh* and *awe*, *cot* and *caught*, *tot* and *taught*, *la* and *law*, *Don* and *dawn* (and this describes the usage of one of us [J.G.]). However, many American speakers of English pronounce each of these pairs of words identically, and regardless of which sound they use in those words, there is no *contrast* to be found there. This is a case in which a distinction is being lost in such dialects, from a historical point of view, but the notion of a phoneme is not, fundamentally, a historical notion; it is used to described the state of a language at a particular moment.

A little reflection will lead us to the conclusion that we can identify exactly how many phonemes a particular language (or really, a particular dialect of a language) possesses. A speaker who contrasts *Don* and *dawn* might have 39 phonemes, for example, and another speaker whose English is identical except that he pronounces those two words the same way will have only 38 phonemes. No one but a linguist is likely ever to notice.

We should not leave the impression that the natural direction of change is towards a decrease in the number of phonemes: it is just as easy for a change in a language to *increase* the number of phonemes. We'll give an example that is a bit more complicated, but interesting in its own right. For a large number of American speakers, the vowel sounds in the word *pat* and the word *pan* are not at all the same, and linguists represent the vowel of *pat* as [æ] and the vowel of *pan* as [$e^{ə}$]. For most Americans, these sounds do not contrast, and there are no minimal pairs distinguished only by these two sounds. However, in a few dialects, some minimal pairs have emerged, and these two vowels in those dialects are now in contrast, leading to a situation in which a new phoneme has arrived in the language. The first pair of words in which this contrast appeared is the pair of words consisting of *can* (as in *a tin can*), with a [$e^{ə}$] and the auxiliary verb *can* (as in *Yes I can!*), with [æ].

29. Haas was one of the relatively few women actively involved in linguistic teaching and research during the period we are looking at. She was herself a student of Sapir's, and she was the doctoral advisor of many influential linguists of the next generation, including William Bright and Karl Teeter. She was a president of the LSA, a fellow of the American Academy of Arts and Sciences, and received a number of honorary PhDs.

30. The best source on Morris Swadesh's life and influence is an essay by Dell Hymes, published as an appendix to Swadesh 1971.

31. Swadesh 1934: 117.

32. Ibid.

33. Ibid., 122.

34. Ibid., 123.

35. See Newman 1946a, 1946b. Hymes's essay in Swadesh 1971: 233 presents some of the finer grained detail of the development of phonology and morphophonology in Sapir's circle. Newman made a field trip in 1930 to work on Yokuts, and published a paper in the *International Journal of American Linguistics* in 1932 on this material. By 1936, he had developed it into a dissertation, which he wrote while a student of Sapir's at Yale. Newman worked at the Institute of Human Relations from 1932 to 1937, the same time that Swadesh was there, developing his work on Nootka. In 1934–36, the three of them—Sapir, Newman, and Swadesh—were working on English together.

Newman's work on Yokuts has remained a classic linguistic analysis, and it has been thoroughly reviewed and reanalyzed by successive theoretical perspectives over the course of the following decades, as much by generative phonologists as by structuralist phonologists.

36. Although Newman 1932 and Newman 1944a have the same title, the first is just a very short overview of a much more substantive analysis developed in the second.

37. Newman 1946a: 226. The asterisk in front of a form had been used in linguistics to mark a reconstructed, unobserved from in an earlier stage of a language.

38. Ibid., 227.

39. Newman 1946a refers to the "formal machinery" of Yawelmani's grammar, and its "automatic regularity" (222).

40. Both of these citations are from Newman 1946a: 227.

41. Ibid., 225.

42. Ibid.

43. Sapir 1929: 207.

44. Ibid.

45. Ibid.

46. Ibid.

47. Leonard Bloomfield (1922) expressed a similar view in a review that he wrote of Sapir (1921):

> We are casting off our dependence on psychology, realizing that linguistics, like every science, must study its subject-matter in and for itself, working on fundamental assumptions of its own; that only on this condition will our results be of value to related sciences (especially, in our case, to psychology) and in the light of these related sciences in the outcome more deeply understandable. In other words, we must study people's habits of language—the way people talk—without bothering about the mental processes that we may conceive to underlie or accompany these

> habits. We must dodge this issue by a fundamental assumption, leaving it to a separate investigation, in which our results will figure as data alongside the results of the other social sciences. . . . Like the rest of us, Dr. Sapir still pays tribute to aprioristic speculation which steals upon us in the guise of psychology; as his own approach is scientific, these false generalizations stand out from the rest of the discussion. Dr. Sapir has less of them than his predecessors; whoever is interested in the progress of our science will welcome his book as a step forward. (92–93)

These words show a consciousness of linguistics as an autonomous profession, based in the first place in a divergence of both methods and goals from the neighboring disciplines and practices.

48. The private correspondence of Fannie Bloomfield is largely in German: http://mms.newberry.org/html. The Bloomfield family did not speak Yiddish at home, which was not unusual among the Jewish families in Bielitz, and unlike Edward Sapir, Leonard Bloomfield did not study Yiddish in his linguistic work. See Hall 1990: 3–4; we have relied on this reference for a number of the points in this section.

49. Marie Bloomfield's life was brief and ended unhappily; it is difficult to know, from our remove today, precisely what effect it had on her older brother Leonard, though it is clear it affected him very much. Marie was an undergraduate student at Barnard College, in New York City, after her parents had passed away. She entered into a serious relationship with Margaret Mead, also an undergraduate there. Mead was involved in at least one other intimate relationship at the same time, and Banner 2010 presents evidence that strongly suggests that it was Marie's unhappiness in this complex relationship that led to her suicide in February of 1923. The reader will recall that we have already encountered Margaret Mead, who would be a student of Franz Boas's and ultimately the most celebrated anthropologist of her generation. About a year later, Mead met Edward Sapir, some 18 years older than she, and became involved romantically with him. This was shortly after the death of Sapir's first wife, and not too long before he moved to the University of Chicago. We have no evidence that either Leonard Bloomfield or Edward Sapir were aware of the unhappy links that joined the two of them through Margaret Mead and Marie Bloomfield. If they did, however, it is not hard to imagine how such knowledge might have impeded establishing a working relationship at Chicago.

50. Despres 1987: 5.

51. Ibid. See also Sayers 1987.

52. Despres 1987: 5.

53. This was Charles Hockett's recollection. From his introductory remarks in Bloomfield and Hockett 1970: 1.

54. Hall 1990: 8.
55. Ibid.
56. Moulton 1970: 512.
57. Lane 1955: 184.
58. Chevalier 1997; Jakobson and Waugh 1979: 266.
59. Hall 1990: 15.
60. Hockett 1987: 411.
61. Emeneau 1988: 757.
62. Ibid., 755.
63. Bloomfield 1929: 268–70.
64. Hall 1990: 14.
65. Bailey (1992: 797) wrote that the

> outlook [of twentieth-century linguistics] . . . was shaped by Americans steeped in the neogrammarian ideas that so influenced intellectual inquiry in our field at the time of the founding of the Society. . . . An English intellectual has recently asserted that "we live in the shadow of a Renaissance as brilliant and dominating as the Italian Renaissance." By this he means that the Anglo-American scientific enterprise has been shaped by "the explosion of genius which burst in Germany and Austria in the eighteenth and nineteenth centuries" (Annan 1984: 165). Hockett has emphasized the same sources for American philology and linguistics . . . "it is hard for us today," wrote Hockett (Bloomfield and Hockett 1970: 536), "to realize how strong and deep was the tradition of German scholarship in American universities before the first World War, and how enormous a blow that tradition received at the time of the War."

66. Much of the information here is from a letter from E. A. Esper written to Charles Hockett, and quoted nearly in full in a concluding chapter of Bloomfield and Hockett 1970.
67. Esper 1968: 113–14.
68. On Meyer and Weiss, see also Levelt 2014.
69. Weiss 1919: 632.
70. Weiss 1919.
71. Bloomfield 1931: 219.
72. Despres 1987: 7.
73. Murray 1994: 121.
74. Sturtevant had recently moved to Yale University, after having been dismissed by Columbia University as part of a policy waged by the president there, Nicholas Murray Butler.
75. Hall 1990: 67.

76. Pike 1989.

77. Bloomfield and Hockett 1970: 541–42.

78. Fries 1970: 197.

79. Bloomfield 1914: 312.

80. Ibid.

81. Ibid., 322.

82. Ibid., 323.

83. Ibid., 60.

84. Ibid.

85. Ibid., 61.

86. Ibid., 54.

87. Ibid., 55.

88. Ibid., 221.

89. Bolling 1917.

90. Ibid., 53. It is worthwhile to pause and think for a moment about Bolling's criticism. Bolling was focused on a problem from a perspective that is not at all obvious to us today, but we can see that he explicitly expected a scientific analysis of a sound change to be explained by a cause that precedes the effect, and that is distinct from the effect; that is, after all, part of what is meant by a sufficient cause. For us today, certain kinds of things can be done mechanically, by computers, which seem to "just happen of themselves—automatically." But we live in a different world from Bolling's; we take computers for granted.

91. Trager 1942: 143, 145.

92. We must imagine that he was referring to Swadesh 1934.

93. Hill 1980: 73.

94. Bloomfield 1934: 32. In this same review, Bloomfield defends in the strongest possible terms the Humean view that the only sensible understanding of the term *cause* is as a shorthand for correlation: he in effect rejected the notion that the aim of science is to uncover a hidden truth in how things really are. Bloomfield's use of the term *correlation* is much the same as our use today, but in chapter 7 we will see European structuralists using the term in the Aristotelian tradition.

95. Bloomfield 1934: 35.

96. Bloomfield 1930: 554–55 for all quotations from this article.

97. Bloomfield 1933: 161.

98. But of course the reader of this book knows full well that there was nothing like total independence, and we can trace the genealogy of thoughts and thinkers that related the two groups.

99. This quotation and the two that follow in this paragraph are from Bloomfield 1926: 153.

100. Saussure 1972: 355.

101. Bloomfield 1926: 155.

102. Ibid., 153.

103. Matthews 1993: 13.

104. Bloch 1949: 88.

105. Bloch 1933: 91–92.

106. Matthews 1993: 118.

107. Ibid., 122.

108. We will discuss the "eclipsing stance" in greater length in volume 2. The phrase alludes to a famous passage from Voegelin and Voegelin 1963: 12. Hockett's remarks are in Hockett 1980:106. Hockett's remarks cited in the text continued:

> Should my mentors, back in the 1930s, have insisted that I work my way through Whitney? Perhaps so, and perhaps that would have made me a better scholar. On the other hand, possibly I would not yet have been mature enough to tune my twentieth-century ears to his nineteenth-century voice. Our receptivities really do change. I'm sure you won't think me facetious if I offer, as another example, the fact that when I tried Milne's Winnie the Pooh first, during my adolescence, it was unspeakably dull, but when later I picked it up to read to my own children it had become poignant magic. (Hockett 1980: 106)

109. Letter from Hockett to John Goldsmith, February 7, 1991.

110. The reader may see that there is both an issue of authority here, and a concern about cross-discipinary relations. The philosopher of science would like to be in a position to legislate what counts as science, across the whole range of academic disciplines; the non-philosopher who is doing his work as he sees fit can either shrug and declare that he really does not care what the philosopher of science decides about what a science can do, or else he has to engage in a full-fledged dialog with someone who does not really know his discipline.

111. Bloomfield 1936: 90.

112. Bloomfield 1938: 90, 95.

113. Murray Emeneau (1988) is an important contribution to this subject.

114. Bloomfield 1933. See, for example, the discussion of base forms there on pp. 211–12.

115. Ibid., 112.

116. See Silverstein 2014 on a number of pertinent points regarding Sapir and Bloomfield.

117. Harris 1973: 255.

118. Bloomfield 1930: 553.

119. Harris 1973:255.

120. Pike 1989: 217–18.

121. Malkiel 1980: 83.

122. Gleason 1988: 4.

123. This quotation and the ones following are from Harris 1973: 252–54.

124. Bloomfield 1925: 1.

125. The situation was undoubtedly more complex than this on the American front. The *American Journal of Philology* was already publishing linguistic work, but it had no competition, and E. A. Esper recalled that he perceived hostility between the journal's editor, Charles Miller, and Bolling as one of the motivations for creating the journal *Language*. Cited in Bloomfield and Hockett 1970: 538.

126. On the history of the LSA, see Murray 1991.

127. Falk 2014.

128. Voegelin and Voegelin 1963: 20. There is much to reflect on in how the American linguists felt that their approach was home-grown, in light of our observations concerning the deep roots of American linguistics in German linguistics.

129. McDavid 1980: 7. We will see in volume 2 what the *Schrecklichkeit* was that McDavid was talking about. The term was used around the time of World War II to refer to the policy of terror and atrocities targeting civilians that was adopted by the German army.

130. This quotation, which continues just below, is from Carroll 1980: 36–37.

Chapter Seven

1. The importance of understanding generative grammar against the background of the developments in philosophy and logic in the period from 1900 to 1940 is emphasized in Tomalin 2006, as we will note a number of times over the course of chapters 7 and 8.

2. Dummett 1993: 1–2.

3. A philosopher who had left the Earth during the first manned space flight and had no contact with academic discussions since 1957 would be surprised to see the matching and mixing of philosophers in this chapter. The last 20 years has seen quite a good deal of rethinking of intellectual currents. Benoist observed quite rightly:

> That phenomenology and analytic philosophy go back to the same historic and conceptual sources is a situation that is starting to be recognized. . . . The common crucible in question is what is starting to be called "Austrian philosophy." The problem is probably not so much a problem of nationality as of philosophical attitude. Carnap was born in Wuppertal, Schlick was from Berlin, and Brentano from the Rhine. It is only a small step from there to conclude that the Vienna Circle and what preceded it were essentially Prussian, or at least German. But if we go back before the

> Vienna Circle, there was in the 19th century a properly Austrian tradition of philosophy which was opposed to what has been called "German idealism," a tradition that goes back to the great shadows cast by the two founding fathers who sketched opposite but complementary orientations, and not totally disconnected: Bolzano and Brentano. They both certainly intended to move out of the wake of the Kantian tradition in order to better link to a Leibnizian inspiration (in the one case) and an Aristotelian inspiration (in the other), all the while adapting the great progress being made in logic and psychology during their lifetimes. . . . It is in this context that Husserl became a philosopher, even if we must not lose sight of his mathematical background . . . and his frequent turn to the British empiricists. (1997: 7–8)

4. A considerable literature exists on the placement of Heidegger in the broader philosophical context; we will not consider his role any further in this book.

5. That may seem like a woefully inadequate description of phenomenology, but doing better is not an easy task. Phenomenology aims to bring to light the character of experience, of subjective experience, which is not hidden but which nonetheless is very hard to see.

6. Benoist 2002. Aurora 2015 is a recent publication drawing linguists' attention to Husserl's importance for their discipline.

7. Kronecker was a strong defender of a version of finitism, which was a precursor of intuitionism, one of the topics of the next chapter. Kronecker's view was that for us to legitimately speak of a mathematical object, we need an explicit algorithm that can run (as we say today) in finite time that produces the object to us. Weierstrass was deeply involved in developing a rigorous foundation for analysis at that point, a rigor which impressed Husserl deeply. Weierstrass aimed to found all of mathematics on natural numbers: positive whole numbers, and thereby removing geometry or spatiality from analysis, and unhooking it from Kantian spatial intuition (recall that the term we translate as *intuition* does not mean for Kant what we mean in everyday English).

8. See Fisette 2009 for details on this period.

9. Ibid., 176.

10. On the connections between mathematics and phenomenology in Husserl, see also Atten 2006b.

11. Atten 2006a: 5.

12. Gray 2008: 204.

13. Lindenfeld 1981: 245.

14. Ibid.

15. S.-Y. Kuroda's discussions of Husserl and Marty are the only places where Husserl's work is juxtaposed to work in generative grammar, notably in Ku-

roda 1973. Kuroda noted that his reading of Husserl and Marty was stimulated by a course he had taken on the history of linguistics taught by Noam Chomsky. Readers of Kuroda's analysis will see that Kuroda placed a great deal of emphasis on some changes in Husserl's presentation of the Port-Royal *grammaire générale*, and he suggested that Husserl is much more sympathetic to that work in the second edition. We do not see a major change, though Husserl explicitly was responding to comments made by Marty after the first edition of Husserl's book, a point that Kuroda discusses in some detail.

Kuroda argued that Marty was right (in any event, more right than Husserl where the two disagreed) and that he was more right because he understood that there could be language-dependent underlying structures, which Kuroda understood in the sense of Chomsky's *Aspects* model. Kuroda explained (101–2) that in Marty's conception he saw a clear connection between the study of the *a priori* in language and the empirical domain of transformational grammar. Marty went beyond Husserl in elaborating his concepts of inner speech-form and a syntax-semantics that is not based on logic.

16. See Casadio 2002 for a very accessible introduction to this development.

17. Zahavi and Stjernfelt 2002: 16

18. Husserl 1900b [2001]: 2:499–500.

19. Lenci and Sandu 2004.

20. See Husserl's Fourth Logical Investigation, entitled *The Distinction between Independent and Dependent Meaning and the Idea of Pure Grammar*, in Husserl 1900b [2001]: 47–76. See also Bundgaard 2004, Gardies 1986: 18 and throughout, and Lenci and Sandu 2004.

21. We do not wish to suggest that Carnap originated this notion. In 1903, Andrew Ingraham discussed the sentence "The gostak distims the doshes," making the point that the sentence is manipulable grammatically even if one does not know what gostaks are, and C. K. Ogden and I. A. Richards made the sentence famous in Ogden and Richards 1923: 46.

22. Husserl 1900b [2001]: 295.

23. Ibid.

24. Ibid. Getting to a complete understanding of Husserl's view on this matter would require not only understanding the evolution of Husserl's thought, but also what the *Port Royal grammaire générale* aimed to accomplish, what differences there were that gave rise to at least the appearance of a disagreement with Anton Marty over this principle, and also what our best understanding is today about the character of universal grammar and how *a priori* logic serves to explain its structure.

25. Gardies 1986: 21.

26. This corresponds to page 68 of Husserl 1900b [2001], but our translation here is not the same.

27. Husserl 1900b [2001]: 69. Given the mathematical context in which Husserl was writing, the only question for a translator today would be whether Husserl's

term *Mannigfaltigkeit* should be translated as *manifold* or as *set*; at the turn of the century, the term in German was used in both ways, even if later *Menge* became the standard term for *set*. But there is no reason to think that Husserl had a Riemannian manifold in mind, but rather a set (Husserl's friend Georg Cantor used *Mannigfaltigkeit* for *set* as well.) The rest of the quotations here from Husserl follow immediately on the same page. The issues of rupture and continuity in the development of set theory are discussed with considerable insight and care in Ferreirós 2007.

28. Ibid. We lean here on the published translation. Actually, in the original German, Husserl gives the conjunction of two nouns to form a new one, but this is overlooked in the published translation.

29. Jakobson 1973a: 13.

30. Russell 1914: 13.

31. Ibid., 224.

32. Carnap 1963: 13.

33. Bouveresse 2012: 197.

34. We are in agreement with Janik and Toulmin 1973 on this point.

35. Bell 1990: x.

36. Ibid.

37. The understanding of the work of the Vienna positivists has evolved in the last 20 years; by the mid-1990s, Thomas Uebel, a philosopher reconsidering their work, would be able to define himself as a member of the set of "writers working on the rediscovery of the Vienna Circle that has burgeoned in recent years" and to respond to what he called "the simple-minded portrayals of the Vienna Circle current in the potted histories with which many writers nowadays tend to preface their own work" (1996: 416).

38. The term *logical positivism* itself goes back to an influential article by Albert Blumberg and Herbert Feigl in 1931 entitled "Logical Positivism: A New Movement in European Philosophy": the paper offered what its authors term a "short and dogmatic" overview of the ideas of Carnap, Reichenbach, and their colleagues. Albert Blumberg is an interesting if marginal character in the history of logical positivism in the United States. Blumberg was born in Baltimore in 1906, of Lithuanian immigrant parents. He married Dorothy Rose Blumberg in 1933; two years older than Blumberg, she was from the wealthy Oppenheim family of Baltimore. Observation by one of us (J.G.): her family knew my grandfather's family in Baltimore—she was the same age as my father; and I remember meeting the Blumbergs at our house in 1967 as a high school student, and discussing Leibniz's philosophy, and Russell's book on the history of philosophy. Blumberg studied in Vienna, where he was close friends with Herbert Feigl, and wrote a doctoral dissertation under Moritz Schlick. He obtained a position in the philosophy department at Johns Hopkins University and was active in the founding of the journal *Science and Society* in 1936, but he resigned in 1937 to take a full-

time position as district administrative secretary for the American Communist Party, and he and his wife remained active in CPUSA activities and politics till after World War II; both were arrested and tried for their involvement in the Communist Party. Later in life, he turned to less radical political activities, and in 1965, Blumberg was hired by Rutgers University as a professor of philosophy, where he remained until his retirement.

39. Uebel 1996.

40. Carnap 1937: 1.

41. Schlick 1915 [1979]: 153.

42. Cited by Reisch 2005: 91.

43. Vienna is an important place for both psychology and philosophy, and its role in psychology during this period is the focus of Gardner and Stevens 1992.

44. Smith 1986: 346n22.

45. There has arisen in the past few years a controversy as to whether Carnap's published work on space understates the degree to which it drew from Husserl's work; see Rosado Haddock 2008 and Jacquette 2009. This controversy, which is hardly settled as we write, is very relevant to the issues regarding influence that we are interested in in this book.

46. Carnap 1963: 16.

47. Schilpp 1963: 13.

48. Ibid., xiv.

49. Ibid., xlvi.

50. Ibid., xviii.

51. Reichenbach 1938.

52. See Reichenbach 1947; see also Witte 1956.

53. Reichenbach 1947: 252.

54. Ibid. On p. 254: "Language has been misconstrued by traditional grammar, since this grammar has no logical place for the existing linguistic forms of the converse." The next two quotations are from p. 255.

55. This and the next quotation are from ibid., 252.

56. Ibid., 255.

57. Not all responses were equally positive. Witte responded to Reichenbach this way:

> Evidently we cannot but show our appreciation for the well-constructed work. *The real objection however is that this logistic grammar is no grammar.* It has used and classified linguistic phaenomena, but according to a division, a scheme, an opposition-system, that not results [*sic*] from the language-phaenomena. A logical, intellectual scheme is forced upon linguistic phaenomena. The outcome is no grammar, but a logical classification of a number

> of language-phaenomena. . . . Admitting that this so called Logistic Grammar puts a number of grammatical values in a clear light, the method is as mistaken as that of psychological grammar. Language has a right to autonomical consideration, a science of its own, is not a part of Logic, Logistics or Psychology [*sic*]. Linguistics being autonomical all those divisions and notions as parts of speech, cases, verb categories etc. have come to existence. With a certain disdain logicians make use of them, but what difficulties they would have had, if busy Traditional Grammarians had not for years and years laboured to render comprehensible and surveyable this distasteful, stupid, primitive Traditional Grammar! In my imagination I see Mr. Reichenbach standing before the virgin wood of linguistic phaenomena without Traditional Grammar! Imagine him in his College-time with his Logistic Grammar, learning to stammer some French and German, not to speek [*sic*] of Russian and Polish, not to mention Latin. (1956: 97)

Witte plainly did not appreciate Reichenbach's offer to help linguists.

58. Feest 2007. One of the participants at the meeting they organized was Kurt Lewin, who Reichenbach had known for many years; we have already met him as the fourth of the early members of the Gestalt psychology movement (see chapter 5). In the early 1920s, both Lewin and Reichenbach were working on the logic of space and time. On Lewin and Reichenbach, see Padovani 2013 and Heis 2013. Lewin was a member of the Executive Committee of the Society for Empirical Philosophy in 1931 (see Heis 2013), and he published in *Erkenntnis* in 1931 the paper which he had presented to the Society the year before. Heider also recalled this meeting; he remembered that neither Carnap nor Reichenbach were well known at that point. "I especially remember Carnap's talk. At that time he was enthusiastic about the symbolic logic of Bertrand Russell and Alfred North Whitehead and proceeded to explicate the statement 'one and one make two.' Before he finished he had covered two blackboards with symbols. I wondered how many blackboards he would have needed for the addition involved in a simple grocery order, and I learned from his example that it is not always profitable to dig all the way down to core concepts" (1989: 135).

59. Feest 2007: 4.

60. Ibid., 9.

61. Feest 2007. This material is also central to the controversy that we alluded to regarding the degree of influence of Husserl on Carnap.

62. Ibid., 13.

63. Uebel 2008.

64. Ibid., 74.

65. Uebel 2004a.

66. Carnap 1963: 57.

67. Uebel 2004b: 62. Uebel cites "Socializing the University," written in 1918, in Reichenbach 2012: 1.56–57.

68. Uebel 1996: 417.

69. Ayer 1977: 129. Smith (1994: 14) observes that Neurath noted that liberalism was "the dominant political current in Vienna. Its world of ideas stems from the enlightenment, from empiricism, utilitarianism and the free trade movement of England. In Vienna's liberal movement, scholars of world renown occupied leading positions. Here an anti-metaphysical spirit was cultivated, for instance, by men like Theodor Gomperz (who translated the works of J. S. Mill) and by Suess, Jodl, and others."

70. This and the quotation in the next paragraph are drawn from Hahn, Neurath, and Carnap 1973: 302.

71. This and the next two quotations are from ibid., 316–17.

72. Uebel 2004 discusses some aspects of the social tendencies among the central members of the Vienna Circle.

73. Næss 2005: 274; the next quotation is from p. 275.

74. Janik and Toulmin 1973: 215.

75. Feigl 1981: 55.

76. Mises 1939 [1951]: 25.

77. Friedman 2012: 3.

78. This quotation and those that follow are all from Schlick 1930: 53–54.

79. Ibid., 7.

80. Neurath 1930: 207.

81. But see Richardson 1992 for a more nuanced view, exposing neo-Kantian aspects of Carnap's ideas.

82. Reichenbach 1938: 5.

83. Carnap 1928 [1967].

84. Reichenbach 1938: 6.

85. Ibid., 6–7.

86. Cited in Ter Hark 2003:19.

87. Carnap 1928 [1967]: xvii. Lotze had made a similar distinction in the nineteenth century, as had Comte, as we have noted. Lotze distinguished between *Genese* and *Geltung*, source and validity of ideas.

88. For the logical positivists, much of the work of philosophical analysis consisted of the analysis of logic, that is, of valid inference. The emphasis here is on the word *valid*: while the psychologist may be interested in any and all sequences of human thoughts, the philosopher is interested only in those passages from thought to thought that are logically valid. As Reichenbach (1947: 1) summarized the matter:

> Logic has often been defined as the science that deals with the laws of thought. . . . If we want to say that logic deals with thinking, we

> had better say that logic teaches us how thinking should proceed and not how it does proceed. . . . It is rather the results of thinking, not the thinking processes themselves, that are controlled by logic. Logic is the touchstone of thinking, not its propelling force, a regulative of thought more than a motive; it formulates the laws by which we judge thought products to be correct, not laws that we want to impose upon thinking.

For the logician, therefore, and the philosopher who wants to understand the edifice of scientific thought, the goal is not to model or understand the actual thought process that one or more scientists went through in reaching a conclusion. What matters, rather, is the rational reconstruction of that development of thought—a term introduced by Rudolf Carnap in 1928. As Reichenbach wrote,

> When we call logic analysis of thought the expression should be interpreted so as to leave no doubt that it is not actual thought which we pretend to analyze. It is rather a substitute for thinking processes, their rational reconstruction, which constitutes the basis of logical analysis. Once a result of thinking is obtained, we can reorder our thoughts in a cogent way, constructing a chain of thoughts between point of departure and point of arrival; it is this rational reconstruction of thinking that is controlled by logic, and whose analysis reveals those rules which we call logical laws. The two realms of analysis to be distinguished may be called context of discovery and context of justification. The context of discovery is left to psychological analysis, whereas logic is concerned only with the context of justification, i.e., with the analysis of ordered series of thought operations so constructed that they make the results of thought justifiable. (2)

Carnap had written something similar in *The Logical Structure of the World*:

> It must be possible to give a rational foundation for each scientific thesis, but this does not mean that such a thesis must always be discovered rationally, that is, through an exercise of the understanding alone. After all, the orientation and the direction of interests are not the result of deliberation, but are determined by emotions, drives, dispositions, and general living conditions. This does not only hold for philosophy, but also the most rational of sciences, namely physics and mathematics. The decision factor is, however . . . the *justification* of a thesis. (xvii)

89. Ferguson 1962: 285.

90. This and the quotations in the next two paragraphs are from Carnap 1937: 1–2.

91. While this book was translated by Amethe Smeaton (Countess von Zeppelin), the English words "transformation rules" and "formative rules" appeared in Carnap 1934, which was translated by the editor of the journal, William Malisoff. Malisoff was an academic and industrial chemist with deep interests in the philosophy of science. He was also a spy for the Soviet Union.

92. Carnap 1934: 9.

93. This quotation and the next are from ibid., 10.

94. Carnap 1937: 5.

95. Ibid., 7.

96. Ibid., 283.

97. Ibid., 8.

98. In volume 2, we will explore the ways in which Chomsky's *Logical Structure of Linguistic Theory* attempted to address just these questions.

99. Both of these quotations come from Bloomfield 1936: 89.

100. Herskovits came from a similar background as Greenberg did; he had been an undergrad at the University of Chicago, class of 1923, and then got his PhD with Boas in anthropology; he is in Boas's genealogy, in chapter 6.

101. This quotation and the next three are from Greenberg 1986: 6–8.

102. Harris 1951: 16.

103. Harris 1954: 146.

104. Harris 1990 [2002]: 4.

105. Chomsky, interview on December 13,1994, in Barsky 1997. Barsky continues:

> Discussing linguistics and philosophy in Chomsky's work, [linguist Carlos] Otero names German-born philosopher Rudolf Carnap as "the best known representative of the group of logical positivists"; he was to have "a direct and decisive influence on Chomsky's teachers," and was "the only non-American philosopher Chomsky read as a student." Carnap was deeply influenced by the work of Bertrand Russell, and made careful studies of Frege, Whitehead, and Wittgenstein, who were models for Chomsky, as well. (53–54)

106. Email, February 25, 2006.

107. Email from Putnam to J.G., October 5, 2006.

108. Email from Mattick to J.G., October 9, 2006.

109. Email from Ryckman to J.G., October 6, 2006.

110. Hiż 1994: 524–25.

111. Bar-Hillel 1964: 1.

112. Carnap 1937.

113. Bar-Hillel 1964: 2.

114. Ibid., 1–2.

115. This quotation and the four quotations following are from Sellars 1939: 41.

116. Regarding Lewis's impact on his student Quine, Isaac notes, "Quine's disagreement with Lewis should not mask his debt to his teacher, even though Quine himself was not entirely aware of it" (2005: 223).

117. Quine 1970: xxiv.

118. Ibid.

119. Isaac 2005: 206. Isaac also noted,

> At a time when European intellectuals were flooding into the United States, the young Quine helped to ensure that analytic philosophy emerged in the American academy as a live tradition rather than a foreign body of doctrine. . . . Quine was one of a number of young philosophers and social scientists who moved to embrace synchronic and fundamentally ahistorical conceptions of human thought and action. Modelled on the natural sciences, guided methodologically by the tools of mathematical logic, and insistent upon the irrelevance of ethical questions to scientific inquiry, Quine's vision of philosophy eschewed historical or interpretive accounts of philosophical puzzles. (209)

120. Goodman continued to develop the material in that book in his courses at the University of Pennsylvania, and his questions about the nature of scientific induction would have a major impact on the thinking of an undergraduate in his class, Noam Chomsky, at the end of the 1940s.

121. Quine 1940: 5.

122. Ibid., 6.

123. Ibid.

124. Chomsky 1957: 51.

125. Quine 1940: 287.

126. Ibid., 7–8.

127. Friedman 1991: 506.

128. Rorty 1990: 173.

Chapter Eight

1. See Grattan-Guinness 1981 for discussion of this period in logic.

2. See Mancosu, Zach, and Badesa 2004 for an excellent analysis of the material covered in this section. See also Shapiro 2005.

3. Fraenkel and Bar-Hillel wrote in *Foundations of Set Theory*,

> In view of the severe restrictions that intuitionism imposes on mathematics it is not surprising that only a handful of mathema-

> ticians have been willing to accept the intuitionistic principles as far as the daily practice of mathematics is concerned. On the other hand, the study of formal properties of intuitionistic logic and mathematics has enjoyed popularity ever since Heyting's formalization in the thirties. The school of Hilbert and some other trends show already for some time full understanding for the basic attitude of intuitionism, as is apparent from the work of Clifford Spector and others. Moreover, there is the remarkably stimulating influence of intuitionism which, mainly in connection with recursion theory, has suggested a number of improvements in classical analysis. . . . In conclusion one can say that the fierce disputes bewteen formalists and intuitionists belong to the past. Although both sides stick to their fundamental principles, a mutual appreciation has developed, which has already begun to bear fruit. (1958: 274)

This Bar-Hillel is the same Yehoshua Bar-Hillel we have met several times, and he will be an important figure in volume 2. The situation, both historically and ethically, is complex; see Velupillai 2011 for a very striking account.

4. We will not discuss this in this book, but there are many excellent books on this subject, such as Li and Vitányi 2008.

5. See Demopoulos and Clark 2005 as well as other papers in Shapiro 2005. We have already discussed Bertrand Russell and Gottlob Frege in the context of their epistemological contributions.

6. There is an additional worry; what sort of objects is it that this principle of induction is a principle *of*? On the face of it, it would seem to be about *properties*; we just said that we use the principle in order to extend the property to all numbers. But what a property is turned out to be a very ticklish question.

7. The problem of *induction* in the empirical sciences is the question (often attributed to David Hume's discussion in his *Treatise of Human Nature*) of how we can go from a finite number of observations to a general statement which can lead to an infinite number of predictions; the problem of induction in the mathematical sciences is much the same: how can we build a proof that has only a finite number of steps, each finite in length, from which we can draw an infinite number of conclusions?

8. Posy 2005. Hesseling 2003: 52ff suggests that the source of the term *intuitionism*—and *formalism*, which Brouwer took as its opposite—is more complex than this. Very few writers in the last 50 years write as if they were aware of the sense of the word *intuition* in the philosophical context, judging from context, and examples of this go back even further. For example, Alan Turing (1936: 249) wrote, "All arguments which can be given [about a certain kind of computability] are bound to be, fundamentally, appeals to intuition, and for this reason rather

unsatisfactory mathematically." Brouwer could not have disagreed more, and it does seem that Turing is using the term *intuition* in a rather casual sense.

By contrast, Rodin 2014 takes the nature of mathematical intuition as a central question, with a sophisticated reading both of Kant and of how his notions evolved during the period we are looking at here. Rodin defends the non-Kantian position that mathematical intuition changes over time; for example, he sees Lobachevsky's generalization of Euclidian geometry as an extension of the geometric intuition that had governed mathematical discourse for 2,000 years.

9. Brouwer wrote, "And in the construction of these sets neither the ordinary language nor any symbolic language can have any other role than that of serving as a non-mathematical auxiliary, to assist the mathematical memory or to enable different individuals to build up the same set," in Brouwer 1913: 86. See also Posy 1974, who refers to Brouwer's position as "constructivity of the left."

10. Two good works that provide an introduction to this area are Yates 1968 and Magee 2001.

11. Despite its importance for our story at this point, the fact is that intuitionism is hardly ever defended by anyone today, at least as it was formulated during the 1920s. That remark is not intended to slight the significance of those working on constructivist approaches to mathematics, such as the late Errett Bishop; his constructivism shared much with earlier intuitionists, but developed mathematically in a different direction.

12. Van Dalen 1990: 18.

13. Ibid.

14. Detlefsen 2005.

15. The proponent of hard logic, the mathematician who wanted to make sure that a result was trustworthy regardless of whether there was a conscious human making sure that each step had been checked for quality—that person was one who wanted to develop the notion of the algorithm. Algorithms are completely explicit statements of the steps that must be taken to accomplish the algorithm's goal, so explicit that a machine can follow them.

Not all mathematicians felt the attraction of thinking of mathematics that way. Henri Poincaré, a Frenchman who was one of the world's greatest mathematicians during the time we are looking at, wrote about that view of mathematics—at least as he saw it:

> Thus it will be readily understood that in order to demonstrate a theorem, it is not necessary or even useful to know what it means . . . we might imagine a machine where we should put in axioms at one end and take out theorems at the other, like that legendary machine in Chicago where pigs go in alive and come out transformed into hams and sausages. It is no more necessary for the mathematician than it is for these machines to know what he is doing. (1908: 157)

16. Wagner 1998; Dauben 1990.

17. Hilbert 1918 [1970]: 9. Ferreirós 2009 points out that Hilbert was himself a logicist, or very close to being one, throughout the first two decades of the twentieth century. He saw arithmetic as part of logic, though geometry was not. On Hilbert and Brouwer, much has been written; we have found Posy 2005 especially helpful here.

18. Hilbert 1927 [1967]: 464–65.

19. See the excellent account in Corry 1997.

20. Hilbert 1899.

21. See Mancosu 1999 for a more nuanced account of the intellectual context which Hilbert was instrumental in shaping and which influenced him in turn.

22. See Mancosu 1999: 320, whose translation we give here.

23. Ibid.

24. Ibid.

25. Ibid.

26. Ibid.

27. Ibid.

28. On Gödel, see Cassou-Noguès 2007.

29. See Janik and Toulmin 1973.

30. Atten 2006b.

31. This should not be read as a claim that Gödel was an intuitionist.

32. See Cassou-Noguès 2007; citations about Gödel whose source is not indicated are from this book.

33. Cassou-Noguès 2007: 55, quoting archives. On teleology, for Gödel: "The theological view of the world is separate from that of the sciences, in that everything has a sense . . . , that is to say, everything is done on purpose (*beabsichtigt ist*)" (Cassou-Noguès 2007: 184). He also wrote, "What I call the theological view of the world is the idea according to which the world and everything in the world has a sense and a reason." Letter to his mother, Gödel 1995: 4.439.

34. Wang 1990: 32.

35. Cassou-Noguès 1997: 85.

36. Gödel, in Wang 1996: 233

37. Godel 1995: 3.353.

38. Wang 1996: 169; the next quote is from page 166.

39. Atten 2015.

40. Wang 1996: 170.

41. Cassou-Noguès 2007: 95, citing Gödel 1995: 2.268. Parsons 1995 has an interesting discussion of Gödel's view in this regard, and he underscores the importance of remembering that in the context of discussions of Kant's view and Gödel's views, we need to bear in mind that the term "intuition" is rarely a good translation of the German word "*Anschauung*."

42. *G* takes the form *(for-all x)not(x proves y)*, where *y* is the Gödel number of *G*, and where *x* is a sequence of expressions in the logical system that forms a valid proof and whose conclusion is *y*: so it says, no matter what valid sequence of logical steps you take, you will not be able to provide a valid proof that the statement whose Gödel number *y* is true. We can see that that means that there is no proof within the system that *y* is true.

Figuring out how to make that happen is complex, but it can be done. Still, if you can prove that statement, then you have a valid sequence of steps whose conclusion is the statement whose Gödel number is *y*, and so *G* is false. And similarly, proving *G* false amounts to providing a valid sequence of steps that does form a proof of *G* (i.e., of the statement whose Gödel number is *y*), and so we have a proof of *G*.

Making the proof work is done in two steps, because it is not at all obvious how to get a statement to make a statement about itself. The process of getting to *G* includes a stage in which we build a function *F* (which takes a single argument) whose Gödel number is *f*, and which takes the form, *(for-all x) not(x proves z)*. So far, nothing is self-referential, but then we place the number *f* into the position of *z*, and the amazing thing is that the sentence becomes self-referential because the Gödel number has changed from what it was before (*f*) to the correct Gödel number of the new statement that has *f* instead of *z*.

Worse yet, we humans can read the sentence and understand what it means, and we can see that it is in fact *true*. So Gödel's conclusion is that there are true statements within any logical system that has a certain formally definable kind of logical consistency, is recursively enumerable, and is rich enough to contain arithmetic; there is a statement in the system which can neither be proven nor disproven within that system.

Gödel's construction is complex, and it requires that the underlying logical form include first order logic, which is to say, the ability to quantify over variables.

43. The reader will recall our discussion in chapter 1, and the long discussion on this at the end of chapter 4. On mechanism, see also Gerring 2010.

44. See Gandy 1988 on the confluence of ideas in 1936, and their background. Robin Gandy was Alan Turing's only graduate student.

45. See Siegmund-Schultze 2001.

46. Church 1932: 346.

47. Cardone and Hindley 2006. Barendregt 1981 is an excellent and standard introduction to the lambda calculus.

48. See volume 2 for further discussion of the integration of semantics into American linguistics.

49. Soare 1996 has become in recent years the standard place to refer for discussion of this point.

50. Church 1935: 332–33.

51. While there is self-reference, there is no recursion here: one of us (J.G.) is Bob Soare's colleague, at the University of Chicago.

52. Daylight 2011: 1756.

53. The term has come back into vogue in recent years among scholars who are interpreting what it is that Chomsky has meant by the term in such publications as Hauser, Chomsky, and Fitch 2002. There they write of recursion that it "provide[s] the capacity to generate an infinite range of expressions from a finite set of elements" (1569), and they "suggest that the FLN [the faculty of language in the narrow sense]—the computational mechanism of recursion—is recently evolved and unique to our species . . . we propose in this hypothesis that FLN comprises only the core computational mechanisms of recursion as they appear in narrow syntax and the mappings to the interfaces" (1572). The discussion does not address the difference between recursion, on the one hand, and other computational methods of generating an infinite set of strings or structures, and the next sentence does not seem at all correct: "By allowing us to communicate an endless variety of thoughts, recursion is clearly an adaptive computation"—recursion, being a particular way of dealing with infinite sets of generated forms, can make very special hardware demands compared with iterative computations (1574). "Even novel capacities such as recursion are implemented in the same type of neural tissue as the rest of the brain and are thus constrained by biophysical, developmental, and computational factors shared with other vertebrates" (1574). And they end with these questions:

> Why did humans, but no other animal, take the power of recursion to create an open-ended and limitless system of communication? Why does our system of recursion operate over a broader range of elements or inputs (e.g., numbers, words) than other animals? One possibility, consistent with current thinking in the cognitive sciences, is that recursion in animals represents a modular system designed for a particular function (e.g., navigation) and impenetrable with respect to other systems. During evolution, the modular and highly domain-specific system of recursion may have become penetrable and domain-general. This opened the way for humans, perhaps uniquely, to apply the power of recursion to other problems. This change from domain-specific to domain-general may have been guided by particular selective pressures, unique to our evolutionary past, or as a consequence (by-product) of other kinds of neural reorganization. Either way, these are testable hypotheses, a refrain that highlights the importance of comparative approaches to the faculty of language. (1578)

From our perspective and for the reasons that we develop in this book, Hauser, Chomsky, and Fitch 2002 fail their reader, and perhaps themselves, in not distinguishing quite different senses of the term *recursion*. Neither Hauser nor Fitch are computer scientists, and while Chomsky is not either, his involvement in the

1950s in the development of our understanding of formal methods of computation of strings was deep enough for readers to take for granted that he was using the term *recursion* in a knowledgeable way, which it does not appear to us that they were.

54. We discussed in chapter 4 the way in which the mathematician Ada Lovelace had seen this coming nearly a hundred years earlier; writing about the Analytical Engine, Babbage's dream that was not realized, she noted, "In enabling mechanism to combine together general symbols in successions of unlimited variety and extent, a uniting link is established between the operations of matter and the abstract mental processes of the most abstract branch of mathematical science" (Hyman 1989: 273). On Turing's psychology, see Shanker 2005.

55. Petzold 2008.

56. See Urquhart 2008, a presentation of Post's life and work. There is little discussion in the literature of Post's impact on linguistics. Just a few scholars over the last 15 years have discussed the roots of generative grammar in Post's work, notably Geoffrey Pullum (together with Barbara Scholz), in Pullum and Scholz 2005 and most recently Pullum 2013, and Tomalin 2006, which is reviewed in Scholz and Pullum 2007.

57. Post used the male-gendered pronouns where we might not today.

58. One thing that disappeared from the Turing machine in this process was Turing's decision to keep the odd-numbered boxes on his tape restricted to a different use than the even-numbered boxes. Turing kept the odd-numbered boxes as a scratch pad, so to speak, for his machine to keep tabs on work it had done. Post did not, and there is hardly a description of Turing's machine today that remembers this fact about Turing's exposition and architecture.

59. Post 1944: 285.

60. Ibid., 286.

61. Rosenbloom 1950: 162–63.

62. Gardies 1986 offers a good discussion of some of the topics here, sensitive both to the philosophical origins and to the more contemporary linguistic questions. See also Feferman and Feferman 2004. Tomalin 2006 discusses this material, emphasizing the importance of it for understanding the roots of contemporary linguistics.

63. There is much more to be said about the genealogy than we present here; see McMahon 1983 for a lengthy discussion of the question as to whether Montague invented Montague grammar, and where should we see the real beginning of categorial grammar. He suggests that Aristotle would not be a bad candidate, but that if we stay closer to the present, Husserl would be an excellent candidate, especially as seen through the work of Leśniewski. He also suggests that Yehoshua Bar-Hillel's role included drawing Richard Montague's attention to categorial grammar in the first place.

64. Smith 1988.

65. See especially Smith and Mulligan 1982 on the logic of mereology and its history.

66. Leśniewski 1929, cited in Gardies 1986.

67. Leśniewski 1992: 19, cited in Casadio 2002.

68. Hiż connected this to Tarski's work as well: "Tarski, in his book on the concept of truth . . . followed Leśniewski's theory of semantic categories and gave a recursive definition of a sentence in the language of the theory of classes, which is the model of what nowadays is called a generative grammar: a recursive definition of the concept of a sentence in a language. Tarski undoubtedly influenced both Harris and, following him, Chomsky. But Tarski maintained that this kind of definition is impossible for the concept of sentence in a natural language, because words are ambiguous and because a natural language contains its own metalanguage. According to him, this last-mentioned property precludes any consistent semantics of a natural language" (1998: 74).

69. Wood 1993.

70. Ajdukiewicz 1967b: 207.

71. The paper was not published in English until 1967 (as Ajdukiewicz 1967a and 1967b), though a manuscript of an earlier translation seems to have circulated in the 1950s; we have an underground version of it that was circulated at the University of Chicago. It appeared in German (as Ajdukiewicz 1935) in 1935. When we look at syntactic models in the 1940s and 1950s in volume 2, we will look at Bar-Hillel's efforts—largely successful—to bring new interest to this paper of Ajdukiewicz. Bar-Hillel also noted in a footnote that there was an English translation that "appeared at the College of the University of Chicago, March 1951." Perhaps he was responsible for the translation? It's very possible, but we will likely never know.

72. Essentially Kotarbiński's term.

73. Ajdukiewicz 1967a: 640.

74. Ibid.

75. Ibid., 641.

76. The example in question is translated differently in the two English translations, Ajdukiewicz 1967a and 1967b.

77. Cited in Wood 1993.

78. On Tarski and many more logicians, see the highly readable Feferman and Feferman 2004.

79. Hiż 1998: 75.

80. Ibid.

81. We can see one harbinger of this change in the temporality of mathematics in the development of the delta/epsilon interpretation of differentiability and continuity in analysis. The reader may recall that a function $f(x)$ is said to be continuous at a point p in modern parlance if for every value of epsilon ε, there is a value of delta δ (which depends on ε, normally), so that when the distance between

p and x is less than δ, the difference between the value of $f(x)$ and $f(x+\delta)$ is less than ε. Mathematics allows the mathematician to bargain with it: tell me how close you want your function to be to a target value, and I will tell you how close your input variable has to be to p. This back-and-forth is reminiscent of long division, but it now is part of the mathematical proof itself. The "delta/epsilon" conception of what had been called "infinitesmals" was first developed by Augustin-Louis Cauchy, a distinguished French mathematician, and Bernard Bolzano, a Czech philosopher greatly influential after his death, but was put on firm foundations by Karl Weierstrass, who was Edmund Husserl's dissertation advisor, and one of Georg Cantor's teachers as well.

82. In a sense, it is the opposite of the epsilon/delta strategy alluded to in the preceding note.

Chapter Nine

1. The work of Patrick Sériot has been central in the development of Western scholarship studying the larger historical and intellectual context of Trubetzkoy and Jakobson, and our understanding is heavily indebted to his work. See notably *Structure et totalité* (Sériot 1999) or the English version (Sériot 2014). We are grateful to him for his comments on this chapter as well. Flack 2007 provides an excellent overview of recent perspectives. See also Comtet 1995.

2. Trubetzkoy's book is more often referred to by its German title—*Grundzüge der Phonologie*—or just by that phrase's first word. We have chosen, with some regret, to use the English translation here.

3. Our understanding of the the work of Trubetzkoy, Jakobson, and the Prague linguists has been reshaped by the translation of a good number of original materials in Russian, and by a number of critical analyses. In addition to the specific citations in this chapter, we would like to note the important background information that comes from the following sources: Trubetzkoy 1950, 1975, 1991, 1996; Florovsky et al. 1921; Sériot 1999; and Toman 1995. Most of our reading of Trubetzkoy's letters has been through Sériot's translation, Trubetzkoy 2006. Jakobson published an earlier edition with an introduction in English (Trubetzkoy 1975), but none of the letters were translated in his edition. Sériot's edition is a translation into French, and most of our references to the individual letters should be understood as references to Sériot's translation.

For critical essays, we are especially indebted to: Adamski 1992; Bassin, Glebov, and Laruelle 2015; Laruelle 2006a, 2006b, 2008; Pomorska 1987; Sériot 1993, 1994, 1999, 2014; Tchougounnikov 2009; Toman 1995, 1997b.

4. See Sériot 1999, 2014.

5. Glebov 2003: 22.

6. Sériot 1999: 9.

7. Jakobson and Pomorska 1983: 2–3.

8. Introduction to Sériot 1999; see also Toman 1995: chap. 10.

9. Trubetzkoy 1969: 311.

10. Trubetzkoy would remain in touch both with Bloomfield and Tesnière. In his final book, *Principles*, the Hopi data are drawn from Bloomfield, and both Sapir and Whorf are cited. See his letters (#95, #142 in Trubetzkoy 2006) to Jakobson; in the latter he refers to his classmate from Leipzig. As far as we can tell, Trubetzkoy was not influenced by Bloomfield; two years after Bloomfield's *Language* had come out, he wrote to Jakobson, "it is hard to read without knowing the language, when every third word has to be looked up. . . . Perhaps there is something useful in it," he remarked. When they spent time together in Leipzig, they undoubtedly spoke German.

11. See the letter to Doroszewski November 27, 1931, cited by Albano Leoni 2013; Trubetzkoy 1975: 273. We are reminded of an observation of Sigmund Freud in an essay on forgetting:

> But the provoking part of it all is the fact that there is scarcely anything to which I am so hostile as the thought of being some one's protégé. What we see of this sort of thing in our country spoils all desire for it, and my character is little suited to the rôle of a protected child. I have always entertained an immense desire to "be the strong man myself." (1901 [1915]: 157)

12. Fortunatov was a generation older than Trubetzkoy. He was born in 1848, and was of the group of young people who were drawn to the linguistics community in Leipzig in the early 1870s—he attended lectures by Curtius and Leskien there, as well as those of Bréal in Paris. He died in 1914. Jakobson recalled later that Shakhmatov was "a great scholar, a great friend of the young people and perhaps the most noble man that I have ever known" (Jakobson and Georgin 1978: 13). It was Shakhmatov that allowed the Moscow Linguistic Circle to be established. See Toman 1995: 57–58.

13. For an analysis of the genealogy of ideas linking Trubetzkoy, Jakobson, and the Eurasianists, see Gasparov 1987.

14. See note 4 of letter 60 in Trubetzkoy 1975.

15. Trubetzkoy 1975: 3–4, cited in Torbakov 2008: 9.

16. This paragraph is based on Chevalier 1997.

17. Quoted in Chevalier 1997: 33, citing Tesnière 1936.

18. After World War II, Menghin finished his career peacefully in Argentina, as an archeologist. See Demoule 2014: 221.

19. Sériot 1999 is a major source on Eurasianism, especially in connection with linguistics. The development of the Eurasianist identity was the work of a number

of Russians and Ukrainians: Nikolai Trubetzkoy, Peter Savitsky, George Vernadsky, Pyotr Suvchinsky, Georgi Florovsky. Vernadsky joined the History Department at Yale University in 1927, and remained there until his retirement in 1956. See also Trubetzkoy 1920; Laruelle 2001, 2006a, 2008. A reader who is aware of events in contemporary Russia will have heard of Eurasianism in the twenty-first century. Eurasianism has indeed been reborn in recent years, notably under the direction of Alexander Dugin (1962–), as a strongly anti-American political force in Russia. Lev Gumilev (1912–92) and Alexander Panarin (1940–2002), notably under the direction of of Edouard Limonov (1943–) and Alexander Dugin, have developed the National Bolshevic Party, and the Eurasian Youth Movement. These movements are founded on antidemocratic, antiliberal, and unsurprisingly, anti-American principles. They do recognize a connection to the Eurasianism of Trubetzkoy and Savitsky.

20. Trubetzkoy 2006: 39, 47–51.

21. Trubetzkoy 1975: letters 3 and 4.

22. Trubetzkoy 1920, 1996.

23. Sériot 1999: 52.

24. Trubetzkoy 1996: 118.

25. Ibid., cited in Sériot 1999: 50.

26. Peter Savitsky was five years younger than Trubetzkoy and an important figure in the development of modern geography. A Russian who had studied at the Polytechnic Institute in St. Petersburg, he emigrated to Prague in 1922, was the director of the Russian lycée there, and also participated as a member of the Prague Linguistic Circle. Suvchinsky was a music critic, and Florovsky was a theologian and intellectual historian. For an overview of the movement, see Riasanovsky 1964.

27. Florovsky et al. 1921 [1996]: 1–2.

28. Trubetzkoy 1921: 88–89, given in French in Sériot 2014: 48. See also Trubetzkoy 1921: 81–100.

29. Joseph de Maistre lived through the experience of the French Revolution. De Maistre was from Savoie (at the time, an independent state), and he became an exile from his home after the Revolution of 1789 in Paris. He followed the same route in exile as Trubtezkoy, but in the opposite direction: he spent 14 years in St Petersburg as the ambassador of Victor-Emmanuel the first. Much as was true for Edmund Burke, who was the other critical figure in the conservative movement of that time, it was his radical opposition to the Revolution of 1789 which was the starting point and always the crucial element in his thinking.

30. Trubetzkoy did not mince his words—not that we would expect him to. In 1920, he wrote,

> How should one fight against the nightmare of an ineluctable Europeanization? It would seem at first sight that the only possi-

bility is a general uprising against the Romano-Germans. If humanity, not that one that the Romano-Germans like to talk about but the real humanity, constituted for the most part by Slavs, Chinese, Hindus, Arabs, Blacks and other peoples, who all, regardless of the color of their skin, moan under the yoke that the Germano-Romans impose on them and spend their national energy on extracting primary materials needed by European factories, if all this humanity could unite in a common struggle against its oppressors, it would succeed sooner or later in removing the much hated yoke and eliminating these predators and all their culture from the surface of the earth. (In French in Sériot 1999: 42n5; in English in Sériot 2014: 33n33)

31. Adamski 1992: 63.

32. Presentation of "L'élément panslave dans la culture russe," in Trubetzkoy 1996: 8–9.

33. The concept of *sobornost* (translated often as "togetherness") was a central idea of Russian religious philosophy. It reflects both a celebration of the collective nature of human society and the fascination of philosophers with the Orthodox Church (*sobor* = cathedral). Glebov 2003: 29.

34. Cited in Sériot 1993: 6. As Sériot observes, "Trubetzkoy despised democracy and had great expectations of countries with a single party that embodied the living idea of the people and the nation (Fascist Italy and Soviet Russia)" (1993: 89). On this, see also Laruelle 2001.

35. A word is in order about the term *Turanian*, rarely used in scholarship today; we are more accustomed to say "Ural-Altaic." The term is due to Lorenzo Hervás, who used it in his *Catalogue of the Languages of the Known Nations* (1800–1805) (Sériot 2014: 41). In Iranian mythology, "Tur" was given "Turkestan" as his territory. The term was picked up by at least some linguists; Max Müller used it, distinguishing between the Aryan languages (our Indo-European), the Semitic languages, and the "Turanian," which he divided into a northern and a southern group.

36. Florovsky et al. 1921 [1996]: 101.

37. In an paper written by Trubetzkoy in 1936, now appearing in Trubetzkoy 1996: 213.

38. Trubetzkoy 1923. Both of the quotations in this paragraph are found in the French version on p. 117. Toman 1995 discusses it as well (203–6).

39. Trubetzkoy 1996: 188–89.

40. The Soviet intelligence agency developed and pursued the strategy that has come to be known since then as "false flag operations"—operations in which foreign nationals are induced to engaged in a project after they have been approached by an intelligence officer who identifies himself as a representative of a *different*

(that is, false) country or interest. The story of the demise of Sidney Reilly, the most famous spy of the twentieth century, is closely connected to this. See also Sériot 1999: 37.

41. Laruelle 2006b.

42. Sériot translated Trubetzkoy 1996: 192–213 into French.

43. Glebov 2003: 23.

44. Trubetzkoy 1996: 200. It also appears in English in Trubetzkoy 1991.

45. Sériot 1999: 55.

46. Trubetzkoy 1991, in Sériot 1999: 50.

47. One Russian thinker who was critical of a mechanistic account was Nikolai Danilevsky (1822–85), who was an early influence on Trubetzkoy. It was Danilevsky who first developed the notion of "Romano-German" influence on Russia, which had its distinct Slavic character, the idea that so appealed to the Eurasianists. Glebov 2015 has been an important source for our presentation in this area, and also Vucinich 1989.

His refutation of a principle of external causality permitted him to defend the notion that change is oriented in a goal-like fashion, and this was pursued by Lev Berg (1876–1950), whose work was later cited by Roman Jakobson in the context of refuting the application of mechanistic darwinism by Neogrammarians.

48. Citations from Jakobson and Pomorska 1983: 64–65.

49. Sériot 2014; Jakobson 1927 [1962]: 17.

50. Jakobson 1927 [1962]: 110. Jakobson's allusion to nomogenesis concerned the anti-Darwinist biologist Lev Berg. As Popov (2008: 367) observes:

> However in the 1920s orthogenesis turned out to be at the foreground of Russian evolutionary biology. Several scientists claimed on the support of this view on evolution, and the special concepts of directed evolution were elaborated, which turned out to be significant for the following development of Russian biology. Two Russian biologists—ichthyologist and geographer Lev Berg and palaeontologist Dmitry Sobolev—elaborated them independently from each other. The development of the directed evolution concept was an essential part of biology in Russia in the XXth century.

51. Jakobson 1927 [1962]: 110.

52. But it is obvious why Jakobson thought the way he did: he was interested in his own quest and his own questions, and he did not feel he had the time to figure out what other people's questions in the past might have been. See Jakobson 1928b [1962]: 5.

53. He makes that point in the passage we are discussing here, Jakobson 1927 [1962]: 109, but this is just one of many places where Jakobson expresses this view.

54. Ibid., 109.

55. Ibid., 17.

56. Jakobson 1928a [1962]: 1.

57. Jakobson 1927 [1962]: 17.

58. Jakobson 1928b [1962]: 6.

59. Trubetzkoy 1939a: 244–45.

60. See Jakobson's foreword to Trubetzkoy 1975: 9.

61. As Jakobson noted; see also letter 41 from Trubetzkoy 1975.

62. The original is in Trubetzkoy 1975: 117, including this quotation and the immediately following quotation.

63. Letter 130, dated May 1934, from Trubetzkoy to Jakobson in Trubetzkoy 1975.

64. Trubetzkoy 1939b.

65. Albano Leoni 2013. Albano Leoni pushes hard the question as to whether Trubetzkoy and Jakobson *really* understood and totally accepted Bühler's ideas. A person who is brimming with his own new ideas never totally understands someone else's thought and certainly never totally accepts someone else's point of view.

66. Trubetzkoy 1975, letter 60, dated May 27, 1930.

67. This is from Trubetzkoy 1936: 8, and the next two quotations are from the previous page.

68. Bühler 1934: 52.

69. This paragraph and the following two draw directly on Jakobson's prose.

70. Letter 73, dated January 8, 1931; letter 74, dated January 29, 1931; letter 130, dated May 1934; and letter 111, dated October 10, 1935, in Trubetzkoy 1975.

71. Trubetzkoy's letter 137, dated June 25, 1935, 363–34. This is a fascinating letter that Trubetzkoy wrote to Jakobson, and to which we will return.

72. Jakobson 1971b: 555. Roman Jakobson cast this motto on Terence's famous quotation from his play Heauton: "Homo sum, humani nihil a me alienum puto", or "I am a human being, nothing human is alien to me."

73. Vanchura's widow in Jakobson and Pomorska 1983: 177.

74. The presence of the Jakobson family in Moscow was a sign in itself of its affluence and social status. Since 1743, the Jews of the Russian Empire were in principle not permitted to travel within the country and were in particular confined to a zone on the periphery of the country, with only a number of exceptions. See Poliakov 1965.

75. We base this principally on the autobiographical remarks published by Jakobson: Jakobson 1997 and Jakobson, Pomorska, and Hrushovski 1980.

76. The Kagan, Brik, and Jakobson families were "Russian Jews originally from Courlande [an area on the Baltic, just south of Riga], affluent members of the bourgeoisie who traveled a great deal and took their children with them abroad, to Venice, Paris, Germany. As children they learned French and German. The house was filled with books and musical instruments. The parents were very interested in the course of painting. It was an unusual milieu, open to the circulation of foreign cultures and new ideas, and an ease with regard to adaptation" (Robel 1978: 35).

77. Overall the best source on this period is Jakobson's recounting (Jakobson 1997). A romantic tale of the Russian Revolution could find no finer, or more complex, story to tell than that of the circle of friends that included Jakobson, Mayakovsky, and Elsa and Lili Brik. Mayakovsky was a flamboyant *artiste*; when he met Elsa's sister Lili, who was married to Ossip Brik, he fell madly in love with Lili. Elsa Brik's career as a writer in France is well-known, where she was known as Elsa Triolet; she was very close to Louis Aragon, with whom she lived, and both were deeply involved in the communist movement and the Third International.

78. Jakobson 1997: 54, 59ff. The group would become the center of an important artistic and poetic movement in Moscow. Jakobson was the heart of this group, and also close to Kasimir Malevich, one of the founders, along with Piet Mondrian and Vassily Kandinsky, two of the most influential contributors to abstract painting.

79. On Husserl in relation to Jakobson, see Holenstein 1975 and 1976, and also Albano Leoni 2013.

80. See also Toman 1995: 28–29; this is an important reference on this era. Toman notes that Jakobson recalled Chelpanov in a statement in 1974; Chelpanov introduced him to Gestalt psychology, but more important, in this seminar, was "Husserl's psychology, a discipline that impressed and influenced students in Moscow in those days very much" (21). Holenstein 1976 notes that Jakobson presented the material on Steinthal and on Husserl in this seminar. In 1911, Chelpanov himself had visited the major psychology labs in the United States—the University of Chicago, Harvard, Yale, Columbia, Cornell, and so on.

81. Dennes 1997. In Moscow in 1914, Shpet published an influential book on Husserl's philosophy. See also Seifrid 2005.

82. This and the following quotations are from Jakobson and Pomorska 1983: 11 On Husserl and Jakobson, see also Albano Leoni 2015.

83. Jakobson 1971a: 527–28. See also Gordin 2006. The Circle was officially recognized in March 1915, though the name "Moscow Linguistic Circle" seems to have first appeared in the fall of 1918, according to Toman 1995: 48. See also Costantini 2010.

84. See also Toman 1995: chap. 3.

85. Holenstein 1975.

86. Jakobson 1973a: 14–15, which includes Jakobson's remark about Pos just below.

87. Trubetzkoy 1975.

88. Jakobson 1975: viii.

89. Jakobson 1982: 7.

90. Holton 1998.

91. Jakobson wrote, "Niels Bohr repeatedly insisted on the deep links that at present tie together physics and linguistics, to whose interrelation both of us de-

voted a joint MIT seminar at the end of the 1950s. The 'exigencies of relativistic invariance,' in Bohr's favored term, were intently discussed with respect to the search for and structure of the ultimate constituents of both the physical and the linguistic universe, the 'elementary quanta,' as they were termed in physics, and were picked up from physics by linguists" (1982: 8). We say that Jakobson was a "broker of ideas," but what we want to say is that he was a *passeur*, a French word which carries other connotations, such as being a go-between and a courier.

92. Jakobson 1997: 38.

93. Ibid., 45.

94. Our primary sources for information on this period are Lukes 1996 and Gulyás 2007.

95. "Constructive Efforts: The American Red Cross and YMCA in Revolutionary and Civil War Russia, 1917–24," a PhD dissertation (2012) written by Jennifer Ann Polk at the University of Toronto, gives a thorough account of the period.

96. Jakobson reminisced about these times in Prague in his *My Futurist Years* (1997). He recalled leaving the mission, saying, "even when I was working for the Mission, I was fed up with my position there" (88). He was friends with a certain Levin, who was Hillerson's aide, and Levin told him that Lenin's foreign minister, Georgy Chicherin, had instructed the first Envoy to ask Jakobson to work on a freelance basis with him. Jakobson recalls working for him until 1928, suggesting that he was no longer a real member of the staff during those years.

Curiously, Jakobson appears for just a brief moment as a very marginal character in another book, *Stalin's Romeo Spy*, by Emil Draitser (Evanston, IL: Northwestern University Press, 2010), a book about Dmitri Bystrolyotov, a Russian spy, who was in Prague in the mid-1920s. The author notes in footnote 6 of chapter 5 that Jakobson "is listed as member of the Soviet Trade Mission staff (as a correspondent) in 1924 (CSAstaff)." Bystrolyotov was a member of the same social circle as Jakobson, a fact that plays a role in the account given by Draitser of Dystrolyotov's espionage work in Prague.

That the Hillerson mission was not a real Red Cross mission was instantly an open secret in Prague. Alice Masaryk, who was Tomáš Masaryk's daughter as well as chair of the Czechoslovak Red Cross, published an open letter to the Soviet foreign minister, Georgy Chicherin, on the day that the mission arrived in Prague, explaining the ways in which the Soviet government had abused the Red Cross as a cover for inappropriate political purposes (as it had in Poland). Nonetheless, the mission settled into the Hotel Imperial in Prague. The Czech government was quite aware of the efforts that Hillerson's mission was making to establish an intelligence network in Prague (Lukes 1996: 8). Hillerson's mission relied on the Czech government's telegraph system to send encrypted messages to Moscow, but unbeknownst to the Russians, the Czech government was able to decypher the messages and follow the mission's activities. As Lukes noted, "Hillerson's main mission in Prague had to do with espionage, sabotage, and money and weapons transfers" (10), and it used its central location in Prague to bolster Soviet military efforts in Ukraine and Poland.

By January of the following year, 1921, the government in Prague had declared Hillerson *persona non grata*, and relations between Prague and Moscow failed to improve. After 1925, "the primary targets for the Soviet agents [GPU and GRU, the Soviet secret services] were the Russian émigrés. They tried to infiltrate, then create unrest, in their associations" (Gulyás 2007: 214).

Jan Nylund wrote that "Roman Jakobson came to Prague as a member of a mission whose task was to repatriate Russian prisoners of war. The mission was much criticized both because of its capacity of being a Soviet institution and because all the members of the mission were Jews, for anti-Semitism was widespread in Prague at this time. Jakobson left the mission after a few weeks and started to study at Charles University, though he was employed by the Soviet diplomatic representation until 1928. As late as 1929 Jakobson was still suspected of being a spy. A daily newspaper, *Národní listy*, wrote: 'Nobody is so naive in the whole of Czechoslovak republic as not to see quite clearly that Mr. Jakobson's Slavic activity in Prague is nothing but a disguise under which Mr. Jakobson fulfils his true mission—the mission of a communist agent.' Jakobson's connections with the German University in Prague gave him the opportunity to submit a thesis in 1930 to receive a doctorate" (Nylund 2013: 167).

Lesley Chamberlain notes that "Jakobson's life as an émigré was subject to the same political tensions as all arrivals from Russia. He came to Prague in July 1920 with an attachment to the Soviet Red Cross Mission in Prague, 'repatriating former Russian prisoners of war who had been stranded in Czechoslovakia since Austro-Hungarian times.' This made him look suspect in the eyes of the White community and their disapproval caused him to resign." Chamberlain quotes Jakobson, who wrote, "You ask me what I'm doing in Prague. I don't know if you know it or not, but in September [1920] I was strongly attacked here for my participation in the Red Cross Mission . . . the professors vacillated whether I was a bandit or a scholar or an unlawful mongrel; in the cabaret they were singing little songs about me---all of this was not very witty. The situation was complex, but it seems to me my fate is to tightrope-walk in inconceivable situations. As a result I have left work (without tears or cursing) and entered university scholarship and so on'" (Jakobson 1997: 117).

Chamberlain continues: "But after he resigned he went hungry and by the end of summer 1921 Jakobson was back in the Soviet mission in Prague, part-time, as 'a free-lance worker' until 1928. Nabokov claimed he was a spy" (ibid., 198–99). We will return to Nakobov and Jakobson in volume 2.

97. Trubetzkoy 1975.

98. Jakobson 1997: 89: "I already knew that I would leave the Mission. I could have left earlier if I had not had a lack of faith in my own abilities. For example, I couldn't bring myself to take the Ph.D. exam for a long time, even though I needed a Czech doctorate."

99. Halle 1979.

100. Jakobson 1962a: 633.

101. Toman 1997a.

102. Toman 1995: chaps. 4 and 5; Fronek 1988, who notes that Mathesius cites Masaryk's remarks on linguistics in his *Versuch einer concreten Logik*, and who emphasizes Masaryk's published ideas about static and dynamic perspectives on language, long before Saussure had developed his ideas. See also Percival (n.d.) and Fontaine 1997. Jakobson was also influenced by Marty; see Albertazzi, Libardi, and Poli 1995: 95.

103. Mathesius (1983: 11) wrote, "It is fair to state that the difference between static and dynamic linguistic problems was first clearly envisaged by the present writer when he was reading, during his university studies, T. G. Masaryk's remarks on linguistics in his *Versuch einer concreten Logik* (Masaryk 1887)."

104. Jakobson and Georgin 1978: 16.

105. Jakobson 1963a.

106. The intellectual universe was smaller and denser to a point which may be astonishing for us today. Roman Jakobson, who knew everyone of intellectual interest in that dense world, recalls that the name Einstein gave to his new theory, "relativity," came from Jost Winteler (1846–1929), a Swiss linguist and phonetician, a colleague of Saussure, in whose house a young student at the Polytechnic of Zurich named Albert Einstein had a room and with whom he shared conversations of an evening. Jakobson 1982.

107. Schrijnen 1928: 252.

108. Cited in Fontaine 1997: 155.

109. Jakobson 1928b [1962]: 4.

110. Jakobson, Pomorska, and Hrushovski 1980: 19.

111. Jakobson 1963a [1971]: 523.

112. Ibid.

113. Toman 1995: 152.

114. Percival n.d., 4.

115. Recall Paul's objection to calling the Neogrammarians a party in chapter 2.

116. Toman 1995: 155. We draw heavily on Toman's discussion here and immediately below.

117. Ibid., 156, who also cites Jakobson's "La scuola linguistica di Praga," which appears in Jakobson's *Selected Writings*, vol. 2, 539–46, and which originally appeared in *La Cultura, an Italian Review* in 1933.

118. Hajičová 2006: 457.

119. The late Stephen Rudy was the foremost authority on Jakobson's scholarship, and his documentation forms the heart of the Jakobson archive at the Massachusetts Institute of Technology library. Rudy's information tells us that Jakobson defended a dissertation entitled *Über den Versbau der serbokroatischen Volksepen* at the German-language University of Prague. A short paper bearing

the same name appears in volume 4 of Jakobson's *Selected Writings*, the volume devoted to Slavic epic studies. It is perhaps a summary of the most significant part of the dissertation, but it appears to be too short to be a dissertation. Patrick Sériot in correspondence has expressed skepticism that Jakobson would have been at the German (as opposed to the Czech) university. This is an interesting question, though not central to our point, and we leave this matter for others to clarify.

120. Letter 137 from Trubetzkoy, dated January 25, 1935, in Trubetzkoy 2006.

121. See Rudy 1990 and the Roman Jakobson Bibliografie site, http://comenius-bibl.wz.cz/Jakobson.html.

122. We have based these remarks on Viel 1984, the work of Stephen Rudy, and the site in the previous note. Between 1915 and 1945, there are 237 entries: 126 are devoted to poetry, folklore, art, philosophy, and quite general questions; 84 are popular or journalistic items; 27 are linguistic, and of these, 17 are about phonology. Of these 17, 3 are collective contributions for meetings. Between 1930 and 1938, there are 6 entries, as many as between 1926 and 1929, and there are 5 between 1939 and 1945.

The contents of Jakobson's *Selected Writings* indicates much the same. Jakobson collected what he took to be his phonological work in the first volume, and he added in 1962 a retrospective account. There are 39 entries in the volume. For 1926, there are 11 pages; for 1927, 42 unpublished pages; for 1928, there are two very short papers, and a book of 100 pages. Between 1930 and 1938, there are eight entries in the volume, often quite short or devoted to general topics:

- "Die Betonung und ihre Rolle in der Wort- und Syntagmaphonologie," Jakobson 1931a.
- "Über die phonologischen Sprachbünde," Jakobson 1931d.
- "La linguistique eurasienne," Jakobson 1931b.
- "Phonemic Notes on Standard Slovak," Jakobson 1931c.
- "Phoneme and Phonology," Jakobson 1932a.
- "Orthographe du vieux russe," Jakobson 1937b.
- "Über die Beschaffenheit der prosodischen Gegensätze," Jakobson 1937c.
- "On Ancient Greek Prosody," Jakobson 1937a.

During the period from 1930 to 1938—that is to say, the period during which Trubetzkoy was writing *The Principles of Phonology*—Jakobson's interests do not appear to us to be focused on phonology in the same way: 114 pages on phonology in eight years, and not on a central question in phonology.

As we step back and look at the dynamic of the work and the publication of Trubetzkoy and Jakobson, we see a burst of intellectual energy by Jakobson at the beginning of the Prague Circle years, with the development of an idea of phonological features and a commitment to a uniformitarian view of oppositions: oppositions are either correlative, in which case they are striking and theoretically interesting, or they are disjunctions. This work energized Trubetzkoy, and launched his theoretical work, but without Trubetzkoy ever buying in to the uniformitarian view of oppositions. Quite the opposite, in fact: for Trubetzkoy, much of the interest was in the *different* types of oppositions that could be explored.

It is all the more astonishing how intellectually productive Jakobson's years were during which he was virtually on the run and a political refugee, immediately following his departure from Czechoslovakia.

See the website comenius-bibl.wz.cz, which is empty between 1929 and 1941, except for *Kindersprache* in its different versions.

123. Jakobson 1962b: 147, based on Sériot's French translation.

124. Sériot 1993: 91. The chapter appears in Jakobson, Pomorska, and Hrushovski 1980, and Jakobson and Pomorska 1983: 85. In Jakobson's words (Jakobson and Pomorska 1983: 85):

> In the thirties I published a number of studies proving the existence of a vast "Eurasian linguistic alliance," which encompassed Russian, the other languages of Eastern Europe, and the majority of the Uralic and Altaic languages, all of which make use of the phonemic opposition of palatalized and nonpalatalized consonants. I also characterized in passing the Circumbaltic languages, which possess a phonemic opposition of two types of word intonation.

125. See, for example, Holms 2009. See also Kyllingstad 2014.

126. See Jangfeldt 1997: 142.

127. Jakobson and Halle 1964.

128. Toman 1997a.

129. Jakobson 1939 [1985]: 133.

130. Cited in Toman 1997b: 244.

131. Jangfeldt 1997. On this period, see Baecklund-Ehler 1977.

132. Jakobson cited Husserl 1900a, 1900b.

133. Jangfeldt 1997: 144; Cassirer 1981: 282.

134. All of the remarks of Ivanov's are found in Ivanov 1983: 48–49.

135. He passed away on October 7, 2017, as we finished this chapter.

136. Perhaps the only real exception to this is Michael Silverstein, our colleague at the University of Chicago.

137. From an article in Czech dated October 31, 1929, and cited in Jakobson 1971a: 711.This is undoubtedly the single most cited sentence of Jakobson's publications.

138. Recall the emphasis Husserl put on the character of the relation between a part and a whole, analyzed by the logic of mereology.

139. Trubetzkoy (1969: 67) remarked, "the determination of the phonemic content of a phoneme presupposes its **prior** classification in the system of distinctive oppositions existing in the language in question. . . . The phonemic inventory of

a language is actually only a corollary of the system of distinctive oppositions. It should always be remembered that in phonology the major role is played, not by the phonemes, but by the distinctive oppositions."

140. Cited in Fontaine 1997: 60.

141. The source of this passage is complicated, and it does not appear that Jakobson published it in an accessible location. It can be found in an article written by Peter Savitsky entitled "L'Eurasie révélée par la linguistique," published in *Le Monde Slave* 1931:364-70, which can be found at http://crecleco.seriot.ch/textes/Savickij31.html. On p. 370, Savitsky wrote that this was a text read by Jakobson to phonologists under the title of "Les unions linguistiques, spécialement phonologiques" on 20 December 1930.

142. This and the following quotation are from Jakobson 1928a [1962]: 1–2.

143. A more technical presentation of the ideas gently alluded to here would involve not energy but rather *action*, a quantity that is more general than the familiar terms of Newtonian physics such as acceleration and force. In classical cases, the action is the difference between the kinetic and the potential energy, and the world evolves in such a way that this amount (integrated over a given time period) is minimized (or better, that a very small change in the world's evolution has no first-order effect on the that integral). As physics has advanced, the precise characterization of action has evolved, but the idea has remained central that the variation on this integral that is called the action is zero—which amounts, roughly speaking, to the statement that a physical system finds an evolution in time that minimizes the total action. A good introduction to this can be found in *The Lazy Universe* by Jennifer Coopersmith (Oxford University Press, 2017).

144. Jakobson 1960.

145. This and the next two quotations are from Jakobson 1973a: 16–17.

146. See Holenstein 1976: 98.

147. Rota 1989 offers an interesting interpretation of Husserl's *Fundierung*, influenced considerably by his reading of Heidegger, in our opinion.

148. Jakobson 1927.

149. Jakobson 1963b [1971]: 582.

150. Holenstein 1976: 99–100.

151. Sériot 1999: 9.

152. Ibid., 22, letter of May 17, 1932.

153. Trubetzkoy 2001: 255.

154. Jakobson 1975: xii.

155. Trubetzkoy 2006: 176, letter to Jakobson, dated July 18, 1929.

156. See Fontaine 1997.

157. Jakobson and Georgin 1978: 15.

158. Jakobson 1929. The most easily accessible source for this is at http://crecleco.seriot.ch/textes/theses29.html.

159. Jakobson, Pomorska, and Hrushovski 1980: 44.

160. Jakobson and Pomorska 1983: 41.

161. Ibid., 25.

162. Jakobson, Pomorska, and Hrushovski 1980: 46; Jakobson and Pomorska 1983: 44–45.

163. On the French/German differences see Trautmann-Waller 2004 and Bergounioux 1997.

164. See Jakobson and Waugh 1979, and Saussure 1968: 100.

165. Saussure 1995: 64.

166. This quotation, and its continuation just below, is from Trubetzkoy 1933: 234.

167. Gadet 1995.

168. Ibid., 230.

169. Trubetzkoy 1938 [1969] : 4.

170. Trubetzkoy 2001: 15.

171. Trubetzkoy wrote at the beginning of his chapter on oppositions, "all these standpoints and principles of classification are valid not only for phonological systems but for any other system of oppositions as well. They contain nothing that is specifically phonological. . . . In [the preceding chapter] we operated with purely logical concepts. We must now combine these logical concepts with acoustic and articulatory, that is, with *phonetic*, concepts" (1938 [1969]: 90–91).

172. Trubetzkoy wrote in a letter to Jakobson in 1935, "It is correct that stylistic variants perform a certain unique function and must be isolated. However, it is false that phonology should study all such phenomena, and I was unable to dissuade [Laziczius]" (Trubetzkoy 2001: 245). Sometimes the opposition's point of view sheds the greatest light. Grammont, the defender of phonetics against whom we can think of Trubetzkoy as inveighing, spoke with dripping irony about phonologists: "Certain 'fonologists' take the term 'foneme' to mean 'the phonological unit' not insofar as it is emitted by the speaking subject, but insofar as it is understood by the listener. This is a surely a joke, for the great majority of listeners are totally incapable of saying what it is that they have heard and to be aware of it" (1933: 77).

173. Toman cites Karchevsky 1927 in connection with this idea. It may be that this idea was indeed first emphasized by Karchevsky, and not Trubetzkoy. As Toman points out, Karchevsky referred to Cassirer's *Substanz- und Funktionsbegriff* (1910) in the context of this discussion.

174. Trubetzkoy 1969: 67–68.

175. Jakobson 1975: xiii.

176. Henning Anderson (1989) looks at a number of approaches to markedness that were being discussed during this period, and notes that Jakobson does not include in his English translation of the Trubetzkoy's letter on markedness cases where Trubetzkoy allows for two opposite members to be equally active, but only in a case where three members are involved.

177. Pos 1938: 246.

178. Toman 1995: 19, 146.

179. Jakobson noted, in Jakobson and Pomorska 1983: 24–25:

> In the light of such considerations, my research on the phonological evolution of Russian and other Slavic languages led me toward the end of the twenties to recognize a special type of phonological relation which I have designated by the logical term *correlation*. This concept proved to be fruitful both for the description of sound systems and for the explanation of their historical evolution as well.

180. This quotation, and its continuation just below, are from Trubetzkoy 1969: 35.

181. In German: *Mal*. He attributes the term to Bühler on p. 10 of *Principles*.

182. Trubetzkoy 1969: 77.

183. This and the following quotation are from Jakobson and Waugh 1979: 92–93.

184. It may be that we read Jakobson's account in too naive a way. In the conversations with his wife that became *Dialogues* (Jakobson and Pomorska 1983), she begins one of the chapters asking him how the concept of the *mark* arose (93). In response, Jakobson explains how "from the beginning" his thinking had been in this area, and he refers to a book by Paul Verrier, a specialist of English verse, on "marked time" which had influenced his thought about rhythm and prosody. Jakobson's suggestion here is at best obscure, as far as we can see. He then skips ahead, with no transition, to refer to the letter from Trubetzkoy, and how enthusiastically he responded to it. There is more than one way to interpret this tale; it may be that Jakobson expects the careful reader to see that he had delved into the notion before Trubetzkoy had brought it up.

185. This and the two following quotations are from Jakobson and Pomorska 1983: 95.

186. Chapter 10 of Jakobson and Pomorska 1983 is devoted to markedness.

187. A third, and completely different, interpretation was sometimes suggested in the context of Gestalt psychology, where one of the characteristics of perception that was studied was the way in which the visual field will always be analyzed in terms of shapes and figures lying against a background (foreground vs. background). This does not have a natural connection, it seems to us, to Trubetzkoy's vision.

Chapter Ten

1. Readers of Michel Foucault will recognize that there is, in our remarks here, a conversation of a sort going on with Foucault. His concept of *epistémé* and his

notion of archeology of knowledge have been influential. To continue this secondary conversation the reader is referred to Foucault 1969.

2. Dennett 1991.

3. See Goldsmith 1990.

4. Alas, the story is probably apocryphal; it may have been Lacan's invention.

5. In some disciplines, command of several languages is taken for granted, but that is becoming less and less true in general.

6. Actually, we know that that is not quite right; linguistic sounds consist of a finite number of simultaneous discrete sequences of sounds.

7. One can occasionally find discussions of the role of free will in which the author seems to assume an equivalence between free will and random decisions, which seems to us an unreasonable assumption to make, but which makes sense in the context we are describing.

8. The principle that "free will blocks the chain of mechanism" is not one that will be found anywhere; it is a principle that seems to be at work as we read what many people write, and it explains what conclusions they draw regarding what is mechanical and what is not.

9. Davies 1978.

10. Hull made this point clearly, as we saw earlier.

11. There are linguists, it appears to us, who make an effort to have a positivist position in the sense just described, and who would like then to conclude that the theoretical concepts match objects in the object of study, that is, in language or knowledge of language. They want to have their cake and eat it too.

12. Some accounts attribute to him the view that analyzing linguistic data must be performed with a strict sequence of methods, with phonemics preceding morphophonemics, itself preceding morphology, and so on. Quite to the contrary, he consistently argued that such an ordering was impossible and unmotivated.

References

Adamski, Darius. 1992. "La personnologie du Prince Nicolas Troubetzkoy et ses développements dans l'oeuvre de Roman Jakobson." *Langages* 107: 62–68.

Ajdukiewicz, Kazimierz. 1935. "Die syntaktische Konnexität." *Studia Philosophica* 1: 1–27.

——. 1967a. "On syntactical coherence [only part 1 translated]." Translation of Ajdukiewicz 1935. *Review of Metaphysics* 20: 635–47.

——. 1967b. "Syntactic connection." Translation of Ajdukiewicz 1935. In *Polish Logic 1920–1939*, edited by S. McCall, 207–31. Oxford: Clarendon.

Albano Leoni, Federico. 2013. "Bühler et le Cercle linguistique de Prague." *Verbum* 31: 89–114.

——. 2015. "Les parties et le tout: Jakobson, Husserl et la phonologie." *Histoire Épistémologie Langage* 37: 27–42.

Albertazzi, L., M. Libardi, and R. Poli. 1996. *The School of Franz Brentano*. Berlin: Springer.

Allegre, Claude. 1985. *Ecume de la terre*. Paris: Fayard.

Alter, Stephen G. 2005. *William Dwight Whitney and the Science of Language*. Baltimore: Johns Hopkins University Press.

Amsterdamska, Olga. 1987. *Schools of Thought: The Development of Linguistics from Bopp to Saussure*. Dordrecht: Reidel.

Anderson, Henning. 1989. "Markedness theory—the first 150 years." In Olga Miseka Tomić, *Markedness in Synchrony and Diachrony*, 11–46. Berlin: De Gruyter.

Angell, James Rowland. 1907. "The province of Functional Psychology." *Psychological Review* 14: 61–91.

——. 1936. "James Rowland Angell." In *A History of Psychology in Autobiography*, vol. 3, edited by Carl Allanmore Murchison, 1–38. Worcester, MA: Clark University Press.

Annan, Noël. 1984. *Leslie Stephen: The Godless Victorian*. London: Weidenfeld and Nicolson.

Anschütz, G. 1911. "Über die Methoden der Psychologie." *Archiv für die gesamte Psychologie* 20: 414–98.

Arens, Katherine. 1989. *Structures of Knowing: Psychologies of the Nineteenth Century.* Dordrecht: Kluwer.

Ash, Mitchell G. 1989. "Max Wertheimer's university career in Germany." *Psychology Research* 51: 52–57.

———. 1995. *Gestalt Psychology in German Culture, 1890–1967: Holism and the Quest for Objectivity.* Cambridge: Cambridge University Press.

Aster, E. von. 1908. "Die psychologische Beobachtung und experimentelle Untersuchung von Denkvorgangen." *Zeitschrift für Psychologie* 49: 56–107.

Atten, Mark van. 2006a. *Brouwer Meets Husserl: On the Phenomenology of Choice Sequences.* Berlin: Springer.

———. 2006b. "Mathematics." In *A Companion to Heidegger*, edited by Hubert L. Dreyfus and Mark A. Wrathall, 585–99. Malden, MA: Blackwell.

———. 2015. *Essays on Gödel's Reception of Leibniz, Husserl and Brouwer.* Berlin: Springer.

Aurora, Simone. 2015. "A forgotten source in the history of linguistics: Husserl's *Logical Investigations.*" *Bulletin d'analyse phénoménologique* 11, no. 5. http://philpapers.org/rec/AURAFS.

Auroux, Sylvain. 1994. *La révolution technologique de la grammatisation.* Liège, Belgium: Mardaga.

Auroux, Sylvain, Gilles Bernard, and Jacques Boulle. 2000. "Le développement du comparatisme indo-européen." In *Histoire des idées linguistiques*, vol. 3, edited by Sylvain Auroux, 155–71. Liège, Belgium: Mardaga.

Ayer, A. J. 1977. *Part of My Life.* London: Collins.

Bacon, Francis. 1620. *Novum Organum.* 1889 edition, edited by Thomas Fowler. Oxford: Clarendon.

Baecklund-Ehler, Astrid. 1977. "Roman Jakobson's cooperation with Scandinavian linguists." In *Roman Jakobson: Echoes of His Scholarship*, edited by Daniel Armstrong and C. H. Van Schooneveld, 21–27. Lisse, Netherlands: Peter de Ridder.

Bailey, Richard W. 1992. "Hans Kurath." *Language* 68: 797–808.

Balkan, David. 1966. "Behaviorism and American urbanization." *History of the Behavioral Sciences* 2: 5–28.

Banner, Lois W. 2010. *Intertwined Lives: Margaret Mead, Ruth Benedict, and Their Circle.* New York: Vintage.

Barendregt, Hendrik Pieter. 1981. *The Lambda Calculus: Its Syntax and Semantics.* Amsterdam: North-Holland.

Bar-Hillel, Yehoshua. 1964. *Language and Information.* Reading, MA: Addison-Wesley.

Barrett, William, and Henry D. Aiken, eds. 1962. *Philosophy in the Twentieth Century: An Anthology.* New York: Random House.

Barsky, Robert. 1997. *Noam Chomsky: A Life of Dissent*. Cambridge, MA: MIT Press.

Bassin, Mark, Sergey Glebov, and Marlene Laruelle. 2015. *Between Europe and Asia: The Origins, Theories, and Legacies of Russian Eurasianism*. Pittsburgh: University of Pittsburgh Press.

Baudouin de Courtenay, Jan. 1870. "Some general remarks on linguistics and language." Reprinted in Baudouin de Courtenay 1972, 49–80.

——. 1889. "On the tasks of linguistics." Reprinted in Baudouin de Courtenay 1972, 125–43.

——. 1895. *Versuch einer Theorie phonetischer Alternationen*. Strasbourg: Trübner. Translated as "An attempt at a theory of phonetic alternations." Reprinted in Baudouin de Courtenay 1972, 144–212.

——. 1972. *A Baudouin de Courtenay Anthology*. Edited by Edward Stankiewicz. Bloomington: Indiana University Press.

Bell, David. 1990. *Husserl*. Cambridge: Cambridge University Press.

Ben-David, Joseph, and Randall Collins. 1966. "Social factors in the origins of a new science: The case of psychology." *American Sociological Review* 31: 451–65.

Benedict, Ruth. 1943. "Franz Boas: An obituary." *The Nation* 156: 15–16.

Beneš, Tuška. 2004. "Comparative linguistics as ethnology: In search of Indo-Germans in Central Asia, 1770–1830." *Comparative Studies of South Asia, Africa and the Middle East* 24: 117–32.

——. 2008. *In Babel's Shadow: Language, Philology, and the Nation in Nineteenth-Century Germany*. Detroit: Wayne State University Press.

Benoist, Jocelyn. 1997. *Phénoménologie, sémantique, ontologie: Husserl et la tradition logique autrichienne*. Paris: Presses Universitaires de France.

——. 2002. "Husserl et la fascination du 'formel.'" In *Les philosophes et la science*, edited by Pierre Wagner, 679–711. Paris: Gallimard.

Benveniste, Emile 1954. "Tendances récente en linguistique générale." Reprinted in *Problèmes de linguistique generale*, 1976, 3–17. Paris: Gallimard.

——. 1964. "Ferdinand de Saussure à l'Ecole des hautes études." *Sciences historiques et philologiques (École Pratique des Hautes Études), Annual report for 1964/1965*: 20–34.

Bergounioux, Gabriel. 1997. "La Société de Linguistique de Paris (1876–1914)." *Bulletin de la Société de linguistique de Paris* 92: 1–26.

——. 2002. "La sélection des langues: darwinisme et linguistique." *Langages* 146: 7–19.

Berlin, Isaiah. 1980. *Personal Impressions*. New York: Viking.

Bernard, Claude. 1865. *Introduction à l'étude de la médecine expérimentale*. Paris: Baillière et fils.

Binet, Alfred. 1903. *Etude expérimentale de l'intelligence*. Paris: Schleicher frères.

Bjork, Daniel W. 1983. *The Compromised Scientist: William James in the Development of American Psychology*. New York: Columbia University Press.

Blackmore, John. 1972. *Ernst Mach: His Life, Work, and Influence*. Berkeley: University of California Press.

———. 1998. "Franz Brentano and the University of Vienna Philosophical Society 1888–1938." In *The Brentano Puzzle*, edited by Roberto Poli, 73–92. Aldershot: Ashgate.

Bloch, Bernard. 1949. "Leonard Bloomfield." *Language* 25: 87–98.

Bloomfield, Leonard. 1914. *An Introduction to the Study of Language*. New York: Henry Holt.

———. 1922. "Review of *Language*, by Edward Sapir." *Classical Weekly* 15: 142–43.

———. 1924. "Review of Saussure 1922." *Modern Language Journal* 8: 317–19.

———. 1925. "Why a linguistic society?" *Language* 1: 1–5.

———. 1926. "A set of postulates for the science of language." *Language* 2: 153–64.

———. 1929. "Review of 'Liebich, B. (1928): Konkordanz Panini-Candra.'" *Language* 5: 267–76.

———. 1930. "Linguistics as a science." *Studies in Philology* 27: 553–57.

———. 1933. *Language*. New York: Henry Holt.

———. 1934. "Review of 'Havers, W. (1931): Handbuch der Erklärenden Syntax.'" *Language* 10: 292.

———. 1936. "Language or ideas?" *Language* 12: 89–95.

———. 1939. "Menomini morphophonemics." *Travaux du Cercle Linguistique de Prague* 8: 103–6.

———. 1943. Untitled note. *Language* 19: 198.

Bloomfield, Leonard, and Charles F. Hockett. 1970. *A Leonard Bloomfield Anthology*. Bloomington: Indiana University Press.

Bloomfield, Maurice. 1884. "On the probability of the existence of phonetic law." *American Journal of Philology* 5, no. 2: 178–85.

———. 1888. "The origin of the recessive accent in Greek." *American Journal of Philology* 9: 1–41.

———. 1908. *The Religion of the Veda; the Ancient Religion of India (from Rig-Veda to Upanishads)*. New York: Putnam.

Blumberg, Albert E., and Herbert Feigl. 1931. "Logical positivism." *Journal of Philosophy* 28: 281–96.

Blumenthal, Arthur. 1970. *Language and Psychology*. Hoboken, NJ: Wiley.

———. 1975. "A reappraisal of Wilhelm Wundt." *American Psychologist* 30: 1081–88.

———. 1979. "The founding father we never knew." *Contemporary Psychology* 24: 547–50.

Blust, Mark. 1996. "The Neogrammarian hypothesis and pandemic irregularity." In *The Comparative Method Reviewed: Regularity and Irregularity in Lan-*

guage, edited by Mark Durie and Malcolm Ross, 136–56. Oxford: Oxford University Press.

Boas, Franz. 1888. *The Central Eskimo*. Washington, DC: The Smithsonian Institution.

——. 1912. "Changes in the bodily form of descendants of immigrants." *American Anthropologist* 14: 530–62.

——. 1934. *Aryans and Non-Aryans*. New York: Information Service Associates.

——. 1940. *Race, Language, and Culture*. New York: Macmillan.

Boas, Franz, Maurice Fishberg, Ellsworth Huntington, and Max J. Kohler. 1934. *Aryan and Semite: With Particular Reference to Nazi Racial Dogmas*. New York: B'nai B'rith.

Boeckx, Cedric, and Massimo Piatelli-Palmarini. 2005. "Language as a natural object—Linguistics as a natural science." *Linguistic Review* 22: 447–66.

Bolling, George. 1917. "Review of Bloomfield's *Introduction to the Study of Language*." *The Classical Weekly* 10. Reprinted in Bloomfield and Hockett 1970, 50–54.

Boole, George. 1854. *An Investigation of the Laws of Thought*. Reprinted, Cambridge: Cambridge University Press, 2009.

——. 1997. *Selected Manuscripts on Logic and Its Philosophy*. Edited by Ivor Grattan-Guinness and Gerard Bornet. Berlin: Springer.

Bopp, Franz. 1820. *Analytical Comparison of the Sanskrit, Greek, Latin and Teutonic Languages*. Annals of Oriental Literature. London: Longman.

——. 1856. *Vergleichende Zergliederung des Sanskrits und der mit ihm verwandten Sprachen*. Published in English as *Comparative Grammar of the Sanskrit, Zend, Greek, Latin, Lithuanian, Gothic, German, and Sclavonic Languages*, 1885. London: Williams and Norgate.

Boring, Edwin G. 1929a. *A History of Experimental Psychology*. New York: D. Appleton–Century.

——. 1929b. "The psychology of controversy." *Psychological Review* 36: 97–121.

——. 1942. "Human nature vs. sensation: William James and the psychology of the present." *American Journal of Psychology* 55: 310–27.

——. 1948. "Masters and pupils among the American psychologists." *American Journal of Psychology* 61: 527–34.

——. 1952. "Edwin Garrigues Boring." In *A History of Psychology in Autobiography*, vol. 4, edited by E. G. Boring, H. Werner, R. M. Yerkes, and H. S. Langfeld, 27–52. Worcester, MA: Clark University Press.

——. 1953. "A history of introspection." *Psychological Bulletin* 50: 169–89.

——. 1961. *Psychologist at Large*. New York: Basic Books.

——. 1963a. "Eponym as placebo." In Boring 1963b, 5–25.

——. 1963b. *History, Psychology and Science: Selected Papers*. New York: Wiley.

Boudewijnse, Gert-Jan A. 1999. "The rise and fall of the Graz School." *Gestalt Theory* 21: 140–58.

Boudon, Raymond. 1986. *L'idéologie*. Paris: Le Seuil.
——. 1990. *L'art de se persuader*. Paris: Le Seuil.
Bouquet, Simon. 1997. *Introduction à la lecture de Saussure*. Paris: Payot.
Bourdieu, Pierre. 1975. "The specificity of the scientific field and the social conditions of the progress of reason." *Social Science Information* 14, no. 6: 19–47.
——. 1980. "Le mort saisit le vif." *Actes de la recherche en sciences sociales* 32: 3–14.
——. 1984. *Homo academicus*. Paris: Editions de Minuit.
——. 2001. *Science de la science et réflexivité*. Paris: Raisons d'agir.
Bouveresse, Jacques. 2012. *Essais VI*. Marseille: Editions Agone.
Bras, Gérard. 2018. *Les voies du peuple*. Paris: Editions Amsterdam.
Brentano, Franz. 1874. *Psychologie vom empirischen Standpunkte*. Leipzig: Duncker and Humblot. Translated as *Psychology from an Empirical Standpoint*, 1995. London: Routledge.
——. 1982. *Deskriptive Psychologie*. Hamburg: Meiner.
Bridgman, Percy Williams. 1927. *The Logic of Modern Physics*. New York: Macmillan.
Bringmann, Wolfgang G., and Gustav A. Ungerer. 1980. "The Foundation of the Institute for Experimental Psychology at Leipzig University." *Psychological Research* 42: 5–18.
Brooks, John I. 1998. *The Eclectic Legacy*. Newark: University of Delaware Press.
Brouwer, Luitzen Jan. 1905. "Life, Art, and Mysticism." In translation in *Notre Dame Journal of Formal Logic* 38 (1996): 389–429.
Bruce, Darryl. 1998. "The Lashley-Hull debate revisited." *History of Psychology* 1: 69–84.
Brugmann, Karl. 1861. *Grundriss der vergleichenden Grammatik der indogermanischen Sprachen*. Translated as *Elements of the Comparative Grammar of the Indo-Germanic Languages*, vol. 1, 1886. New York: Westermann.
Bruner, Jerome. 1983. *In Search of Mind*. New York: Basic Books.
Buckley, Kerry W. 1989. *Mechanical Man: John Broadus Watson and the Beginnings of Behaviorism*. New York: Guilford.
Buffon, Georges-Louis Leclerc de. 1749–1789. *Histoire naturelle, générale et particulière, avec la description du Cabinet du Roy*. Paris: Imprimerie royale.
Bühler, Karl. 1908. "Antwort auf die von W. Wundt erhobenen Einwande gegen die Methode der Selbstbeobachtung an experimentell erzeugten Erlebnissen." *Archiv für die gesamte Psychologie* 1: 93–123.
——. 1919. *Abriss der geistigen Entwicklung des Kindes*. Leipzig: Quelle & Meyer.
——. 1931. "Phonetik und Phonologie." Reprinted in *Phonometrie*, vol. 2, edited by Eberhard Zwirner and Kennosuke Ezawa, 1968, 190–222. Basel: Karger.
——. 1934. *Sprachtheorie*. Reprinted in English as *Theory of Language*, 2011. Amsterdam: John Benjamins.

Bundgaard, Peer F. 2004. "The ideal scaffolding of language: Husserl's fourth logical investigation in the light of cognitive linguistics." *Phenomenology and the Cognitive Sciences* 3: 49–80.

Burridge, Kate. 2013. "Nineteenth-century study of sound change from Rask to Saussure." In *The Oxford Handbook of the History of Linguistics*, edited by Keith Allen, 141–65. Oxford: Oxford University Press.

Buss, Mareike, Lorella Ghiotti, and Ludwig Jaeger. 2003. "Notes sur l'accentuation lituanienne." In *Cahiers de L'Herne "Saussure"* no. 76, edited by Simon Bouquet, 505–26. Paris: L'Herne.

Cahan, David. 2003. *From Natural Philosophy to the Sciences*. Chicago: University of Chicago Press.

Caldwell, William. 1898. "Professor Titchener's view of the self." *Psychology Review* 5: 401–8.

——. 1899. "The postulates of a structural psychology." *Psychology Review* 6: 187–91.

Campbell, Lyle, and William J. Poser. 2008. *Language Classification, History and Method*. Cambridge: Cambridge University Press.

Cao, Tian Yu. 1997. *Conceptual Development of 20th Century Field Theories*. Cambridge: Cambridge University Press.

Cardone, Felice, and J. Roger Hindley. 2009. "Lambda-calculus and combinators in the 20th century." In Gabbay and Woods, 723–817.

Carnap, Rudolf. 1928. *Der Logische Aufbau der Welt*. Translated as *The Logical Structure of the World*, 1967. Berkeley: University of California Press.

——. 1934. "On the character of philosophical problems." *Philosophy of Science* 1: 5–19.

——. 1937. *Logische Syntax der Sprache*. Translated as *The Logical Syntax of Language*, 2014. London: Routledge.

——. 1963. "Intellectual biography." In *The Philosophy of Rudolf Carnap*, edited by Paul Schilpp, 3–86. La Salle, IL: Open Court.

Carroll, John B. 1980. "The tale of a theoretical and applied linguistic psychologist." In Davis and O'Cain, 31–52.

Casadio, Claudia. 2002. *Logic for Grammar*. Rome: Bulzoni.

Cassirer, Ernst. 1910. *Substanzbegriff und Funktionsbegriff: Untersuchungen über die Grundfragen der Erkenntniskritik*. Berlin: Bruno Cassirer.

——. 1950. *The Problem of Knowledge*. New Haven, CT: Yale University Press.

Cassirer, Toni. 1981. *Mein Leben mit Ernst Cassirer*. Hildesheim, Germany: Gerstenberg.

Cassou-Noguès, Pierre. 2007. *Les démons de Gödel*. Paris: Le Seuil.

Chamberlain, Houston Stewart. 1899. *Die Grundlagen des neunzehnten Jahrhunderts*. Munich: F. Bruckmann.

Chamberlain, Lesley. 2007. *Lenin's Private War: The Voyage of the Philosophy Steamer and the Exile of the Intelligentsia*. New York: St. Martin's Press.

Chater, Nick, Alex Clark, John Goldsmith, and Amy Perfors. 2015. *Empiricism and Language Learnability*. Oxford: Oxford University Press.

Chevalier, Jean-Claude. 1997. "Trubetzkoy, Jakobson et la France, 1919–1939." *Cahiers de l'ILSL* 9: 31–43.

Chomsky, Noam. 1957. *Syntactic Structures*. Janua Linguarum, Series Minor 4. The Hague: Mouton.

——. 1979. *Language and Responsibility*. New York: Pantheon.

——. 2002. *On Nature and Language*. Cambridge: Cambridge University Press.

Chomsky, Noam, and Morris Halle. 1968. *The Sound Pattern of English*. New York: Harper and Row.

Chomsky, William. 1945. "How the study of Hebrew grammar began and developed." *Jewish Quarterly Review* 35, no. 3: 281–301.

Church, Alonzo. 1932. "A set of postulates for the foundation of logic." *Annals of Science* 33: 346–66.

——. 1935. "An unsolvable problem of elementary number theory (abstract)." *Bulletin of the American Mathematical Society* 41: 332–33.

Coffa, J. Alberto. 1991. *The Semantic Tradition from Kant to Carnap*. Cambridge: Cambridge University Press.

Cohen, David. 1979. *J. B. Watson: The Founder of Behaviorism*. London: Routledge and Kegan Paul.

Cohen, I. Bernard. 1952. "Orthodoxy and scientific progress." *Proceedings of the American Philosophical Society* 96: 505–12.

——. 1960. *The Birth of a New Physics*. New York: Norton.

Collins, Randall. 1998. *The Sociology of Philosophy: A Global Theory of Intellectual Change*. Cambridge, MA: Harvard University Press.

Collitz, Hermann. 1925. "The scope and aims of linguistic science." *Language* 1: 1–13.

Comte, Auguste. 1830. *Cours de philosophie positive*. Paris: Bachelier.

——. 1996. *Philosophie des sciences (Anthologie de textes et de discours)*. Paris: Gallimard.

Comtet, Roger. 1995. "L'école phonologique de Leningrad et l'école phonologique de Moscou." *Histoire Épistémologie Langage* 17: 183–209.

Coneybeare, William D. and William Phillips. 1822. *Outline of the Geology of England and Wales*. London: William Phillips.

Corry, Leo. 1997. "David Hilbert and the axiomatization of physics (1894–1905)." *Archive for History of Exact Sciences* 51: 83–198.

Costantini, Michel. 2010. "Reconnaissance de dette: Jakobson et Lévi-Strauss." *Actes sémiotiques* 113. http://epublications.unilim.fr/revues/as/1764.

Cowan, Robert. 2010. *The Indo-German Identification: Reconciling South Asian Origins and European Destinies, 1765–1885*. Rochester, NY: Camden House.

Culicover, Peter W., and Andrzej Nowak. 2003. *Dynamical Grammar: Minimalism, Acquisition, and Change*. Oxford: Oxford University Press.

Cuvier, Georges. 1817. *Le règne animal distribué d'après son organisation: pour servir de base à l'histoire naturelle des animaux et d'introduction à l'anatomie comparée*. Paris: Chez Déterville.

Danziger, Kurt. 1979a. "The positivist repudiation of Wundt." *Journal of the History of the Behavioral Sciences* 15: 205–30.

———. 1979b. "The social origins of modern psychology: Positivist sociology and the sociology of knowledge." In *Psychology in Social Context: Towards a Sociology of Psychological Knowledge*, edited by A. R. Buss, 27–45. New York: Irvington.

———. 1980a. "The history of introspection reconsidered." *Journal of the History of the Behavioral Sciences* 16: 241–62.

———. 1980b. "Wundt's psychological experiment in the light of his philosophy of science." *Psychological Research* 42: 109–22.

———. 1983. "Origins and basic principles of Wundt's *Völkerpsychologie*." *British Journal of Social Psychology* 22: 303–13.

———. 2001. "Wundt and the temptations of psychology." In *Wilhelm Wundt in History: The Making of a Scientific Psychology*, edited by Robert W. Rieber and David K. Robinson, 69–94. Berlin: Springer.

Darnell, Regna. 1990. *Edward Sapir: Linguist, Anthropologist, Humanist*. Berkeley: University of California Press.

———. 1998. "Camelot at Yale: The construction and dismantling of the Sapirian synthesis, 1931–1939." *American Anthropologist* 100: 361–72.

Darnell, Regna, and Dell Hymes. 1986. "Edward Sapir's six-unit classification of American Indian languages: The search for time perspective." In *Studies in the History of Western Linguistics*, edited by Theodora Bynon and F. R. Palmer, 202–44. Cambridge: Cambridge University Press.

Darwin, Charles. 1859. *On the Origin of Species by Means of Natural Selection, or the Preservation of Favoured Races in the Struggle for Life*. London: John Murray.

———. 1871. *The Descent of Man, and Selection in Relation to Sex*. London: John Murray.

Dauben, Joseph W. 1990. *George Cantor: His Mathematics and Philosophy of the Infinite*. Princeton, NJ: Princeton University Press.

Davies, Anna Morpurgo. 1978. "Analogy, segmentation and the early Neogrammarians." *Transactions of the Philological Society* 71: 36–60.

———. 2006. "Saussure and Indo-European linguistics." In *The Cambridge Companion to Saussure*, edited by Carol Sanders, 9–29. Cambridge: Cambridge University Press.

Davis, Boyd H., and Raymond K. O'Cain, eds. 1980. *First Person Singular*. Vol. 1. Amsterdam: John Benjamins.

Davis, Martin. 1958. *Computability and Unsolvability*. New York: McGraw-Hill.

Daylight, Edgar G. [Karel Van Oudheusden]. 2011. "Dijkstra's rallying cry for generalization: The advent of recursion in programming, 1950s–1960s." *Computer Journal* 54: 1756–72.

De Mauro, Tullio. 1972. "Notes." In Saussure 1972, 405–77.

Demetz, Peter. 1998. *Prague in Black and Gold: Scenes from the Life of a European City*. New York: Hill and Wang.

Demopoulos, William, and Peter Clark. 2005. "The logicism of Frege, Dedekind, and Russell." In Shapiro, 129–65.

Demoule, Jean-Paul. 1999. "Destin et usages des Indos-Européens." *Mauvais Temps* 5: 63–74.

———. 2003. "Les pierres et les mots: Freud et les archéologues." *Alliage*, no. 52.

———. 2014. *Mais où sont passés les Indo-Européens?* Paris: Le Seuil.

Dennes, Maryse. 1997. "Husserl et l'école russe de phénoménologie." *Slavica Occitania* 4: 153–75.

Dennett, Daniel. 1991. *Consciousness Explained*. Boston: Little, Brown.

Despres, Leon. 1987. "My recollections of Leonard Bloomfield." In *Leonard Bloomfield: Essays on His Life and Work*, edited by Robert A. Hall Jr., 3–13. Amsterdam: John Benjamins.

Detlefsen, Michael. 2005. "Formalism." In Shapiro, 236–317.

Dewey, John. 1884. "The new psychology." *Andover Review* 2: 278–89.

———. 1910. "William James." *Journal of Philosophy* 7: 505–8.

Dummett, Michael. 1993. *Origins of Analytic Philosophy*. Cambridge, MA: Harvard University Press.

Dwight, Benjamin W. 1859. *Modern Philology: Its Discoveries, History and Influence*. Vol. 1. New York: A. S. Barnes and Burr.

Edgerton, Franklin. 1928. "Maurice Bloomfield: 1855–1928." *Language* 48: 193–99.

Ehrenfels, Christian von. 1890. "Über 'Gestaltqualitäten.'" 1890. *Vierteljahrsschrift für wissenschaftliche Philosophie* 14: 242–92.

Ellis, Willis D. 1938. *A Source Book of Gestalt Psychology*. New York: Harcourt Brace.

Emeneau, Murray B. 1943. "Franz Boas as a linguist." *Memoirs of the American Anthropological Association* 61: 35–38.

———. 1988. "Bloomfield and Pāṇini." *Language* 64: 755–60.

Erasmus, Desiderius. 1992. *Collected Works of Erasmus*. Edited by James F. Brady and John C. Olin. Toronto: University of Toronto Press.

Espagne, Michel. 1998. "Wilhelm Wundt. La 'psychologies des peuples' et l'histoire culturelle." *Revue Germanique Internationale* 10. http://rgi.revues.org/688.

Esper, Erwin Allen. 1968. *Mentalism and Objectivism in Linguistics: The Sources of Leonard Bloomfield's Psychology of Language*. New York: American Elsevier.

Evans, Rand B. 1972. "E. B. Titchener and his lost system." *Journal of the History of the Behavioral Sciences* 8: 168–80.

Fabian, Reinhard, ed. 1986. *Christian von Ehrenfels: Leben und Werk.* Amsterdam: Rodopi.

——. 1996. "Christian von Ehrenhels (1859–1932)." In Albertazzi et al., 161–74.

Falk, Julia S. 2014. Ninetieth Anniversary of the Linguistic Society of America: A Commemorative Symposium, January 4. *The LSA Linguistic Institutes.*

Fechner, Gustav. 1860. *Elemente der Psychophysik.* Leipzig: Breitkopf und Härtel. In translation in *The Classical Psychologists*, edited by Benjamin Rand, 1912. Boston: Houghton Mifflin.

Feest, Uljana. 2007. "Science and experience/Science of experience: Gestalt psychology and the anti-metaphysical project of the *Aufbau*." *Perspectives on Science* 15: 1–25.

Feferman, Anita Burdman, and Solomon Feferman. 2004. *Alfred Tarski: Life and Logic.* Cambridge: Cambridge University Press.

Feigl, Herbert. 1981. *Inquiries and Provocations: Selected Writings, 1929–1974.* Berlin: Springer.

Ferguson, Charles. 1962. "Review of *The Sound Pattern of Russian* by Morris Halle." *Language* 38: 284–98.

Ferreirós, José. 2007. *Labyrinth of Thought: A History of Set Theory and Its Role in Modern Mathematics.* Berlin: Springer. First edition 1999. Barcelona: Birkhäuser.

——. 2009. "Hilbert, logicism, and mathematical existence." *Synthese* 170: 33–70.

Ferry, Luc, Jean-Pierre Person, and Alain Renaut. 1979. *Philosophies de l'université: l'idealisme allemand et la question de l'université.* Paris: Payot.

Feuer, Lewis S. 1969. *The Conflict of Generations.* New York: Basic Books.

——. 1975. *Ideology and the Ideologists.* New York: Harper and Row.

Fichte, Johann Gottlieb. 1808. "Reden an die deutsche Nation." In *Philosophische Bibliothek*, vol. 24, edited by Alexander Aichele. Hamburg: Meiner.

Fisette, Denis. 2009. "Stumpf and Husserl on phenomenology and descriptive psychology." *Gestalt Theory* 31: 175–90.

Flack, Patrick. 2007. "Roman Jakobson and the transition of German thought to the structuralist paradigm: Towards a new historiography of structuralism." *Acta Structuralica* 1: 1–15.

Florovsky, Georges, P. N. Savitsky, Nikolai Trubetzkoy, and Pierre Souvtchinsky. 1921. *Exodus to the East: Forebodings and Events: An Affirmation of the Eurasians.* Reprint, edited by Ilya Vinkovetzky and Charles Schlacks Jr., 1996. Idyllwild, CA: Charles Schlacks Jr.

Focht, Mildred. 1935. *What Is Gestalt-Theory?* New York: George Grady Press.

Fodor, Jerry. 1983. *Modularity of Mind.* Cambridge, MA: MIT Press.

Fontaine, Jacqueline. 1997. "S. Karcevski et R. Jakobson, grammairiens de la langue russe à l'époque du Cercle linguistique de Prague." *Cahiers de L'ILSL*, no. 9: 75–86.

Forster, Michael. 2010. *After Herder: Philosophy of Language in the German Tradition*. Oxford: Oxford University Press.

Foucault, Michel. 1969. *L'archéologie du savoir*. Translated as *The Archaeology of Knowledge*. 2nd ed. Classics Series, 2002. London: Routledge.

Fought, John, ed. 1999. *Leonard Bloomfield: Critical Assessments of Leading Linguists*. London: Routledge.

Fraenkel, A. A., and Yehoshua Bar-Hillel. 1958. *Foundations of Set Theory*. Amsterdam: North-Holland.

Frege, Gottlob. 1879. *Begriffsschrift: Eine der arithmetischen nachgebildete Formelsprache des reinen Denkens*. Halle: Louis Nebert. Reprinted in translation in Heijenoort 1967, 1–82.

———. 1884. *Die Grundlagen der Arithmetik*. Published in translation as *The Foundations of Arithmetic*, 1968. Evanston, IL: Northwestern University Press.

———. 1997. *The Frege Reader*. Edited by Michael Beaney. Oxford: Blackwell.

Freud, Sigmund. *Zur Psychopathologie des Alltagsleben*. 1901. In English, *Psychopathology of Everyday Life*, 1915. New York: Macmillan.

Friedman, Michael. 1991. "The re-evaluation of logical positivism." *Journal of Philosophy* 88: 505–19.

———. 2000. *A Parting of the Ways: Carnap, Cassirer, and Heidegger*. Chicago: Open Court.

———. 2012. "Scientific philosophy from Helmholtz to Carnap and Quine." In *Rudolf Carnap and the Legacy of Logical Empiricism*, edited by Richard Creath, 1–11. Berlin: Springer.

Fries, Charles Carpenter. 1952. *The Structure of English: An Introduction to the Construction of English Sentences*. New York: Harcourt Brace.

———. 1970. "The 'Bloomfield School.'" In *Trends in European and American Linguistics: 1930–1960*, edited by Christine Mohrmann, Alf Sommerfelt, and Joshua Whatmough, 196–224. Utrecht: Spectrum.

Fronek, Josef. 1988. "T. G. Masaryk and Prague School linguistic structuralism." In *On Masaryk*, edited by Josef Novák, 81–96. Amsterdam: Rodopi.

Gabbay, Dov M., and John Woods, eds. 2009. *Logic from Russell to Church*. Volume 5 of *Handbook of the History of Logic*. New York: Elsevier.

Gadet, Françoise. 1995. "Jakobson sous le pavillon saussurien." *Linx* 7: 449–59.

Gall, Franz Joseph, and Johann Gaspar Spurzheim. 1810. *Anatomie et physiologie du système nerveux en général, et du cerveau en particulier, avec des observations sur la possibilité de reconnoître plusieurs dispositions intellectuelles et morales de l'homme et des animaux, par la configuration de leurs têtes*. Paris: F. Schoell.

Gandy, Robin. 1991. "The confluence of ideas in 1936." In *The Universal Turing Machine: A Half-Century Survey*, edited by Rolf Herken. Oxford: Oxford University Press.

Gardies, Jean-Louis. 1986. *Rational Grammar.* Munich: Philosophia.

Gardner, Howard. 1985. *The Mind's New Science.* New York: Basic Books.

Gardner, Sheldon, and Gwendolyn Stevens. 1992. *Red Vienna and the Golden Age of Psychology, 1918–1938.* New York: Praeger.

Gasparov, Boris. 1987. "The ideological principles of the Prague school phonology." In *Language, Poetry and Poetics*, edited by Krystyna Pomorska, Elzbieta Chodakowska, Hugh McLean, and Brent Vine, 49–78. Berlin: De Gruyter.

———. 2013. *Beyond Pure Reason: Ferdinand de Saussure's Philosophy of Language and Its Early Romantic Antecedents.* New York: Columbia University Press.

Gengerelli, Joseph A. 1976. "Graduate school reminiscences: Hull and Koffka." *American Psychologist* 31: 685–88.

Gérard, René. 1963. *L'orient et la pensée romantique allemande.* Paris: Marcel Didier.

Gerring, John. 2010. "Causal mechanisms: Yes, but. . . ." *Comparative Political Studies* 43: 1499–526.

Gibbs, Josiah W. 1853. "Remarks on the preceding essay." *Journal of the American Oriental Society* 3: 469–72.

———. 1857. *Philological Studies with English Illustrations.* New Haven, CT: Durrie and Peck.

Gilbert, William. 1600. *De Magnete.* London: Peter Short.

Gleason, H. A., Jr. 1955. *Introduction to Descriptive Linguistics.* New York: Henry Holt.

———. 1988. "Theories in conflict." Unpublished manuscript.

Glebov, Sergey. 2003. "Science culture and empire: Eurasianism as a modernist movement." *Slavic and East European Information Resources* 4: 13–31.

———. 2015. "N. S. Trubetskoi's Europe and Mankind and Eurasianist Antievolutionism." In Bassin et al., 48–67.

Gobineau, Arthur. 1853. *Essai sur l'inégalité des races humaines.* Paris: F. Didot frères.

Gödel, Kurt. 1995. *Collected Works.* Edited by S. Feferman. Oxford: Oxford University Press.

Godel, Robert. 1957. *Les sources manuscrites du Cours de linguistique générale de Ferdinand de Saussure.* Geneva: Droze.

Godfrey, J. J. 1967. "Sir William Jones and Père Coeurdoux: A philological footnote." *Journal of the American Oriental Society* 87: 57–59.

Goldschmidt, Walter Rochs. 1959. *The Anthropology of Franz Boas: Essays on the Centennial of His Birth.* Menasha, WI: American Anthropological Association.

Goldsmith, John A. 1990. *Autosegmental and Metrical Phonology*. Oxford: Blackwell.

——. 2011. "The syllable." In *The Handbook of Phonological Theory*, edited by John Goldsmith, Alan Yu, and Jason Riggle, 162–96. Oxford: Wiley Blackwell.

Goldstein, Doris S. 1968. "'Official philosophies' in modern France: The example of Victor Cousin." *Journal of Social History* 1: 259–79.

Goldstein, Jan. 2005. *The Post-Revolutionary Self: Politics and Psyche in France, 1750–1850*. Cambridge, MA: Harvard University Press.

——. 2013. "Neutralizing Freud: The lycée philosophy class and the problem of the reception of psychoanalysis in France." *Critical Inquiry* 40: 40–82.

Gordin, Michael D. 2006. "'Le Premier Cercle': le Kruzhok de Heidelberg et la nationalisation de la chimie russe." *Sociologie du Travail* 48: 286–307.

Gossett, Thomas. 1963. *Race: The History of an Idea in America*. Oxford: Oxford University Press.

Gould, Stephen Jay. 1981. *The Mismeasure of Man*. London: Penguin.

Grafton, Anthony. 2004. "Review of *In No Man's Land*." *New York Review of Books*, February 26.

Grammont, Maurice. 1912. "Review of *Mélanges linguistiques offerts à M. Ferdinand de Saussure*." *Revue des langues romanes* 55: 387–89.

——. 1933. "Review of *Eléments de phonologie française*, by Georges Gougenheim." *Revue des langues romanes* 67: 574–76.

Grattan-Guinness, I. 1981. "On the development of logics between the two world wars." *American Mathematical Monthly* 88: 495–509.

——. 2000. *The Search for Mathematical Roots, 1870–1940: Logics, Set Theories and the Foundations of Mathematics from Cantor Through Russell to Gödel*. Princeton, NJ: Princeton University Press.

Gray, Jeremy. 2008. *Plato's Ghost: The Modernist Transformation of Mathematics*. Princeton, NJ: Princeton University Press.

Greenberg, Joseph. 1960. "A quantitative approach to the morphological typology of language." *International Journal of American Linguistics* 26: 178–94.

——. 1986. "On being a linguistic anthropologist." *Annual Review of Anthropology* 15: 1–24.

Grimm, Jacob. 1848. *Geschichte der deutschen Sprache*. 3rd ed., 1868. Leipzig: S. Hirzel.

Gross, Alan G. 1998. "Do disputes over priority tell us anything about science?" *Science in Context* 11: 161–79.

Grout, Lewis. 1853. "An essay on the phonology and orthography of the Zulu and kindred dialects in Southern Africa." *Journal of the American Oriental Society* 3: 421–68.

Guillin, Vincent. 2004. "Théodule Ribot's ambiguous positivism: Philosophical and epistemological strategies in the founding of French scientific psychology." *Journal of the History of the Behavioral Sciences* 40: 165–81.

Gulyás, László. 2007. *Beneš: Statesman or Charlatan? The Plans and the Reality 1908–1948*. Toronto: Corvinus.

Gundlach, Horst. 2014. "Max Wertheimer, *Habilitation* Candidate at the Frankfurt Psychological Institute." *History of Psychology* 17: 134–48.

Hackenberg, Timothy D. 1995. "Jacques Loeb, B. F. Skinner, and the legacy of prediction and control." *Behavior Analyst* 18: 225–36.

Haeckel, Ernst. 1879. *Les preuves du transformisme: réponse à Virchow*. Paris: G. Baillière.

Hahn, Hans, Otto Neurath, and Rudolf Carnap. 1929. *Wissenschaftliche Weltauffassung: Der Wiener Kreis*. Vienna: Artur Wolf.

———. 1973. "The scientific conception of the world: The Vienna Circle." In *Empiricism and Sociology*, edited by M. Neurath and R. S. Cohen, 299–318. Dordrecht: Reidel.

Hajičová, Eva. 2006. "Old linguists never die, they only get obligatorily deleted." *Computational Linguistics* 32: 457–69.

Hall, Robert A., Jr. 1949. "In Memoriam Leonard Bloomfield (April 1877–April 18, 1949)." *Lingua* 2: 117–23.

———. 1990. *A Life for Language*. Amsterdam: John Benjamins.

Halle, Morris. 1962. "Phonology in generative grammar." *Word* 18: 1–3, 54–72.

———. 1979. "Roman Jakobson." In *International Encyclopedia of the Social Sciences*, vol. 18, 335–41. New York: Free Press.

Hannush, Mufid James. 1987. "John B. Watson remembered: An interview with James B. Watson." *Journal of the History of the Behavioral Sciences* 23: 137–52.

Harris, Marvin. 1968. *The Rise of Anthropological Theory: A History of Theories of Culture*. New York: Thomas Y. Crowell.

Harris, Zellig. 1951. *Methods in Structural Linguistics*. Chicago: University of Chicago Press.

———. 1954. "Distributional structure." *Word* 10: 146–62.

———. 1973. "Review of Charles F. Hockett (Ed.), *A Leonard Bloomfield Anthology*." *International Journal of American Linguistics* 39: 252–55.

———. 1990. "The background of transformation and metalanguage analysis." Reprinted in *The Legacy of Zellig Harris: Language and Information in the 21st Century*, edited by Bruce Nevin, 2002, 1–14. Amsterdam: John Benjamins.

Harvey, David Allen. 2014. "The lost Caucasian civilization: Jean-Sylvain Bailly and the roots of the Aryan myth." *Modern Intellectual History* 11: 279–306.

Hauser, Marc D., Noam Chomsky, and W. Tecumseh Fitch. 2002. "The faculty of language: What is it, who has it, and how did it evolve?" *Science* 298 (November): 1569–79.

Heidbreder, Edna. 1933. *Seven Psychologies*. New York: D. Appleton–Century.

Heider, Fritz. 1989. "Fritz Heider." In *A History of Psychology in Autobiography*, vol. 8, edited by Gardner Lindzey, 127–56. Stanford, CA: Stanford University Press.

Heijenoort, Jean van, ed. 1967. *From Frege to Gödel*. Cambridge, MA: Harvard University Press.

Heis, Jeremy. 2013. "Ernst Cassirer, Kurt Lewin, and Hans Reichenbach." In *The Berlin Group and the Philosophy of Logical Empiricism*, edited by Nikolay Milkov and Volker Peckhaus, 67–94. Berlin: Springer.

Helmholtz, Hermann. 1885. *On the Sensations of Tone*. New York: Dover.

Helson, Harry. 1925. "The psychology of 'Gestalt.'" *American Journal of Psychology* 36: 342–70.

Hendel, Charles W. 1949. "Ernst Cassirer." In *The Philosophy of Ernst Cassirer*, edited by Paul A. Schilpp, 55–60. New York: Tudor.

Henle, Mary. 1984. "Robert M. Ogden and Gestalt psychology in America." *Journal of the History of the Behavioral Sciences* 10: 9–19.

Herrnstein, R. J. 1972. "Nature as nurture: Behaviorism and the instinct doctrine." *Behaviorism* 1: 23–52.

Hesseling, Dennis E. 2003. *Gnomes in the Fog: The Reception of Brouwer's Intuitionism in the 1920s*. Basel: Birkhäuser.

Hilbert, David. 1899. *Grundlagen der Geometrie*. Reprinted in 2015. Berlin: Springer.

——. 1918. "Axiomatic thinking." Published in *Philosophia Mathematica*, vols. 1–7, issue 1–2 (1970): 1–12.

——. 1927. "The foundations of mathematics." Reprinted in Heijenoort 1967, 464–79.

Hill, Archibald A. 1980. "How many revolutions can a linguist live through?" In Davis and O'Cain, 69–76.

——. 1991. "The LSA and North American linguistics, 1950–1968." *Historiographia Linguistica* 18: 49–152.

Hindeland, Michael J. 2006. "Edward Bradford Titchener: A pioneer in perception." *Journal of the History of the Behavioral Sciences* 7, no. 1: 23–28.

Hiż, Henry. 1998. "Between logic and linguistics." In *First Person Singular III: Autobiographies by North American Scholars in the Language Sciences*, edited by E. F. K. Koerner, 71–80. Amsterdam: John Benjamins.

Hobbes, Thomas. 1651. *The Leviathan*. Critical edition by Noel Malcolm, 2014. Oxford: Oxford University Press.

Hobsbawm, Eric. 1962. *The Age of Revolution 1789–1848*. New York: World Publishing.

——. 1975. *The Age of Capital, 1848–1875*. New York: Vintage.

——. 1987. *The Age of Empire, 1875–1914*. London: Weidenfeld and Nicholson.

Hockett, Charles F. 1942. "A system of descriptive phonology." *Language* 18: 3–21.

——. 1949. "Two fundamental problems in phonemics." *Studies in Linguistics* 7: 29–51.

——. 1958. *A Course in Modern Linguistics*. New York: Macmillan.

——. 1980. "Preserving the heritage." In Davis and O'Cain, 99–110.

——. 1987. "Letters from Bloomfield to Michelson and Sapir." *Historiographia Linguistica* 14: 39–60.

Holenstein, Elmar. 1975. "Jakobson and Husserl: A contribution to the genealogy of structuralism." *Human Context* 7: 61–83. Reprinted in *Edmund Husserl: The Web of Meaning*, vol. 4, edited by Rudolf Bernet, Donn Welton, and Gina Zavota, 2005, 11–48. London: Routledge.

——. 1976. *Roman Jakobson's Approach to Language*. Bloomington: Indiana University Press.

Holmboe, C. A. 1855. "Respecting an alphabet suited to the languages of Southern Africa." *Journal of the American Oriental Society* 5: 427–29.

Holms, Linda. 2009. "Norwegian academics and their contribution to international reconciliation and peace: The Institute for Comparative Research in Human Culture." In *Academics as Public Intellectuals*, edited by Sven Eliasson and Ragnvald Kalleberg, 87–100. Cambridge, UK: Cambridge Scholars Publishing.

Holton, Gerald. 1998. "Einstein and the cultural roots of modern science." *Daedalus* 127, no. 1: 1–44.

Honeybone, Patrick. 2005. "Brugmann, Karl." In *Key Thinkers in Linguistics and the Philosophy of Language*, edited by Siobhan Chapman and Christopher Routledge, 45–48. Edinburgh: Edinburgh University Press.

Hoyningen-Huene, Paul. 1995. "Two letters of Paul Feyerabend to Thomas S. Kuhn on a draft of *The Structure of Scientific Revolutions*." *Studies in History and Philosophy of Science* 26: 353–87.

Hull, Clark L. 1935. "The conflicting psychologies of learning—a way out." *Psychological Review* 42: 491–516.

——. 1952. "Autobiography." In *A History of Psychology in Autobiography*, vol. 4, edited by E. G. Boring, H. S. Langfeld, H. Werner, and R. M. Yerkes, 143–62. Worcester, MA: Clark University Press.

——. 1984. *Mechanisms of Adaptive Behavior*. Edited by Abram Amsel and Michael E. Rashotte. New York: Columbia University Press.

Hull, Clark L., Arthur I. Gladstone, John M. Felsinger, and Harry G. Yamaguchi. 1947. "A proposed quantification of habit strength." *Psychological Review* 54: 237–54.

Humboldt, Alexander von. 1849. *Cosmos: A Sketch of a Physical Description of the Universe*. Vol. 2. London: Longman.

——. 1858. *Cosmos: A Sketch of a Physical Description of the Universe*. Vol. 1. New York: Harper.

Humboldt, Wilhelm. 1836. *Über die Kawi-Sprache auf der Insel Java*. Berlin: Duemmler.

——. 1999. *On Language*. Cambridge: Cambridge University Press.

Hume, David. 1738. *Treatise on Human Nature*. London: John Noon.

Husserl, Edmund. 1900a. *Logical Investigations*. Vol. 1. 2nd ed. 1913. London: Routledge. Translation of *Logische Untersuchungen*. Reprinted by Routledge, 2001.

——. 1900b. *Logical Investigations.* Vol. 2. 2nd ed. 1913. London: Routledge. Translation of *Logische Untersuchungen.* Reprinted, London: Routledge, 2001.

——. 1919. "Remiscences of Franz Brentano." Reprinted in translation in McAlister 1976, 47–55.

Isaac, Joel. 2005. "W. V. Quine and the origins of analytic philosophy in the United States." *Modern Intellectual History* 2: 205–34.

——. 2012. *Working Knowledge: Making the Human Sciences from Parsons to Kuhn.* Cambridge, MA: Harvard University Press.

Ivanov, Vjačeslav. 1983. "Roman Jakobson: The future." In *A Tribute to Roman Jakobson 1896–1982*, edited by Morris Halle, 47–57. Berlin: De Gruyter.

Jacquette, Dale. 2009. "Review of Guillermo E. Rosado Haddock, 'The Young Carnap's Unknown Master: Husserl's Influence on *Der Raum* and *Der Logische Aufbau der Welt.*'" *History and Philosophy of Logic* 30: 194–200.

Jakobson, Roman. 1927. "Remarques sur l'évolution phonologique du russe." Reprinted in SW1, 7–116.

——. 1928a. "The concept of the sound law and the teleological criterion." Reprinted in SW1, 1–2.

——. 1928b. "Proposition au premier congrès international de linguistes." Reprinted in SW1, 3–6.

——. 1929. "Theses presented to the First Congress of Slavic Philologists in Prague, 1929." Reprinted in *The Prague School: Selected Writings, 1929–1946*, edited by P. Steiner, 3–31. Austin: University of Texas Press, 1983.

——. 1930. "Problems of language in Masaryk's writings." Reprinted in *T. G. Masaryk in Perspective*, edited by M. Čapek and K. Hrubú, 1981, 63–82. New York: SVU Press.

——. 1931a. "Die Betonung und ihre Rolle in der Wort- und Syntagmaphonologie." Reprinted in SW1, 117–36.

——. 1931b. "La linguistique eurasienne." In Russian. Reprinted in SW1, 144–201.

——. 1931c. "Phonemic notes on Standard Slovak." Reprinted in SW1, 221–30.

——. 1931d. "Über die phonologischen Sprachbünde." Reprinted in SW1, 137–43.

——. 1931e. "Les unions phonologiques des langues." *Le Monde Slave* 8: 371–78.

——. 1932a. "Phoneme and Phonology." Reprinted in SW1, 231–33.

——. 1932b. "Zur Struktur des russischen Verbums." Reprinted in SW2, 3–15.

——. 1937a. "On Ancient Greek prosody." Reprinted in SW1, 262–71.

——. 1937b. "Orthographe du vieux russe." Reprinted in SW1, 247–53.

——. 1937c. "Über die Beschaffenheit der prosodischen Gegensätze." Reprinted in SW1, 254–61.

——. 1939. "The Czech part in Church Slavonic culture." Reprinted in SW6, 129–52.

——. 1941. *Kindersprache, Aphasie und allgemeine Lautgesetze*. Uppsala: Almqvist & Wiksell. Published in English translation as *Child Language, Aphasia, and Phonological Universals*. The Hague: Mouton, 1968.

——. 1944. "Franz Boas' approach to language." *International Journal of American Linguistics* 10: 188–95.

——. 1960. "Parts and wholes in language." Reprinted in SW2, 280–84.

——. 1962a. "Retrospect." Reprinted in SW1, 631–52.

——. 1962b. *Selected Writings: Word and Language*. Vol. 1. Abbreviated SW1. Paris: Mouton.

——. 1963a. "Efforts toward a means-ends model of language in interwar continental linguistics." *Trends in Modern Linguistics* 2:104–8. Utrecht: Spectrum. Reprinted in SW2, 522–26.

——. 1963b. "Implications of language universals for linguistics." In *Universals of Language*, edited by Joseph Greenberg, 208–18. Cambridge, MA: MIT Press. Reprinted in SW2, 580–92.

——. 1971a. "An example of migratory terms and institutional models (On the fiftieth anniversary of the Moscow Linguistic Circle)." Reprinted in SW2, 527–38.

——. 1971b. *Selected Writings: Word and Language*. Abbreviated SW2. Paris: Mouton.

——. 1973a. *Main Trends in the Science of Language*. London: Allen and Unwin.

——. 1973b. "Relations entre la science du langage et les autres sciences." In *Essaies de linguistique générale II*. Paris: Éditions de Minuit. Translated as *Main Trends in the Science of Language*. New York: Harper and Row, 1973.

——. 1975. "Foreword." In *N. S. Trubetzkoy's Letters and Notes*, edited by Roman Jakobson. The Hague: Mouton.

——. 1982. "Einstein et la science du langage." *Le débat* 20: 131–42.

——. 1985. *Selected Writings: Early Slavic Paths and Crossroads*. Abbreviated SW6. Berlin: Mouton.

——. 1997. *My Futurist Years*. Edited by Bengt Jangfeldt and Stephen Rudy. New York: Marsilio.

Jakobson, Roman, and R. Georgin. 1978. "Entretien avec Roman Jakobson." In *Jakobson*, Cahiers Cistre 5, 11–27. Lausanne: L'Age d'Homme.

Jakobson, Roman, and Morris Halle. 1964. "The term Canaan in Medieval Hebrew." In *For Max Weinreich on His Seventieth Birthday*, 147–72. The Hague: Mouton.

Jakobson, Roman, and Krystyna Pomorska. 1980. *Dialogues*. Paris: Flammarion.

——. 1983. *Dialogues*. Cambridge, MA: MIT Press.

Jakobson, Roman, Krystyna Pomorska, and Benjamin Hrushovski. 1980. "Dialogue on time in language and literature." *Poetics Today* 2: 15–27.

Jakobson, Roman, and Linda Waugh. 1979. *The Sound Shape of Language*. Bloomington: Indiana University Press.

James, William. 1890. *The Principles of Psychology*. 2 vols. New York: Dover.

——. 1895. "The knowing of things together." *Psychological Review* 2: 105–24.

Jangfeldt, Bengt. 1997. "Roman Jakobson in Sweden 1940–41." *Cahiers de L'ILSL* 9: 141–49.

Janik, Allan, and Stephen Toulmin. 1973. *Wittgenstein's Vienna*. New York: Simon and Schuster.

Jespersen, Otto. 1922. *Language: Its Nature, Development and Origin*. London: Allen and Unwin.

Jones, William. 1798. "Third anniversary discourse: On the Hindus." *Asiatick Researches* 1: 415–31.

Joos, Martin. 1957. *Readings in Linguistics: The Development of Descriptive Linguistics in America 1925–1956*. Chicago: University of Chicago Press.

Joseph, John. 2002. *From Whitney to Chomsky: Essay in the History of American Linguistics*. Amsterdam: John Benjamins.

——. 2012. *Saussure*. Oxford: Oxford University Press.

Jussieu, Antoine Laurent de. 1824. "Méthode naturelle des végétaux." In *Dictionnaire des sciences naturelles*. Vol. 30. Strasbourg: F. G. Levrault.

Kant, Immanuel. 1781. *Kritik der reinen Vernunft*. Translated as *Critique of Pure Reason*. The Cambridge Edition of the Works of Immanuel Kant, 1998. Cambridge: Cambridge University Press.

Karchevsky, Serge. 1927. *Système du verbe russe, Essai de linguistique synchronique*. Prague: Plamja.

Kemmer, Suzanne. 2013. "Observations on the 'Neogrammarian Manifesto.'" http://www.ruf.rice.edu/~kemmer/Found/ostbrugessay.html.

King, D. Brett, and Michael Wertheimer. 2005. *Max Wertheimer and Gestalt Theory*. Piscataway, NJ: Transaction Publishers.

Kiparsky, Paul. 1974. "From Paleogrammarians to Neogrammarians." In *Studies in the History of Linguistics: Traditions and Paradigms*, edited by Dell Hymes, 331–45. Bloomington: Indiana University Press.

Koerner, E. F. K. 1975. "European structuralism: Early beginnings." In *Current Trends in Linguistics*, vol. 13, edited by Thomas A. Sebeok, 717–827. Paris: Mouton.

——. 1980. "L'importance de William Dwight Whitney pour les jeunes linguistes de Leipzig et pour Ferdinand de Saussure." *Lingvisticae Investigationes* 4: 379–94.

——. 1987. "On the problem of 'Influence.'" In *Papers in the History of Linguistics: Proceedings of the Third International Conference on the History of the Language Sciences 1984*, edited by Hans Aarsleff, Louis G. Kelly, and Hans-Josef Niederehe, 13–28. Amsterdam: John Benjamins.

——. 1990. "The place of Friedrich Schlegel in the development of historical-comparative linguistics." In *Leibniz, Humboldt, and the Origins of Compar-*

ativism, edited by Tullio De Mauro and Lia Formigari, 239–61. Amsterdam: John Benjamins.

——, ed. 1991. *First Person Singular*. Vol. 2. Amsterdam: John Benjamins.

Koffka, Kurt. 1922. "Perception: An introduction to the Gestalt-Theorie." *Psychological Bulletin* 19: 531–85.

——. 1935. *Principles of Gestalt Psychology*. New York: Harcourt Brace.

Köhler, Wolfgang. 1920. *Die physische Gestalten in Ruhe und stationären Zustand: Eine naturphilosophische Untersuchung*. Braunschweig: Friedr. Vieweg & Sohn.

——. 1929. *Gestalt Psychology*. New York: Liveright.

——. 1959. "Gestalt psychology today." *American Psychologist* 14: 727–34.

——. 1969. *The Task of Gestalt Psychology*. Princeton, NJ: Princeton University Press.

Kreiser, Lothar. 2001. *Gottlob Frege: Leben–Werk–Zeit*. Hamburg: Felix Meiner.

Kripke, Saul, 2013. "The Church–Turing 'thesis' as a special corollary of Gödel's completeness theorem." In *Computability: Turing, Gödel, Church, and Beyond*, edited by B. Jack Copeland, Carl J. Posy, and Oron Shagrir, 77–104. Cambridge, MA: MIT Press.

Kroeber, Alfred Louis. 1984. "Reflections on Edward Sapir, scholar and man." In *Edward Sapir: Appraisals of His Life and Word*, edited by E. F. K. Koerner, 131–39. Amsterdam: John Benjamins.

Krueger, Robert G., and Clark Hull. 1931. "An electro-chemical parallel to the conditioned reflex." *Journal of General Psychology* 5: 262–69.

Kuroda, S.-Y. 1973. "Edmund Husserl, *Grammaire Générale et Raisonnée* and Anton Marty." *Foundations of Language* 10: 169–75.

Kusch, Martin. 1995. *Psychologism: A Case Study in the Sociology of Philosophical Knowledge*. London: Routledge.

——. 2005. *Psychological Knowledge: A Sociological History and Philosophy*. London: Routledge.

Kyllingstad, Jon Røyne. 2014. *Measuring the Master Race: Physical Anthropology in Norway, 1890–1945*. Cambridge, UK: Open Book Publishers.

Labov, William. 1981. "Resolving the Neogrammarian controversy." *Language* 57: 267–309.

——. 1994. *Principles of Linguistic Change: Internal Factors*. Oxford: Blackwell.

——. 2001. *Principles of Linguistic Change: Social Factors*. Oxford: Blackwell.

——. 2007. "Transmission and diffusion." *Language* 83: 344–87.

Lagemann, Ellen Condliffe. 1989. *The Politics of Knowledge: The Carnegie Corporation, Philanthropy, and Public Policy*. Middletown, CT: Wesleyan University Press.

Laks, Bernard. 2007. "Genetics and language: Comparatism and genealogy." In *Origins and Evolution of Languages: Models, Theories, Paradigms*, edited by

S. Cleuziou, J.-P. Demoule, Pierre Encrevé, and Bernard Laks, 157–92. London: Equinox.

——. 2012. "La phonotactique saussurienne: système et loi de valeur." *Langages* 185: 91–109.

——. 2013. "Why is there variation instead of nothing?" *Language Sciences* 39: 31–53.

Lane, George S. 1955. "Carl Darling Buck." *Language* 31: 181–89.

Lanman, Charles R., ed. 1897. *The Whitney Memorial Meeting: A Report of that Session of the First American. Congress of Philologists, Which Was Devoted to the Memory of the Late Professor William Dwight Whitney, of Yale University: Held at Philadelphia, Dec. 28, 1894.* Boston: Ginn.

Larson, Cedric A., and John J. Sullivan. 1965. "Watson's relation to Titchener." *Journal of the History of the Behavioral Sciences* 1: 338–54.

Laruelle, Marlène. 2001. "Les idéologies de la 'Troisième Voie' dans les années 1920: le mouvement eurasiste russe." *Revue d'histoire* 70: 31–46.

——. 2006a. "The Eurasianist ideology and the Eurasian history: Empire as the *natural* solution for the post-Soviet space." Unpublished manuscript.

——. 2006b. "Regards sur la réception du racialisme allemand chez les panslavistes et les eurasistes russes." *Revue germanique internationale* 3: 145–55.

——. 2008. *Russian Eurasianism: An Ideology of Empire.* Baltimore: Johns Hopkins University Press.

Lashley, Karl. 1923. "The behavioristic interpretation of consciousness I." *Psychological Review* 30: 237–72.

——. 1931. "Cerebral control versus reflexology: A reply to Professor Hunter." *Journal of General Psychology* 5: 3–19.

Lashley, Karl, and M. Wade. 1946. "The Pavlovian theory of generalization." *Psychological Review* 53: 72–87.

Leahey, Thomas H. 1981. "The mistaken mirror: On Wundt's and Titchener's psychologies." *Journal of the History of the Behavioral Sciences* 17: 273–82.

Lehmann, Winfred P., ed. and trans. 1967. *A Reader in Nineteenth-Century Historical Indo-European Linguistics.* Bloomington: Indiana University Press.

Lemov, Rebecaa. 2005. *World as Laboratory: Experiments with Mice, Mazes, and Men.* New York: Hill and Wang.

Lenci, Alessandro, and Gabriel Sandu. 2004. "Logic and linguistics in the twentieth century." In *Development of Modern Logic*, edited by Leila Haaparanta, 775–847. Oxford: Oxford University Press.

Leśniewski, Stanisław. 1929. "Grundzüge eines neuen Systems der Grundlagen der Mathematik." *Fundamenta Mathematicae* 14: 1–81.

——. 1992. *Collected Works.* 2 vols. Dordrecht: Kluwer.

Levelt, Wilhelm. 2014. *A History of Psycholinguistics: The Pre-Chomskyan Era.* Oxford: Oxford University Press.

Lewis, Herbert S. 2001. "The passion of Franz Boas." *American Anthropologist* 103: 447–67.

Li, Ming, and Paul M. B. Vitányi. 2008. *An Introduction to Kolmogorov Complexity and Its Applications*. 3rd ed. Berlin: Springer.

Lindenfeld, David. 1981. *The Transformation of Positivism: Alexius Meinong and European Thought, 1880–1920*. Berkeley: University of California Press.

Linn, Mary S., Tessie Naranjo, Sheilah Nicholas, Inée Slaughter, Akira Yamamoto, and Ofelia Zepeda. 2002. "Awakening the languages: Challenges of enduring language programs; Field reports from fifteen programs from Arizona, New Mexico and Oklahoma." In *Indigenous Languages across the Community*, edited by Barbara Burnaby and Jon Reyhner, 105–26. Washington, DC: ERIC Clearinghouse.

Loeb, Jacques. 1912. *The Mechanistic Conception of Life*. Chicago: University of Chicago Press.

Lowie, Robert H. 1944. "Franz Boas 1858–1942." *Journal of American Folklore* 57: 59–64.

Lukes, Igor. 1996. *Czechoslovakia between Stalin and Hitler: The Diplomacy of Edward Beneš in the 1930s*. Oxford: Oxford University Press.

Mach, Ernst. 1897. *Contributions to the Analysis of the Sensations*. Chicago: Open Court.

——. 1901. *Die Mechanik in ihrer Entwicklung*. 4th ed. Leipzig: F. A. Brockhaus.

——. 1905. *Erkenntnis und Irrtum*. Leipzig: Barth. Translated as *Knowledge and Error*, trans. T. J. McCormack and P. Fouldes, 1976. Dordrecht: Reidel.

Magee, Glenn. 2001. *Hegel and the Hermetic Tradition*. Ithaca, NY: Cornell University Press.

Malkiel, Yakov. 1980. "Early years in America." In Davis and O'Cain, 79–98.

Mancosu, Paolo. 1999. "Between Russell and Hilbert: Behmann on the foundations of mathematics." *Bulletin of Symbolic Logic* 5: 303–30.

Mancosu, Paolo, Richard Zach, and Calixto Badesa. 2004. "The development of mathematical logic from Russell to Tarski: 1900–1935." In *The Development of Modern Logic*, edited by Leila Haaparanta, 318–470. Oxford: Oxford University Press.

Mannheim, Karl 1923. "The problem of generations." Reprinted in *Essays on the Sociology of Knowledge*, vol. 5, edited by Paul Kecskemeti, 1952, 276–320. Oxford: Oxford University Press.

——. 1938. *Ideology and Utopia*. London: Routledge and Kegan Paul.

Marchand, Suzanne L. 2009. *German Orientalism in the Age of Empire: Religion, Race, and Scholarship*. Cambridge: Cambridge University Press.

Marthelot, Perrine. 2012. *Karl Bühler: du contexte à la signification*. Paris: Armand Colin.

Masaryk, Tomáš. 1887. *Versuch einer concreten Logik*. Vienna: Konegen.

Matthews, P. H. 1993. *Grammatical Theory in the United States from Bloomfield to Chomsky.* Cambridge: Cambridge University Press.

McAlister, Linda L. 1976. *The Philosophy of Brentano.* Atlantic Highlands, NJ: Humanities Press.

McCumber, John. 2007. *Reshaping Reason: Toward a New Philosophy.* Bloomington: Indiana University Press.

McDavid, Raven. 1980. "Linguistics, through the Kitchen Door," in Davis and O'Cain, 4–20.

McDougall, William. 1912. *Psychology: The Study of Behaviour.* New York: Henry Holt.

——. 1929. "Fundamentals of psychology: Behaviorism examined." In *The Battle of Behaviorism: An Exposition and an Exposure*, edited by John B. Watson and William McDougall, 40–85. New York: W. W. Norton.

McMahon, William E. 1983. "Did Montague invent 'Montague Grammar'?" *Histoire Épistémologie Langage* 5: 117–37.

Meillet, Antoine. 1926. *Linguistique historique et linguistique générale.* Paris: Champion.

——. 1966. "F.d.S." Reprinted in *Portraits of Linguists: A Biographical Source Book for the History of Western Linguistics, 1746–1963*, vol. 2, edited by T. A. Sebeok, 92–99. Bloomington: Indiana University Press.

Menabrea, Federico Luigi. 1842. "Notions sur la machine analytique de M. Charles Babbage." *Bulletin de la bibliothèque universelle de Genève* 41: 352–76.

——. 1884. *Comptes-rendus hebdomadaires de l'Académie des sciences* 99. Paris.

Mendeleev, Dmitri. 1889. "The periodic law of the chemical elements." *Journal of the Chemical Society* 55: 634–56.

Merchant, Jason. 2001. *The Syntax of Silence.* Oxford: Oxford University Press.

Merz, J. T. 1903. *A History of European Thought in the Nineteenth Century.* Vol. 2. Edinburgh: W. Blackwood.

Metcalf, George J. 1974. "The Indo-European hypothesis in the sixteenth and seventeenth centuries." In *Studies in the History of Linguistics: Traditions and Paradigms*, edited by Dell Hymes, 233–57. Bloomington: Indiana University Press.

Mill, John Stuart. 1865. *Auguste Comte and Positivism.* 2nd ed., 1866. London: N. Trübner.

Mises, Richard von. 1939. Reprinted as *Positivism: A Study in Human Understanding*, 1951. Cambridge, MA: Harvard University Press.

Monod, Gabriel. 1876. "Du progrès des études historiques en France depuis le XVIème siècle." *Revue historique* 1: 5–38.

Moore, Jay. 1999. "The basic principles of behaviorism." In *The Philosophical Legacy of Behaviorism*, 41–68. Dordrecht: Kluwer.

Morawski, J. G. 1986. "Organizing knowledge and behavior at Yale's Institute of Human Relations." *Isis* 77: 219–42.

Moulton, William. 1970. "Leonard Bloomfield as a Germanist." In *A Leonard Bloomfield Anthology*, edited by Charles F. Hockett, 512–23. Bloomington: Indiana University Press.

Mugdan, Joachim. 1984. *Jan Baudouin de Courtenay (1845–1929): Leben und Werk*. Munich: Wilhelm Fink.

Müller, Max. 1862. *The Science of Language*. New York: Charles Scribner.

———. 1864. *Lectures on the Science of Language*. London: Longman.

———. 1868. *On the Stratification of Language*. London: Longman.

———. 1887. *The Science of Thought*. New York: Charles Scribner.

Mulligan, Kevin. 1988. "On structure: Bühler's linguistic and psychological examples." In *Karl Bühler's Theory of Language/Karl Bühlers Sprachtheorie: Proceedings of the Conference held at Kirchberg, August 26, 1984 and Essen, November 21–24, 1984*, vol. 2, edited by Achim Eschbach, 203. Philadelphia: John Benjamins.

Mulligan, Kevin, and Barry Smith. 1985. "Franz Brentano on the ontology of mind." *Philosophy and Phenomenological Research* 45: 627–44.

———. 1988. "Mach and Ehrenfels: The foundations of Gestalt theory." In *Foundations of Gestalt Theory*, edited by Barry Smith, 124–57. Munich: Philosophia.

Murchison, Carl Allanmore, ed. 1961. *A History of Psychology in Autobiography*. Vol. 2. New York: Russell and Russell.

Murray, Stephen O. 1991. "The first quarter century of the Linguistic Society of American, 1924–1949." *Historiographia Linguistica* 18: 1–48.

———. 1994. *Theory Groups and the Study of Language in North America*. Amsterdam: John Benjamins.

Musolff, Andreas. 1997. "Bühler, Karl." In *Handbook of Pragmatics*, edited by Jef Verschueren, Jan-Ola Östman, Jan Blommaert, and Chris Bulcaen, 1–15. Amsterdam: John Benjamins.

Næss, Arne. 2005. "Logical empiricism and the uniqueness of the Schlick seminar: A personal experience with consequence." In *The Selected Works of Arne Næss*, edited by Alan Drengson, 1765–83. Amsterdam: Springer Netherlands.

Nerlich, Brigitte, and David D. Clarke. 2001. "Mind, meaning and metaphor: The philosophy and psychology of metaphor in 19th-century Germany." *History of the Human Sciences* 14: 39–61.

Neurath, Otto. 1930. "Protokollsatz." *Erkenntnis* 3: 204–14.

Newman, Stanley S. 1932. "The Yawelmani dialect of Yokuts." *International Journal of American Linguistics* 7: 85–89.

———. 1946a. "The Yawelmani dialect of Yokuts." In *Linguistic Structures of Native America*, edited by H. Hoijer, 222–48. New York: Johnson.

———. 1946b. *Yokuts Language of California*. New York: Viking.

Newmeyer, Frederick J. 1986. *Linguistic Theory in America: The First Quarter-Century of Transformational Generative Grammar.* 2nd ed. New York: Academic Press.

Newton, Isaac. 1687. *Philosophiae Naturalis Principia Mathematica.* London: Benjamin Motte.

Nicolas, Serge. 2013. *Histoire de la psychologie.* Paris: Dunod.

Nicolas, Serge, and Ludovic Ferrand. 2002. "Alfred Binet and higher education." *History of Psychology* 5: 264–83.

Nicolas, Serge, and David J. Murray. 1999. "Théodule Ribot (1839–1916), founder of French psychology: A biographical introduction." *History of Psychology* 2: 277–301.

Norman, Frederick. 1929. "'Indo-European' and 'Indo-Germanic.'" *Modern Language Review* 24: 313–21.

Novák, Josef. 1988. "Masaryk and the Brentano school." In *On Masaryk: Texts in English and German*, edited by Josef Novák, 27–38. Amsterdam: Rodopi.

Nylund, Jan H. 2013. "The Prague School of Linguistics and its influence on New Testament language studies." In *The Language of the New Testament*, edited by Stanley E. Porter and Andrew W. Pitts, 155–224. Leiden: Brill.

Ogden, C. K., and I. A. Richards. 1923. *The Meaning of Meaning.* New York: Harcourt Brace.

Oldroyd, David R. 2003. "The earth sciences." In *From Natural Philosophy to the Sciences*, 88–128. Chicago: University of Chicago Press.

Olender, Maurice. 1992. *The Languages of Paradise: Race, Religion, and Philology in the Nineteenth Century.* Cambridge, MA: Harvard University Press.

Osthoff, Hermann, and Karl Brugmann. 1878. "Introduction." In *Morphologische Untersuchungen auf dem Gebiete der indogermanischen Sprachen*, vol. 1, 3–20. Leipzig: S. Hirzel.

Padovani, Flavia. 2013. "Genidentity and topology of time: Kurt Lewin and Hans Reichenbach." In *The Berlin Group and the Philosophy of Logical Empiricism*, vol. 273, edited by Nikolay Milkov and Volker Peckhaus, 97–122. Berlin: Springer.

Paul, Harry W. 1972. *The Sorcerer's Apprentice: The French Scientist's Image of German Science, 1840–1919.* Gainesville: University of Florida Press.

Paul, Hermann. 1886. "Review of H. Schuchardt, *Über die Lautgesetze.*" *Literaturblatt für germanische und romanische Philologie* 7: 1–6.

Peckhaus, Volker. 1999. "19th century logic between philosophy and mathematics." *Bulletin of Symbolic Logic* 5: 433–50.

Pedersen, Holger. 1931. *The Discovery of Language: Linguistic Science in the 19th Century.* Bloomington: Indiana University Press.

Percival, Keith. n.d. "On the genealogy of linguistic structuralism."

Petzold, Charles. 2008. *The Annotated Turing.* Hoboken, NJ: Wiley.

Pickering, Mary. 1993. *August Comte: An Intellectual Biography*. Vol. 1. Cambridge: Cambridge University Press.

Pietarinen, Ahti-Veikko. 2009. "Significs and the origins of analytic philosophy." *Journal of the History of Ideas* 70: 467–90.

Pike, Kenneth. 1989. "Recollections of Bloomfield." *Historiographia Linguistica* 16: 217–23.

Poincaré, Henri. 1908. *Science et méthode*. Paris: Flammarion.

Poliakov, Leon. 1965. *The History of Anti-Semitism*. New York: Vanguard.

Pomorska, Krystyna, ed. 1987. *Language, Poetry and Poetics*. Berlin: De Gruyter.

Popov, Igor. 2008. "Orthogenesis versus Darwinism: The Russian case." *Revue d'histoire des sciences* 61: 367–97.

Popper, Karl. 1935. *Logik der Forschung: Zur Erkenntnistheorie der moderner Naturwissenschaft*. Berlin: Springer.

Pos, Hendrick. 1938. "La notion d'opposition en linguistique." In *Onzième congrès international de psychologie*, edited by H. Piéron and I. Myerson, 246–47. Paris: Alcan.

Posy, Carl. 1974. "Brouwer's constructivism." *Synthese* 27: 125–59.

——. 2005. "Intuitionism and philosophy." In Shapiro, 318–55.

Pullum, Geoffrey K. 2013. "The central question in comparative syntactic metatheory." *Mind and Language* 28: 492–521.

Pullum, Geoffrey K., and Barbara C. Scholz. 2005. "Contrasting applications of logic in natural language syntactic description." In *Logic, Methodology and Philosophy of Science: Proceedings of the Twelfth International Congress*, edited by Petr Hájek, Luis Valdés-Villanueva, and Dag Westerståhl, 481–503. London: King's College Publications.

Quine, Willard Van Orman. 1940. *Mathematical Logic*. Cambridge, MA: Harvard University Press.

——. 1971. [No title.] In *PSA 1970: In Memory of Rudolf Carnap; Proceedings of the 1970 Biennial Meeting, Philosophy of Science Association*, vol. 8, edited by Roger C. Buck and Robert S. Cohen, xxii–xxv. Boston Studies in the Philosophy of Science. Dordrecht: Reidel.

Radwanska-Williams, Joanna. 1994. *A Paradigm Lost: The Linguistic Thought of Mikołaj Kruszewski*. Amsterdam: John Benjamins.

Ratzel, Friedrich. 1882. *Anthropogeographie*. Vol. 1. Stuttgart: J. Engelhorn.

Reck, Erich H., and Steve Awodey. 2004. "Frege's lectures on logic and their influence." In *Frege's Lectures on Logic: Carnap's Jena Notes, 1910–1914*, edited and translated by Erich H. Reck and Steve Awodey, 17–42. Chicago: Open Court.

Reed, Edward. 1997. *From Soul to Mind: The Emergence of Psychology from Erasmus Darwin to William James*. New Haven: Yale University Press.

Reichenbach, Hans. 1938. *Experience and Prediction*. Chicago: University of Chicago Press.

——. 1947. *Elements of Symbolic Logic*. New York: Macmillan.

——. 2012. *Selected Writings, 1909–1953*. Edited by Maria Reichenbach and Robert S. Cohen. Berlin: Springer.

Reisch, George A. 2005. *How the Cold War Transformed Philosophy of Science: To the Icy Slopes of Logic*. Cambridge: Cambridge University Press.

Renan, Ernest. 1862. *De la part des peuples sémitiques dans l'histoire de la civilisation: discours d'ouverture des cours de langue hébraïque, chaldaïque et syriaque au Collège de France*. Paris: Michel-Lévy frères.

Riasanovsky, Nicholas. 1964. "Prince N. S. Trubetskoi's 'Europe and Mankind.'" *Jahrbücher für Geschichte Osteuropas* 13: 207–20.

Ribot, Théodule. 1870. *La psychologie anglaise contemporaine: la méthode expérimentale*. Paris: J.-B. Baillière.

——. 1877. "Philosophy in France." *Mind* 2: 366–86.

——. 1879. *La psychologie allemande contemporaine*. 2nd ed., 1885. Paris: Librairie G. Baillière.

——. 1881. *Les maladies de la mémoire*. Paris: Librairie G. Baillière.

——. 1883. *Les maladies de la volonté*. Paris: Librairie G. Baillière.

Richards, Robert. 2002. "The linguistic creation of man: Charles Darwin, August Schleicher, Ernst Haeckel, and the missing link in 19th-century evolutionary theory." In *Experimenting in Tongues: Studies in Science and Language*, edited by Matthias Doerres, 21–48. Stanford, CA: Stanford University Press.

Richardson, Alan W. 1992. "Logical idealism and Carnap's construction of the world." *Synthese* 93: 59–92.

——. 1998. *Carnap's Construction of the World: The Aufbau and the Emergence of Logical Empiricism*. Cambridge: Cambridge University Press.

Ricoeur, Paul. 1974. "Science et idéologie." *Revue philosophique de Louvain* 72: 328–56.

Rieber, Robert W., and David K. Robinson. 2001. *Wilhelm Wundt in History: The Making of a Scientific Psychology*. Berlin: Springer.

Riskin, Jessica. 2016. *The Restless Clock: A History of the Centuries-Long Argument over What Makes Living Things Tick*. Chicago: University of Chicago Press.

Ritchie, Benbow F. 1964. *Edward Chace Tolman*. Biographical Memoirs, vol. 37. Washington, DC: National Academy of Sciences.

Robel, Leon. 1978. "Les années de formation." In *Jakobson. Cahiers Cistre* 5, 34–39. Lausanne: L'Age d'Homme.

Robins, Robert H. 1990. "Leibniz and Wilhelm von Humboldt and the history of comparative linguistics." In *Leibniz, Humboldt, and the Origins of Comparitivism*, edited by Tullio De Mauro and Lia Formigari, 85–102. Amsterdam: John Benjamins.

Rodin, Andrei. 2014. *Axiomatic Method and Category Theory*. Berlin: Springer.

Rorty, R. 1990. *Philosophy and the Mirror of Nature*. Oxford: Blackwell.

Rosado Haddock, Guillermo E. 2008. *The Young Carnap's Unknown Master: Husserl's Influence on "Der Raum" and "Der Logische Aufbau der Welt."* Aldershot, UK: Ashgate.

Rosenbloom, Paul. 1950. *The Elements of Mathematical Logic*. New York: Dover.

Ross, Dorothy. 1972. *G. Stanley Hall: The Psychologist as Prophet*. Chicago: University of Chicago Press.

Rota, Gian-Carlo. 1989. "*Fundierung* as a logical concept." *The Monist* 72: 70–77.

Rousselot, Jean-Pierre. 1897. *Principes de phonétique expérimentale*. Paris: Didier.

Royce, Josiah. 1892. *The Spirit of Modern Philosophy: An Essay in the Form of Lectures*. Boston: Houghton Mifflin.

Rudy, Stephen. 1990. *Roman Jakobson: A Bibliography of His Writings*. Berlin: De Gruyter.

Ruëgg, Walter. 2004. *A History of the University in Europe*. Vol. 3, *Universities in the Nineteenth and Early Twentieth Centuries (1800–1945)*. Cambridge: Cambridge University Press.

Russell, Bertrand. 1903. *Principles of Mathematics*. Cambridge: Cambridge University Press.

——. 1914. *Our Knowledge of the External World*. London: Allen and Unwin.

——. 1945. *A History of Western Philosophy*. New York: Simon and Schuster.

Saint-Hilaire, Geoffroy. 1818. *Philosophie anatomique*. Paris: Méquignon-Marvis.

Sapir, Edward. 1921. *Language*. New York: Harcourt Brace.

——. 1929. "The status of linguistics as a science." *Language* 5: 207–14.

——. 1949. *Selected Writings in Language, Culture, and Personality*, edited by David G. Mandelbaum. Berkeley: University of California Press.

——. 2011. *The Psychology of Culture: A Course of Lectures*. Edited by Judith T. Irvine. Berlin: De Gruyter.

Sarton, G., J. R. Mayer, J. P. Poule, and Sadi Carnot. 1929. "The discovery of the law of conservation of energy." *Isis* 13: 18–44.

Saussure, Ferdinand de. 1879. *Mémoire sur le système primitive des voyelles dans les langues indo-européennes*. Leipzig: Teubner.

——. 1916. *Cours de linguistique générale*. Edited by Charles Bally and Albert Sechehaye, with the collaboration of Albert Riedlinger. Lausanne: Payot.

——. 1960. "Souvenirs de F. de Saussure concernant sa jeunesse et ses études." *Cahiers Ferdinand de Saussure* 17: 12–25.

——. 1968. *Cours de linguistique générale*. Edited by Rudolph Engler. Wiesbaden: Otto Harrassowitz.

——. 1972. *Cours de linguistique générale*. Edited by Tullio De Mauro. Paris: Payot.

——. 1978 [1874]. "Essai pour réduire les mots du grec du latin et de l'allemand à un petit nombre de racines." *Cahiers Ferdinand de Saussure* 32: 76–101.

——. 1995. *Phonétique: Il manoscritto di Harvard*. Edited by Maria Pia Marchese. Padua: Unipress.

——. 2002a. *Ecrits de linguistique générale*. Edited by Simon Bouquet and Rudolf Engler. Paris: Gallimard.

——. 2002b. *Théorie des sonantes: Il manoscritto du Ginevra (BPU Ms Fr 3955)*. Edited by Maria Pia Marchese. Florence: Università degli studi di Firenze.

——. 2006. *Writings in General Linguistics*. Edited by Simon Bouquet, Rudolf Engler, Carol Sanders, and Matthew Pires. Oxford: Oxford University Press.

Sayers, Frances Clarke. 1987. "The small mythologies of Leonard Bloomfield." In *Leonard Bloomfield: Essays on His Life and Work*, edited by Robert A. Hall Jr., 15–23. Amsterdam: John Benjamins.

Schilpp, Paul Arthur. 1949. *The Philosophy of Ernst Cassirer*. New York: Tudor.

——. 1963. *The Philosophy of Rudolf Carnap*. Chicago: Open Court.

Schlegel, Friedrich von. 1808. *Über die Sprache und Weisheit der Indier*. Reprinted in 1977. Heidelberg: Mohr and Zimmer.

——. 1849. "On the language and wisdom of the Indians." In *The Aesthetic and Miscellaneous Works of Frederick von Schlegel*. London: Henry G. Bohn.

Schleicher, August. 1863. *Die Darwinsche Theorie und die Sprachwissenschaft: Offenes Sendschreiben an Herrn Dr. Ernst Häckel*. Weimar: H. Boehlau.

——. 1874. *A Compendium of the Comparative Grammar of the Indo-European, Sanskrit, Greek and Latin Languages*. London: Trübner.

Schlick, Moritz. 1915. Die philosophische Bedeutung des Relativsatsprinzips. *Zeitschrift für Philosophie und philosophische Kritik* 159: 129–75. Reprinted in 1979 as "The philosophical significance of the principle of relativity," in *Moritz Schlick Philosophical Papers*, vol. 1. Berlin: Springer.

——. 1930. "Die Wende der Philosophie." *Erkenntnis* 1: 4–11. Translated as "The turning point in philosophy," in *Logical Positivism*, edited by A. J. Ayer, 1959, 53–59. Glencoe, IL: Free Press.

Schmit, Roger. 1992. "Die Idee der logischen Grammatik bei Husserl: Eine begriffgeschichtliche Betrachtung." *Husserl Studies* 9: 31–49.

Schnaitter, Roger. 1999. "Some criticisms of behaviorism." In *The Philosophical Legacy of Behaviorism*, edited by Bruce Thyer, 209–49. Dordrecht: Kluwer.

Scholz, Barbara C., and Geoffrey K. Pullum. 2007. "Tracking the origins of transformational generative grammar." *Journal of Linguistics* 43: 701–23.

Schrijnen, Joseph. 1928. "Le Premier Congrès International de Linguistes à la Haye du 10 à 15 avril 1928. Résumé, donné à la clôture du Congrès." *Anthropos* 23: 259–62.

Schuchardt, Hugo. 1885. *Über die Lautgesetze: Gegen die Junggrammatiker*. Berlin: Oppenheim.

——. 1979. *The Ethnography of Variation: Selected Writings on Pidgins and Creoles*. Cambridge: Cambridge University Press.

Schuhmann, Karl. 1988. "Husserl and Masaryk." In *On Masaryk: Texts in English and German*, edited by Josef Novák, 129–56. Amsterdam: Rodopi.

Searle, John. 1992. *The Rediscovery of the Mind*. Cambridge, MA: MIT Press.

Sebeok, Thomas A. 1987. "Karl Bühler." In *Classics of Semiotics*, 129–45. Berlin: Springer.

Seifrid, Thomas. 2005. *The Word Made Self: Russian Writings on Language, 1860–1930*. Ithaca, NY: Cornell University Press.

Sellars, Roy Wood. 1939. "Positivism in contemporary philosophic thought." *American Sociological Review* 4: 26–42.

Sériot, Patrick. 1993. "La double vie de Troubetzkoy, ou la clôture des systèmes." In *Le gré des langues* 5: 88–115. Paris: L'Harmattan.

——. 1994. "Aux sources du structuralisme: une controverse biologique en Russie." *Etudes de lettres* (janv–mars 1994): 89–103.

——. 1999. *Structure et totalité*. Paris: Presses Universitaires de France.

——. 2014. *Structure and the Whole: East, West and Non-Darwinian Biology in the Origins of Structural Linguistics*. Berlin: De Gruyter.

Shanker, Stuart. 2005. "Turing and the origins of AI." In *Mathematics and the Historian's Craft*, edited by Glen Van Brummelen and Michael Kinyon, 297–328. Berlin: Springer.

Shapin, Steven. 1981. "Of gods and kings: Natural philosophy and politics in the Leibniz–Clarke disputes." *Isis* 72: 187–215.

Shapiro, Stewart. 2005. *The Oxford Handbook of Philosophy of Mathematics and Logic*. Oxford: Oxford University Press.

Shibatani, Masayoshi, and Theodora Bynon. 1995. "Approaches to language typology: A conspectus." In *Approaches to Language Typology*, edited by M. Shibatani and Theodora Bynon, 1–27. Oxford: Oxford University Press.

Sieg, Wilfried. 2008. "On computability." In *Philosophy of Mathematics*, edited by Andrew Irvine, 535–630. Amsterdam: Elsevier.

Siegmund-Schultze, Reinhard. 2001. *Rockefeller and the Internationalization of Mathematics between the Two Wars*. Berlin: Springer.

Sievers, Eduard. 1881. *Grundzüge der Phonetik*. Leipzig: Breitkopf & Härtel.

Silverstein, Michael, ed. 1971. *Whitney on Language*. Cambridge, MA: MIT Press.

——. 2014. "In praise of 'exceptionless': Linguistics among the human sciences at Bloomfield and Sapir's Chicago." *History and Philosophy of the Language Sciences*. https://hiphilangsci.net/2014/11/26/in-praise-of-exceptionless-linguistics-among-the-human-sciences-at-bloomfield-and-sapirs-chicago.

Skidelsky, Edward. 2011. *Ernst Cassirer: The Last Philosopher of Culture*. Princeton, NJ: Princeton University Press.

Smail, Daniel Lord. 2008. *On Deep History and the Brain*. Berkeley: University of California Press.

Smith, Barry. 1982. *Parts and Moments: Studies in Logic and Formal Ontology*. Munich: Philosophia.

——. 1988. "Austrian origins of logical positivism." *Nijhoff International Philosophy Series* 38: 19–53.

——. 1994. *Austrian Philosophy: The Legacy of Franz Brentano*. Chicago: Open Court.

Smith, Barry, and Kevin Mulligan. 1982. "Pieces of a theory." In *Parts and Moments: Studies in Logic and Formal Ontology*, edited by Barry Smith, 15–110. Munich: Philosophia.

Smith, Laurence D. 1986. *Behaviorism and Logical Positivism: A Reassessment of the Alliance*. Stanford, CA: Stanford University Press.

Soare, Robert. 1996. "Computability and recursion." *Bulletin of Symbolic Logic* 2: 284–321.

Sokal, Michael M. 1990. "G. Stanley Hall and the institutional character of psychology at Clark 1889–1920." *Journal of the History of the Behavioral Sciences* 26: 114–24.

——. 1994. "Gestalt psychology in America in the 1920s and 1930s." In *Gestalt Psychology: Its Origins, Foundations and Influence*, edited by Stefano Poggi, 87–119. Florence: Leo S. Olschki.

Stratton, A. W., and A. H. Ewing. 1920. *Studies in Honor of Maurice Bloomfield Professor of Sanskrit and Comparative Philology in the Johns Hopkins University*. New Haven: Yale University Press.

Stumpf, Carl. 1919. "Reminiscences of Brentano." Reprinted in McAlister 1976, 10–46.

Swadesh, Morris. 1934. "The phonemic principle." *Language* 10: 117–29.

——. 1971. *The Origin and Diversification of Language*. New York: Aldine, Atherton.

Sweet, Henry. 1877. *A Handbook of Phonetics*. Oxford: Henry Frowde.

——. 1908. *The Sounds of English: An Introduction to Phonetics*. Oxford: Clarendon.

Sweet, Paul R. 1980. *Wilhelm von Humboldt: A Biography*. Vol. 2. Columbus: Ohio State University Press.

Taylor, Richard. 1843. *Scientific Memoirs*. Vol. 3. London: Richard and John E. Taylor.

Tchougounnikov, Sergueï. 2009. "L'effet 'néo-grammarien' entre la Russie et l'Allemagne." *Slavica Occitania* 29: 115–42.

Teeter, Karl. 1964. "Descriptive linguistics in America: Triviality vs. irrelevance." *Word* 20: 197–206.

Ter Hark, Michel. 2003. *Popper, Otto Selz, and the Rise of Evolutionary Epistemology*. Cambridge: Cambridge University Press.

Tesnière, Lucien. 1936. "Antoine Meillet." *Bulletin de la faculté des lettres de Strasbourg* 2: 1–10.

Tifrit, Ali. 2005. "Marque et syllabe: de la représentation des fricatives du français dans le Modèle à Contours." PhD thesis, Université de Paris X–Nanterre.

Titchener, Edward B. 1898. "The postulates of a structural psychology." *Philosophical Review* 7: 449–65.

——. 1899. "Structural and functional psychology." *Philosophical Review* 8: 290–99.

——. 1925. "Experimental psychology: A retrospect." *American Journal of Psychology* 36: 313–23.

Tocqueville, Alexis de. 1840. *Democracy in America*. New York: Heffner.

Tolman, Edward C. 1932. *Purposive Behavior in Animals and Men*. New York: The Century Co.

Tomalin, Marcus. 2006. *Linguistics and the Formal Sciences: The Origins of Generative Grammar*. Cambridge: Cambridge University Press.

Toman, Jindrich. 1995. *The Magic of a Common Language: Jakobson, Mathesius, Trubetzkoy, and the Prague Linguistic Circle*. Cambridge, MA: MIT Press.

——. 1997a. "Jakobson and Bohemia/Bohemia and the East." *Cahiers de L'ILSL*, no. 9: 229–39.

——. 1997b. "Notes on Roman Jakobson's culture of scholarship." *Acta Linguistica Hafniensa* 29: 89–100.

Torbakov, Igor. 2008. *Rethinking the Nation: Imperial Collapse, Eurasianism, and George Vernadsky's Historical Scholarship*. Occasional Papers no. 302. Washington, DC: Woodrow Wilson International Center for Scholars.

——. 2015. "Becoming Eurasian: The intellectual odyssey of Georgii Vladimirovich Vernadsky." In Bassin et al., 113–36.

Trager, George L. 1942. "The phoneme 'T': A study in theory and method." *American Speech* 17: 144–48.

Trautmann, Thomas R. 2006. *Languages and Nations: The Dravidian Proof in Colonial Madras*. Berkeley: University of California Press.

Trautmann-Waller, Cécile. 2004. *Quand Berlin pensait les peuples—Anthropologie, ethnologie et psychologie (1850–1900)*. Edited by Cécile Trautmann-Waller. Paris: CNRS Editions.

Trubetzkoy, Nikolai. 1920. "Europe and mankind." Reprinted in Trubetzkoy 1991, 1–64.

——. 1921. "The upper and lower stories of Russian culture." Reprinted in Trubetzkoy 1991, 81–100.

——.1923. "The Tower of Babel and the confusion of Tongues." In English translation in Trubetzkoy 1991, 147–60. In French translation in Trubetzkoy 1996, 115–26.

——. 1927a. "Le nationalisme pan-eurasien." In Trubetzkoy 1996, 181–91.

——. 1927b. "Sur le problème de la connaissance de la Russie par elle-même." Reprinted in Trubetzkoy 1996, 175–80.

——. 1933. "La phonologie actuelle." *Journal de psychologie normale et pathologique* 30: 227–46.

——. 1935. "On racism." Reprinted in Trubetzkoy 1991, 277–88.

——. 1936. "D'une théorie des oppositions phonologiques." *Journal de psychologie normale et pathologique* 33: 5–18.

——. 1939a. *Grundzüge der Phonologie*. Prague: Travaux du Cercle linguistique de Prague.
——. 1939b. "Zur allgemeinen Theorie der phonologischen Vokalsysteme." *Travaux du Cercle linguistique de Prague* 1: 39–67.
——. 1950. *The Common Slavic Element in Russian Culture*. Edited by Leon Stilman. New York: Columbia University Department of Slavic Languages.
——. 1969. *Principles of Phonology*. Translation of 1939 edition. Berkeley: University of California Press.
——. 1975. *N. S. Trubetzkoy's Letters and Notes*. Edited by Roman Jakobson, H. Baran, O. Ronen, and Martha Taylor. The Hague: Mouton.
——. 1991. *The Legacy of Genghis Khan and Other Essays on Russia's Identity*. Edited by Anatoly Liberman. Ann Arbor: Michigan Slavic Publications.
——. 1996. *L'Europe et l'humanité*. Edited by Patrick Sériot. Liège, Belgium: P. Mardaga.
——. 2001. *Studies in General Linguistics and Language Structure*. Edited by Anatoly Liberman. Durham, NC: Duke University Press.
——. 2006. *Correspondence avec Roman Jakobson et autres écrits*. Edited by Patrick Sériot. Lausanne: Payot.
Turing, Alan M. 1936–37. "On computable numbers, with an application to the *Entscheidungsproblem*." *Proceedings of the London Mathematical Society* 42: 230–65.
Uebel, Thomas E. 1996. "Anti-foundationalism and the Vienna Circle's revolution in philosophy." *British Journal for the Philosophy of Science* 47: 415–40.
——. 2004a. "Carnap, the Left Vienna Circle, and neopositivist antimetaphysics." In *Carnap Brought Home: The View from Jena*, edited by Steve Awodey and Carstein Klein, 247–77. Chicago: Open Court.
——. 2004b. "Education, enlightenment and positivism: The Vienna Circle's scientific world-conception revisited." *Science and Education* 13: 41–66.
——. 2008. "Writing a revolution: On the production and early reception of the Vienna Circle's manifesto." *Perspectives on Science* 16: 70–102.
Uriagereka, Juan. 1998. *Rhyme and Reason*. Cambridge, MA: MIT Press.
Urquhart, Alasdair. 2008. "Emil Post." In Gabby and Wood, 617–66.
Velupillai, K. Vela. 2011. "Freedom, anarchy and conformism in academic research." *Interdisciplinary Journal of Economics and Business Law* 1: 76–101.
Verburg, P. A. 1949. "The background to the linguistic conceptions of Bopp." *Lingua* 2: 438–68.
Viel, Michel. 1984. *La notion de marque chez Trubetzkoy et Jakobson: un épisode de la pensée structurale*. Paris: Didier.
Voegelin, Carl F., and Florence M. Voegelin. 1963. "On the history of structuralizing in 20th century America." *Anthropological Linguistics* 5: 12–37.
Vonk, Frank. 2004. "La Sprachtheorie de Karl Bühler d'un point de vue linguistique." *Les dossiers de HEL*. http://dossierhel.hypotheses.org.

Vucinich, Alexander. 1989. *Darwin in Russian Thought*. Berkeley: University of California Press.

Wagemans, Johan, James H. Elder, Michael Kubovy, Stephen E. Palmer, Mary A. Peterson, Manish Singh, and Rüdiger von der Heydt. 2012. "100 Years of visual Gestalts." *Psychological Bulletin* 138: 1172–217.

Wagner, Pierre. 1998. *La machine en logique*. Paris: Presses Universitaires de France.

Wang, Hao. 1974. *From Mathematics to Philosophy*. London: Routledge and Kegan Paul.

———. 1990. *Reflections on Kurt Gödel*. Cambridge, MA: MIT Press.

———. 1996. *A Logical Journey: From Gödel to Philosophy*. Cambridge, MA: MIT Press.

Watson, John B. 1913. "Psychology as the behaviorist views it." *Psychological Review* 20: 158–77.

———. 1919. *Psychology from the Standpoint of a Behaviorist*. Philadelphia: J. B. Lippincott.

———. 1929. "Behaviorism—The Modern Note in Psychology." Remarks delivered publicly in 1924. In John B. Watson and William McDougall, *The Battle of Behaviorism*. New York: W. W. Norton.

———. 1930. *Behaviorism*. 2nd ed. New York: W. W. Norton.

Watson, John B., and Karl Lashley. 1913. "Notes on the development of a young monkey." *Journal of Animal Behavior* 3: 114–39.

Weidman, Nadine M. 1999. *Constructing Scientific Psychology: Karl Lashley's Mind-Brain Debates*. Cambridge: Cambridge University Press.

Weimer, Walter B. 1974. "The history of psychology and its retrieval from historiography: I. The problematic nature of history." *Science Studies* 4: 235–58.

Weiss, Albert Paul. 1919. "The relation between physiological psychology and behavioral psychology." *Journal of Philosophy, Psychology, and Scientific Methods* 16: 626–34.

Wellek, Albert. 1968. "The impact of German immigration on the development of American psychology." *Journal of the History of the Behavioral Sciences* 4: 207–29.

Wertheimer, Max. 1922. "Untersuchungen zur Lehre von der Gestalt I." *Psychologische Forschung* 1: 47–58.

———. 1923. "Untersuchungen zur Lehre von der Gestalt II." *Psychologische Forschung* 4: 301–50.

———. 1938. "Gestalt theory." In *A Source Book of Gestalt Psychology*, edited by Willis D. Ellis, 71–88. New York: Harcourt Brace.

———. 1945. *Productive Thinking*. New York: Harper.

Wettersten, John. 1988. "Külpe, Bühler, Popper." In *Karl Bühler's Theory of Language*, edited by Achim Eschbach, 327–47. Amsterdam: John Benjamins.

Whewell, William. 1832. "Review of Lyell's *Principles of Geology*, V. 2." *Quarterly Review* 47: 103–32.

——. 1839. "Excerpts from the Anniversary Address of the Rev. Wm. Whewell before the Geological Society of London." *American Journal of Science and Arts* 37: 238.

——. 1858. *History of Scientific Ideas: Being the First Part of the Philosophy of the Inductive Sciences*. London: John W. Parker and Son.

Whitehead, Alfred North. 1916. "The organization of thought." *Science* 22: 409–19.

Whitehead, Alfred North, and Bertrand Russell. 1910–1913. *Principia Mathematica*. 3 vols. Cambridge: Cambridge University Press.

Whitfield, Steven J. 2010. "Franz Boas: The anthropologist as public intellectual." *Society* 47: 430–38.

Whitman, Charles O. 1898. "Animal behavior." *Biological Lectures Delivered at the Marine Biological Laboratory (Woods Hole)* 6: 285–338.

Whitney, William Dwight. 1865. "On the relation of vowels and consonants." *Journal of the American Oriental Society* 8: 357–73. Reprinted in Silverstein 1971, 198–214.

——. 1867a. "Languages and dialects." *North American Review* 104: 30–64.

——. 1867b. *Language and the Study of Language*. New York: Scribner. Reprinted in Silverstein 1971, 7–110.

——. 1872. "Steinthal on the origin of language." *North American Review* 114: 272–308. Reprinted in Silverstein 1971, 133–69.

——. 1874. "The relation of vowel and consonant." In *Oriental and Linguistic Studies*, second series, 277–300. New York: Scribner.

——. 1875. *Life and Growth of Language*. New York: Appleton.

——. 1879. *Sanskrit Grammar*. Cambridge, MA: Harvard University Press.

Whorf, Benjamin Lee. 1956. *Language, Thought, and Reality: Selected Writings of Benjamin Lee Whorf*. Cambridge, MA: MIT Press.

Winston, Andrew S. 2001. "Cause into function: Ernst Mach and the reconstruction of explanation in psychology." In *The Transformation of Psychology: Influences of 19th-Century Philosophy, Technology, and Natural Science*, edited by Christopher D. Green, Marlene Shore, and Thomas Teo, 107–31. Washington, DC: American Psychological Association.

Witte, A. J. J. de. 1956. "Towards a New Grammar." *Synthese* 10: 91–101.

Wolf, Theta H. 1973. *Alfred Binet*. Chicago: University of Chicago Press.

Wood, Mary McGee. 1993. *Categorial Grammars*. London: Routledge.

Wozniak, Robert H. 1997. *Theoretical Roots of Early Behaviorism: Functionalism, the Critique of Introspection, and the Nature and Evolution of Consciousness*. London: Routledge.

Wright, Cory, and William Bechtel. 2007. "Mechanisms and psychological explanation." In *Philosophy of Psychology and Cognitive Science*, edited by Paul Thagard, 31–79. Amsterdam: Elsevier.

Wundt, Wilhelm. 1894. "Über psychische Kausalität und das Prinzip des psychophysischen Parallelismus." *Philosophische Studien* 10: 1–124.

——. 1896. *Grundriss der Psychologie*. Leipzig: Engelmann. Translated as *Outlines of Psychology*, 1897. New York: Stechert.

Yates, Frances A. 1939. "Giordano Bruno's conflict with Oxford." *Journal of the Warburg Institute* 2: 227–42.

——. 1968. "The hermetic tradition in Renaissance science." In *Art, Science, and History in the Renaissance*, edited by Charles Singleton, 255–74. Baltimore: Johns Hopkins Press.

Zahavi, Dan, and Frederick Stjernfelt. 2002. *One Hundred Years of Phenomenology: Husserl's Logical Investigations Revisited*. Dordrecht: Kluwer.

Zenkovsky, V. V. 1953. *Russian Thinkers and Europe*. Ann Arbor. MI: J. W. Edwards.

Zipf, G. K. 1936. *The Psycho-biology of Language: An Introduction to Dynamic Philology*. Cambridge, MA: MIT Press.

Zuckerman, Harriet. 1977. *Scientific Elite: Nobel Laureates in the United States*. New York: Free Press.

Index